Bacterial–Epithelial Cell Cross-Talk
Molecular mechanisms in pathogenesis

An emerging theme in molecular and cellular microbiology has been the ability of many pathogens to usurp the host cell and eventually colonize the host. Microbial pathogens have evolved different ways of interacting with their hosts and possess virulence factors that interfere with or stimulate a variety of host-cell physiological responses. This interaction between bacteria and host is not unidirectional – both pathogens and host cells engage in a signaling cross-talk. Research focused on this cross-talk reveals not only novel aspects of bacterial pathogenesis but also key information about epithelial biology, with broader implications in the prevention and treatment of infectious diseases. Written by leading researchers in the field, this book provides a valuable overview for graduate students and researchers. It details these remarkable host–pathogen interactions, uniquely providing a comprehensive understanding of the host–bacterial interactions that occur at mucosal surfaces, including the gastrointestinal, respiratory, and urogenital tracts.

Beth A. McCormick is Associate Professor of Pediatrics and Microbiology at Harvard Medical School and Associate Professor in the Department of Pediatric Gastroenterology and Nutrition at Massachusetts General Hospital.

Over the past decade, the rapid development of an array of techniques in the fields of cellular and molecular biology have transformed whole areas of research across the biological sciences. Microbiology has perhaps been influenced most of all. Our understanding of microbial diversity and evolutionary biology, and of how pathogenic bacteria and viruses interact with their animal and plant hosts at the molecular level, for example, has been revolutionized. Perhaps the most exciting recent advance in microbiology has been the development of the interface discipline of cellular microbiology, a fusion of classic microbiology, microbial molecular biology, and eukaryotic cellular and molecular biology. Cellular microbiology is revealing how pathogenic bacteria interact with host cells in what is turning out to be a complex evolutionary battle of competing gene products. Molecular and cellular biology are no longer discrete subject areas but vital tools and an integrated part of current microbiological research. As part of this revolution in molecular biology, the genomes of a growing number of pathogenic and model bacteria have been fully sequenced, with immense implications for our future understanding of microorganisms at the molecular level.

Advances in Molecular and Cellular Microbiology is a series edited by researchers active in these exciting and rapidly expanding fields. Each volume focuses on a particular aspect of cellular or molecular microbiology and provides an overview of the area, as well as examines current research. This series will enable graduate students and researchers to keep up with the rapidly diversifying literature in current microbiological research.

ADVANCES IN MOLECULAR AND CELLULAR MICROBIOLOGY

Published titles

1. *Bacterial Adhesion to Host Tissues.* Edited by Michael Wilson 0-521-80107-9
2. *Bacterial Evasion of Host Immune Responses.* Edited by Brian Henderson & Petra Oyston 0-521-80173-7
3. *Dormancy in Microbial Diseases.* Edited by Anthony Coates 0-521-80940-1
4. *Susceptibility to Infectious Diseases.* Edited by Richard Bellamy 0-521-81525-8
5. *Bacterial Invasion of Host Cells.* Edited by Richard Lamont 0-521-80954-1
6. *Mammalian Host Defense Peptides.* Edited by Deirdre Devine & Robert Hancock 0-521-82220-3
7. *Bacterial Protein Toxins.* Edited by Alistair Lax 0-521-82091-X
8. *The Dynamic Bacterial Genome.* Edited by Peter Mullany 0-521-82157-6
9. *Salmonella Infections.* Edited by Pietro Mastroeni & Duncan Maskell 0-521-83504-6
10. *The Influence of Cooperative Bacteria on Animal Host Biology.* Edited by Margaret J. McFall-Ngai, Brian Henderson & Edward G. Ruby 0-521-83465-1
11. *Bacterial Cell-to-Cell Communication.* Edited by Donald R. Demuth & Richard Lamont 0-521-84638-7

Forthcoming titles

12. *Phagocytosis of Bacteria and Bacterial Pathogenicity.* Edited by Joel Ernst & Olle Stendahl 0-521-84569-6

Advances in Molecular and Cellular Microbiology 13

Bacterial–Epithelial Cell Cross-Talk

Molecular Mechanisms in Pathogenesis

EDITED BY
BETH A. McCORMICK
Massachusetts General Hospital and Harvard Medical School

CAMBRIDGE UNIVERSITY PRESS
Cambridge, New York, Melbourne, Madrid, Cape Town, Singapore, São Paulo

Cambridge University Press
The Edinburgh Building, Cambridge CB2 2RU, UK

Published in the United States of America by Cambridge University Press, New York

www.cambridge.org
Information on this title: www.cambridge.org/9780521852449

First published 2006

Printed in the United Kingdom at the University Press, Cambridge

A catalog record for this publication is available from the British Library

ISBN-13 978-0-521-85244-9 hardback
ISBN-10 0-521-85244-7 hardback

Contents

Part III Host cell signaling by bacteria

Part IV Exploitation of host niches by pathogenic bacteria: mechanisms and consequences

Color plates appear between pages 56 and 57.

Contributors

Christopher M. Bailey
Bacterial Pathogenesis and Genomics Unit
Division of Immunity and Infection
The Institute for Biomedical Research
University of Birmingham
Edgbaston
UK

Joseph T. Barbieri
Medical College of Wisconsin
Milwaukee WI 53226
USA

Helen J. Betts
Bacterial Pathogenesis and Genomics Unit
Division of Immunity and Infection
The Institute for Biomedical Research
University of Birmingham
Edgbaston
UK

Patrice Boquet
Laboratoire de Bactériologie
Centre Hospitalier Universitaire de Nice
Hôpital de l'Archet
Nice
France

Bobby J. Cherayil
Massachusetts General Hospital and Harvard Medical School
Boston, MA
USA

René Clément
Unité INSERM
Faculté de Médecin
Nice
France

Andrew T. Gewirtz
Epithelial Pathobiology Division
Department of Pathology and Laboratory Medicine
Emory University
Atlanta GA
USA

Marisa I. Gómez
Departments of Pediatrics
College of Physicians and Surgeons
Columbia University
New York
USA

W. Vallen Graham
Department of Pathology
University of Chicago
Chicago IL 60637
USA

Wolf-Dietrich Hardt
Institute of Microbiology
ETH Zurich
Zurich
Switzerland

Christof R. Hauck
Department of Cell Biology
University of Konstanz
Konstanz
Germany

Gail A. Hecht
Department of Medicine, Section of Digestive Diseases and Nutrition
University of Illinois at Chicago
Chicago IL 60612-7323
USA

Ian R. Henderson
Bacterial Pathogenesis and Genomics Unit
Division of Immunity and Infection
The Institute for Biomedical Research
University of Birmingham
Edgbaston
UK

Luce Landraud
Laboratoire de Bactériologie
Centre Hospitalier Universitaire de Nice
Hôpital de l'Archet
Nice
France

Anthony T. Maurelli
Department of Microbiology and Immunology
F. Hébert School of Medicine
Uniformed Services University of the Health Sciences
Bethesda MD 20814-4799
USA

Beth A. McCormick
Pediatric Gastroenterology Unit
Massachusetts General Hospital and Harvard Medical School
Charlestown MA 021291
USA

D. Scott Merrell
Department of Microbiology and Immunology
Uniformed Services University of the Health Sciences
Bethesda MD 20814-4799
USA

Randall J. Mrsny
Welsh School of Pharmacy
Cardiff University
Cardiff
UK

Mark J. Pallen
Bacterial Pathogenesis and Genomics Unit
Division of Immunity and Infection

The Institute for Biomedical Research
University of Birmingham
Edgbaston
UK

Alice S. Prince
College of Physicians and Surgeons
Columbia University
New York
USA

Donnie Edward Shifflett
Department of Medicine, Section of Digestive Diseases and Nutrition
University of Illinois at Chicago
Chicago IL 60612-7323
USA

Samuel Tesfay
Department of Medicine, Section of Digestive Diseases and Nutrition
University of Illinois at Chicago
Chicago IL 60612-7323
USA

Jerrold R. Turner
Department of Pathology
University of Chicago
Chicago IL 60637
USA

Matam Vijay-Kumar
Epithelial Pathobiology Division
Department of Pathology and Laboratory Medicine
Emory University
Atlanta GA
USA

Brit Winnen
Institute of Microbiology
ETH Zurich
Zurich
Switzerland

Glenn M. Young
University of California
Davis, CA 95616
USA

Part I Introduction to the host and bacterial pathogens

CHAPTER 1

Overview of the epithelial cell

W. Vallen Graham and Jerrold R. Turner

Bacteria must overcome multiple obstacles in order to achieve successful pathogenesis. In many cases, this requires bypassing the first line of host defense: the barrier provided by epithelial surfaces of the integument and the gastrointestinal, respiratory, and urinary tracts. To overcome these barriers, pathogenic organisms frequently initiate mechanisms that exploit essential cellular processes of the epithelium. These cellular processes are therefore critical to our understanding of bacterial pathogenesis. Their description is the goal of this chapter.

GASTROINTESTINAL TRACT

The gastrointestinal epithelium forms a critical interface between the internal milieu and the lumen. The latter should be considered the external environment, since the gut is essentially a tube running through the body that communicates with the external environment at each end. Thus, like the skin, the barrier formed by the gastrointestinal epithelium is critical in preventing noxious luminal contents from accessing the internal tissues. In contrast to the skin, the gastrointestinal tract must also support passive paracellular and active transcellular transport of nutrients, electrolytes, and water. The barrier formed by the gastrointestinal epithelium must therefore be highly regulated and selectively permeable. Consistent with this, barrier permeability and epithelial transport function vary at individual sites within the gastrointestinal tract according to regional differences in the specific nutrients and ions transported.

Bacterial–Epithelial Cell Cross-Talk: Molecular Mechanisms in Pathogenesis, ed. Beth A. McCormick. Published by Cambridge University Press.

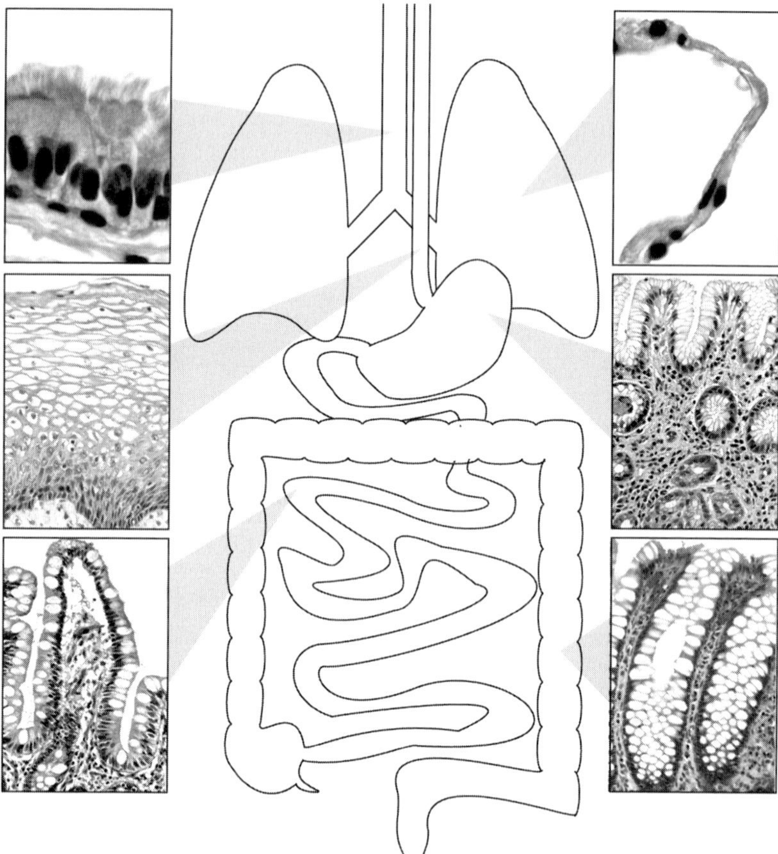

Figure 1.1 The human respiratory and gastrointestinal tracts. Selected histological sections are displayed. Counterclockwise from the top right are sections including epithelium from lung alveolus, gastric fundus, colon, ileum, esophagus, and bronchus. See text for detailed descriptions.

The oral cavity and esophagus are lined by stratified squamous epithelium (Figure 1.1), much like the skin. Keratinization, a normal feature of the skin, is abnormal in the esophagus and typically occurs only as a response to injury or in association with neoplasia. The esophageal squamous epithelium can be divided into three layers. The basal layer houses the stem cells and their progeny, which, in the normal state, extend up to three cell layers above the basement membrane. Normally this represents 10–15% of overall epithelial height, but it can increase dramatically during active inflammation, as often accompanies esophageal infection. Above the basal layer are the post-mitotic prickle and functional layers. Squamous cells become progressively more glycogenated and flattened as they pass through these layers on their way to the luminal surface. A variety of specialized cell types, including Langerhans cells, which process and present antigens, and endocrine cells, are intermixed with the squamous epithelium. Modified salivary glands are present within the esophageal submucosa. These salivary glands secrete mucus for lubrication, growth factors to augment epithelial cell growth and to aid in repair, and, in the distal esophagus, bicarbonate to neutralize gastric acid. The esophageal squamous mucosa ends at a sharply demarcated site, the Z-line, which marks the transition to the columnar epithelium of the stomach.

The stomach is a distensible saccular portion of the gastrointestinal tract in which food is mixed with gastric acid and digestive enzymes. This mixture, termed chyme, exits the gastric antrum through the pylorus at a controlled rate. The cardia is the small section of the stomach just distal to the Z-line. Like the cardia, the mucosae of the antrum and pylorus, which are the distal portions of the stomach, have long gastric pits and relatively short glands lined by foveolar, or mucus-producing epithelium. This epithelium is preferentially colonized by *Helicobacter pylori*. Thus, the antrum is most frequently affected and the optimal site of diagnostic biopsy for *H. pylori* infection. The fundus, which is a dome-shaped region projecting upward and to the left of the gastroesophageal junction, and the body of the stomach are lined by acid-producing, or oxyntic, mucosa with short pits and an expanded glandular compartment (Figure 1.1). The superficial areas of oxyntic mucosa are lined by foveolar cells that secrete neutral mucins, but the deep pits and glands contain parietal, chief, and endocrine cells. Parietal cells secrete acid as well as intrinsic factor, which is necessary for the absorption of vitamin B12 in the terminal ileum, while chief cells secrete pepsinogen. These areas are typically colonized by *H. pylori* only after loss of parietal cells and resultant increases in gastric pH. Chronic gastritis involving any portion of the stomach can also lead to metaplasia, with the appearance of small intestinal-like mucin-secreting goblet cells. Unlike gastric foveolar mucus cells, these cells express acid mucins, have a well-developed brush border, and are not readily colonized by *H. pylori*.

Despite its name, the small intestine is the longest portion of the gastrointestinal tract, extending for a length of approximately 6 m in adults. The small intestine is divided into three functionally distinct sections: duodenum, jejunum, and ileum. The primary function of each of these sections is to absorb luminal contents, although the specific molecules absorbed differ by region. For example, although 90% of absorption occurs within the first 100 cm of the small intestine, the jejunum is the principal site of absorption for Na^+ cotransported monosaccharides and amino acids. The ileum is essential to the enterohepatic recirculation of bile salts and the absorption of certain vitamins, including B12. Moreover, the terminal ileum contains abundant lymphoid tissue, termed Peyer's patches. These lymphoid follicles are a favored site of *Yersina* infection. Since the Peyer's patches actively sample luminal antigens, including those provided by bacteria, the terminal ileum is an effective location for this function; luminal bacterial load averages 1 000 000 bacteria per 1 ml in the distal ileum as opposed to 100 bacterial organisms per 1 ml in the duodenum and jejunum.

Small-intestinal nutrient absorption is facilitated by tissue architecture (Figure 1.1). The surface area of the small intestine is increased greatly by grossly visible mucosal folds, smaller finger-like protrusions, or villi, and a dense microvillus brush border. These combine to provide an overall luminal surface area that is approximately 600-fold greater than the corresponding serosal surface area. Thus, the loss of villi and mucosal flattening that accompany some infections and generalized bacterial overgrowth syndromes can result in significant malabsorption due simply to loss of absorptive surface area.

In addition to regional specialization along the length of the small intestine, there is significant specialization from the regenerative basally located crypt to the tip of the villus. The epithelial stem cells reside in the crypt, where they proliferate and give rise to specialized enteroendocrine cells and Paneth cells, which remain in the crypt. The stem cells also differentiate into goblet cells, M-cells, and undifferentiated crypt cells. The goblet cell is named for its large mucin vacuole, which compresses the cytoplasm into a shape reminiscent of a wine goblet. Goblet cells are present throughout the crypt–villus axis (Figure 1.1). M-cells, or follicle-associated epithelial cells, are most prominent in the distal ileum overlying Peyer's patches, but they can be found throughout the intestines. M-cells are specialized for the bulk endocytic sampling of luminal contents and are found in the flattened dome epithelium overlying lymphoid follicles. The most prominent feature of the M-cell is its shape, described as having microfolds, from which the 'M' is derived. The basal membrane of the M-cell is retracted from the basement membrane to form a cleft into which lymphocytes and macrophages migrate. Apically derived endocytic vesicles are rapidly released into this cleft by exocytosis, with some luminal particles arriving in the cleft space within 10 min of apical endocytosis. This highly efficient transport pathway is exploited as a route into and across the epithelium by some infectious organisms.

The undifferentiated crypt cells remain mitotically active while they are in the crypt and, thus, are somewhat intermediate between stem cells and terminally differentiated cells. Despite their name, undifferentiated crypt cells possess significant functional differentiation and effectively transport Cl^- into the intestinal lumen. This creates an osmotic gradient that draws water into the lumen. The latter is essential for normal homeostasis and is present in an exaggerated form in many diarrheal diseases, including cholera. As these undifferentiated crypt cells migrate towards the villus, they lose their ability to divide and modify their gene-expression profile to become specialized for absorption. This includes expression of absorptive nutrient and ion transporters, brush-border digestive enzymes, and kinases unique to villus epithelium (Clayburgh, Rosen *et al.*, 2004; Llor *et al.*, 1999). Ultimately, these absorptive cells become senescent and are shed from the villus tip. This occurs via an orderly process that allows the epithelium to maintain barrier function despite cell shedding (Madara 1990; Rosenblatt *et al.*, 2001).

The colon, which begins at the ileocecal valve and continues to the rectum, is approximately 1 m long in adults. The main functions of the colon are absorption of water and electrolytes and storage of waste products. Transport is accomplished by crypt and surface epithelial cells, which are functionally similar to those in the small-intestinal crypt and villus, although villi are not present (Figure 1.1). Bacterial load in the colonic lumen can be quite high, with typical yields of 10^{11} organisms per 1 ml. These commensal organisms typically coexist peacefully with the host and may serve beneficial roles (Abreu *et al.*, 2005; Fukata *et al.*, 2005; Neish *et al.*, 2000; Rakoff-Nahoum *et al.*, 2004). The presence of these commensal organisms may also prevent overgrowth of pathogenic organisms. For example, incomplete sterilization of the lumen by antibiotic treatment may allow overgrowth of *Clostridium difficile*, which release toxins that glucosylate and inactivate rho (Just *et al.*, 1995). This results in a severe and sometimes fatal colitis characterized by dense adherent pseudomembranes composed of purulent material.

RESPIRATORY TRACT

The proximal larynx is lined by stratified squamous epithelium. Although this is similar to that present in the esophagus, glycogenation is less prominent in the laryngeal epithelium. This gives way to pseudostratified columnar epithelium, although patches of squamous epithelium continue to be present distally. The columnar epithelium can vary in thickness from only a few cells to many cells. Small round reserve cells characterize the basal layer, while the most superficial layers are covered by ciliated epithelial cells and variable numbers of mucin-producing goblet cells. Together, these surface cells provide an important line of defense, as the cilia beat continually and propel a thin layer of mucin with entrapped foreign material upward and out of the airway. This mucociliary clearance is critical, as demonstrated by patients with Kartagener syndrome, in whom ciliary immotility results in recurrent infection (Eliasson *et al.*, 1977).

The larynx leads to the trachea and bronchi. These large-bore airways are encircled by cartilaginous rings, which prevent their collapse. The trachea and bronchi are lined by ciliated columnar epithelium similar to that seen in the distal larynx. *Corynebacterium diphtheriae* can colonize these sites and release a toxin that adenosine diphosphate (ADP)-ribosylates elongation factor 2, resulting in epithelial cell death. The mucosal necrosis and sloughing that follow result in accumulation of mucus, cellular debris, and inflammatory cells that obstruct the airways and kill by asphyxiation.

As the bronchi branch into smaller bronchioles, the pseudostratified epithelium becomes simpler, assuming a single-layer cuboidal morphology lined by ciliated and non-ciliated epithelial cells. The latter, termed Clara cells, secrete surfactant and serve as progenitor cells in repair after injury. The bronchioles terminate in the functional unit of the lung, the alveolus, where gas exchange takes place. Alveoli are lined by a thin layer of type 1 squamous and type 2 cuboidal pneumocytes (Figure 1.1). The terminally differentiated type 1 cells cover over 90% of the alveolar surface and provide a thin separation between the airspace and blood that allows for efficient gas exchange. Type 2 cells synthesize and secrete surfactant and also serve as alveolar stem cells, giving rise to additional type 2 cells as well as type 1 cells. In the presence of infection, for example with *Streptococcus pneumoniae*, the most common cause of community-acquired acute pneumonia, the alveoli may become filled with fluid and inflammatory cells, including neutrophils and macrophages. Infection spreads rapidly within contiguous alveoli, leading to consolidation of large areas of lung parenchyma and loss of the surface area necessary for gas exchange.

URINARY TRACT

The renal collecting ducts empty into the renal pelvis. The latter is lined by urothelium, which is histologically similar in the renal pelvis, ureters, and bladder. This epithelium is also termed transitional, as its appearance is intermediate between that of non-keratinizing stratified squamous epithelium, as seen in the esophagus, and pseudostratified columnar epithelium, like that of the bronchi. The thickness of the urothelium varies significantly, from only two to three cells thick in the renal pelvis to seven or more cells in the contracted bladder. As the bladder distends, the urothelium thins to two to three cells that are elongated parallel to the basement membrane.

Like other stratified epithelia, proliferation occurs in the basal zone, which is populated by small cuboidal cells. The surface is comprised of terminally differentiated cells. In the bladder, these surface cells are termed umbrella cells due to their large ellipsoid shape. Like deeper cells, the umbrella cells become flattened and difficult to appreciate in the distended bladder. They are often absent in urothelial carcinoma, making their presence a somewhat reassuring, although non-diagnostic, feature in bladder biopsies. Bacterial infections of the bladder and urethra include *Escherichia, Proteus, Klebsiella,* and *Enterobacter* and are more common in women than men due to the shorter length of the female urethra. Such infections can extend in a retrograde fashion to involve ureters, renal pelvis, and renal parenchyma in ascending pyelonephritis.

VECTORIAL TRANSPORT

Despite the differences in specialized cellular functions and morphology, all of these tissues are lined by a polarized epithelium that provides protection from the environment. Therefore, for the purpose of understanding cellular architecture, the structure of these polarized epithelia can be generalized. At the cellular level, polarization allows for asymmetric morphology as well as distinct molecular composition and function of membrane domains. The free surface, which faces the lumen, is referred to as apical, while the surfaces in contact with the basement membrane and adjacent cells are referred to as basal and lateral, respectively. The basal and lateral membrane domains typically function as a single unit, termed the basolateral domain. Specific proteins and lipids can be targeted to apical and basolateral membrane domains. For example, Na^{+} nutrient cotransporters are recruited to the apical membrane in transporting epithelia (Harris *et al.*, 1992). Similarly, the GM1 ganglioside used as a receptor for cholera toxin is also primarily apical (Lencer *et al.*, 1995; Wolf *et al.*, 1998). Other proteins, such as β1 integrins and interferon gamma (IFNγ) and tumor necrosis factor alpha (TNFα) receptors, predominantly localize to the basolateral surface in polarized epithelia (Boll *et al.*, 1991; Madara and Stafford, 1989; Moller *et al.*, 1994). Finally, some proteins, such as the polymeric immunoglobulin receptor, have more complex trafficking patterns. The polymeric immunoglobulin receptor is trafficked to the basolateral membrane first, where it picks up immunoglobulin cargo; then the receptor is endocytosed and moved to the apical membrane, where it releases the cargo (Casanova *et al.*, 1990, 1991; Mostov *et al.*, 1986).

In each case, the appropriately polarized distribution of proteins and lipids is necessary for proper epithelial function. For example, as noted above,

Na^+ nutrient cotransporter proteins are apically targeted in intestinal and renal tubular epithelium (Harris *et al.*, 1992; Turner *et al.*, 1996; Wright *et al.*, 1997). In the case of glucose absorption, glucose is transported into the cell against its concentration gradient by SGLT1, the apical intestinal Na^+/glucose cotransporter by Na^+ cotransport using the Na^+ gradient, from high outside the cell to low inside the cell, as the driving force (Hediger *et al.*, 1987). This Na^+ gradient is maintained by the Na^+/K^+ ATPase. The Na^+/K^+ ATPase is localized to the basolateral membrane along with the facilitated glucose transporter GLUT2 (Dominguez *et al.*, 1992; Miyamoto *et al.*, 1992). Thus, after SGLT1 transports Na^+ and glucose across the apical membrane, glucose and Na^+ exit the cytoplasm by crossing the basolateral membrane. This allows for directional net absorption and generates an ionic gradient that draws in water from the lumen. Similar mechanisms move water and solutes across the alveoli (Nielson and Lewis, 1990). Clearly, none of this would be possible in the absence of polarized membrane domains.

Membrane polarization is also essential for secretion. The cystic fibrosis transmembrane conductance regulator (CFTR) is a transporter responsible for apical cyclic adenosine monophosphate (cAMP)-activated anion conductance (Anderson *et al.*, 1991; Kartner *et al.*, 1991; Tabcharani *et al.*, 1991). Cystic fibrosis is linked to mutations in CFTR, and these mutations are responsible for the poorly hydrated viscous luminal accumulations that cause intestinal, pancreatic, and airway obstruction (Gregory *et al.*, 1990; Riordan *et al.*, 1989). In normal subjects without CFTR mutations, cholera toxin activates CFTR by elevating intracellular cAMP levels (Breuer *et al.*, 1992). This results in net transport of Cl^- ions into the intestinal lumen, followed by massive fluid accumulation (Field, 2003). Thus, cholera takes advantage of normal cellular processes to cause massive diarrhea, which spreads infection in areas with poor sanitation.

MEMBRANE TRAFFIC

In addition to transporters that allow solutes to cross apical and basolateral membrane domains, epithelia also utilize vesicular routes in order to traffic materials and to respond to extracellular stimuli. The trafficking of immunoglobulin A (IgA) from basolateral to apical surfaces by polymeric immunoglobulin receptor is a particularly well characterized example of such transport. As noted above, the polymeric immunoglobulin receptor is initially trafficked to the basolateral membrane. Here, it binds polymeric immunoglobulin, most prominently IgA (Casanova *et al.*, 1990, 1991; Mostov *et al.*, 1986). The binding of IgA triggers internalization of the complex and trafficking of the endocytic cargo to the apical surface. A proteolytic cleavage event at this surface results in release of IgA at the apical, or luminal, surface along with a portion of the receptor, termed the secretory component (Aroeti *et al.*, 1993; Casanova *et al.*, 1991; Song *et al.*, 1994). Although this finely tuned process plays an important role in mucosal immune protection, it has apparently not been able to escape the attention of *Streptococcus pneumoniae*. These bacteria exploit the presence of uncleaved polymeric immunoglobulin receptor at the apical membrane to bind to the secretory component portion of the receptor and invade host cells (Zhang *et al.*, 2000). The binding of *S. pneumoniae* to polymeric immunoglobulin receptor drives transcytosis in reverse, from apical to basolateral, thereby effectively transporting bacteria across the epithelium (Brock *et al.*, 2002). This pathway is clinically relevant, as the pattern of polymeric immunoglobulin receptor expression, greater in nasopharyngeal than distal respiratory epithelium, likely also explains the more frequent colonization of upper airways by *S. pneumoniae* (Zhang *et al.*, 2000).

In general, endocytosis in epithelial and other cells occurs by two main pathways, which are mediated by clathrin and caveolin. Clathrin-mediated endocytosis is best characterized in processes of macromolecule and integral membrane protein internalization. In this pathway, surface receptors bind cargo and initiate a signaling cascade. Receptor-mediated endocytosis of low-density lipoprotein (LDL) and transferrin are classic examples of endocytosis by clathrin-coated vesicles (Anderson *et al.*, 1978; Hewlett *et al.*, 1994). The clathrin coat is readily recognized by electron microscopy as a regular lattice of clathrin molecules that covers the cytoplasmic face of the membrane at sites of internalization (Pearse, 1976). Like other cellular processes, bacterial pathogens have evolved means of using clathrin-mediated endocytosis to their own advantage. For example, Shiga and Shiga-like toxins enter the cell through clathrin-mediated endocytosis (Sandvig and van Deurs, 1996, 2000). These toxins ultimately cross intracellular membranes to block ribosomal activity, resulting in inhibition of protein synthesis and cell death (Obrig *et al.*, 1987; Reisbig *et al.*, 1981).

The second major endocytic pathway occurs through caveolae, literally little caves, which were initially described morphologically (Palade and Bruns, 1968). Extensive work has defined cholesterol- and glycosphingolipid-rich microenvironments at these sites that often also contain proteins of the caveolin family (Dupree *et al.*, 1993; Lisanti *et al.*, 1993; Sargiacomo *et al.*, 1993). Although the biophysical structure of these lipid domains remains controversial, data do show that a variety of receptors and signaling molecules are concentrated in caveolae (Penela *et al.*, 2003). Thus, in addition to internalization of macromolecules such as micronutrients, chemokines, and hormones, caveolae have been proposed to be platforms for signal transduction (Penela *et al.*, 2003). Like clathrin-mediated endocytosis, bacterial pathogens have developed means of co-opting caveolin-mediated endocytosis. For example, cholera toxin binds to ganglioside GM1, which is concentrated in caveolae. This allows the toxin to enter the cell through caveolin-mediated endocytosis (Duncan *et al.*, 2002; Lencer *et al.*, 1995). Similarly, entry of *Chlamydia trachomatis* is associated with caveolin-mediated endocytosis (Duncan *et al.*, 2002; Norkin *et al.*, 2001).

CYTOSKELETON

The cytoskeleton serves as a structural framework for epithelial shape, supports membrane traffic, and plays an essential role in development and maintenance of polarity (Figure 1.2). In addition to actin, tubulin, and intermediate filaments, the epithelial cytoskeleton is composed of a complex array of accessory proteins that regulate filament assembly and movement. The intermediate filaments form a stable fiber system that surrounds the nucleus and extends to the plasma membrane (Brown *et al.*, 1983). In polarized epithelial cells, microtubules provide an important scaffold for cell shape. The microtubule organizing center, located near the nucleus, coordinates nucleation of microtubule polymerization in a defined polarized fashion (Murphy and Stearns, 1996). This polarity allows for directional transport of vesicles to and from specific membranes. In addition to roles in regulating placement of organelles and vesicular targeting, microtubules can also be used to enhance bacterial invasion. For example, *Campylobacter jejuni* have been shown to invade intestinal epithelial cells in a microtubule-dependent, actin-independent manner (Kopecko *et al.*, 2001; Oelschlaeger *et al.*, 1993). In ciliated respiratory epithelial cells, the core of each cilium is a precisely organized ring of microtubules. Dynein, a microtubule-based motor protein, causes the cilia to flex, resulting in the ciliary beating that is necessary for mucociliary clearance (Camner *et al.*, 1975). Patients with primary ciliary dyskinesia, including many with Kartagener syndrome, have mutations that shorten dynein arms, resulting in immotile cilia (Camner *et al.*, 1975). As discussed above, these patients are at increased risk for respiratory infection. Affected males also tend to be sterile, due to inefficiencies of sperm motility, another microtubule and dynein-dependent process.

In contrast to microtubules, actin filaments or microfilaments are found in close association with the plasma membrane of polarized epithelial cells. These filaments are present at apical, lateral, and basal membranes, but they

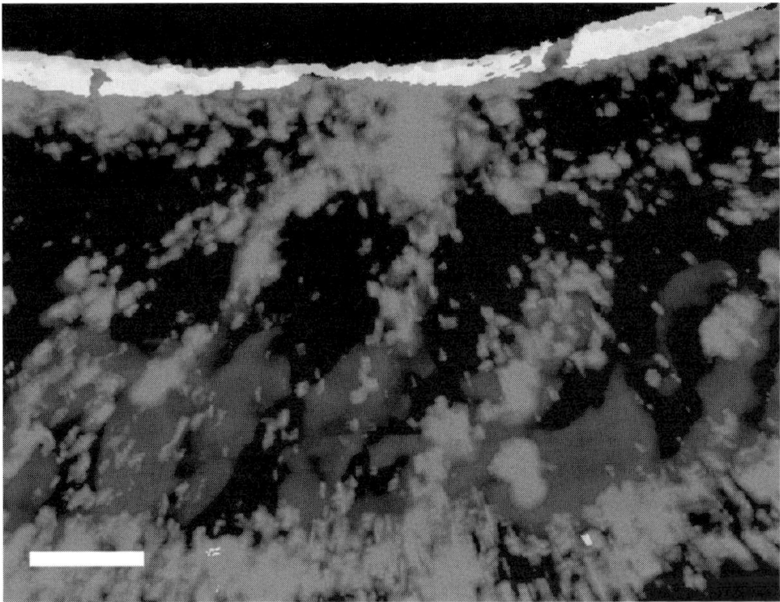

Figure 1.2 Three-dimensional reconstruction of human jejunal epithelium stained for tubulin (red), f-actin (green), and nuclei (blue). Note the prominent actin ring (green) concentrated at the level of the apical junctional complex. Actin is also present within the microvilli and along basolateral membranes but cannot be appreciated in this image due to the high concentration of actin at the perijunctional ring. Microtubules (red) extend throughout the cell and play a significant role in vesicular trafficking. Bar = 5 μm. See also Color plate 1.

are particularly concentrated at sites of cell–cell or cell–substrate contact. In intestinal epithelia, actin cores also form the substructure of microvilli. Moreover, a contractile circumferential actin belt encircles each epithelial cell at the level of the tight and adherens junctions. Contraction of this actomyosin belt can be regulated during normal tissue function and can be disrupted in disease. Microfilaments also play important roles in endocytosis, and the small GTPases of the rho family that regulate actin are frequent targets of bacterial pathogens. For example, *Salmonella* use the type III secretion system to inject a toxin that serves as a guanine exchange factor for cdc42 and rac (Hardt *et al.*, 1998; Lu and Walker, 2001; Parrello *et al.*, 2000). This induces actin-dependent membrane ruffling, leading ultimately to bacterial internalization. As discussed above, *Clostridium difficile* toxins glucosylate and inactivate rho (Hippenstiel *et al.*, 1997; Just *et al.*, 1994, 1995). This key regulator of actin structure is also targeted by *Clostridium botulinum* C3 toxin, which inactivates rho via ribosylation (Aktories *et al.*, 1992; Narumiya *et al.*, 1990; Wilde *et al.*, 2000).

INTERCELLULAR JUNCTIONS

Many cells function effectively as individual units. Examples include red blood cells, immune cells, and fibroblasts, which at most make occasional transient junctions with other cells. In contrast, epithelia are ineffectual as independent cells. Their function requires that they form an intact surface with a sustained network of junctions joining each cell to its neighbors and, for the basal cell layer, the matrix of the basement membrane. In the absence of these junctions, epithelia would be unable to form a barrier. Thus, efforts to achieve vectorial transport would be unsuccessful, as any transport could easily be negated by diffusion in the reverse direction. Additionally, the mere presence of a barrier is an essential component of epithelial structure. These barriers provide protection from external hazards, including airborne toxins, luminal bacteria, and noxious chemicals. Thus, as one might guess, bacteria have evolved to exploit, evade, and disrupt intercellular junctions.

Interactions between individual cells and the basement membrane are typically mediated by integrins. This family of proteins can interact with many matrix proteins, including laminin and collagen, as well as other cell types by generating a diverse array of molecules with differing alpha- and beta-chain compositions (Albelda *et al.*, 1994; Fuchs *et al.*, 1997; Giancotti and Ruoslahti 1999; Humphries 1990; Luscinskas and Lawler 1994). β1 integrins link the extracellular matrix to the intracellular actin cytoskeleton and can also serve as receptors for *Yersinia enterocolitica* and *Yersinia pseudotuberculosis* invasion (Isberg and Leong, 1990). In this case, since β1 integrins are localized to basolateral membrane domains, some disruption of the apical junctional complex must occur before *Yersinia* can access β1 integrins. Adjacent cells are also bound to one another by desmosomes, which are linked to intermediate filaments, thus providing a semi-rigid framework, and, in many cases, by gap junctions, which facilitate intercellular signaling by allowing diffusion of small molecules between the cytosol of adjacent cells.

Despite the importance of these structures to epithelial function, the components of the apical junctional complex may well be the most critical. This has not escaped the attention of pathogens, such as *Porphyromonas gingivalis* and *Bacteroides fragilis*, that attack the apical junctional complex by secreting proteases that degrade critical components of the complex (Katz *et al.*, 2000; Koshy *et al.*, 1996; Wu *et al.*, 1998).

Assembly and placement of the apical junctional complex are linked closely to the establishment of epithelial polarity. As implied by the name, the apical junctional complex is located at the most apical region of the lateral membrane. The complex consists of the tight junctions, which define the border between apical and basolateral membrane domains, and the adherens junctions, which are located just basal to the tight junctions. The principal proteins of the adherens junction mediate Ca^{+2}-dependent homotypic intercellular adhesions and are termed cadherins (Hatta and Takeichi, 1986). Several distinct cadherins can be expressed, and this differential expression forms the basis of self-recognition by different embryonic cell types (Takeichi, 1988). Epithelia express E-cadherin. The cytoplasmic tail of E-cadherin interacts directly with cytoplasmic scaffolding and signaling proteins, including β-catenin, and indirectly with the actin cytoskeleton (Itoh *et al.*, 1997; McCrea *et al.*, 1991; Ozawa *et al.*, 1990; Yonemura *et al.*, 1995). Thus, adherens junctions perform essential roles in signal transduction in addition to their function in maintenance of intercellular contact. Cadherins can also be used as landmarks by bacteria. For example, *Listeria monocytogenes* use E-cadherin to specifically target human epithelial cells (Mengaud, Lecuit *et al.*, 1996; Mengaud, Ohayon *et al.*, 1996). The bacterial protein internalin binds a specific sequence in E-cadherin that differs in other species (Lecuit *et al.*, 1999). This binding triggers lipid raft-mediated internalization of the E-cadherin–*Listeria* complex, thus facilitating bacterial invasion (Seveau *et al.*, 2004).

Although adherens junctions are thought to provide much of the strength that supports intercellular junctions, they are quite porous and do not form a significant barrier to paracellular diffusion. Thus, sealing of the paracellular space falls to the tight junction. This intercellular seal of variable permeability has been described as a "gate," allowing differential passage of water, ions, and other molecules (Diamond, 1977; Gumbiner, 1987). As noted above, tight junctions also physically separate apical and basolateral membranes, forming a "fence" that prevents diffusion of protein and lipid components of the plasma membrane, thereby maintaining distinct apical and basolateral membrane domains (Diamond, 1977; Gumbiner, 1987).

Although many tight-junction proteins have now been identified, their precise roles in assembly and maintenance of tight-junction structure are poorly understood. At least three classes of transmembrane proteins have been described (Furuse *et al.*, 1998). These include the claudin family, whose members appear to define the charge-selective permeability of tight junctions in different tissues, as demonstrated elegantly by in vitro (Van Itallie *et al.*, 2003) and in vivo (Simon *et al.*, 1999) studies. Specific claudin isoforms are targeted by *Clostridium perfringens* enterotoxin, which binds to and targets claudins 3 and 4 for degradation, resulting in loss of barrier function in epithelia that express these claudin isoforms (Sonoda *et al.*, 1999).

In contrast to claudins, the role of another tetramembrane-spanning protein, occludin, is less well defined. Numerous in vitro studies have suggested that occludin plays an important role in tight-junction function, including the observation that peptides derived from the second extracellular loop can disrupt epithelial barrier function (Nusrat *et al.*, 2005; Wong and Gumbiner, 1997). However, this and other in vitro observations suggesting the functional importance of occludin are challenged by the report that occludin knockout mice do not have readily apparent barrier defects (Saitou *et al.*, 2000). Nonetheless, several in vivo analyses suggest that occludin internalization is a reliable marker of tight junction disruption by enteropathogenic *Escherichia coli* (EPEC) and other inflammatory stimuli (Clayburgh *et al.*, 2005; Shifflett *et al.*, 2005). The precise function of occludin remains to be determined.

The third major family of described transmembrane tight junction proteins, junctional adhesion molecules, can localize to tight junctions as well as desmosomes. At these sites they appear to play a host of roles, including mediating transepithelial inflammatory cell migration (Zen *et al.*, 2004) and resealing of tight junctions after injury (Liu *et al.*, 2000). Finally, a long list of peripheral membrane proteins, including ZO-1, ZO-2, ZO-3, MUPP-1, and cingulin, have been described (Citi *et al.*, 1988; Haskins *et al.*, 1998; Jesaitis and Goodenough 1994; Stevenson *et al.*, 1986). These are able to maintain numerous binding interactions with one another and other proteins, leading to the assembly of a protein complex at the tight junction.

BARRIER REGULATION

The structure of the tight junction is critically dependent on an intact ring of actin and myosin, the perijunctional actomyosin ring, at the site of the apical junctional complex. Mere depolymerization of this actomyosin ring, as induced by toxin-mediated rho inactivation, is sufficient to cause massive disruption of barrier function (Hecht *et al.*, 1988; Nusrat *et al.*, 1995). Less massive regulation of tight-junction permeability also occurs in response to physiological stimuli. For example, activation of Na^+-nutrient cotransporters causes increases in paracellular permeability that allow paracellular amplification of transcellular nutrient and water absorption at appropriate times, e.g. after a nutrient-rich meal (Madara and Pappenheimer, 1987). In vivo and in vitro studies have shown that initiation of Na^+/glucose cotransport results in activation of a signaling cascade that leads to myosin light chain kinase (MLCK)-mediated myosin II regulatory light chain (MLC) phosphorylation, perijunctional actomyosin ring contraction, and increased paracellular permeability (Berglund *et al.*, 2001; Turner *et al.*, 1997). The distal portion of this pathway, that following MLCK activation, is also used by EPEC and enterohemorrhagic *Escherichia coli* (EHEC) (Philpott *et al.*, 1998; Spitz *et al.*, 1995; Yuhan *et al.*, 1997). Like initiation of Na^+/glucose cotransport (Turner *et al.*, 1997), EPEC infection results in MLC phosphorylation and increased paracellular permeability. In each case, MLCK inhibitors are able to partially or completely prevent increases in paracellular permeability, suggesting an essential role for MLCK in this process (Turner *et al.*, 1997; Yuhan *et al.*, 1997; Zolotarevsky *et al.*, 2002). Although the specific bacterial effectors responsible for EPEC-induced MLC phosphorylation have not been defined, it is clear that the presence of an intact type III secretion system is required, raising the possibility that the effector is injected directly into the epithelial cytoplasm (Simonovic *et al.*, 2000). In the case of EPEC, it also appears likely that upstream activation of protein kinase C zeta (PKC-ζ) contributes to MLCK activation (Tomson *et al.*, 2004).

Advances in our understanding of the biology of and, potentially, developing therapy for these barrier defects have been limited by the lack of effective and specific inhibitors of MLCK. This deficiency has now been overcome by the development of stable peptide analogs of a highly specific membrane-permeant MLCK inhibitor, termed PIK (Owens *et al.*, 2005; Zolotarevsky *et al.*, 2002). This inhibitor has been used to specifically probe the role of MLCK in intestinal epithelial barrier dysfunction. As noted above, epithelial barrier dysfunction is seen in a variety of infectious diseases of the intestines. As expected, PIK prevents much of the loss of barrier function that accompanies EPEC infection in vitro (Zolotarevsky *et al.*, 2002).

Epithelial barrier function is also compromised in inflammatory diseases such as Crohn's disease (Clayburgh, Shen *et al.*, 2004). In this case, it is thought that barrier disruption is mediated by local release of T helper type 1 (Th1) cytokines, including IFNγ and TNFα. Indeed, mucosal levels of these cytokines are elevated during the active phase of inflammatory bowel disease. Consistent with a role in barrier dysfunction, these cytokines are able to disrupt barrier function in cultured epithelial monolayers, and in vivo antagonism of IFNγ or TNFα can diminish disease severity and restore barrier function (Brown *et al.*, 1999; Ferrier *et al.*, 2003; Musch *et al.*, 2002; Suenaert *et al.*, 2002). To define the mechanisms of this cytokine-dependent barrier dysfunction, MLC phosphorylation was assessed in cultured monolayers after treatment with IFNγ and TNFα. Marked increases in MLC phosphorylation accompanied functional barrier loss (Zolotarevsky *et al.*, 2002). The MLCK

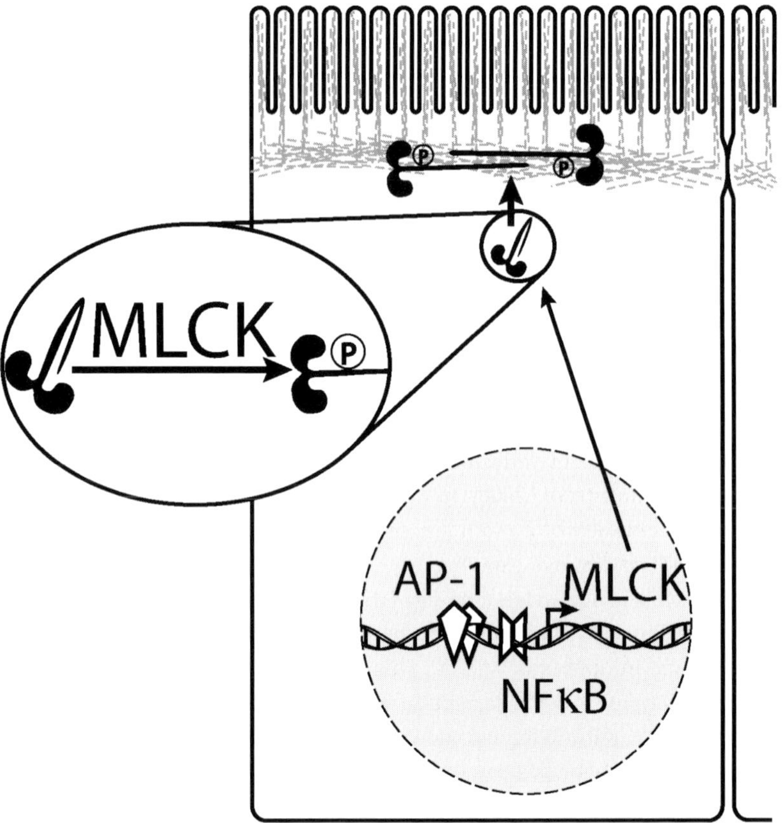

Figure 1.3 Model for epithelial barrier dysfunction secondary to myosin light chain kinase (MLCK) upregulation. Extracellular stimuli trigger a signaling pathway that activates activating protein 1 (AP-1) and/or nuclear factor kappa B (NFκB) transcription factors. These increase transcription of MLCK mRNA, thereby increasing the cytoplasmic pool of MLCK protein. MLCK induces myosin II regulatory light chain (MLC) phosphorylation and a conformational change in myosin. This activates myosin, which leads to contraction of the actomyosin ring and decreased tight-junction barrier function. Although the involvement of this pathway in cytokine-induced barrier dysfunction has been demonstrated, the role of MLCK transcriptional activation in bacterial pathogenesis has not been reported.

inhibitor PIK reduced MLC phosphorylation and normalized barrier function (Zolotarevsky *et al.*, 2002). Thus, MLC phosphorylation is necessary for this in vitro cytokine-dependent barrier dysfunction. Other studies suggest that this may also be the mechanism of acute in vivo cytokine-dependent barrier loss (Clayburgh *et al.*, 2005).

To better understand the mechanisms by which IFNγ and TNFα cause increases in MLC phosphorylation, expression of MLCK, the responsible kinase, was assessed following cytokine treatment. The data show that in cultured epithelial monolayers, MLCK protein expression is upregulated by cytokine treatment and that this correlates with elevated MLC phosphorylation and barrier dysfunction (Ma *et al.*, 2005; Wang *et al.*, 2005). Both IFNγ and TNFα are required for this upregulation of MLCK expression, but the role of IFNγ seems to be in priming the cells to respond to TNFα (Wang *et al.*, 2005). In the absence of IFNγ priming, TNFα is unable to induce changes in barrier function, MLC phosphorylation, or MLCK expression (Wang *et al.*, 2005). Some authors have suggested that these effects might be mediated by nuclear factor kappa B (NFκB), a canonical signaling pathway activated by TNF-α (Ma *et al.*, 2004, 2005). However, other data demonstrate that inhibition of NFκB does not reduce cytokine-dependent barrier dysfunction, MLC phosphorylation, or MLCK upregulation (Wang *et al.*, 2005). The latter observations are consistent with in vitro and in vivo studies suggesting a protective role for epithelial NFκB following cytokine exposure (Chen *et al.*, 2003; Soler *et al.*, 1999).

To better define the mechanisms by which these cytokines and other stimuli, including pathogens, regulate MLCK expression, 4 kilobases of DNA sequence upstream of the MLCK transcriptional start site was cloned (Graham *et al.*, 2004). When fused to a reporter gene and transfected into intestinal epithelial cells, this construct responded to IFNγ and TNFα like endogenous MLCK (Graham *et al.*, 2004). In silico sequence analysis identified numerous possible transcription factor binding sites, including four putative activating protein 1 (AP-1) binding sites and two putative NFκB binding sites. Several of these sites were confirmed to be functional protein binding sites based on electrophoretic mobility shift assay (Graham *et al.*, 2005). Cotransfection with constitutively active AP-1 and NFκB showed that activation of either transcription factor pathway was able to upregulate MLCK transcription (Graham *et al.*, 2005). Mutational analysis confirmed that these AP-1 and NFκB sites were indeed necessary for MLCK transcriptional activation (Figure 1.3). Further analysis showed clearly that in these differentiated epithelial monolayers, AP-1 activation is responsible for the observed increase in MLCK transcription after cytokine exposure (Graham *et al.*, 2005). Together, these data indicate that AP-1 induces MLCK upregulation in response to IFNγ and TNFα, but they also suggest that other stimuli, including infection by *H. pylori* or EPEC, may upregulate MLCK expression via NFκB activation (Graham *et al.*, 2005; Keates *et al.*, 1997; Munzenmaier *et al.*, 1997; Savkovic *et al.*, 1997; Sharma *et al.*, 1998). Thus, although this has not yet been demonstrated, it may be another example of bacterial exploitation of host regulatory mechanisms.

CONCLUSIONS

Epithelia perform many specialized functions that are essential to homeostasis as well as host defense. In many cases, these are exploited effectively by bacteria in their efforts to achieve infection and overcome the epithelial barrier.

ACKNOWLEDGMENTS

We are grateful for the excellent technical assistance of Edwina Witkowski in the preparation of Figure 1.2.

Supported by the National Institutes of Health (DK061931 and DK068271), the Crohn's and Colitis Foundation of America, and the University of Chicago Digestive Disease Center (P30 DK42086).

REFERENCES

Abreu, M. T., Fukata, M., and Arditi, M. (2005). TLR signaling in the gut in health and disease. *J. Immunol.* **174**, 4453–4460.

Aktories, K., Mohr, C., and Koch, G. (1992). *Clostridium botulinum* C3 ADP-ribosyltransferase. *Curr. Top. Microbiol. Immunol.* **175**, 115–131.

Albelda, S. M., Smith, C. W., and Ward, P. A. (1994). Adhesion molecules and inflammatory injury. *FASEB J.* **8**, 504–512.

Anderson, M. P., Gregory, R. J., Thompson, S., *et al.* (1991). Demonstration that CFTR is a chloride channel by alteration of its anion selectivity. *Science* **253**, 202–205.

Anderson, R. G., Vasile, E., Mello, R. J., Brown, M. S., and Goldstein, J. L. (1978). Immunocytochemical visualization of coated pits and vesicles in human fibroblasts: relation to low density lipoprotein receptor distribution. *Cell* **15**, 919–933.

Aroeti, B., Kosen, P. A., Kuntz, I. D., Cohen, F. E., and Mostov, K. E. (1993). Mutational and secondary structural analysis of the basolateral sorting signal of the polymeric immunoglobulin receptor. *J. Cell Biol.* **123**, 1149–1160.

Berglund, J. J., Riegler, M., Zolotarevsky, Y., Wenzl, E., and Turner, J. R. (2001). Regulation of human jejunal transmucosal resistance and MLC phosphorylation by Na^+-glucose cotransport. *Am. J. Physiol. Gastrointest. Liver Physiol.* **281**, G1487–1493.

Boll, W., Partin, J. S., Katz, A. I., Caplan, M. J., and Jamieson, J. D. (1991). Distinct pathways for basolateral targeting of membrane and secretory proteins in polarized epithelial cells. *Proc. Natl. Acad. Sci. U. S. A.* **88**, 8592–8596.

Breuer, W., Kartner, N., Riordan, J. R., and Cabantchik, Z. I. (1992). Induction of expression of the cystic fibrosis transmembrane conductance regulator. *J. Biol. Chem.* **267**, 10 465–10 469.

Brock, S. C., McGraw, P. A., Wright, P. F., and Crowe, J. E., Jr (2002). The human polymeric immunoglobulin receptor facilitates invasion of epithelial cells by *Streptococcus pneumoniae* in a strain-specific and cell type-specific manner. *Infect. Immun.* **70**, 5091–5095.

Brown, D. T., Anderton, B. H., and Wylie, C. C. (1983). The organization of intermediate filaments in normal human colonic epithelium and colonic carcinoma cells. *Int. J. Cancer* **32**, 163–169.

Brown, G. R., Lindberg, G., Meddings, J., *et al.* (1999). Tumor necrosis factor inhibitor ameliorates murine intestinal graft-versus-host disease. *Gastroenterology* **116**, 593–601.

Camner, P., Mossberg, B., and Afzelius, B. A. (1975). Evidence of congenitally nonfunctioning cilia in the tracheobronchial tract in two subjects. *Am. Rev. Respir. Dis.* **112**, 807–809.

Casanova, J. E., Breitfeld, P. P., Ross, S. A., and Mostov, K. E. (1990). Phosphorylation of the polymeric immunoglobulin receptor required for its efficient transcytosis. *Science* **248**, 742–745.

Casanova, J. E., Apodaca, G., and Mostov, K. E. (1991). An autonomous signal for basolateral sorting in the cytoplasmic domain of the polymeric immunoglobulin receptor. *Cell* **66**, 65–75.

Chen, L. W., Egan, L., Li, Z. W., *et al.* (2003). The two faces of IKK and NF-kappaB inhibition: prevention of systemic inflammation but increased local injury following intestinal ischemia-reperfusion. *Nat. Med.* **9**, 575–581.

Citi, S., Sabanay, H., Jakes, R., Geiger, B., and Kendrick-Jones, J. (1988). Cingulin, a new peripheral component of tight junctions. *Nature* **333**, 272–275.

Clayburgh, D. R., Rosen, S., Witkowski, E. D., *et al.* (2004). A differentiation-dependent splice variant of myosin light chain kinase, MLCK1, regulates epithelial tight junction permeability. *J. Biol. Chem.* **279**, 55 506–55 513.

Clayburgh, D. R., Shen, L., and Turner, J. R. (2004). A porous defense: the leaky epithelial barrier in intestinal disease. *Lab. Invest.* **84**, 282–291.

Clayburgh, D. R., Barrett, T. A., Tang, Y., *et al.* (2005). Epithelial myosin light chain kinase-dependent barrier dysfunction mediates T cell activation-induced diarrhea in vivo. *J. Clin. Invest.* **115**, 2702–2715.

Diamond, J. M. (1977). Twenty-first Bowditch lecture: The epithelial junction: bridge, gate, and fence. *Physiologist* **20**, 10–18.

Dominguez, J. H., Camp, K., Maianu, L., and Garvey, W. T. (1992). Glucose transporters of rat proximal tubule: differential expression and subcellular distribution. *Am. J. Physiol.* **262**, F807–812.

Duncan, M. J., Shin, J. S., and Abraham, S. N. (2002). Microbial entry through caveolae: variations on a theme. *Cell. Microbiol.* **4**, 783–791.

Dupree, P., Parton, R. G., Raposo, G., Kurzchalia, T. V., and Simons, K. (1993). Caveolae and sorting in the trans-Golgi network of epithelial cells. *EMBO J.* **12**, 1597–1605.

Eliasson, R., Mossberg, B., Camner, P., and Afzelius, B. A. (1977). The immotile-cilia syndrome: a congenital ciliary abnormality as an etiologic factor in chronic airway infections and male sterility. *N. Engl. J. Med.* **297**, 1–6.

Ferrier, L., Mazelin, L., Cenac, N., *et al.* (2003). Stress-induced disruption of colonic epithelial barrier: role of interferon-gamma and myosin light chain kinase in mice. *Gastroenterology* **125**, 795–804.

Field, M. (2003). Intestinal ion transport and the pathophysiology of diarrhea. *J. Clin. Invest.* **111**, 931–943.

Fuchs, E., Dowling, J., Segre, J., Lo, S. H., and Yu, Q. C. (1997). Integrators of epidermal growth and differentiation: distinct functions for beta 1 and beta 4 integrins. *Curr. Opin. Genet. Dev.* **7**, 672–682.

Fukata, M., Michelsen, K. S., Eri, R., *et al.* (2005). Toll-like receptor-4 is required for intestinal response to epithelial injury and limiting bacterial translocation in a murine model of acute colitis. *Am. J. Physiol. Gastrointest. Liver Physiol.* **288**, G1055–1065.

Furuse, M., Fujita, K., Hiiragi, T., Fujimoto, K., and Tsukita, S. (1998). Claudin-1 and -2: novel integral membrane proteins localizing at tight junctions with no sequence similarity to occludin. *J. Cell Biol.* **141**, 1539–1550.

Giancotti, F. G., and Ruoslahti, E. (1999). Integrin signaling. *Science* **285**, 1028–1032.

Graham, W. V., Wang, F., Wang, Y., and Turner, J. R. (2004). Transcriptional activation of myosin light chain kinase mediates TNFalpha-induced barrier dysfunction. *Gastroenterology* **126**, abstract.

Graham, W. V., Wang, F., Wang, Y., *et al.* (2005). AP1-dependent transcriptional activation of myosin light chain kinase (MLCK) mediates IFNgamma-TNFalpha-induced barrier dysfunction. *FASEB J.* **19**, abstract.

Gregory, R. J., Cheng, S. H., Rich, D. P., *et al.* (1990). Expression and characterization of the cystic fibrosis transmembrane conductance regulator. *Nature* **347**, 382–386.

Gumbiner, B. (1987). Structure, biochemistry, and assembly of epithelial tight junctions. *Am. J. Physiol.* **253**, C749–758.

Hardt, W. D., Chen, L. M., Schuebel, K. E., Bustelo, X. R., and Galan, J. E. (1998). *S. typhimurium* encodes an activator of Rho GTPases that induces membrane ruffling and nuclear responses in host cells. *Cell* **93**, 815–826.

Harris, D. S., Slot, J. W., Geuze, H. J., and James, D. E. (1992). Polarized distribution of glucose transporter isoforms in Caco-2 cells. *Proc. Natl. Acad. Sci. U. S. A.* **89**, 7556–7560.

Haskins, J., Gu, L., Wittchen, E. S., Hibbard, J., and Stevenson, B. R. (1998). ZO-3, a novel member of the MAGUK protein family found at the tight junction, interacts with ZO-1 and occludin. *J. Cell Biol.* **141**, 199–208.

Hatta, K. and Takeichi, M. (1986). Expression of N-cadherin adhesion molecules associated with early morphogenetic events in chick development. *Nature* **320**, 447–449.

Hecht, G., Pothoulakis, C., LaMont, J. T., and Madara, J. L. (1988). *Clostridium difficile* toxin A perturbs cytoskeletal structure and tight junction permeability of cultured human intestinal epithelial monolayers. *J. Clin. Invest.* **82**, 1516–1524.

Hediger, M. A., Coady, M. J., Ikeda, T. S., and Wright, E. M. (1987). Expression cloning and cDNA sequencing of the Na^+/glucose co-transporter. *Nature* **330**, 379–381.

Hewlett, L. J., Prescott, A. R., and Watts, C. (1994). The coated pit and macropinocytic pathways serve distinct endosome populations. *J. Cell Biol.* **124**, 689–703.

Hippenstiel, S., Tannert-Otto, S., Vollrath, N., *et al.* (1997). Glucosylation of small GTP-binding Rho proteins disrupts endothelial barrier function. *Am. J. Physiol.* **272**, L38–43.

Humphries, M. J. (1990). The molecular basis and specificity of integrin-ligand interactions. *J. Cell. Sci.* **97 (Pt 4)**, 585–592.

Isberg, R. R. and Leong, J. M. (1990). Multiple beta 1 chain integrins are receptors for invasin, a protein that promotes bacterial penetration into mammalian cells. *Cell* **60**, 861–871.

Itoh, M., Nagafuchi, A., Moroi, S., and Tsukita, S. (1997). Involvement of ZO-1 in cadherin-based cell adhesion through its direct binding to alpha catenin and actin filaments. *J. Cell Biol.* **138**, 181–192.

Jesaitis, L. A. and Goodenough, D. A. (1994). Molecular characterization and tissue distribution of ZO-2, a tight junction protein homologous to ZO-1 and the Drosophila discs-large tumor suppressor protein. *J. Cell Biol.* **124**, 949–961.

Just, I., Fritz, G., Aktories, K., *et al.* (1994). *Clostridium difficile* toxin B acts on the GTP-binding protein Rho. *J. Biol. Chem.* **269**, 10 706–10 712.

Just, I., Selzer, J., Wilm, M., *et al.* (1995). Glucosylation of Rho proteins by *Clostridium difficile* toxin B. *Nature* **375**, 500–503.

Kartner, N., Hanrahan, J. W., Jensen, T. J., *et al.* (1991). Expression of the cystic fibrosis gene in non-epithelial invertebrate cells produces a regulated anion conductance. *Cell* **64**, 681–691.

Katz, J., Sambandam, V., Wu, J. H., Michalek, S. M., and Balkovetz, D. F. (2000). Characterization of *Porphyromonas gingivalis*-induced degradation of epithelial cell junctional complexes. *Infect. Immun.* **68**, 1441–1449.

Keates, S., Hitti, Y. S., Upton, M., and Kelly, C. P. (1997). *Helicobacter pylori* infection activates NF-kappa B in gastric epithelial cells. *Gastroenterology* **113**, 1099–1109

Kopecko, D. J., Hu, L., and Zaal, K. J. M. (2001) Campylobacter jejuni: microtubule-dependent invasion. *Trends Microbiol.* **9**, 389–396.

Koshy, S. S., Montrose, M. H., and Sears, C. L. (1996). Human intestinal epithelial cells swell and demonstrate actin rearrangement in response to the metalloprotease toxin of *Bacteroides fragilis*. *Infect. Immun.* **64**, 5022–5028.

Lecuit, M., Dramsi, S., Gottardi, C., *et al.* (1999). A single amino acid in E-cadherin responsible for host specificity towards the human pathogen *Listeria monocytogenes*. *EMBO J.* **18**, 3956–3963.

Lencer, W. I., Moe, S., Rufo, P. A., and Madara, J. L. (1995). Transcytosis of cholera toxin subunits across model human intestinal epithelia. *Proc. Natl. Acad. Sci. U. S. A.* **92**, 10 094–10 098.

Lencer, W. I., Hirst, T. R., and Holmes, R. K. (1999). Membrane traffic and the cellular uptake of cholera toxin. *Biochim. Biophys. Acta* **1450**, 177–190.

Lisanti, M. P., Tang, Z. L., and Sargiacomo, M. (1993). Caveolin forms a hetero-oligomeric protein complex that interacts with an apical GPI-linked protein: implications for the biogenesis of caveolae. *J. Cell Biol.* **123**, 595–604.

Liu, Y., Nusrat, A., Schnell, F. J., *et al.* (2000). Human junction adhesion molecule regulates tight junction resealing in epithelia. *J. Cell Sci.* **113 (Pt 13)**, 2363–2374.

Llor, X., Serfas, M. S., Bie, W., *et al.* (1999). BRK/Sik expression in the gastrointestinal tract and in colon tumors. *Clin. Cancer Res.* **5**, 1767–1777.

Lu, L. and Walker, W. A. (2001). Pathologic and physiologic interactions of bacteria with the gastrointestinal epithelium. *Am. J. Clin. Nutr.* **73**, 1124S–1130S.

Luscinskas, F. W., and Lawler, J. (1994) Integrins as dynamic regulators of vascular function. *FASEB J.* **8**, 929–938.

Ma, T. Y., Iwamoto, G. K., Hoa, N. T., *et al.* (2004). TNF-alpha-induced increase in intestinal epithelial tight junction permeability requires NF-kappa B activation. *Am. J. Physiol. Gastrointest. Liver Physiol.* **286**, G367–376.

Ma, T. Y., Boivin, M. A., Ye, D., Pedram, A., and Said, H. M. (2005). Mechanism of TNF-α modulation of Caco-2 intestinal epithelial tight junction barrier: role of myosin light-chain kinase protein expression. *Am. J. Physiol. Gastrointest. Liver Physiol.* **288**, G422–430.

Madara, J. L. (1990). Maintenance of the macromolecular barrier at cell extrusion sites in intestinal epithelium: physiological rearrangement of tight junctions. *J. Membr. Biol.* **116**, 177–184.

Madara, J. L. and Pappenheimer, J. R. (1987). Structural basis for physiological regulation of paracellular pathways in intestinal epithelia. *J. Membr. Biol.* **100**, 149–164.

Madara, J. L. and Stafford, J. (1989). Interferon-gamma directly affects barrier function of cultured intestinal epithelial monolayers. *J. Clin. Invest.* **83**, 724–727.

McCrea, P. D., Turck, C. W., and Gumbiner, B. (1991). A homolog of the armadillo protein in Drosophila (plakoglobin) associated with E-cadherin. *Science* **254**, 1359–1361.

Mengaud, J., Lecuit, M., Lebrun, M., *et al.* (1996). Antibodies to the leucine-rich repeat region of internalin block entry of *Listeria monocytogenes* into cells expressing E-cadherin. *Infect. Immun.* **64**, 5430–5433.

Mengaud, J., Ohayon, H., Gounon, P., Mege, R. M., and Cossart, P. (1996). E-cadherin is the receptor for internalin, a surface protein required for entry of *L. monocytogenes* into epithelial cells. *Cell* **84**, 923–932.

Miyamoto, K., Takagi, T., Fujii, T., *et al.* (1992). Role of liver-type glucose transporter (GLUT2) in transport across the basolateral membrane in rat jejunum. *FEBS Lett.* **314**, 466–470.

Moller, P., Koretz, K., Leithauser, F., *et al.* (1994). Expression of APO-1 (CD95), a member of the NGF/TNF receptor superfamily, in normal and neoplastic colon epithelium. *Int. J. Cancer* **57**, 371–377.

Mostov, K. E., de Bruyn Kops, A., and Deitcher, D. L. (1986). Deletion of the cytoplasmic domain of the polymeric immunoglobulin receptor prevents basolateral localization and endocytosis. *Cell* **47**, 359–364.

Munzenmaier, A., Lange, C., Glocker, E., *et al.* (1997). A secreted/shed product of *Helicobacter pylori* activates transcription factor nuclear factor-kappa B. *J. Immunol.* **159**, 6140–6147.

Murphy, S. M. and Stearns, T. (1996). Cytoskeleton: microtubule nucleation takes shape. *Curr. Biol.* **6**, 642–644.

Musch, M. W., Clarke, L. L., Mamah, D., *et al.* (2002). T cell activation causes diarrhea by increasing intestinal permeability and inhibiting epithelial Na^+/K^+-ATPase. *J. Clin. Invest.* **110**, 1739–1747.

Narumiya, S., Morii, N., Sekine, A., and Kozaki, S. (1990). ADP-ribosylation of the rho/rac gene products by botulinum ADP-ribosyltransferase: identity of the enzyme and effects on protein and cell functions. *J. Physiol. (Paris)* **84**, 267–272.

Neish, A. S., Gewirtz, A. T., Zeng, H., *et al.* (2000). Prokaryotic regulation of epithelial responses by inhibition of IkappaB-alpha ubiquitination. *Science* **289**, 1560–1563.

Nielson, D. W. and Lewis, M. B. (1990). Effects of amiloride on alveolar epithelial PD and fluid composition in rabbits. *Am. J. Physiol.* **258**, L215–219.

Norkin, L. C., Wolfrom, S. A., and Stuart, E. S. (2001). Association of caveolin with *Chlamydia trachomatis* inclusions at early and late stages of infection. *Exp. Cell. Res.* **266**, 229–238.

Nusrat, A., Giry, M., Turner, J. R., *et al.* (1995). Rho protein regulates tight junctions and perijunctional actin organization in polarized epithelia. *Proc. Natl. Acad. Sci. U. S. A.* **92**, 10 629–10 633.

Nusrat, A., Brown, G. T., Tom, J., *et al.* (2005). Multiple protein interactions involving proposed extracellular loop domains of the tight junction protein occludin. *Mol. Biol. Cell* **16**, 1725–1734.

Obrig, T. G., Moran, T. P., and Brown, J. E. (1987). The mode of action of Shiga toxin on peptide elongation of eukaryotic protein synthesis. *Biochem. J.* **244**, 287–294.

Oelschlaeger, T. A., Guerry, P., and Kopecko, D. J. (1993). Unusual microtubule-dependent endocytosis mechanisms triggered by *Campylobacter jejuni* and *Citrobacter freundii*. *Proc. Natl. Acad. Sci. U. S. A.* **90**, 6884–6888.

Owens, S., Graham, W. V., Siccardi, D., Turner, J. R., and Mrsny, R. J. (2005). A strategy to identify stable membrane-permeant peptide inhibitors of myosin light chain kinase. *Pharm. Res.* **22**, 703–709.

Ozawa, M., Ringwald, M., and Kemler, R. (1990). Uvomorulin-catenin complex formation is regulated by a specific domain in the cytoplasmic region of the cell adhesion molecule. *Proc. Natl. Acad. Sci. U. S. A.* **87**, 4246–4250.

Palade, G. E. and Bruns, R. R. (1968). Structural modulations of plasmalemmal vesicles. *J. Cell Biol.* **37**, 633–649.

Parrello, T., Monteleone, G., Cucchiara, S., *et al.* (2000). Up-regulation of the IL-12 receptor beta 2 chain in Crohn's disease. *J. Immunol.* **165**, 7234–7239.

Pearse, B. M. (1976). Clathrin: a unique protein associated with intracellular transfer of membrane by coated vesicles. *Proc. Natl. Acad. Sci. U. S. A.* **73**, 1255–1259.

Penela, P., Ribas, C., and Mayor, F., Jr (2003). Mechanisms of regulation of the expression and function of G protein-coupled receptor kinases. *Cell. Signal.* **15**, 973–981.

Philpott, D. J., McKay, D. M., Mak, W., Perdue, M. H., and Sherman, P. M. (1998). Signal transduction pathways involved in enterohemorrhagic *Escherichia coli*-induced alterations in T84 epithelial permeability. *Infect. Immun.* **66**, 1680–1687.

Rakoff-Nahoum, S., Paglino, J., Eslami-Varzaneh, F., Edberg, S., and Medzhitov, R. (2004). Recognition of commensal microflora by toll-like receptors is required for intestinal homeostasis. *Cell* **118**, 229–241.

Reisbig, R., Olsnes, S., and Eiklid, K. (1981). The cytotoxic activity of *Shigella* toxin: evidence for catalytic inactivation of the 60 S ribosomal subunit. *J. Biol. Chem.* **256**, 8739–8744.

Riordan, J. R., Rommens, J. M., Kerem, B., *et al.* (1989). Identification of the cystic fibrosis gene: cloning and characterization of complementary DNA. *Science* **245**, 1066–1073.

Rosenblatt, J., Raff, M. C., and Cramer, L. P. (2001). An epithelial cell destined for apoptosis signals its neighbors to extrude it by an actin- and myosin-dependent mechanism. *Curr. Biol.* **11**, 1847–1857.

Saitou, M., Furuse, M., Sasaki, H., *et al.* (2000). Complex phenotype of mice lacking occludin, a component of tight junction strands. *Mol. Biol. Cell* **11**, 4131–4142.

Sandvig, K. and van Deurs, B. (1996). Endocytosis, intracellular transport, and cytotoxic action of Shiga toxin and ricin. *Physiol. Rev.* **76**, 949–966.

Sandvig, K. and van Deurs, B. (2000). Entry of ricin and Shiga toxin into cells: molecular mechanisms and medical perspectives. *EMBO J.* **19**, 5943–5950.

Sargiacomo, M., Sudol, M., Tang, Z., and Lisanti, M. P. (1993). Signal transducing molecules and glycosyl-phosphatidylinositol-linked proteins form a caveolin-rich insoluble complex in MDCK cells. *J. Cell Biol.* **122**, 789–807.

Savkovic, S. D., Koutsouris, A., and Hecht, G. (1997). Activation of NF-kappaB in intestinal epithelial cells by enteropathogenic *Escherichia coli*. *Am. J. Physiol.* **273**, C1160–1167.

Seveau, S., Bierne, H., Giroux, S., Prevost, M. C., and Cossart, P. (2004). Role of lipid rafts in E-cadherin- and HGF-R/Met-mediated entry of *Listeria monocytogenes* into host cells. *J. Cell. Biol.* **166**, 743–753.

Sharma, S. A., Tummuru, M. K., Blaser, M. J., and Kerr, L. D. (1998). Activation of IL-8 gene expression by *Helicobacter pylori* is regulated by transcription factor nuclear factor-kappa B in gastric epithelial cells. *J. Immunol.* **160**, 2401–2407.

Shifflett, D. E. Clayburgh, D. R. Koutsouris, A., Turner, J. R., and Hecht, G. A. (2005). Enteropathogenic *E. coli* disrupts tight junction barrier function and structure in vivo. *Lab. Invest.* **85**, 1308–1324.

Simon, D. B., Lu, Y., Choate, K. A., *et al.* (1999) Paracellin-1, a renal tight junction protein required for paracellular Mg^{2+} resorption. *Science* **285**, 103–106.

Simonovic, I., Rosenberg, J., Koutsouris, A., and Hecht, G. (2000). Enteropathogenic *Escherichia coli* dephosphorylates and dissociates occludin from intestinal epithelial tight junctions. *Cell. Microbiol.* **2**, 305–315.

Soler, A. P., Marano, C. W., Bryans, M., *et al.* (1999). Activation of NF-kappaB is necessary for the restoration of the barrier function of an epithelium undergoing TNF-alpha-induced apoptosis. *Eur. J. Cell. Biol.* **78**, 56–66.

Song, W., Bomsel, M., Casanova, J., *et al.* (1994). Stimulation of transcytosis of the polymeric immunoglobulin receptor by dimeric: IgA an autonomous signal for basolateral sorting in the cytoplasmic domain of the polymeric immunoglobulin receptor. *Proc. Natl. Acad. Sci. U. S. A.* **91**, 163–166.

Sonoda, N., Furuse, M., Sasaki, H., *et al.* (1999). *Clostridium perfringens* enterotoxin fragment removes specific claudins from tight junction strands: evidence for direct involvement of claudins in tight junction barrier. *J. Cell. Biol.* **147**, 195–204.

Spitz, J., Yuhan, R., Koutsouris, A., *et al.* (1995). Enteropathogenic *Escherichia coli* adherence to intestinal epithelial monolayers diminishes barrier function. *Am. J. of Physiol. Gastrointest. Liver Physiol.* **268**, G374–G379.

Stevenson, B. R., Siliciano, J. D., Mooseker, M. S., and Goodenough, D. A. (1986). Identification of ZO-1: a high molecular weight polypeptide associated with the tight junction (Zonula Occludens) in a variety of epithelia. *J. Cell. Biol.* **103**, 755–766.

Suenaert, P., Bulteel, V., Lemmens, L., *et al.* (2002). Anti-tumor necrosis factor treatment restores the gut barrier in Crohn's disease. *Am. J. Gastroenterol.* **97**, 2000–2004.

Tabcharani, J. A., Chang, X. B., Riordan, J. R., and Hanrahan, J. W. (1991). Phosphorylation-regulated Cl- channel in CHO cells stably expressing the cystic fibrosis gene. *Nature* **352**, 628–631.

Takeichi, M. (1988). The cadherins: cell–cell adhesion molecules controlling animal morphogenesis. *Development* **102**, 639–655.

Tomson, F. L., Koutsouris, A., Viswanathan, V. K., *et al.* (2004). Differing roles of protein kinase C-zeta in disruption of tight junction barrier by enteropathogenic and enterohemorrhagic *Escherichia coli. Gastroenterology* **127**, 859–869.

Turner, J. R., Lencer, W. I., Carlson, S., and Madara, J. L. (1996). Carboxy-terminal vesicular stomatitis virus G protein-tagged intestinal Na^+-dependent glucose cotransporter (SGLT1): maintenance of surface expression and global transport function with selective perturbation of transport kinetics and polarized expression. *J. Biol. Chem.* **271**, 7738–7744.

Turner, J. R., Rill, B. K., Carlson, S. L., *et al.* (1997). Physiological regulation of epithelial tight junctions is associated with myosin light-chain phosphorylation. *Am. J. Physiol. Cell Physiol.* **273**, C1378–C1385.

Van Itallie, C. M., Fanning, A. S., and Anderson, J. M. (2003). Reversal of charge selectivity in cation or anion-selective epithelial lines by expression of different claudins. *Am. J. Physiol. Renal Physiol.* **285**, F1078–1084.

Wang, F., Graham, W. V., Wang, Y., *et al.* (2005). Interferon-gamma and tumor necrosis factor-alpha synergize to induce intestinal epithelial barrier dysfunction by up-regulating myosin light chain kinase expression. *Am. J. Pathol.* **166**, 409–419.

Wilde, C., Genth, H., Aktories, K., and Just, I. (2000). Recognition of RhoA by *Clostridium botulinum* C3 exoenzyme. *J. Biol. Chem.* **275**, 16 478–16 483.

Wolf, A. A., Jobling, M. G., Wimer-Mackin, S., *et al.* (1998). Ganglioside structure dictates signal transduction by cholera toxin and association with caveolae-like membrane domains in polarized epithelia. *J. Cell Biol.* **141**, 917–927.

Wong, V. and Gumbiner, B. M. (1997). A synthetic peptide corresponding to the extracellular domain of occludin perturbs the tight junction permeability barrier. *J. Cell Biol.* **136**, 399–409.

Wright, E. M., Hirsch, J. R., Loo, D. D., and Zampighi, G. A. (1997). Regulation of Na^+/glucose cotransporters. *J. Exp. Biol.* **200**, 287–293.

Wu, S., Lim, K. C., Huang, J., Saidi, R. F., and Sears, C. L. (1998). *Bacteroides fragilis* enterotoxin cleaves the zonula adherens protein, E-cadherin. *Proc. Natl. Acad. Sci. U. S. A.* **95**, 14 979–14 984.

Yonemura, S., Itoh, M., Nagafuchi, A., and Tsukita, S. (1995). Cell-to-cell adherens junction formation and actin filament organization: similarities and differences between non-polarized fibroblasts and polarized epithelial cells. *J. Cell. Sci.* **108 (Pt 1)**, 127–142.

Yuhan, R., Koutsouris, A., Savkovic, S. D., and Hecht, G. (1997). Enteropathogenic *Escherichia coli*-induced myosin light chain phosphorylation alters intestinal epithelial permeability. *Gastroenterology* **113**, 1873–1882.

Zen, K., Babbin, B. A., Liu, Y., *et al.* (2004). JAM-C is a component of desmosomes and a ligand for CD11b/CD18-mediated neutrophil transepithelial migration. *Mol. Biol. Cell* **15**, 3926–3937.

Zhang, J.-R., Mostov, K. E., Lamm, M. E., *et al.* (2000). The polymeric immunoglobulin receptor translocates pneumococci across human nasopharyngeal epithelial cells. *Cell* **102**, 827–837.

Zolotarevsky, Y., Hecht, G., Koutsouris, A., *et al.* (2002). A membrane-permeant peptide that inhibits MLC kinase restores barrier function in in vitro models of intestinal disease. *Gastroenterology* **123**, 163–172.

CHAPTER 2

Evolution of bacterial pathogens

Anthony T. Maurelli

INTRODUCTION

The evolution of bacterial pathogens is essentially the story of how all life evolves. It is a story of mutation and selection, of adaptation and survival, of incremental genetic changes and quantum leaps in genome content. All organisms evolve, but the evolution of microbes is the best studied. The short generation times of microbes and the ability to grow individual populations to large numbers allow researchers to study rare events over many generations – something that is next to impossible with larger, more complex organisms such as fruit flies, mice, and humans.

This chapter begins with a brief overview of the principles of mutation and selection in bacteria. This background will prepare the reader for the subsequent sections that will discuss the various forms of horizontal gene transfer (HGT) and how they each contribute to bacterial evolution. Specific examples are presented to give the reader insight into the enormous power of genetic selection as well as the great diversity of pathways that bacteria take in adapting to their environment. We also highlight some recurrent themes in bacterial pathogenesis. The chapter concludes with a discussion of a new paradigm of bacterial pathogen evolution that involves the loss of gene function as an adaptation to colonization of the host. The concept of pathoadaptation is introduced and expanded to include selection for gene loss as the pathogen improves its fitness within the host niche.

Bacterial–Epithelial Cell Cross-Talk: Molecular Mechanisms in Pathogenesis, ed. Beth A. McCormick. Published by Cambridge University Press.

BASIC BACTERIAL GENETICS

Principles of mutation and selection

No biological system is perfect, and mistakes happen during growth and in the reproductive process. Errors are made during replication of bacterial genomes. Some errors are corrected and some escape proofreading and repair. This error rate is the basis for spontaneous mutation in bacteria. In addition, environmental influences can increase this rate of spontaneous mutation. Ionizing radiation (ultraviolet, X-rays), chemical mutagens, and other biological entities can increase the rate of spontaneous mutation in a population of bacteria. Alterations at the nucleic-acid level (genotypic changes) may be silent or may be manifested as a visible, measurable change in the organism (phenotypic changes). Thus, mutations may change the DNA sequence of a gene without altering the amino-acid sequence of the protein it encodes (silent mutation) or may result in an altered codon, but one that has no detectable effect on protein function (neutral mutation).

We tend to think of mutations as leading to loss of function, and in the majority of cases they do. Single base-pair changes, insertions, deletions, and rearrangements can all potentially alter the structure of a gene and cause loss of gene function. It is also true that these same types of mutation can result in gain of function. Thus, mutations can revert to restore original gene function, e.g. a base-pair change back to the wild-type base or precise excision of an insertion from a gene. Mutations can also create novel functions for an existing protein, e.g. an alteration that extends the substrate range of an enzyme. Gene duplication also allows flexibility in the evolutionary process, as changes can arise in the duplicated gene and be selected for without loss of the original function, which is maintained in the first gene copy. Regardless of whether a genetic alteration creates a loss of function or gain of function, in the final analysis successful mutations are those that impart some selective advantage to the organism, e.g. better fitness in a particular environmental niche. These mutations tend to be maintained within and spread through the population. Mutations that result in a selective disadvantage to the bacterium will be outcompeted by strains expressing the wild-type gene and lost. Neutral mutations will arise and persist, but in the absence of selection the mutant genotype will be restricted to a subset of the population.

Horizontal gene transfer in bacteria

Evolution via spontaneous mutation is a slow process that proceeds by trial and error. Small changes are made and, if favorable to the bacterium, the mutation is maintained in the genome. Clearly the number of mutational events that result in favorable changes is much smaller than the number of events that are not favored, e.g. 1 : 40 000 in the *lac* system (Roth *et al.*, 2003). In addition, beneficial mutations are rare; multiple beneficial mutations in the same organism are even rarer. Clearly, then, spontaneous mutation alone is not the answer to the question of how bacterial pathogens evolved.

If an organism could acquire new or better genes from another organism, then the course of evolution would be vastly accelerated. The extraordinary adaptive capacity of bacteria is due in part to their ability to acquire genes from other bacteria by means of HGT (Ochman *et al.*, 2000). HGT not only allows bacteria to obtain novel genes that have themselves already been tested by selection for providing a benefit to the bacterium, but also allows the recipient strain to acquire large blocks of DNA in a single transfer event. Thus, genes for novel multi-gene metabolic pathways or complex surface structures can be inherited and dramatically alter the ability of the bacteria to colonize new niches in a single step. In addition to chromosomal DNA, the mobile genetic elements that are spread by HGT include plasmids, bacteriophages, and transposable elements, including insertion sequences and transposons.

There are three basic processes of gene transfer in bacteria: transformation, transduction, and conjugation. Transformation is the direct uptake of naked DNA (plasmids or fragments of genomic DNA) from the external medium into a competent recipient. The overall contribution of transformation to bacterial evolution is limited by several factors. Not all bacteria are "competent" for transformation, i.e. capable of taking up DNA directly from the external medium. Although transformation has now achieved wide application as a tool for gene transfer in the research laboratory, its prevalence in nature is restricted to only a few naturally competent bacteria. Furthermore, some of these naturally transformable bacteria only recognize DNA with specific uptake sequences for transformation. Another limitation of transformation is the relatively small size of DNA that can be transformed.

Transduction (specialized and generalized) is gene transfer mediated by bacteriophage. This method of gene transfer is limited by the host range of the bacteriophage.

Conjugation involves direct cell-to-cell contact between donor and recipient bacteria and can transfer chromosomal DNA and plasmid DNA. The size of DNA transferred by conjugation ranges from small 1–2-kilobase pair (kb) plasmids to the entire bacterial chromosome. Even mega-plasmids of up to several hundred kilobase pairs can be transferred by conjugation.

In most cases of HGT, stable inheritance of the newly acquired genetic material requires recombination of the donor DNA into the genome of the recipient. Recombination can occur at homologous or non-homologous DNA sequences. Donor chromosomal DNA is generally recombined into the recipient by the RecA-mediated general recombination machinery at homologous sites in the recipient chromosome. Stable inheritance of bacteriophage DNA is achieved by site-specific (e.g. bacteriophage λ) or random (e.g. bacteriophage μ) integration into the bacterial chromosome or by autonomous replication as a plasmid (e.g. bacteriophage P1). In the case of HGT of plasmids, recombination is not required for stable inheritance if the plasmid is capable of autonomous replication in the cytoplasm of the recipient. Some plasmids are self-transmissible and thus possess greater potential for spread through a population than do non-conjugative plasmids. Similarly, some transposable elements are conjugative and transmitted more readily through a population.

CONTRIBUTIONS OF HORIZONTAL GENE TRANSFER TO PATHOGEN EVOLUTION: QUANTUM LEAPS

This section illustrates some examples of how the various mobile genetic elements described above have shaped the genomes of bacterial pathogens.

Plasmids

Plasmids are linear or circular DNA molecules that replicate autonomously in the cytoplasm of a host bacterium. All three processes of gene transfer are capable of transferring plasmids from donor to recipient, but the real power of plasmids in the arena of evolution lies in their potential for self-transmission by conjugation. Conjugative plasmids can transfer themselves to a broad range of recipient bacteria and can even be transferred across species boundaries. Add to this the ability to combine multiple genes on a single transferable genetic element and it is clear that plasmids are one of the most potent drivers of bacterial evolution.

One of the earliest demonstrations of how plasmids contribute to bacterial virulence came from the seminal work of H. Willy Smith and his colleagues on diarrhea in piglets. They showed that colonization factors, enterotoxins, and hemolysin production were all encoded on transmissible plasmids in toxigenic strains of *Escherichia coli* (Smith and Halls, 1967; Smith and Linggood, 1971a,b). Bovine and human isolates of diarrheogenic *E. coli* were also shown to carry plasmids that encode these virulence factors. Smith and Linggood further showed that toxigenic *E. coli* strains were more likely to harbor a plasmid than were *E. coli* strains isolated from asymptomatic animals, thus making the first strong link between plasmid carriage and pathogenicity. Although colonization factors and toxins were encoded on separate but compatible plasmids, isolates were found that carried the gene for heat-stable enterotoxin alone and others that carried both heat-stable and heat-labile enterotoxin genes on the same plasmid. So *et al.* (1979) later showed that the gene for heat-stable toxin was part of a transposable element, thus linking together two mobile genetic elements (see also Transposable elements, p. 39). Later studies confirmed that genes for both colonization factors and enterotoxins could be found on the same plasmid (Echeverria *et al.*, 1986).

Resistance to antibiotics imparts a powerful selective advantage to a bacterial pathogen in the presence of the myriad antibiotics used to treat infectious diseases. Shortly after the recognition of multiple antibiotic-resistant strains of *Shigella* in Japan in the late 1950s, Akiba *et al.* (1960) suggested that multiple resistance might be transferred from drug-resistant *E. coli* to *Shigella* in the intestinal tracts of patients. Later experiments by Mitsuhashi *et al.* (1960), Watanabe (1963), and Nakaya *et al.* (1960) confirmed and extended these initial observations on transmissible drug resistance, and Mitsuhashi *et al.* (1960) proposed the term "resistance factor" (R-factor) for the transmissible plasmids that encode antibiotic resistance (Watanabe, 1963). Thus, plasmids were shown to contribute to the adaptation of bacterial pathogens to the new antibiotic era by imparting on their hosts the ability to resist the drugs that were increasingly being used to treat the diseases these bacteria caused.

Smith and Linggood concluded their paper on the specific role of enterotoxins and colonization factors in the pathogenesis of diarrhea with the prescient remark that since the transfer factors found on plasmids that encode toxins, colonization factors, and antibiotic resistance determinants can be common to all these plasmids, "it is interesting to speculate whether 'new' enteropathogenic strains of *E. coli* will emerge at a more rapid rate in the present antibiotic era than hitherto" (Smith and Linggood, 1971a).

There are many other examples of plasmids that encode virulence determinants that allow pathogenic bacteria to colonize new niches and cause damage to the host. Genes for invasion of colonic epithelial cells and cell-to-cell spread of *Shigella* spp. are found on a large 220-kb plasmid (Sansonetti, 1993; Sansonetti *et al.*, 1982). A 70-kb virulence plasmid enables *Yersinia* spp. to resist phagocytosis by macrophages and to inhibit the respiratory burst (Cornelis *et al.*, 1998). One of the plasmids associated with virulence in *Bacillus anthracis* encodes the biosynthetic genes for the poly-γ-D-glutamic acid capsule that protects the bacillus from phagocytosis (Makino *et al.*, 1989). A plasmid in *Agrobacterium tumifaciens* endows this organism with the ability to colonize and promote crown gall tumor formation in plants (Van *et al.*, 1974). Interestingly, *Agrobacterium* pathogenesis is mediated by transfer of the bacteria's Ti plasmid to the plant cell, a process that has been exploited widely to introduce recombinant DNA into plants.

Other plasmid genes encode proteins that mediate host damage, including toxins, membrane lytic enzymes (e.g. hemolysins), and signal transduction molecules (e.g. for induction of cellular apoptosis). A partial list of bacterial toxins encoded by plasmids would include tetanus toxin; *E. coli* hemolysins, enterotoxins, and cytotoxic necrotizing factor; *Staphylococcus aureus* exfoliative toxin; the lethal factor and edema factor of *B. anthracis*; and the insecticidal parasporal crystal toxins of many subspecies of *Bacillus thuringiensis*. The virulence plasmids of *Shigella* and *Yersinia* mentioned above also enable these pathogens to induce apoptosis in macrophages and thus to escape this arm of the host defense. The purpose here is not to provide an exhaustive list of plasmids that have contributed to the evolution of bacterial pathogens but rather to illustrate the broad range of virulence factors that can be encoded on plasmids.

In summary, plasmid acquisition by HGT provides the quantum leap that propels the evolution of bacterial pathogens from a non-pathogenic progenitor. Plasmids provide a genetic platform on which nature can build "virulence cassettes" through the sequential addition of genes that, as a group, contribute to a pathogenic phenotype. Once these plasmids have evolved, their spread can be facilitated by conjugation, whether through self-transmission or mobilization. But the unanswered question is where did these plasmids come from? One can speculate that over the course of evolution, mobile genetic elements are randomly assorted. Linkages (combinations) of genes whose coexpression favors survival in a particular niche are selected. These persist and serve as the building blocks for acquisition of new genes that add to and improve the fitness of the organism. When those genes are integrated, selection of fitness locks them into the evolving plasmid platform, and a virulence plasmid is born.

Bacteriophages

Bacteriophages (phages or bacterial viruses) are ultimate parasites. They depend absolutely on their bacterial host for growth and are incapable of growth independent of their host. There are two classes of phages: virulent and temperate. Virulent phages infect a susceptible bacterium, replicate, and lyse the host cell, liberating progeny phages. Temperate or lysogenic phages can follow the lytic pathway or can persist quietly in the host bacterium in the prophage or lysogenic state. In this quiescent state, the genes of the phage lytic pathway are repressed, and most of the phage genes are not transcribed. Usually the genome of a lysogenic phage is integrated into the chromosome of the host bacteria. This assures inheritance of the phage in daughter cells and maintenance of the phage in the bacterial population as the integrated phage genome is replicated as part of the bacterial chromosome. This lysogenic state may persist indefinitely. A significant consequence of lysogeny for the bacterium is when expression of genes encoded by the lysogenic phage contributes to a change in the visible phenotype of the bacterial host. This condition is known as lysogenic conversion.

Years before the discovery of plasmids, bacteriophages were known to play an important role in making a bacterium a pathogen. The production of diphtheria toxin by strains of *Corynebacterium diphtheriae* was first shown to be linked to the presence of a lysogenic β-phage in these strains by Freeman (1951), Groman (1953), Barksdale and Pappenheimer (1954), and others in the 1950s. The gene for diphtheria toxin is encoded by the β-phage. Strains of *C. diphtheriae* that do not carry this bacteriophage are unable to synthesize the toxin and are not associated with disease. Other examples of toxins that are synthesized from phage-encoded genes in bacterial lysogens include the Shiga toxins of enterohemorrhagic *E. coli*, cholera toxin of *Vibrio cholerae*, enterotoxin A of *Staphylococcus aureus*, and the botulinum toxins of *Clostridium botulinum* (Brussow *et al.*, 2004).

Another example of how lysogenic conversion contributes to bacterial pathogen evolution is the synthesis or modification of cell-surface structures. Bacteriophage SfV is a lysogenic phage of *Shigella flexneri* that encodes a glucosyl transferase that converts serotype Y (3,4) to serotype 5a (V; 3,4) by addition of glucose to the second rhamnose residue of the O antigen (Huan *et al.*, 1997). Mutants that are deleted for the phage-encoded glucosyl transferase show a marked survival defect in ligated rabbit ileal loops. Glucosylation of the O antigen restores virulence (West *et al.*, 2005). It is well known that O antigen protects bacteria from innate immune defenses such as bile salts, complement-mediated lysis, and antimicrobial peptides. However, glucosylation of O antigen serves a more specific function for *Shigella*. Glucosylation induces transition of the lipopolysaccharide (LPS) to a more compact structure, shortening the distance that it extends beyond the outer membrane. This compact structure optimizes the exposure of the type III secretion system (TTSS) needles that extend from the bacterial cytoplasm across the inner membrane and periplasmic space and through the outer membrane to deliver effectors required for invasion. Thus, lysogenic conversion of *Shigella* permits alteration of the “shield” (O antigen) while maximizing the effectiveness of the "sword" (TTSS needle) and making the bacterium a more efficient pathogen (West *et al.*, 2005).

A dramatic example of lysogenic conversion in bacterial pathogenesis was the discovery of a bacteriophage associated with invasive meningococci (Bille *et al.*, 2005). *Neisseria meningitidis* is a common inhabitant of the human nasopharynx; in a small proportion of people who are colonized, the bacteria invade the bloodstream and can go on to cause meningitis. A filamentous phage is present in the chromosome of the disease-causing bacteria. The precise factors that this phage contributes to the invasive capacity of *N. meningitidis* are not known, but the observation that the 8-kb phage can excise and be secreted from the lysogenic host suggests the potential for spread and conversion of commensal *N. meningitidis* into invasive pathogens (Bille *et al.*, 2005).

This example of lysogenic conversion in *N. meningitidis* illustrates the potential of virulence gene spread by HGT mediated by bacteriophage. However, one should keep in mind that the spread of bacteriophage genes is limited by the host range of the bacteriophage. Thus, although the potential certainly exists for this filamentous phage to spread within a population of commensal *N. meningitidis,* it is unlikely that the phage will spread to other members of the normal flora in the nasopharynx. Extension of the host range of a bacteriophage requires either a mutation in the phage attachment protein in order to allow it to recognize a receptor on a new bacterial host or a mutation in (or acquisition of) the receptor protein on the surface of the bacterium that permits attachment of the phage. A second obstacle that needs to be overcome is the stable integration of the phage genome into the bacterial chromosome. Although this step generally takes place at specific sequences in the bacterial chromosome, phage integration at sites with limited homology to the normal phage-attachment site does occur.

Other ways in which bacteriophage contribute to bacterial pathogen evolution include enhancing bacterial resistance to serum and phagocytes, transduction of genes for resistance to antibiotics, and promoting transmission, colonization/adhesion, and sometimes invasion (Wagner and Waldor, 2002). Table 2.1 lists some of these virulence phenotypes associated with lysogenic conversion.

Phage integration and excision is an ongoing part of bacterial evolution. The remnants of genomes for dozens of bacteriophage have been uncovered in the genome sequences of many bacterial pathogens. In a review of the published bacterial genome sequences to date, Casjens (2003) found that 51 of the 82 analyzed genomes carry prophages with the prophages constituting 10–20% of a bacterium's genome. Even if many of these prophages

Table 2.1 *Virulence-associated phenotypes encoded by bacteriophages*

Bacterial host	Phage	Protein
Corynebacterium diphtheriae	β-Phage	Diphtheria toxin
Clostridium botulinum	Phage C1	Neurotoxin
Enterohemorrhagic *Escherichia coli*	H-19B, 933	Shiga toxins
Salmonella enterica	GIFSY-2, Fels-1	Superoxide dismutase
	Fels-1	Neuraminidase
		Superoxide dismutase
	ϕ 34, P22	Glucosylation
Shigella flexneri	SfII, SfV, SfX	Glucosyl transferase
	Sf6	O-antigen acetylase
Staphylococcus aureus	ϕ 13	Staphylokinase enterotoxin A
	ϕ ETA	Exfoliative toxin A
	ϕ PVL	Leukocidin
Streptococcus pyogenes	H4489A	Hyaluronidase
	T12	Toxin type A
Vibrio cholerae	CTXϕ	Cholera toxin

are defective and undergoing mutational decay, they still may express genes that contribute to the phenotype of the bacterial host. Bacteriophages are also major contributors to differences between individual strains within a species. For example, a comparison of enterohemorrhagic *E. coli* O157:H7 (5.5 million base pairs [Mbp]) with the laboratory strain *E. coli* K-12 (4.6 Mbp) revealed two highly conserved genomes that share a core backbone of 4.1 Mbp. Much of the additional genomic content of *E. coli* O157:H7 is strain-specific sequence attributable to the presence of a multitude of lambda-like phages (Ohnishi *et al.*, 2001).

How did bacteriophages acquire the genes that they carry into bacterial hosts that contribute to pathogen evolution? The origin of a given toxin gene or virulence factor that is carried on a lysogenic bacteriophage is difficult to reconstruct. Intact genes may have come from an ancestral bacterial source. Alternatively, new genes may have formed in the endless mixing of bacteriophages with host bacteria. The genomes of today's prophages are the result of numerous recombination events that occurred over time, and the dynamics of phage-host DNA exchanges in bacterial evolution are just beginning to be explored. What is clear, however, is that lysogenic conversion has had an enormous impact on the evolution of pathogenic bacteria.

Transposable elements

Transposable elements are discrete segments of DNA that are capable of moving (transposing) from one replicon to another or to a different position on the same replicon. For example, a transposable element can move from a chromosome and insert into a plasmid, and vice versa, or move between a bacteriophage and a plasmid or a chromosome. The mechanism of transposition differs depending on the transposable element. Replicative transposition generates a new copy of the transposable element, which hops into a new site while leaving a copy of itself inserted in the original location. By contrast, the "cut-and-paste" or conservative mechanism moves the entire transposable element intact to a new site without creating another copy. Regardless of the mechanism involved, a common feature of all transposable elements is that transposition is independent of the general (i.e. RecA-dependent) homologous recombination system. The target sites for insertion generally are random, but some transposable elements have preferred "hot spots" for insertion while others are site-specific.

The movement of transposable elements creates a variety of genetic alterations. Insertion into a gene normally inactivates the gene. Precise excision restores the gene to its original state. Imprecise excision can result in deletion of DNA sequences adjacent to the original site of insertion and may create gene fusions. Transposable elements also mediate genomic rearrangements such as inversions and duplications.

The two classes of transposable elements are insertion sequences and transposons. Insertion sequences, or IS elements, are small segments of DNA (< 2 kb) that contain no genes other than those required for transposition. Transposons are larger and more complex and contain genes in addition to those that mediate transposition. They may also contain IS elements as part of their genetic structure.

Antibiotic resistance is probably the best example of how transposable elements have contributed to bacterial pathogen evolution. Hedges and Jacob (1974) described the first transposon, Tn*3*, which encodes ampicillin resistance. In the ensuing years, numerous transposons encoding resistance to a wide range of antibiotics were discovered, including Tn*1* and Tn*2* (ampicillin resistance), Tn*5* and Tn*6* (kanamycin resistance), Tn*7* (trimethoprim and streptomycin/spectinomycin resistance), Tn*9* (chloramphenicol resistance), Tn*10* (tetracycline resistance), and Tn*917* (erythromycin resistance); see Berg *et al.* (1989) for an excellent summary of transposons and their uses as genetic tools to modify bacteria. Transposons are found in Gram-positive as well as Gram-negative bacteria. Some transposons encode multiple drug resistance, as in the case of Tn*1545*, a 25-kb transposon that encodes resistance to tetracycline, erythromycin, and kanamycin. An interesting form of transposon first discovered in a Gram-positive organism is Tn*916*. This transposon not only encodes genes for transposition but also is capable of mediating its own transfer by conjugation. Thus, the class of conjugative transposons combines the power of HGT with random insertion in the genome of the recipient organism (for a review, see Salyers *et al.*, 1995).

The ability of antibiotic resistance genes to transpose between DNA replicons independent of general recombination functions explains in large part the rapid evolution and wide dissemination of multiple antibiotic-resistance plasmids described earlier. Examination of the gene organization of R-factors revealed that they are composites of plasmid sequences with cassettes of antibiotic-resistance genes contained within transposons. For example, R100 (NR1) expresses multiple antibiotic resistance encoded by Tn*10*, a Tn*9*-related transposon, and a transposable element encoding mercury resistance (Womble and Rownd, 1988). Indeed the presence of identical resistance genes in a wide variety of plasmids present in bacteria of diverse phylogenetic origins clearly underlines the power of transposition coupled with HGT and strong genetic selection (in the form of global application of antibiotics) in the modern-day phase of bacterial evolution.

Transposons that encode virulence factors also have contributed to bacterial pathogen evolution. The best known example is that of the heat-stable toxin STa of enterotoxigenic *E. coli* (ETEC), which is found on a transposon, Tn*1681* (So *et al.*, 1979). ETEC strains that infect pigs also synthesize a heat-stable toxin, which is encoded on a transposon, Tn*4521* (Lee *et al.*, 1985) although this toxin shares no homology with STa from the human ETEC. In both cases, the transposons can be found on plasmids that carry colonization factors and antibiotic-resistance genes.

Pathogenicity islands

Pathogenicity islands (PAIs) are a relatively new concept in our understanding of bacterial pathogen evolution (Groisman and Ochman, 1996; Hacker and Kaper, 2000; Schmidt and Hensel, 2004). A PAI is a block of genes present in a bacterial pathogen that is missing from a related but non-pathogenic reference strain. The island contains a gene or genes that are known or suspected to be involved with virulence of the pathogen.

Furthermore, the genes in the island have an atypical base composition relative to the reference genome, i.e. the G+C content of the DNA in the island is markedly different from the genomic region surrounding the island. Genes within the PAI also show a different codon usage compared with genes in the rest of the genome. PAIs are large (10–200 kb) and are frequently inserted within or adjacent to tRNA genes. Some PAIs are contained within plasmids or bacteriophages. PAIs usually contain markers of genetic mobility. PAIs are often flanked by direct repeats and contain IS elements or parts of IS elements. They carry cryptic or functional mobility genes such as resolvases and transposases usually associated with transposable elements, and integrases, which show similarities to the enzymes of lysogenic bacteriophages. Indeed, the presence of these genes and other DNA sequences with homology to bacteriophages underscores the mosaic architecture of PAIs and reflects the multistep evolution of this genetic element.

The presence of mobility genes and DNA structures (e.g. direct repeats) that favor gene rearrangements accounts for the instability of certain PAIs. Thus, some PAIs are intrinsically unstable and undergo excision or deletion at a rate higher than the normal mutation frequency.

The contribution of PAIs to bacterial pathogen evolution is highlighted in Table 2.2. This table presents only a small sampling of the many PAIs identified in Gram-negative and Gram-positive pathogens of humans, animals, and plants. A more complete listing of PAIs and the virulence properties that they encode can be found in Hacker and Kaper (2000) and Schmidt and Hensel (2004). It is clear even from this limited list that PAIs have evolved to provide bacteria with a broad range of properties that permit colonization of their hosts. Perhaps the most widespread virulence property encoded by PAIs is the dedicated secretion system used to transport pathogenic factors from the bacterial cytosol to the external medium and even into the target eukaryotic cell. The most prominent of these systems is the TTSS (Hueck, 1998). In a typical TTSS, more than 20 genes contribute to synthesis of the needle structure and the inner and outer membrane channels through which it passes to translocate proteins out of the bacterium. All of the genes for the proper functioning of this complex macromolecular structure, including the structural components of the secretion apparatus, the secreted effector proteins and their chaperones, and regulatory genes that control proper expression of the TTSS genes, are contained within the PAI. Thus, PAIs are self-contained virulence cassettes.

How did PAIs arise? The IS elements and transposon sequences in PAIs likely played a role in the original formation of PAIs and probably continue to play a role in their evolution. Similarly, the presence of fragmented and

Table 2.2 *Virulence-associated phenotypes encoded by pathogenicity islands*

Bacteria	Pathogenicity island	Virulence properties
Bordetella pertussis	*ptx-ptl* locus	Type IV secretion system for secretion of pertussis toxin
Uropathogenic *Escherichia coli*	PAI I_{536}–PAI III_{536}	Hemolysin, P-fimbriae, S-fimbriae
Enteropathogenic *E. coli*	LEE	Type III secretion system, invasion
Enterohemorrhagic *E. coli*	LEE	Type III secretion system, invasion
Helicobacter pylori	*cag* PAI	Type IV secretion system, *cag* antigen
Legionella pneumophila	*icm/dot* region	Type IV secretion system
Listeria monocytogenes	LIPI-1, LIPI-2	Phospholipase, listeriolysin O, ActA, internalin, sphingomyelinase
Salmonella spp.	SPI-1–SPI-5	Type III secretion system and effectors, invasion, apoptosis, survival in monocytes
Shigella spp.	Virulence plasmid	Type III secretion system and effectors, invasion, apoptosis
	SHI-1–SHI-3	Enterotoxin, protease, aerobactin synthesis and transport
Vibrio cholerae	VPI	TCP-adhesin, regulator
Yersinia spp.	HPI	Yersiniabactin synthesis, hemin uptake
	Yop regulon	Type III secretion system and effectors (Yops)

HPI, high pathogenicity island; LEE, locus of enterocyte effacement; LIPI, *Listeria* pathogenicity island; PAI, pathogenicity island; SHI, *Shigella* pathogenicity island; SPI, *Salmonella* pathogenicity island; TCP, toxin co-regulated pilus; VPI, *Vibrio* pathogenicity island.

intact bacteriophage genes in PAIs argues for their contribution in the formation of PAIs. In fact, the continual exchange of genetic elements between phages, bacterial genomes, and various other mobile DNAs also explains the sometimes fuzzy distinction between phages, plasmids, and PAIs and the chimeric nature of some extant phages. The mobility genes within PAIs suggest the potential for HGT and another means of evolution by quantum leaps. However, it is clear that although some PAIs are inherently unstable due to the presence of these mobile elements and expression of transposases, other PAIs are quite stable. This observation reflects the normal process of evolution wherein once a genetic element has been acquired by an organism, if expression of the genes within that element is beneficial to the organism then selection will favor mutations that stabilize the genetic element. Therefore, although some "young" PAIs are unstable, more "mature" PAIs have already deleted or otherwise inactivated the genes that provided the PAI the mobility that it used to move into a new population (Hacker and Kaper, 2000).

EVOLUTION OF BACTERIAL PATHOGENS FROM COMMENSALS

The challenge of colonizing a new niche

The minimalist view of bacterial evolution is that change is driven by the need to survive in a competitive and sometimes hostile environment. One bacterium grows and divides to become two bacteria. The organisms that can successfully execute this simple equation persist. When variants that are better fit for a particular environment arise within a population, they will tend to be selected for and eventually become the dominant species in the population or even replace the less fit ancestors entirely. Another pathway is when organisms evolve to occupy a new niche. One can view the evolution of bacterial pathogens from normally harmless commensals in this context. One should also keep in mind that we tend to take a host-centric view of host–pathogen interactions. In our view damage to the host defines the pathogen, but the damage the pathogen produces may be secondary to the principal objective of the bacterium: to grow and divide and, in most cases, to do so in a new host.

The challenge facing the organism that attempts to colonize a new niche is two-fold: to adapt to the nutrient limitations of the new environment and to compete successfully with any indigenous microbial populations in the new niche. If the new niche is a site normally devoid of any microbes then the absence of competition makes the task simpler. When the new niche is another biological system, an additional challenge is evading or resisting innate and acquired host defenses against colonization. Microbes and their mammalian hosts have evolved together over many thousands of years and present a microcosm of coevolution and symbiosis. Thus, mammals possess normal flora microbial populations that protect the host by outcompeting potential pathogens. And yet some bacterial pathogens have evolved to exploit new niches in the mammalian host where normal flora are not present. Intracellular pathogens such as *Shigella* and *Salmonella* that have acquired the means of invading host epithelial cells can grow unimpeded in this normally sterile site. Pathogens that can penetrate into sterile tissue spaces have no normal flora to compete with but need to avoid the highly evolved weapons of innate and acquired immunity. Here, too, pathogens have evolved to cope with these defenses. Thus, *Mycobacterium tuberculosis* and *Legionella pneumophila* infect the lung and survive when ingested by alveolar macrophages. Encapsulated bacteria such as *Bacillus anthracis* that penetrate into the bloodstream resist phagocytosis and can grow to very large numbers. Some pathogens such as *Neisseria gonorrhoeae* (Seifert, 1996) and *Borrelia hermsii* (Barbour and Restrepo, 2000) have overcome the problem of the acquired immune response of the host by evolving mechanisms for continually changing their antigenic profiles while colonizing the host. There are many more examples of pathogens of humans, animals, and plants and the mechanisms they employ to grow in the host. However, in the final analysis, these pathogens all represent new clones of pre-existing populations that have adapted to and successfully competed to colonize new niches within the host.

Pathoadaptation: making a better pathogen

Although pathogen evolution progresses through mutation and gene acquisition via HGT, the newly acquired phenotypes continue to undergo selection once the pathogen has moved to colonize a new niche. This process is known as pathoadaptation. Pathoadaptive mutations are genetic modifications that enhance fitness of the pathogen in the novel (host) environment. One example of pathoadaptation is at the level of virulence gene expression. In many pathogens, newly acquired virulence genes, whether present on a plasmid or PAI, appear to have been brought under the control of a regulator that was already present in the genome of the pathogen's ancestor. In *Shigella* spp., expression of the genes encoding invasion effectors and the TTSS that exports these effectors is tightly regulated by growth temperature (Dorman and Porter, 1998). The genes are expressed at 37° C, the temperature of the human host, and repressed at lower temperatures (Maurelli *et al.*, 1984).

This temperature regulation is controlled by two transcriptional activators, VirF and VirB, encoded on the virulence plasmid of *Shigella*. One of these activators is in turn controlled by a chromosomally encoded repressor, H-NS (Hromockyj *et al.*, 1992; Maurelli and Sansonetti, 1988). The gene for H-NS is conserved in nearly all enterobacteriaceae, pathogen and commensal, and is clearly not specifically a virulence gene (Bertin *et al.*, 1999). The "hard wiring" of plasmid-encoded virulence genes of *Shigella* into a pre-existing regulatory circuit is probably a result of an adaptation to regulate virulence gene expression, such that it occurs only when the pathogen is inside its host. In this way, *Shigella* conserves energy by not synthesizing a complex surface structure for invading mammalian cells when it is not within the mammalian host. Temperature then serves as an inducing signal to the bacterium to de-repress expression of these genes so that the pathogen can effectively compete in the host environment.

A variation on this theme is the global regulation of virulence genes in enteropathogenic *E. coli*. A plasmid-encoded activator, Per, is responsible for upregulation of virulence genes in the LEE locus (locus of enterocyte effacement) as well as the *bfp* genes for a bundle-forming pilus that the organism uses to facilitate attachment to mammalian cells. The LEE locus is a chromosomal PAI that encodes a TTSS, secreted proteins, and intimin. The *bfp* genes are found on the same plasmid as the *per* gene (Mellies *et al.*, 1999). Furthermore, a LEE-encoded regulator (Ler) is part of the Per-mediated regulatory cascade but also regulates expression of virulence genes encoded outside the LEE (Elliott *et al.*, 2000). Thus, the evolution of regulatory circuits that allow coordinated control of multiple unlinked virulence genes is an important part of pathoadaptation and occurs after acquisition of these genes by the new pathogen.

Genome reduction: the flip side to quantum leaps

Acquisition of virulence genes is not the only mechanism by which pathogens evolve. Pathogenic bacteria also evolve from commensal bacteria by loss of gene function. Thus, pathoadaptive mutation via gene loss complements the pathway of bacterial pathogen evolution by gain-of-function mutation and gene acquisition. This model of evolution starts with the premise that genes required for fitness in one niche may actually inhibit fitness in another environment. For example, an organism evolves as a commensal within a certain niche in the host, and selective pressures result in an organism that colonizes and makes optimal use of the available nutrients in that niche. As the organism acquires new genes that allow it to colonize a new niche, a new set of selective pressures is brought to bear. These forces will favor the emergence of variants that have eliminated or downregulated expression of any gene that is incompatible with growth in the new niche. When the gene in question affects the pathogenicity of the organism, it is called an "anti-virulence gene". Therefore, we define an anti-virulence gene as a gene whose expression is incompatible with virulence. Each newly evolved pathogen adapts to its new lifestyle by modifying the anti-virulence genes in its genome. These alterations consist of any means that eliminate or reduce expression of the anti-virulence gene, including deletion of the gene, point mutations within the gene, and suppression of gene expression. The best example of pathoadaptive mutation by loss of anti-virulence genes is the case of the *cadA* gene in *Shigella* spp., which we discuss below.

It is important to note that pathoadaptative mutation by loss of genes that are incompatible with the pathogen's new lifestyle is distinct from evolution by reduction. In this process, the commitment to an obligate intracellular lifestyle for certain pathogens results in loss of genes not essential to life within the host (Moran and Plague, 2004). Some of the best examples of genome reduction include *Mycobacterium leprae* (Cole *et al.*, 2001), *Coxiella burnetii* (Seshadri *et al.*, 2003), and the Rickettsiae (Andersson *et al.*, 1998). Not only have these organisms deleted genes that are no longer required for their survival, but also many of the superfluous genes that remain are riddled with mutations that eliminate their function. These "pseudogenes" may ultimately be removed by deletion. The smaller genomes of these pathogens are also due in part to the limited opportunities for gene acquisition from other organisms that is caused by growth restricted to within a host.

Shigella: gene acquisition + gene loss = more efficient pathogen

Bacteria of the genus *Shigella* are Gram-negative rods that are the causative agents of bacillary dysentery and shigellosis. The bacteria are highly host-adapted and cause disease only in humans and primates. *Shigella* invade cells of the colonic epithelium, replicate intracellularly, spread from cell to cell, and cause abscesses and ulcerations of the intestinal lining, leading to the bloody mucoid stools characteristic of dysentery. Bacterial invasion and replication also lead to an intense inflammatory response that serves both the host and the pathogen (see Sansonetti *et al.*, 1999 for model).

The four species of *Shigella* (*S. dysenteriae, S. flexneri, S. boydii, S. sonnei*) are related so closely to *E. coli* that they should be included in a single species. The chromosomes of these organisms are largely colinear and the genomes are more than 90% homologous (Jin *et al.*, 2002). However, studies have demonstrated that *Shigella* strains do not form a single subgroup of *E. coli*, as would be expected of a distinct genus, but are instead derived from separate *E. coli* strains (Pupo *et al.*, 1997, 2000). The majority of *Shigella* strains fall into three main clusters within *E. coli*, and seven different *Shigella* lineages have been identified through sequence analysis of a number of chromosomal loci (Lan and Reeves, 2002). Therefore, the shigellae are a group of pathogenic *E. coli*. Acquisition of the large virulence plasmid was the crucial event in the evolution of *Shigella*, and this plasmid is what distinguishes *Shigella* from the non-pathogenic commensal *E. coli*. The plasmid is found in all species of *Shigella* and encodes genes for expression of the hallmarks of *Shigella* virulence: invasion, intracellular replication, intercellular spread, and induction of an inflammatory response. As explained earlier, the clustering of these virulence genes and their low G+C content also suggest that they constitute a PAI within the plasmid. Thus, horizontal transfer of the virulence plasmid to commensal *E. coli* has occurred multiple times, on each occasion giving rise to new *Shigella* clones (Pupo *et al.*, 2000). Further evidence of independent HGT in the evolution of *Shigella* comes from the finding of distinct PAIs in the genomes of *S. flexneri* and *S. boydii* (Purdy and Payne, 2001).

These findings suggest that traits unique to and shared by *Shigella* species are the result of convergent evolution driven by unique selective forces present within the new host niche (i.e. inside colonocytes) that were encountered by each newly evolved *Shigella* clone. These unique *Shigella* traits arose through either gain-of-function mutations (horizontal transfer of the virulence plasmid) or loss-of-function mutations (deletion of traits expressed by ancestral *E. coli* strains). Thus, the convergent evolution of the seven *Shigella* lineages presents a unique opportunity for the identification and study of anti-virulence genes and pathoadaptive mutations as well as the study of pathogen evolution.

As outlined above, ancestral traits that interfere with virulence are lost from the newly evolved pathogen genome early on, as increased fitness of adapted clones fixes these beneficial mutations in the newly or recently evolved pathogen population. Therefore, traits absent in all pathogenic clones of a species but commonly expressed in the closely related commensal ancestor species are strong candidates for pathoadaptive mutations that have arisen by convergent evolution. Evidence in support of this new model of pathogen evolution was first provided by comparison of *Shigella* with its commensal ancestor *E. coli*. Although *Shigella* and *E. coli* share many biochemical traits, some characteristic markers have proven useful over the years to differentiate *Shigella* from *E. coli*. One of these is lysine decarboxylase (LDC) activity, which is encoded by the *cadA* gene in *E. coli*. LDC is expressed in more than 90% of

E. coli isolates. In contrast, none of the *Shigella* clones expresses LDC activity. Moreover, pathogenic enteroinvasive strains of *E. coli* that cause a disease clinically indistinguishable from dysentery caused by *Shigella* also lack LDC activity (Silva *et al.*, 1980). Lack of LDC activity in the Shigellae is consistent with the expected pattern of a pathoadaptive mutation and suggests that *cadA* may be an anti-virulence gene for *Shigella*. Experimental evidence for the anti-virulence nature of the *cadA* gene was demonstrated by examination of the virulence phenotypes of a strain of *S. flexneri* 2a transformed with the *cadA* gene from *E. coli* K-12. The LDC producing *Shigella* is still invasive, but it fails to produce the wild-type level of enterotoxic activity in the rabbit ileal loop and the Ussing chamber assays (Maurelli *et al.*, 1998). Further analysis showed that cadaverine, the product of the decarboxylation of lysine, is an inhibitor of the virulence plasmid-encoded *Shigella* enterotoxins. Cadaverine is also responsible for the block in the ability of an LDC expressing *S. flexneri* to elicit transepithelial migration of polymorphonuclear neutrophils in a polarized tissue culture model system for the inflammatory response (McCormick *et al.*, 1999). This phenotype appears to be due to a failure of the bacteria to escape the phagolysosome after invasion of the polarized cells (Fernandez *et al.*, 2001). Attenuation of these virulence phenotypes is linked to expression of LDC and production of cadaverine in the *S. flexneri* 2a strain transformed with the *cadA* gene from *E. coli* K-12. Therefore, *cadA* behaves as an anti-virulence gene for *Shigella*.

Genomic analysis suggested that the chromosomal region to which *cadA* maps in *E. coli* K-12 had undergone a large deletion (a "black hole") in *S. flexneri* 2a (Maurelli *et al.*, 1998). Subsequent sequence analysis of the *cadA* region of four *Shigella* lineages revealed novel genetic arrangements that were distinct in each strain examined (Day *et al.*, 2001). Insertion sequences, a phage genome, and/or loci from different positions on the ancestral *E. coli* chromosome displaced the *cadA* locus to form distinct genetic linkages unique to each *Shigella* lineage (Figure 2.1). None of these novel gene arrangements was observed in representatives of all *E. coli* phylogenies. It is interesting to note that in the case of *S. boydii*, the *cadA* region is deleted and replaced with a PAI that encodes an iron-uptake system (Purdy and Payne, 2001). However, it is not possible to determine whether acquisition of the pathogenicity island occurred before, coincident with, or after deletion of the *cadA* anti-virulence gene. Collectively, these observations indicate that inactivation of the *cadA* anti-virulence gene occurred independently in each *Shigella* lineage. The convergent evolution of these pathoadaptive mutations demonstrates that following evolution from commensal *E. coli*, strong pressures in host tissues selected *Shigella* clones with increased fitness and virulence

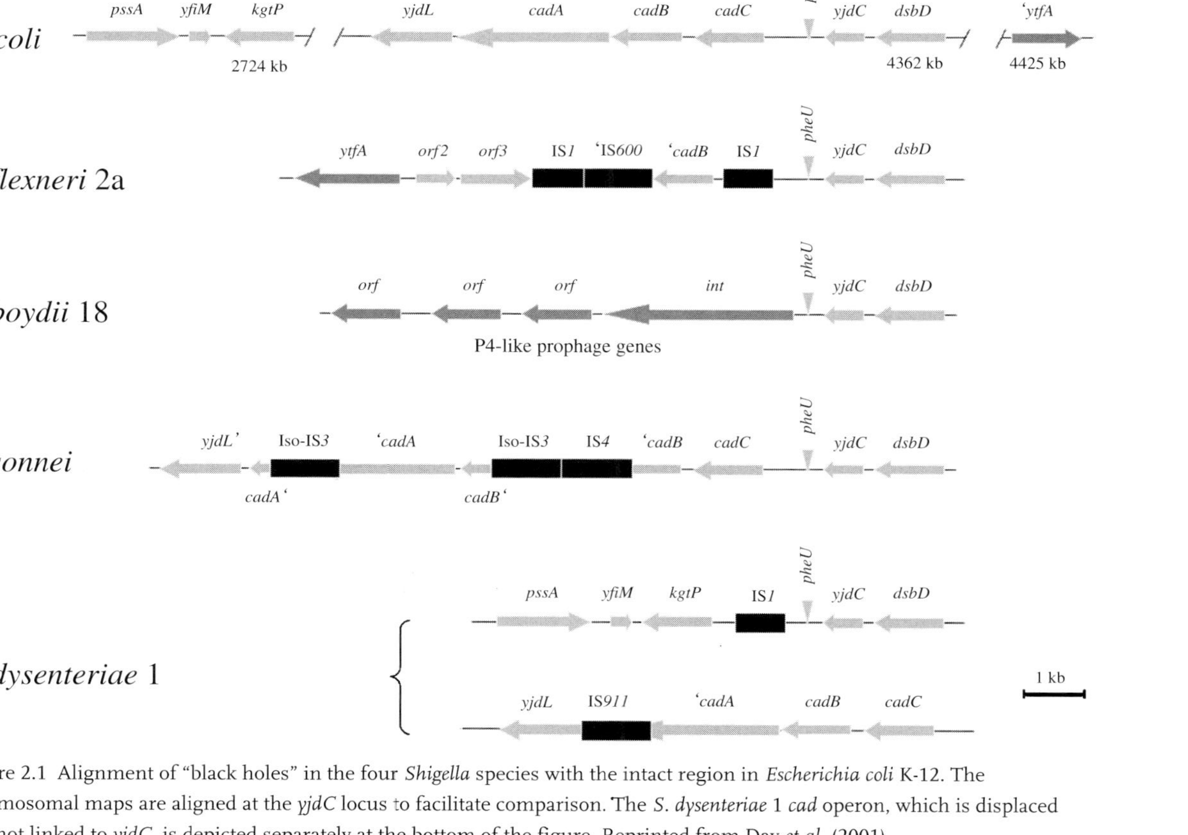

Figure 2.1 Alignment of "black holes" in the four *Shigella* species with the intact region in *Escherichia coli* K-12. The chromosomal maps are aligned at the *yjdC* locus to facilitate comparison. The *S. dysenteriae* 1 *cad* operon, which is displaced and not linked to *yjdC*, is depicted separately at the bottom of the figure. Reprinted from Day *et al.* (2001).

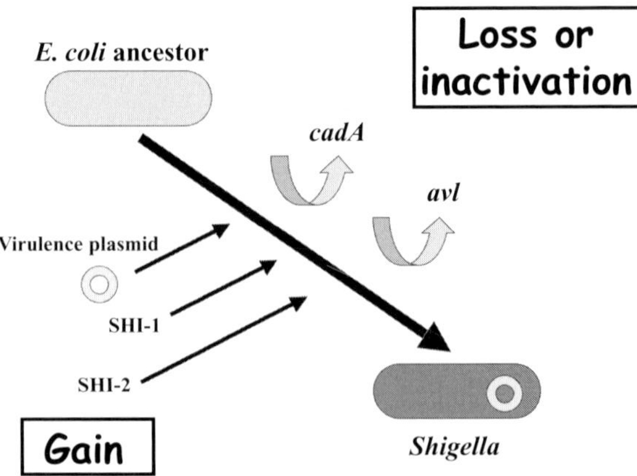

Figure 2.2 Evolution of *Shigella* from a commensal *Escherichia coli* ancestor. Schematic representation of evolution by gene acquisition and gene loss. SHI-1 and SHI-2 are *Shigella* pathogenicity islands. *cadA* is the gene for lysine decarboxylase. *avl* represents any anti-virulence locus.

through the loss of an ancestral trait (i.e. LDC). These observations strongly support the role of pathoadaptive mutation as an important pathway in the evolution of pathogenic organisms.

The *ompT* gene of *E. coli* K-12, which encodes an outer-membrane protease (Grodberg and Dunn, 1988), provides another example of a gene expressed by a non-pathogen that is incompatible with virulence when expressed in a pathogen. OmpT protease activity destroys the surface-expressed IcsA protein of *Shigella* when *icsA* is cloned and expressed in *E. coli* K-12 (Nakata *et al.*, 1993). Introduction and expression of *ompT* from *E. coli* K-12 into *S. flexneri* abolishes plaque-forming ability and the ability to produce conjunctivitis in the Sereny test (Nakata *et al.*, 1993). These two phenotypes are dependent on IcsA expression in the outer membrane. Thus, expression of *ompT* is incompatible with *Shigella* virulence as it destroys a critical outer-membrane protein essential for post-invasion virulence phenotypes. Interestingly, the *ompT* locus is contained within a 21-kb cryptic prophage in *E. coli* K-12 (Lindsey *et al.*, 1989) and the entire prophage is absent in all four species of *Shigella* as well as in enteroinvasive *E. coli* (Nakata *et al.*, 1993). *E. coli* strains that represent lineages from which the different groups of *Shigella* evolved do not contain this cryptic prophage. This observation suggests that the absence of the *ompT* anti-virulence gene from *Shigella* is not the result of a pathoadaptive mutation; rather, the presence of *ompT* may have imposed a limitation on the potential ancestral strains that were capable of giving rise to the successful *Shigella* we see today. Figure 2.2 presents a schematic view of how a pathogen (in this case, *Shigella*) arises from a commensal ancestor by gene acquisition and gene loss.

The evolution of *E. coli* O157:H7 is another good example of bacterial pathogen evolution that involves sequential gain and loss of genes. Acquisition of toxin-encoding phages (described above) was followed by loss of motility and a shift in the type of O antigen expressed. A model detailing how *E. coli* O157:H7 may have evolved from an enteropathogenic *E. coli*-like ancestor has been proposed by Whittam and colleagues (Feng *et al.*, 1998; Wick *et al.*, 2005).

CONCLUSIONS

Although mutation plays an important role in the overall evolution of any species, the greatest contribution to the evolution of bacterial pathogens from their commensal ancestors has been the acquisition of virulence genes in the form of plasmids, bacteriophages, and PAIs. Gene exchange by transduction and conjugation provides the genetic diversity, and pressures to compete successfully within a niche or to colonize a novel niche act to select out the best fit of the new recombinants. The pathogens that we see today are the winners in the genetic lottery. They have acquired the right combination of genes to compete within the host and ensure their transmission to a new host. Once acquired, the process of pathoadaptation acts to "fine tune" the pathogen to its new lifestyle. This process may involve genome reduction, i.e. removal of genes that are no longer needed to produce the nutrients that the pathogen can now obtain from its host. Pathoadaptation also involves the inactivation or removal of anti-virulence genes, i.e. genes that are incompatible with virulence.

Where did the virulence genes found in plasmids, bacteriophages, transposable elements, and PAIs originate? This question is perhaps the most interesting but also the most difficult to answer. However, consider that the number of prokaryotes in the various aquatic and terrestrial ecosystems of the planet is estimated to exceed 150 000 different species. Only about 4000 have been described thus far. Therefore, an enormous untapped reservoir of genetic information exists that may provide some answers to this question. New methods for cultivating previously uncultivable species will contribute to unlocking this treasure of genetic information. More importantly, perhaps, will be the ability to use whole-genome shotgun sequencing and computational genomics in order to sort out the genes of complex microbial populations without the need for isolation of pure cultures (Venter *et al.*, 2004). In the future, our abilities to look back at the tracks of bacterial pathogen evolution will continue to improve and provide answers to these questions.

REFERENCES

Akiba, T., Koyama, K., Ishiki, Y., Kimura, S., and Fukushima, T. (1960). On the mechanism of the development of multiple-drug resistant clones of *Shigella*. *Jpn. J. Microbiol.* **4**, 219–227.

Andersson, S. G., Zomorodipour, A., Andersson, J. O., *et al.* (1998). The genome sequence of *Rickettsia prowazekii* and the origin of mitochondria. *Nature* **396**, 133–140.

Barbour, A. G. and Restrepo, B. I. (2000). Antigenic variation in vector-borne pathogens. *Emerg. Infect. Dis.* **6**, 449–457.

Barksdale, W. L. and Pappenheimer, A. M., Jr (1954). Phage–host relationships in nontoxigenic and toxigenic diphtheria bacilli. *J. Bacteriol.* **67**, 220–232.

Berg, C. M., Berg, D. E., and Groisman, E. A. (1989). Transposable elements and the genetic engineering of bacteria. In *Mobile DNA*, ed. D. E. Berg and M. M. Howe. Washington, DC: American Society for Microbiology, pp. 879–925.

Bertin, P., Benhabiles, N., Krin, E., *et al.* (1999). The structural and functional organization of H-NS-like proteins is evolutionarily conserved in gram-negative bacteria. *Mol. Microbiol.* **31**, 319–329.

Bille, E., Zahar, J. R., Perrin, A., *et al.* (2005). A chromosomally integrated bacteriophage in invasive meningococci. *J. Exp. Med.* **201**, 1905–1913.

Brussow, H., Canchaya, C., and Hardt, W. D. (2004). Phages and the evolution of bacterial pathogens: from genomic rearrangements to lysogenic conversion. *Microbiol. Mol. Biol. Rev.* **68**, 560–602.

Casjens, S. (2003). Prophages and bacterial genomics: what have we learned so far? *Mol. Microbiol.* **49**, 277–300.

Cole, S. T., Eiglmeier, K., Parkhill, J., *et al.* (2001). Massive gene decay in the leprosy bacillus. *Nature* **409**, 1007–1011.

Cornelis, G. R., Boland, A., Boyd, A. P., *et al.* (1998). The virulence plasmid of *Yersinia*, an antihost genome. *Microbiol. Mol. Biol. Rev.* **62**, 1315–1352.

Day, W. A., Jr, Fernandez, R. E., and Maurelli, A. T. (2001). Pathoadaptive mutations that enhance virulence: genetic organization of the *cadA* regions of *Shigella* spp. *Infect. Immun.* **69**, 7471–7480.

Dorman, C. J. and Porter, M. E. (1998). The *Shigella* virulence gene regulatory cascade: a paradigm of bacterial gene control mechanisms. *Mol. Microbiol.* **29**, 677–684.

Echeverria, P., Seriwatana, J., Taylor, D. N., *et al.* (1986). Plasmids coding for colonization factor antigens I and II, heat-labile enterotoxin, and heat-stable enterotoxin A2 in *Escherichia coli*. *Infect. Immun.* **51**, 626–630.

Elliott, S. J., Sperandio, V., Giron, J. A., *et al.* (2000). The locus of enterocyte effacement (LEE)-encoded regulator controls expression of both LEE- and non-LEE-encoded virulence factors in enteropathogenic and enterohemorrhagic *Escherichia coli*. *Infect. Immun.* **68**, 6115–6126.

Feng, P., Lampel, K. A., Karch, H., and Whittam, T. S. (1998). Genotypic and phenotypic changes in the emergence of *Escherichia coli* O157:H7. *J. Infect. Dis.* **177**, 1750–1753.

Fernandez, I. M., Silva, M., Schuch, R., *et al.* (2001). Cadaverine prevents the escape of *Shigella flexneri* from the phagolysosome: a connection between bacterial dissemination and neutrophil transepithelial signaling. *J. Infect. Dis.* **184**, 743–753.

Freeman, V. J. (1951). Studies on the virulence of bacteriophage-infected strains of *Corynebacterium diphtheriae*. *J. Bacteriol.* **61**, 675–688.

Grodberg, J. and Dunn, J. J. (1988). *ompT* encodes the *Escherichia coli* outer membrane protease that cleaves T7 RNA polymerase during purification. *J. Bacteriol.* **170**, 1245–1253.

Groisman, E. A. and Ochman, H. (1996). Pathogenicity islands: bacterial evolution in quantum leaps. *Cell* **87**, 791–794.

Groman, N. B. (1953). The relation of bacteriophage to the change of *Corynebacterium diphtheriae* from avirulence to virulence. *Science* **117**, 297–299.

Hacker, J. and Kaper, J. B. (2000). Pathogenicity islands and the evolution of microbes. *Annu. Rev. Microbiol.* **54**, 641–679.

Hedges, R. W. and Jacob, A. E. (1974). Transposition of ampicillin resistance from RP4 to other replicons. *Mol. Gen. Genet.* **132**, 31–40.

Hromockyj, A. E., Tucker, S. C., and Maurelli, A. T. (1992). Temperature regulation of *Shigella* virulence: identification of the repressor gene *virR*, an analogue of *hns*, and partial complementation by tyrosyl transfer RNA (tRNA1(Tyr)). *Mol. Microbiol.* **6**, 2113–2124.

Huan, P. T., Bastin, D. A., Whittle, B. L., Lindberg, A. A., and Verma, N. K. (1997). Molecular characterization of the genes involved in O-antigen modification, attachment, integration and excision in *Shigella flexneri* bacteriophage SfV. *Gene* **195**, 217–227.

Hueck, C. J. (1998). Type III protein secretion systems in bacterial pathogens of animals and plants. *Microbiol. Mol. Biol. Rev.* **62**, 379–433.

Jin, Q., Yuan, Z., Xu, J., *et al.* (2002). Genome sequence of *Shigella flexneri* 2a: insights into pathogenicity through comparison with genomes of *Escherichia coli* K12 and O157. *Nucleic Acids Res.* **30**, 4432–4441.

Lan, R. and Reeves, P. R. (2002). *Escherichia coli* in disguise: molecular origins of *Shigella*. *Microbes Infect.* **4**, 1125–1132.

Lee, C. H., Hu, S. T., Swiatek, P. J., *et al.* (1985). Isolation of a novel transposon which carries the *Escherichia coli* enterotoxin STII gene. *J. Bacteriol.* **162**, 615–620.

Lindsey, D. F., Mullin, D. A., and Walker, J. R. (1989). Characterization of the cryptic lambdoid prophage DLP12 of *Escherichia coli* and overlap of the DLP12 integrase gene with the tRNA gene *argU*. *J. Bacteriol.* **171**, 6197–6205.

Makino, S., Uchida, I., Terakado, N., Sasakawa, C., and Yoshikawa, M. (1989). Molecular characterization and protein analysis of the cap region, which is essential for encapsulation in *Bacillus anthracis*. *J. Bacteriol.* **171**, 722–730.

Maurelli, A. T. and Sansonetti, P. J. (1988). Identification of a chromosomal gene controlling temperature-regulated expression of *Shigella* virulence. *Proc. Natl. Acad. Sci. U. S. A.* **85**, 2820–2824.

Maurelli, A. T., Blackmon, B., and Curtiss, R., III (1984). Temperature-dependent expression of virulence genes in *Shigella* species. *Infect. Immun.* **43**, 195–201.

Maurelli, A. T., Fernandez, R. E., Bloch, C. A., Rode, C. K., and Fasano, A. (1998). "Black holes" and bacterial pathogenicity: a large genomic deletion that enhances the virulence of *Shigella* spp. and enteroinvasive *Escherichia coli*. *Proc. Natl. Acad. Sci. U. S. A.* **95**, 3943–3948.

McCormick, B. A., Fernandez, M. I., Siber, A. M., and Maurelli, A. T. (1999). Inhibition of *Shigella flexneri*-induced transepithelial migration of polymorphonuclear leucocytes by cadaverine. *Cell Microbiol.* **1**, 143–155.

Mellies, J. L., Elliott, S. J., Sperandio, V., Donnenberg, M. S., and Kaper, J. B. (1999). The Per regulon of enteropathogenic *Escherichia coli*: identification of a regulatory cascade and a novel transcriptional activator, the locus of enterocyte effacement (LEE)-encoded regulator (Ler). *Mol. Microbiol.* **33**, 296–306.

Mitsuhashi, S., Harada, K., and Hashimoto, H. (1960). Multiple resistance of enteric bacteria and transmission of drug-resistance to other strains by mixed cultivation. *Jpn. J. Exp. Med.* **30**, 179–184.

Moran, N. A. and Plague, G. R. (2004). Genomic changes following host restriction in bacteria. *Curr. Opin. Genet. Dev.* **14**, 627–633.

Nakata, N., Tobe, T., Fukuda, I., *et al.* (1993). The absence of a surface protease, OmpT, determines the intercellular spreading ability of *Shigella*: the relationship between the *ompT* and *kcpA* loci. *Mol. Microbiol.* **9**, 459–468.

Nakaya, R., Nakamura, A., and Murata, Y. (1960). Resistance to transfer agents in *Shigella*. *Biochem. Biophys. Res. Commun.* **3**, 654–659.

Ochman, H., Lawrence, J. G., and Groisman, E. A. (2000). Lateral gene transfer and the nature of bacterial innovation. *Nature* **405**, 299–304.

Ohnishi, M., Kurokawa, K., and Hayashi, T. (2001). Diversification of *Escherichia coli* genomes: are bacteriophages the major contributors? *Trends Microbiol.* **9**, 481–485.

Pupo, G. M., Karaolis, D. K., Lan, R., and Reeves, P. R. (1997). Evolutionary relationships among pathogenic and nonpathogenic *Escherichia coli* strains inferred from multilocus enzyme electrophoresis and *mdh* sequence studies. *Infect. Immun.* **65**, 2685–2692.

Pupo, G. M., Lan, R., and Reeves, P. R. (2000). Multiple independent origins of *Shigella* clones of *Escherichia coli* and convergent evolution of many of their characteristics. *Proc. Natl. Acad. Sci. U. S. A.* **97**, 10 567–10 572.

Purdy, G. E. and Payne, S. M. (2001). The SHI-3 iron transport island of *Shigella boydii* 0-1392 carries the genes for aerobactin synthesis and transport. *J. Bacteriol.* **183**, 4176–4182.

Roth, J. R., Kofoid, E., Roth, F. P., *et al.* (2003). Regulating general mutation rates: examination of the hypermutable state model for Cairnsian adaptive mutation. *Genetics* **163**, 1483–1496.

Salyers, A. A., Shoemaker, N. B., Stevens, A. M., and Li, L. Y. (1995). Conjugative transposons: an unusual and diverse set of integrated gene transfer elements. *Microbiol. Rev.* **59**, 579–590.

Sansonetti, P. J. (1993). Molecular mechanisms of cell and tissue invasion by *Shigella flexneri. Infect. Agents Dis.* **2**, 201–206.

Sansonetti, P. J., Kopecko, D. J., and Formal, S. B. (1982). Involvement of a plasmid in the invasive ability of *Shigella flexneri. Infect. Immun.* **35**, 852–860.

Sansonetti, P. J., Arondel, J., Huerre, M., Harada, A., and Matsushima, K. (1999). Interleukin-8 controls bacterial transepithelial translocation at the cost of epithelial destruction in experimental shigellosis. *Infect. Immun.* **67**, 1471–1480.

Schmidt, H. and Hensel, M. (2004). Pathogenicity islands in bacterial pathogenesis. *Clin. Microbiol. Rev.* **17**, 14–56.

Seifert, H. S. (1996). Questions about gonococcal pilus phase- and antigenic variation. *Mol. Microbiol.* **21**, 433–440.

Seshadri, R., Paulsen, I. T., Eisen, J. A., *et al.* (2003). Complete genome sequence of the Q-fever pathogen *Coxiella burnetii. Proc. Natl. Acad. Sci. U. S. A.* **100**, 5455–5460.

Silva, R. M., Toledo, M. R., and Trabulsi, L. R. (1980). Biochemical and cultural characteristics of invasive *Escherichia coli. J. Clin. Microbiol.* **11**, 441–444.

Smith, H. W. and Halls, S. (1967). The transmissible nature of the genetic factor in *Escherichia coli* that controls haemolysin production. *J. Gen. Microbiol.* **47**, 153–161.

Smith, H. W. and Linggood, M. A. (1971a). Observations on the pathogenic properties of the K88, Hly and Ent plasmids of *Escherichia coli* with particular reference to porcine diarrhoea. *J. Med. Microbiol.* **4**, 467–485.

Smith, H. W. and Linggood, M. A. (1971b). The transmissible nature of enterotoxin production in a human enteropathogenic strain of *Escherichia coli. J. Med. Microbiol.* **4**, 301–305.

So, M., Heffron, F., and McCarthy, B. J. (1979). The *E. coli* gene encoding heat stable toxin is a bacterial transposon flanked by inverted repeats of IS1. *Nature* **277**, 453–456.

Van, L. N., Engler, G., Holsters, M., *et al.* (1974). Large plasmid in *Agrobacterium tumefaciens* essential for crown gall-inducing ability. *Nature* 252, 169–170.

Venter, J. C., Remington, K., Heidelberg, J. F., *et al.* (2004). Environmental genome shotgun sequencing of the Sargasso Sea. *Science* **304**, 66–74.

Wagner, P. L. and Waldor, M. K. (2002). Bacteriophage control of bacterial virulence. *Infect. Immun.* **70**, 3985–3993.

Watanabe, T. (1963). Infective heredity of multiple drug resistance in bacteria. *Bacteriol. Rev.* **27**, 87–115.

West, N. P., Sansonetti, P., Mounier, J., *et al.* (2005). Optimization of virulence functions through glucosylation of *Shigella* LPS. *Science* **307**, 1313–1317.

Wick, L. M., Qi, W., Lacher, D. W., and Whittam, T. S. (2005). Evolution of genomic content in the stepwise emergence of *Escherichia coli* O157:H7. *J. Bacteriol.* **187**, 1783–1791.

Womble, D. D. and Rownd, R. H. (1988). Genetic and physical map of plasmid NR1: comparison with other IncFII antibiotic resistance plasmids. *Microbiol. Rev.* **52**, 433–451.

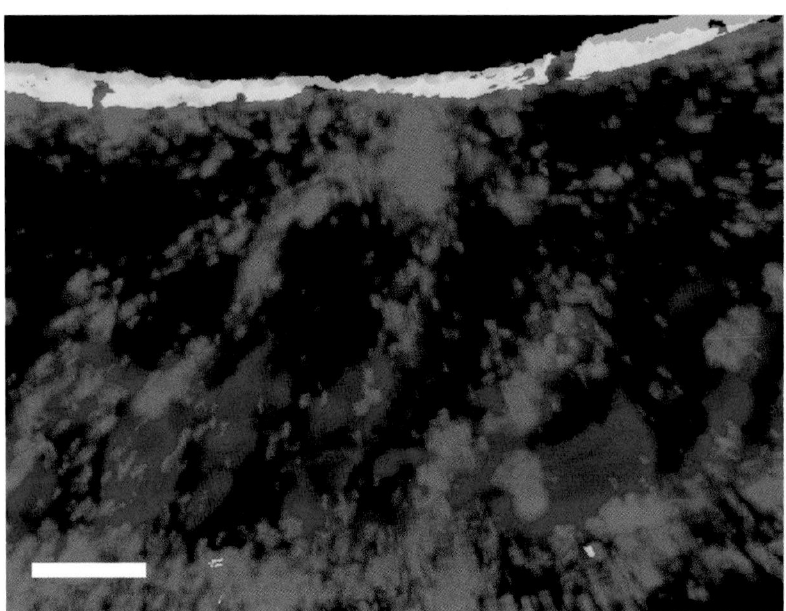

Plate 1 Three-dimensional reconstruction of human jejunal epithelium stained for tubulin (red), f-actin (green), and nuclei (blue). Note the prominent actin ring (green) concentrated at the level of the apical junctional complex. Actin is also present within the microvilli and along basolateral membranes but cannot be appreciated in this image due to the high concentration of actin at the perijunctional ring. Microtubules (red) extend throughout the cell and play a significant role in vesicular trafficking. Bar = 5 μm.

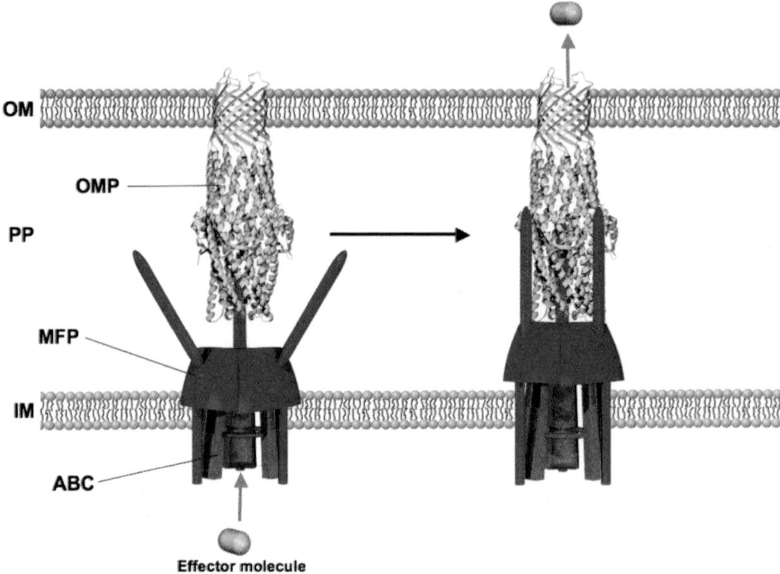

Plate 2 Schematic overview of the type I secretion system. The positions of the outer membrane (OM), periplasm (PP), inner membrane (IM), and the major components of the type I secretion system are depicted. The effector molecule interacts with the ATP-binding cassette (ABC) transporter, triggering a conformational change in the major facilitator protein (MFP), such that it interacts with TolC (OMP) to form a continuous channel across the cell envelope through which the effector molecule is translocated to the exterior of the cell. The structure of TolC was solved to 2.1-Å resolution; the functional unit consists of three TolC monomers oligomerized into a regular structure embedded in the OM and extending into the periplasm. The structures for an MFP and an ABC transporter of a type I secretion systems are yet to be solved, although their putative locations in the cell envelope have been predicted from classical biochemical analyses.

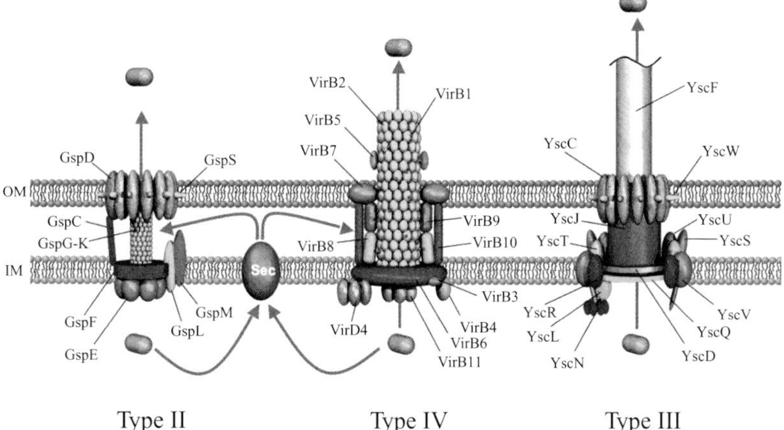

Plate 3 Schematic representation of the multicomponent type II, III, and IV secretion systems. The type II secretion system is exemplified by pullulanase secretion in *Klebsiella oxytoca*, the type III secretion system by Yop secretion in *Yersinia*, and the type IV system by the VirB/VirD system of *Agrobacterium tumefaciens*. The type II and sometimes the type IV secretion systems utilize the Sec-dependent pathway for translocation of effector molecules across the inner membrane (IM). The type III secretion systems, and in the majority of cases the type IV secretion systems, translocate proteins directly across the cell envelope without the need for a periplasmic intermediate. The major structural components of each system are depicted in relation to the known or deduced position in the cell envelope. The position of the outer membrane (OM) and IM are indicated.

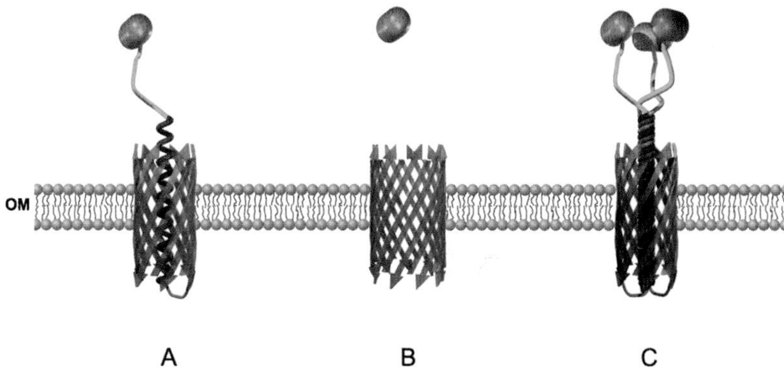

Plate 4 Schematic overview of the type V secretion systems. The secretion pathway across the outer membrane (OM) is depicted for the autotransporter (A), the two-partner (B), and the oligomeric coiled-coil system (C) secretion systems. After translocation through the inner membrane, the β-domains form β-barrel pore-like structures in the OM. After formation of the pore, the effector molecule inserts into the pore and is translocated to the bacterial cell surface, where it may or may not undergo further processing. The linker regions required for secretion of the autotransporters and oligomeric coiled-coil proteins are shaded in red and the autochaperone regions green. Equivalent regions appear to be missing in the two-partner secretion pathway, where the effector molecule and the β-barrel structure are encoded as separate proteins.

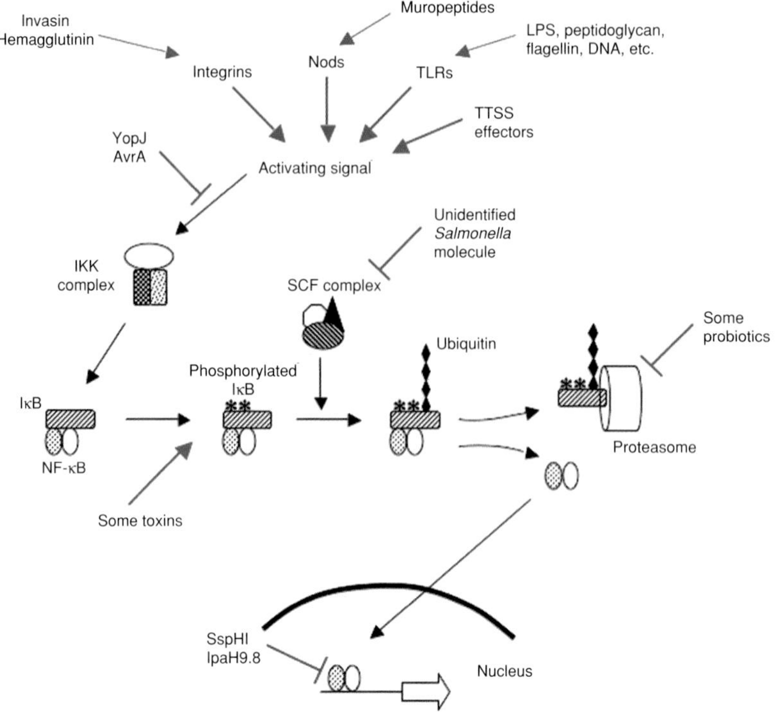

Plate 5 The basic steps in the activation of nuclear factor kappa B (NF-κB) are shown in black. The activating and inhibitory effects of bacterial molecules on this process are indicated in blue.

IκB, inhibitor kappa B; IKK, inhibitor kappa B kinase; LPS, lipopolysaccharide; Nod, nucleotide-binding oligomerization domain; SCF, Sκp1-culin-F-box; TLR, Toll-like receptor; TTSS, type 3 secretion system.

Part II Bacterial cell biology and pathogenesis

CHAPTER 3

Bacterial secretion systems

Helen J. Betts, Christopher M. Bailey,
Mark J. Pallen, and Ian R. Henderson

To survive in any given niche, bacteria must be capable of sensing, interacting with, and responding to their environment. The method and extent to which bacteria interact with their environment are governed to a large degree by the proteinaceous molecules located on the bacterial cell surface or released into the extracellular milieu. Due to differences in cell-envelope architecture, this process of protein secretion is markedly different between Gram-positive and Gram-negative organisms.

GRAM-POSITIVE VERSUS GRAM-NEGATIVE BACTERIA

Gram-positive bacteria possess a single biological membrane termed the cytoplasmic membrane, which is surrounded by a thick cell wall. The majority of proteins targeted for secretion possess an N-terminal amino-acid signal peptide and utilize the Sec-dependent pathway (Holland, 2004). The Sec machinery is composed of several membrane-associated proteins, including an ATPase (SecA), the Sec translocon (SecYEG), which appears to be the basic unit of cellular life forms, several integral membrane proteins (e.g. SecD, SecF), and a signal peptidase that removes the signal peptide during translocation of the proteins across the cytoplasmic membrane (Dalbey and Chen, 2004). In addition to the Sec pathway, several alternative protein-secretion systems have been recognized in Gram-positive organisms, including the Tat (twin arginine translocation) and ESAT-6/WXG-100 pathways (Pallen, 2002; Robinson and Bolhuis, 2004). However, the role of these systems in protein secretion in Gram-positive bacteria is minor in comparison with the Sec-dependent pathway. Once translocated across the cytoplasmic

Bacterial–Epithelial Cell Cross-Talk: Molecular Mechanisms in Pathogenesis, ed. Beth A. McCormick. Published by Cambridge University Press.

membrane, the mature protein either can be released into the extracellular milieu or may remain in contact with the cell wall. In Gram-positive bacteria, five major types of surface protein are recognized: (i) lipoproteins that are attached covalently by their N-terminus to long-chain fatty acids of the cytoplasmic membrane; (ii) proteins attached to the cell surface by S-layer homology domains; (iii) proteins binding to components of the cell wall e.g. choline; (iv) proteins anchored to the cytoplasmic membrane by hydrophobic transmembrane domains; and (v) proteins possessing an LPXTG motif anchored covalently to the cell wall by the enzyme sortase (Ton-That *et al.*, 2004).

In contrast to Gram-positive bacteria, Gram-negative bacteria encompass two lipidaceous membranes, the cytoplasmic (or inner) and outer membranes, which sandwich the peptidoglycan and periplasmic space between them. Protein translocation across the cytoplasmic membrane in Gram-negative bacteria generally follows the same routes as in Gram-positive organisms, and again it relies heavily on the Sec-dependent pathway. However, in contrast to Gram-positive bacteria, protein translocation across the Gram-negative cytoplasmic membrane results in release of the protein into the periplasmic space. Thus, proteins that are destined for the cell surface or extracellular milieu must also cross the additional barrier to secretion formed by the outer membrane. To achieve translocation of proteins to the cell surface and beyond, Gram-negative bacteria have evolved several dedicated secretion systems, some of which bypass the Sec-dependent system and integrate translocation of proteins across both the cytoplasmic and outer membranes in a temporally linked fashion. At least five major highly ordered secretion pathways (types I, II, III, IV and V) have been identified in Gram-negative bacteria, some of which allow proteins to be targeted directly into host cells (Henderson *et al.*, 2004).

TYPE I SECRETION

The type I secretion system (T1SS) of Gram-negative bacteria permits secretion of effector molecules from the cytoplasm to the external milieu in a coupled process such that the molecule bypasses the periplasmic compartment (Delepelaire, 2004). Effectors vary in size (from fewer than 80 to more than 8000 amino acids) and function. Each T1SS requires the presence of three cell-envelope molecules to effect secretion of its cognate effector molecule: a pore-forming outer-membrane protein (OMP), a membrane fusion protein (MFP) that spans the periplasm, and an inner-membrane ATP-binding cassette (ABC) transporter (Figure 3.1) (Binet *et al.*, 1997; Delepelaire, 2004).

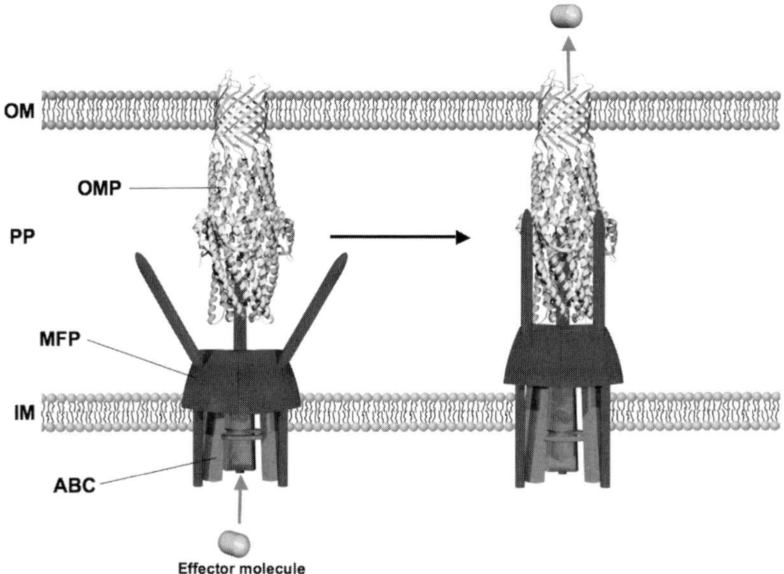

Figure 3.1 Schematic overview of the type I secretion system. The positions of the outer membrane (OM), periplasm (PP), inner membrane (IM), and the major components of the type I secretion system are depicted. The effector molecule interacts with the ATP-binding cassette (ABC) transporter, triggering a conformational change in the major facilitator protein (MFP), such that it interacts with TolC (OMP) to form a continuous channel across the cell envelope through which the effector molecule is translocated to the exterior of the cell. The structure of TolC was solved to 2.1-Å resolution; the functional unit consists of three TolC monomers oligomerized into a regular structure embedded in the OM and extending into the periplasm. The structures for an MFP and an ABC transporter of a type I secretion systems are yet to be solved, although their putative locations in the cell envelope have been predicted from classical biochemical analyses. See also Color plate 2.

The secreted effector molecules often possess a characteristic nine-amino-acid glycine-rich repeat (GGXGXDXXX), which specifically binds calcium ions (Baumann *et al.*, 1993). However, several proteins known to be secreted by the T1SS do not possess such motifs, e.g. HasA of *Serratia marcesens* (Ghigo *et al.*, 1997; Letoffe *et al.*, 1994a), and others possess different types of repeat that also bind calcium, e.g. LapA of *Pseudomonas fluorescens* (Hinsa *et al.*, 2003). Nevertheless, several studies have demonstrated that the repeats are necessary for the function of those molecules that possess them. The functions of the effector molecules vary considerably, from enzymes to adhesins and toxins (Delepelaire, 2004). The prototypical effector of the T1SS family is the *Escherichia coli* α-hemolysin, HlyA (Gentschev *et al.*, 2002). HlyA

Table 3.1 *Examples of type I secretion systems (T1SS)*

Organism	T1SS	Effector	ABC	MFP	OMP
Bordetella pertussis	Cyclolysin	CyaA	CyaB	CyaD	CyaE
Caulobacter crescentus	S-layer protein	RsaA	RsaD	RsaE	RsaF
Erwinia chrysanthemi	Metalloproteases	PrtA, PrtB, PrtC, PrtG	PrtD	PrtE	PrtF
Escherichia coli	Haemolysin A	HlyA	HlyB	HlyD	TolC
E. coli	Colicin V	CvaC	CvaB	CvaA	TolC
Pseudomonas fluorescens	Adhesin	LapA	LapB	LapC	LapE
Serratia marcescens	Haemophore	HasA	HasD	HasE	HasF
S. marcescens	Lipase	LipA	LipB	LipC	LipD

ABC, ATP-binding cassette; MFP, membrane fusion protein; OMP, outer-membrane protein.

is a lipid-modified protein possessing 11 to 17 glycine-rich repeats, which are thought to interact with the host cell. The interaction stimulates insertion of HlyA into the plasma membrane of eukaryotic cells, causing pore formation and the release of cytoplasmic contents. Other functionally characterized members of this family and the associated elements of their T1SS machinery are listed in Table 3.1.

The advent of genomics has led to the identification of a large number of putative T1SSs and associated effector molecules across the breadth of the Proteobacteria (Delepelaire, 2004). As with other secretion systems, the functional characterization of these T1SS effector molecules has lagged behind our understanding of the steps involved in secretion. Thus, the full contribution of the T1SS to the biology of Gram-negative organisms remains enigmatic. However, reconstitution of various T1SS in *E. coli* has revealed that a common feature of the effector molecules, with only a very limited number of exceptions, is the presence of a C-terminal secretion signal (Guzzo *et al.*, 1991; Letoffe *et al.*, 1994b; Wandersman *et al.*, 1987).

The location of the signal sequence at the C-terminus implies that the effector molecule is secreted in a post-translational fashion. The secretion of the T1SS effector molecule begins when the C-terminal signal sequence interacts with the ABC transporter protein (Jarchau *et al.*, 1994). The specific amino acids within the signal sequence that are required for this interaction are poorly understood. It appears that the signal sequences have a high propensity for α-helical structure. Indeed, in the case of HlyA, an amphiphilic α-helical region has been shown to be essential for secretion (Delepelaire, 2004). Furthermore, bioinformatic analyses of the C-terminal regions of putative T1SS effector molecules identified through genome sequencing reveal that the region is rich in several amino acids (LDAVTSIF) and poor in others (KHPMWC) (Delepelaire, 2004). Nevertheless, it appears that the signal sequence is specific and is generally recognized only by its dedicated ABC transporter (Binet and Wandersman, 1995).

The ABC transporter comprises two domains: a transmembrane domain consisting of six transmembrane α-helices and a nucleotide-binding domain extending into the cytoplasm. These molecules function as multimers, which are presumed to form a pore across the inner membrane. Investigations of protein–protein interactions indicate that the nucleotide-binding domain interacts with the C-terminal signal sequence (Benabdelhak *et al.*, 2003). Although the role of ABC transporters in recognition and specificity of the substrate has been demonstrated, ATP is not required for the recognition process itself; indeed, the presence of ATP appears to abrogate the initial interaction (Benabdelhak *et al.*, 2003; Koronakis *et al.*, 1995; Koronakis *et al.*, 1991). Rather, it appears that ATP hydrolysis by the ABC transporter provides the impetus to drive translocation across the cell envelope through the ABC transporter–MFP–OMP complex.

The role of the MFP in secretion is not very well understood. The MFPs appear to possess a conserved structure, comprising a short N-terminal cytoplasmic domain, a transmembrane domain, and a large C-terminal periplasmic domain. Analyses of the primary structure revealed coiled-coiled domains, which are presumed to mediate the oligomerization of the MFP into larger homotrimeric structures that cap the periplasmic domain of the ABC transporter (Thanabalu *et al.*, 1998). The prevailing theory of secretion suggests that interaction of the effector molecule with the ABC transporter triggers a conformational change in the MFP, such that it interacts with the OMP, allowing secretion of the effector molecule to the external milieu (see Figure 3.1) (Sharff *et al.*, 2001). However, this model is somewhat controversial, since investigations of some systems indicate that interaction of the MFP and ABC transporter occur only after the ABC transporter binds the effector molecule, whereas in other systems it appears that the two components interact before substrate binding (Letoffe *et al.*, 1996). Whether these differences correspond to real differences in the mechanisms of secretion or to differences in the experimental procedures remains to be determined. Nevertheless, it is clear that in all cases, binding of the substrate molecule drives the formation of the translocation channel through specific interactions between the ABC transporter and the MFP and between the MFP and the OMP.

The crystal structure of TolC, the OMP of the *E. coli* α-hemolysin secretion system, has contributed greatly to our understanding of the T1SS (Koronakis *et al.*, 2000). TolC exists as a trimer anchored in the outer membrane by a β-barrel structure with α-helices protruding approximately 100 Å into the periplasm. Each monomer contributes four β-strands, to form a 30–35-Å-wide β-barrel structure, and four α-helices, which form a larger super-helix, such that the molecule tapers to 3.5 Å at the end of the periplasmic domain. As 3.5 Å is insufficient to allow passage of folded molecules, it was suggested that TolC opens at the distal end in an iris-like mechanism to generate a maximal opening of 16–20 Å; this is wide enough to allow passage of proteins that have adopted their secondary structure. It has been suggested that the conformational change in MFP that occurs after binding of the effector molecule to the ABC transporter causes each MFP monomer to interact with the α-helices of a TolC monomer, such that a continuous channel is formed across the cell envelope at the appropriate time. The interaction of MFP with TolC is probably responsible for the mechanical force that leads to the opening of the TolC iris (Andersen *et al.*, 2002; Koronakis *et al.*, 2000).

TYPE II SECRETION

Secretion via the type II secretion system (T2SS) is exemplified by pullulanase (PulA) secretion from *Klebsiella oxytoca* (Takizawa and Murooka, 1985). PulA is a starch-hydrolyzing enzyme that forms micelles once it is secreted into the external environment. Since the discovery of the T2SS in *K. oxytoca*, the pathway has been identified in a variety of other bacterial species exporting a range of proteins with diverse functions, some of which appear to be important determinants of bacterial virulence (Table 3.2) (Sandkvist, 2001a). Furthermore, the striking similarities at the genetic and structural levels between the type IV fimbrial systems and T2SSs have led to the suggestion that the fimbrial biogenesis pathway may be considered a T2SS (Bitter *et al.*, 1998). Indeed, the demonstration that the *Vibrio cholerae* type IV fimbrial system was essential for secretion of a soluble colonization factor (TcpF) demonstrated that these two systems are linked closely at a functional level (Kirn *et al.*, 2003). The type IV fimbriae have also been implicated as important virulence determinants involved in cell adhesion, biofilm formation, DNA uptake, and a specialized form of motility termed twitching motility (Mattick, 2002).

Table 3.2 *Examples of type II secretion systems*

Organism	Locus	Secreted effector molecules
Aeromonas hydrophila	*exe*	Aerolysin, amylase, phospholipase, protease
Burkholderia pseudomallei	*gsp*	Phospholipase C
Erwinia chrysanthemi	*out*	Pectinase, cellulase
Escherichia coli H10407 (ETEC)	*gsp*	Heat-labile enterotoxin
E. coli O157	*etp*	StcE metalloprotease
Klebsiella oxytoca	*pul*	Pullulanase
Legionella pneumophila	*lsp*	Msp protease, phospholipase A
Pseudomonas aeruginosa	*xcp*	Exotoxin A, elastase, phospholipase C
Vibrio cholerae	*eps*	Cholera toxin, HAP, chitinase, lipase, neuraminidase
Xanthomonas campestris	*xps*	Polygalacturonate lyase, α-amylase

ETEC, enterotoxigenic *Escherichia coli*; HAP, hemagglutinin/protease.

Secretion of proteins via the T2SS is a two-step process, whereby the effector molecules are translocated across the inner membrane and then dock with the T2SS machinery before secretion across the outer membrane (Figure 3.2). In general, the effector molecules are synthesized with an N-terminal signal sequence, which targets the proteins to the Sec pathway (Filloux, 2004). However, evidence has demonstrated that some effector molecules can cross the inner membrane by another apparatus termed the Tat system (Voulhoux *et al.*, 2001). It appears in both cases that this two-step process allows the effector molecules to adopt a folded state before interaction with the T2SS machinery and translocation across the outer membrane (Filloux, 2004). Interestingly, recognition of the effector molecule by the T2SS machinery appears to be a species-specific process. Thus, the *K. oxytoca* pullulanase is not recognized by the *Pseudomonas aeruginosa* T2SS. However, heterologous secretion has been described for the effector molecules of closely related species (de Groot *et al.*, 1991; Wong *et al.*, 1990). The basis of this discrimination between effector molecules remains unknown but is presumed to relate to specific conformation signals within the effector molecule that are recognized by the cognate T2SS (Filloux, 2004).

To effect secretion across the outer membrane, the T2SS requires 12 to 16 proteins, which are usually encoded in large operons (Filloux, 2004;

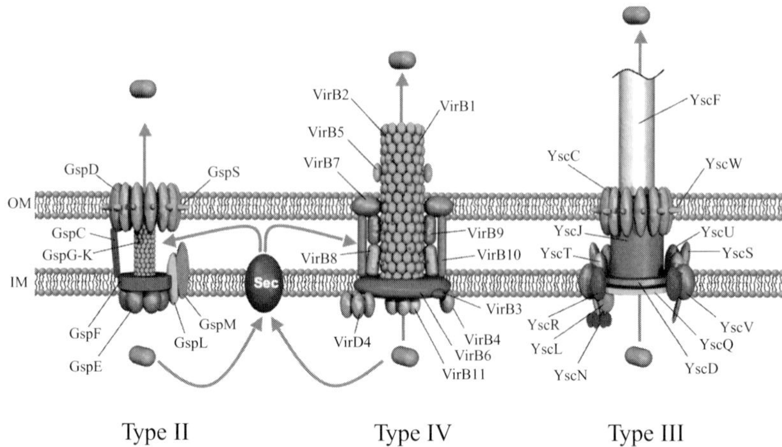

Figure 3.2 Schematic representation of the multicomponent type II, III, and IV secretion systems. The type II secretion system is exemplified by pullulanase secretion in *Klebsiella oxytoca*, the type III secretion system by Yop secretion in *Yersinia*, and the type IV system by the VirB/VirD system of *Agrobacterium tumefaciens*. The type II and sometimes the type IV secretion systems utilize the Sec-dependent pathway for translocation of effector molecules across the inner membrane (IM). The type III secretion systems, and in the majority of cases the type IV secretion systems, translocate proteins directly across the cell envelope without the need for a periplasmic intermediate. The major structural components of each system are depicted in relation to the known or deduced position in the cell envelope. The position of the outer membrane (OM) and IM are indicated. See also Color plate 3.

Sandkvist, 2001b). The genes encoding these proteins have been given the generic designation *gsp*, and a consensus nomenclature has been established for the functional homologues defined by the letters A to O and S (the systems of *P. aeruginosa* are labeled P to Z). Surprisingly for an outer-membrane translocation apparatus, only two of these proteins are located in the outer membrane; the others are variously associated with the inner membrane or the periplasmic space (Figure 3.2). Based on a variety of biochemical and genetic investigations, it is possible to divide the apparatus composing the T2SS into several domains: (i) the inner membrane platform upon which (ii) the pseudopilus is built, (iii) the connector proteins that mediate interaction between the inner-membrane platform, and (iv) the outer-membrane complex.

The inner-membrane platform is thought to be composed of GspE, GspF, GspL, and GspM (Filloux, 2004). Evidence suggests that GspE interacts with the cytoplasmic face of the inner membrane and regulates the secretion process by supplying energy to promote translocation and assembly of the pseudopilins (Sandkvist, 2001b). GspL, an integral inner-membrane protein, is required for the association of GspE with the membrane, and GspM is thought to stabilize GspL within the inner membrane (Sandkvist, 2001b). GspF, a polytopic inner-membrane protein, interacts with GspL and GspE to strengthen the association of GspE with the inner membrane and may form a pore in the cytoplasmic membrane through which the pseudopilins are translocated to the periplasmic space (Py *et al.*, 2001).

The pseudopilins are encoded by *gspG–K* and possess an N-terminal signal sequence that is recognized by the prepilin peptidase GspO. After cleavage of the N-terminal signal, GspO methylates the prepilins before translocation into the periplasmic space (Bleves *et al.*, 1998; Pepe and Lory, 1998). Once translocated across the membrane, the pseudopilins are presumed to assemble into a macromolecular pilus-like structure termed the pseudopilus. Overexpression of the pseudopilins leads to production of a pilus structure reminiscent of the type IV pili. However, in the case of the T2SS, a pilus-like structure cannot be visualized on the cell surface, despite clear evidence for interaction of the pseudopilins within the periplasmic space (Sauvonnet *et al.*, 2000). Based on several studies, it has been suggested that the pseudopilus may grow to push the effector molecule through the outer-membrane complex (Filloux, 2004).

The outer-membrane complex is formed by GspD and GspS. GspD belongs to a family of proteins termed the secretins, homologs of which are also required for type III secretion (Bitter, 2003). GspD is an integral outer-membrane protein predicted to consist largely of transmembrane β-strands. Twelve to 14 subunits form a larger-order homomeric pore-like structure in the outer membrane through which effector molecules are secreted to the external environment (Bitter *et al.*, 1998; Chen *et al.*, 1996; Nouwen *et al.*, 2000). Biochemical and electron microscopy studies have revealed that GspD can adopt a ring-shaped structure with a large central cavity of ~95 Å, which is consistent with translocation of folded molecules across the outer membrane (Nouwen *et al.*, 1999). In some cases, insertion of GspD into the outer membrane depends on a small lipoprotein termed GspS (Hardie *et al.*, 1996; Nouwen *et al.*, 1999). Although its role has not yet been established, it is likely that GspS protects GspD from proteolytic degradation.

The functions of GspC and GspN remain unknown. GspN is absent from many T2SSs but, when present, is thought to play a role similar to that of GspC. The topology of GspC is reminiscent of TonB and possesses a C-terminal PDZ domain (Bleves *et al.*, 1996; Pallen and Ponting, 1997). PDZ domain mediate protein–protein interactions, particularly in the organization of multimolecular complexes. Interestingly, in some homologs the PDZ domain from GspC is replaced by coiled-coiled domains, which are also implicated in protein–protein interactions. Several observations have led to the suggestion that GspC connects the inner-membrane platform with the outer-membrane complex and regulates the secretion of effector molecules by controlling the opening of the pore formed by GspD and by transducing energy through the system (Bleves *et al.*, 1999; Gerard-Vincent *et al.*, 2002; Possot *et al.*, 2000). In addition, GspC may contribute to the species-specific transport of effector molecules by defining the transport of specific substrates through the GspD pore (Bouley *et al.*, 2001; Lindeberg *et al.*, 1996; Possot *et al.*, 2000). In summary, the components of the T2SS interact to form a multiprotein complex spanning both the inner and outer membranes and through which effector molecules are translocated to the extracellular milieu.

TYPE III SECRETION

Type III secretion systems (T3SSs) can be divided into two classes on the basis of function and evolutionary history: the flagellar and non-flagellar systems (Aizawa, 2001; Galan and Collmer, 1999; Gophna *et al.*, 2003; Macnab, 2004; Pallen *et al.*, 2005a; Plano *et al.*, 2001). The flagellar systems are widespread among bacteria and are involved in the biosynthesis of the flagellum, the chief organelle of motility in bacteria. However, in this chapter we will focus almost exclusively on the non-flagellar type III secretion systems (NF-T3SSs), which are characterized by the possession of a translocation apparatus that mediates transfer of bacterial proteins into eukaryotic cells. NF-T3SSs occur in a variety of Gram-negative pathogens of plants, mammals, and insects (Table 3.3). Intriguingly, such systems have also been identified in symbiotic and environmental bacteria, where there they are probably involved in symbiotic or exploitational interactions between bacteria and invertebrate and even protozoal hosts (Pallen *et al.*, 2005a).

The structural components of the T3SS apparatus span both bacterial membranes and the periplasm, allowing effector proteins to be exported in a single step. In non-flagellar systems, the secretion apparatus is supplemented by a lanceolate structure protruding from the apparatus (the “needle”) and a distal translocation apparatus (or “translocon”) that mediates delivery of effectors in the host-cell cytoplasm. Indeed, the NF-T3SS has been referred to as a “molecular syringe” (Ghosh, 2004; Hueck, 1998; Tampakaki *et al.*, 2004).

The Ysc-Yop NF-T3SS from *Yersinia* was the first such system to be characterized and is often treated as the archetype (Bolin *et al.*, 1985). However, thanks largely to genome sequencing, NF-T3SSs have been identified

Table 3.3 *Examples of type III secretion system (T3SS) effectors*

Effector	System	Function/enzymatic activity
YopH	Ysc	Phosphotyrosine phosphatase – disrupts focal adhesions
		Downregulates inflammatory response
YopO	Ysc	Serine threonine kinase
YopT	Ysc	Cleaves RhoA at the C-terminus
YopP/J	Ysc	Disrupts MAP kinase pathways
SptP	Spi-1	Mimics action of Rho-inactivating proteins
SopB	Spi-1	Inositol phosphate phosphatase
SipC	Spi-1	Involved in host-cell actin polymerization
SipA	Spi-1	Inhibits actin depolymerization
SopE	Spi-1	Stimulates Rac1 and CDC42 to promote membrane ruffling
SpiC	Spi-2	Influences vesicular trafficking
EspB	LEE	Recruits α-catenin to actin pedestal
MAP	LEE	Targets mitochondria
EspH, EspF	LEE	Manipulation of actin cytoskeleton

LEE, locus of enterocyte effacement.

in several dozen bacterial species, including many important pathogens. Well-studied human pathogens containing one or more NF-T3SSs include *Salmonella enterica*, pathogenic *E. coli*, *Shigella*, and *P. aeruginosa*, which cause diseases ranging from diarrhea and lung infection to septicemia (He *et al.*, 2004). Commercially important plant pathogens with NF-T3SSs include *Pseudomonas syringae*, *Xanthomonas*, and *Ralstonia solanacearum* (Genin and Boucher, 2004; Jin *et al.*, 2003; Weber *et al.*, 2005). Some NF-T3SSs mediate internalization into eukaryotic cells; for example, T3SS effectors secreted by *S. enterica* and *Shigella* induce engulfment and entry into host cells. However, in other settings, e.g. *E. coli* and *Yersinia*, the bacterial cells remain extracellular and yet are still able to subvert host cellular processes by injecting effectors.

Some species contain more than one NF-T3SS, each playing a distinct role in pathogenesis. For example, *S. enterica* has two such systems, each encoded by a distinct chromosomal locus: Spi-1 and Spi-2 (*Salmonella* pathogenicity islands 1 and 2) (Knodler *et al.*, 2002; Shea *et al.*, 1996). The Spi-1-encoded system is essential for initial attachment and cell entry, while the system encoded by Spi-2 is required for intracellular replication and proliferation. Furthermore, it is not uncommon for an organism to contain a functional NF-T3SS and non-functional remnants of a second system. For instance, enterohemorrhagic *E. coli* O157:H7 (EHEC) possesses the locus for enterocyte effacement (LEE), encoding a fully functional T3SS required for pathogenesis, and also the *E. coli* type III secretion system 2 (ETT2) locus, encoding a degenerate T3SS found in a wide range of *E. coli* strains, including commensals (Pallen *et al.*, 2005a,b; Ren *et al.*, 2004). The identification of ETT2 genes in non-pathogens demonstrates that the presence of T3SS genes cannot be considered a bona fide indicator of virulence. Interestingly, regulators encoded within the ETT2 locus in EHEC are involved in regulatory cross-talk with the LEE (Zhang *et al.*, 2004).

NF-T3SSs consist of five general components (Figure 3.2):

- Secretion apparatus and needle (often together termed the needle complex).
- The translocation apparatus, including proteins that form a pore in the eukaryotic cell plasma membrane.
- Regulators.
- Chaperones.
- Translocated effector proteins.

Each component is discussed briefly here before we provide a more detailed analysis focused on effectors and their roles in cross-talk between bacteria and epithelial cells and other host cells. Due to space constraints, discussion will be confined largely to the five best-studied systems from the four mammalian pathogens *Yersinia*, *S. enterica*, *Shigella*, and *E. coli*. The reader is directed elsewhere for information on NF-T3SSs in phytopathogens (Alfano and Collmer, 2004; Buttner and Bonas, 2002; He, 1998; He and Jin, 2003; Innes, 2003; Jin and He, 2001; Mudgett, 2005; Puhler *et al.*, 2004).

Secretion apparatus

The T3SS apparatus consists of around two dozen proteins, some of which are highly conserved, which form multiprotein structures spanning the inner and outer membranes (Ghosh, 2004; Hueck, 1998). The secretion machinery is thought to be energized by ATP hydrolysis, as proteins with proven ATPase activity (YscN, InvC) have been shown to be essential for type III secretion (Eichelberg *et al.*, 1994; Woestyn *et al.*, 1994).

In *Yersinia*, genes for the T3SS proteins are located on a large virulence plasmid. Three loci, VirA, VirB, and VirC, encode the structural Ysc proteins, the secreted Yop proteins, regulators, and chaperones (Allaoui *et al.*,

1995; Bleves and Cornelis, 2000). The inner-membrane complex consists of YscDFJLNQRSTUV. Many of the inner-membrane proteins contain putative transmembrane domains (Ghosh, 2004; Hueck, 1998). YscC forms an oligomeric complex spanning the outer membrane (Burghout *et al.*, 2004). Homologs of YscC are common to all NF-T3SSs but are absent from flagellar systems. These proteins belong to the secretin family, which also possess features of type II secretion.

The needle forms a hollow extracellular appendage, which has been visualized by electron microscopy in *Yersinia*, *Salmonella*, and *Shigella*, attached to the outer-membrane ring (Cordes *et al.*, 2003; Hoiczyk and Blobel, 2001; Marlovits *et al.*, 2004). The secretion apparatus and the needle mediate movement of effectors from the bacterial cytosol across the cell wall, in preparation for their translocation into host cells. Numerous studies have shown that mutations in genes encoding needle proteins result in an inability to secrete or translocate effectors and thus cause an attenuation of virulence (Kubori *et al.*, 2000; Sukhan *et al.*, 2003; Tamano *et al.*,2000, 2002). The morphology of the structures distal to the secretion apparatus varies from one system to another. Electron micrographs of the needles from the *Yersinia*, *Salmonella* Spi-1, and *Shigella* T3SSs reveals a straight rigid structure that is very short in comparison with the flagellum (Cordes *et al.*, 2003; Hoiczyk and Blobel, 2001; Marlovits *et al.*, 2004). By contrast, pathogenic *E. coli* strains that harbour the LEE produce a small needle, capped by a long filamentous appendage, reminiscent of the flagellar filament, which is termed the EspA pilus (Daniell *et al.*, 2003; Knutton *et al.*, 1998). Phytopathogens also display a long pilus-like structure, the Hrp pilus, which appears to be thinner and more flexible than the EspA pilus.

Regulators and chaperones

Regulation of T3SS gene expression is complex, usually involving a hierarchy of transcriptional regulators, and varies greatly from one system to the next. The reader is referred to a recent review of this complex subject for more information (Francis *et al.*, 2002). Molecular chaperones, which interact with translocator and effector molecules in the bacterial cytosol, play a pivotal role in T3SS, as evidenced by the fact that mutational ablation of chaperone genes usually results in decreased substrate secretion and often attenuation of virulence (Jackson *et al.*, 1998; Neyt and Cornelis, 1999; Tucker and Galan, 2000; Wainwright and Kaper, 1998). Despite their importance, the functions of T3SS chaperones are often unclear. Proven or suspected roles cover a wide range of activities:

- Chaperones may prevent degradation of substrate before secretion.
- Chaperones may act as anti-aggregation factors, partitioning substrates, and preventing their premature association.
- As the internal channel within the needle has a small diameter, which is unlikely to permit secretion of fully folded proteins, chaperones may maintain substrates in a secretion-competent state (Smith and Hultgren, 2001).
- Chaperones may escort the effector or translocator to the secretion apparatus.
- As the number of effectors far outweighs the number of chaperones, some authors have suggested that chaperones determine a hierarchy, so that substrates with a chaperone may be secreted before those without a chaperone (Boyd *et al.*, 2000; Feldman and Cornelis, 2003).
- Some chaperones influence transcriptional regulation of the T3SS (Darwin and Miller, 2000; Francis *et al.*, 2002).

All of these models appear plausible, and most chaperones probably perform a range of functions.

T3SS chaperones are classified according to whether they have a higher propensity to interact with effector proteins (class I chaperones) or translocator proteins (class II chaperones) (Parsot *et al.*, 2003). The structures of several class I chaperones have been elucidated, revealing anti-parallel beta-sheets wrapped around an alpha-helical binding groove forming a new alpha-beta fold (Luo *et al.*, 2001; Stebbins and Galan, 2001). No experimentally determined structure exists for any class II chaperone. However, homology modeling suggests that class II chaperones consist of an array of three tetratricopeptide repeats that fold up to form a hydrophobic binding groove (Pallen *et al.*, 2003). This model has received substantial support from mutagenesis studies (M. S. Francis, H. J. Betts, and M. J. Pallen, unpublished data).

The translocation apparatus

The term "translocator" is reserved for proteins that are required for effector translocation but dispensable for type III protein secretion into the external milieu. In the *Yersinia* Ysc–Yop system, secretion of Yop effectors requires the translocators YopB, YopD, and LcrV (Pettersson *et al.*, 1999). Both YopB and LcrV have been shown to be involved in pore formation in the eukaryotic plasma membrane (Tardy *et al.*, 1999). It has been demonstrated that LcrV interacts with both YopB and YopD and is required for their secretion. Experimental evidence suggests that YopB and YopD are secreted in a complex with each other. YopD may be involved in pore formation via its putative transmembrane domains (Marenne *et al.*, 2003). However, YopD also appears to have additional effector functions (Hakansson *et al.*, 1993).

Enteropathogenic *E. coli* (EPEC) and EHEC along with the mouse pathogen *Citrobacter rodentium*, produce a T3SS encoded by the locus for enterocyte effacement or LEE. As noted above, the LEE-encoded system is characterized by a filamentous extension to the needle, the EspA pilus, through which the translocators EspD and EspB are transported (Chen and Frankel, 2005; Knutton *et al.*, 1998). EspB and EspD work together to produce a pore in the plasma membrane. However, in parallel with YopD, EspB also appears to have an effector function.

Unlike *Yersinia* and *E. coli*, which remain extracellular, *S. enterica* utilizes its two T3SSs to invade and replicate within cells (Galan, 2001). Spi-1 proteins initiate adherence to and invasion of M-cells, while Spi-2 proteins are involved in intracellular survival and replication. Spi-1 translocated SipA, SipB, and SipC act together to form the translocation machinery, with SipB and SipC forming a pore (Miki *et al.*, 2004).

Effectors

In contrast to the largely conserved nature of the secretion and translocation apparatus, the repertoire of effectors secreted via T3SSs varies not only from one system to another but also from strain to strain. Furthermore, genes encoding effector proteins are often encoded outside the locus encoding the structural components (Boyd and Brussow, 2002; Gruenheid *et al.*, 2004). Genomic data-mining exercises have identified many candidate effectors, and thus the number of suspected effectors far outweighs the number that have been characterized experimentally. However, a unifying assumption is that all effectors are engaged in bacterial–host cross-talk, subverting host cellular processes to the advantage of the bacterium.

A variety of enzymatic activities have been observed among effectors (Barz *et al.*, 2000; Bliska *et al.*, 1991; Kaniga *et al.*, 1996; Knight and Barbieri, 1997; Liu *et al.*, 1997; Norris *et al.*, 1998; Yahr *et al.*, 1998) including the following:

- Phosphotyrosine phosphatases, e.g. YopH from *Yersinia* and SptP from *Salmonella*.
- Serine-threonine kinases, e.g. YopO.
- Inositol phosphate phosphatase, e.g. SopB.
- ADP-ribosyltransferases, e.g. ExoS and ExoT from *P. aeruginosa*.
- Adenylate cyclase, e.g. ExoY from *P. aeruginosa*.

Studies of effector function highlight several recurring themes (Lu and Walker, 2001; Espinosa and Alfano, 2004): disruption of the cytoskeleton; disruption of signaling pathways, particularly to evade immune responses; and modulation of vesicular trafficking.

Cytoskeletal effects

As noted, EHEC and EPEC remain extracellular but are still able to use the LEE-encoded T3SS in order to induce cytoskeletal changes in the underlying host cell. The best-characterized effector from this system is the translocated intimin receptor (Tir), which, after passing through the translocation machinery, embeds itself in the plasma membrane (Kenny *et al.*, 1997). The extracellular loop from Tir binds to intimin, an outer-membrane protein expressed on the bacterial cell surface, mediating an intimate bacterial attachment to the host cell. The intracellular components of Tir trigger the accumulation of polymerized actin beneath the attached bacterium to form a pedestal (Campellone and Leong, 2003). The translocator EspB also moonlights as an effector, recruiting alpha-catenin to the actin pedestal. Other LEE-encoded effectors that target the cytoskeleton include MAP, which contributes to filiopodia formation (and also targets mitochondria), EspH, and EspF (which manipulates the actin cytoskeleton to disrupt tight junctions) (Dean and Kenny, 2004; Kenny, 2002; Tu *et al.*, 2003).

T3SS in *Salmonella* leads to actin rearrangements, starting with membrane ruffles and progressing to internalization of the bacterium into membrane-bound *Salmonella*-containing vacuoles (SCVs). SipC is essential for actin polymerization, while SipA potentiates this activity by inhibiting depolymerization (McGhie *et al.*, 2001).

Small GTPases, such as Rho, Rac, and CDC42, control cytoskeletal organization and various other activities in human cells, making them ideal targets for type III effectors (Schwartz, 2004). YopT from *Yersinia* cleaves RhoA at the C-terminus (Iriarte and Cornelis, 1998), while *S. enterica* SopE potentiates the membrane ruffling initiated by SipA and SipC by stimulating Rac1 and CDC42. SptP, ExoS, and YopE modulate Rho functioning by mimicking the action of Rho-inactivating proteins (Aktories *et al.*, 2000). In this capacity, SptP is thought to mediate restoration of cytoskeletal architecture after engulfment of *Salmonella* cells (Zhou and Galan, 2001).

Subversion of the immune system

YopH is a tyrosine phosphatase that disrupts focal adhesions targeting host proteins Fyb and SKAP-HOM (Deleuil *et al.*, 2003). YopH has also been implicated in downregulation of the inflammatory response, by inhibiting release of cytokines involved in macrophage recruitment (Navarro *et al.*, 2005). YopP (also termed YopJ) disrupts MAP kinase pathways, inhibiting IKK-β, the kinase responsible for IκB phosphorylation (Denecker *et al.*, 2001; Ruckdeschel *et al.*, 1998). This prevents activation of NF-κB, which in turn abrogates release of pro-inflammatory cytokines. YopP/J may also trigger macrophage apoptosis, although it is unclear which pathways predominate in this effect.

In contrast to *Yersinia*, *Salmonella* effectors upregulate the inflammatory response, activating MAP kinase and NF-κB and thereby stimulating production of pro-inflammatory cytokines. This presumably induces the inflammatory diarrhea characteristic of salmonellosis (Hobbie *et al.*, 1997). Several effectors are implicated in this phenomenon, including SopE, SopE2, and SopB (Hernandez *et al.*, 2003). The translocator/effector SipB may also contribute to the pro-inflammatory response by inducing macrophage apoptosis.

Vesicular trafficking

Intracellular replication of *Salmonella* is dependent on effectors secreted through the Spi-2 T3SS. SpiC is secreted into the macrophage cytosol and targets at least two different host proteins, TassC and Hook3, presumably influencing vesicular trafficking (Lee *et al.*, 2002; Shotland *et al.*, 2003; Uchiya *et al.*, 1999). In the later stages of infection, SCVs elaborate filamentous extrusions, termed *Salmonella*-induced filaments (Sifs) (Garcia-del Portillo *et al.*, 1993). Although striking in appearance, their function remains unknown. Several Spi-2 effectors contribute to Sif formation or are associated with them, e.g. SifA, SseF, SseG, SseJ, and SopD2 (Kuhle *et al.*, 2004; Stein *et al.*, 1996; Waterman and Holden, 2003). Other Spi-2 effectors interact with cytoskeletal components such as filamin and profilin. The Spi-2 system also contributes to intracellular survival by influencing release of reactive oxygen and reactive nitrogen radicals within the SCV, although the mechanisms remain unclear (Vazquez-Torres and Fang, 2001a,b).

TYPE IV SECRETION

Type IV secretion systems (T4SSs) show several functional similarities to non-flagellar T3SSs, such as one-step secretion across the Gram-negative cell wall and translocation via a needle-like organelle directly into target cells. Furthermore, they share common components, e.g. secretin domains and ATPases. However, in overall composition, structure, and function, the two types of secretion system are clearly distinct, with the T4SS sharing a common ancestor with the bacterial conjugation machinery. Furthermore, in contrast to T3SSs, T4SSs can deliver DNA as well as proteins to a diverse range of organisms (Christie and Vogel, 2000). Several mammalian and plant pathogens have been shown to mediate virulence factors in a T4SS-dependent manner, including *Bordetella pertussis, Legionella pneumophila, H. pylori, Brucella* spp., *Bartonella* spp., and the archetype *Agrobacterium tumefaciens* (Cascales and Christie, 2003; Ding *et al.*, 2003). Curiously, genome sequencing has revealed putative T4SSs in a number of non-pathogens, e.g. *Caulobacter crescentus* and *Rhizobium etli* (Christie and Vogel, 2000).

Based on ancestral lineage, T4SSs are split into two subclasses: type IVA secretion systems include the archetype from *A. tumefaciens* and type IVB secretion systems encompass systems assembled from *Shigella flexneri* IncI ColIb-P9 plasmid Tra homologues (Christie and Vogel, 2000). There are three known types of substrate:

- DNA conjugation intermediates, where single-stranded (ss) DNA is exported in complex with one or more proteins, e.g. the *A. tumefaciens* T-complex;
- multisubunit proteins such as pertussis toxin (PT);
- monomeric proteins such as CagA from *H. pylori* (Christie and Vogel, 2000).

We will focus on the four best-characterized systems – VirB/D4 from *A. tumefaciens*, Ptl from *B. pertussis*, the Cag system from *H. pylori*, and the Dot/Icm system from *L. pneumophila* – with an emphasis on secreted substrates and the subversion of host cellular machinery. The *A. tumefaciens* T4S machinery is thus far the only system to be structurally characterized, although closely related organisms, including *B. pertussis* and *Brucella* spp., are likely to have a similar structure (Burns, 2003) (Figure 3.2). A list of secreted type IV substrates is given in Table 3.4.

The archetypal *Agrobacterium tumefaciens* VirB/D4 system

A. tumefaciens causes crown-gall disease. It utilizes the T4SS to deliver oncogenes to the chromosomes of a variety of dicotyledonous plants, resulting in tumor formation. Tumors produce opines, which are assimilated by bacteria as a food source (Ziemienowicz, 2001). Tumor formation results from translocation of transfer DNA (T-DNA), ss DNA derived from the tumor-inducing plasmid (pTi), into host cells (Van Montagu *et al.*, 1980).

Table 3.4 *Some examples of type IV secretion system effectors*

Effector	System	Function/enzymatic activity
T-strand	VirB/D4	Integrates into host-cell chromosome
A-subunit of pertussis toxin	Pertussis toxin	Adenylate cyclase activity; interferes with host-cell signal-transduction pathways
CagA	Cag	Cortical actin polymerization and cytoskeletal rearrangement
DotA	Dot/Icm	Forms pores in eukaryotic plasma membrane
LidA	Dot/Icm	Function undefined; could induce vesicle recruitment

The VirB/D4 T4SS consists of a secretion apparatus spanning both bacterial membranes and the periplasm. A long transfer (T-) pilus, reminiscent of the Hrp pilus from T3S, is formed on the surface of the organism, allowing the export of substrates into host cells.

The *virB* locus on the Ti plasmid encodes at least 11 VirB proteins, of which ten make up the transporter apparatus spanning the membranes and extending into the extracellular milieu. A core of five proteins, VirB4, VirB7, VirB9, VirB10, and VirB11, is conserved among type IVA systems. VirB2 is the major pilin subunit and in conjunction with the minor VirB5 subunit forms a pilus approximately 3.8 nm in diameter. VirB4 and VirB6 are integral membrane proteins found in the inner membrane. VirB11 is associated with the inner side of the inner membrane. VirB7–B10 are predicted to span the membranes fractionating to both inner and outer regions and forming a translocation channel (Figure 3.2) (Das and Xie, 1998). VirB4 and VirB11 contain nucleotide-binding sites essential for transport, and it is suggested that these proteins project into the bacterial cytosol and energize secretion via ATP hydrolysis. VirB1 is predicted to have glycosidase activity, possibly involved in peptidoglycan hydrolysis, forming a pore for the other VirB proteins to fill (Burns, 2003). Many T4SSs have one or more coupling proteins (CP) acting as a chaperone to deliver substrates to the translocon. VirE1 is required to stabilize VirE2, a protein with ss DNA binding capacity, which acts to protect the transported T-complex from nucleases (Zhao *et al.*, 2001).

The T-DNA region on pTi is flanked by imperfect 25-base-pair direct repeats. VirD1 and VirD2 cleave these sites and excise the lower strand, to yield the single-stranded T-DNA. The VirE2–T-strand–VirD2 complex is delivered to the nucleus, a process mediated by nuclear localization signals. The T-DNA then integrates into the host-cell chromosome by illegitimate recombination.

Pertussis toxin: an exception to the rule?

PT is the sole secreted effector of the *B. pertussis* T4SS and is essential for the pathogenesis of whooping cough. PT is an A-B_5 bacterial exotoxin consisting of five subunits S1–S5, where the S1 subunit is the enzymatically active A component. Following Sec-dependent export to the periplasm, the subunits are assembled into the mature toxin before continuing their journey through the T4SS (Burns, 2003). The toxin is not translocated directly into the host-cell cytoplasm; rather, it is secreted into the extracellular milieu before being taken up. The B oligomer mediates binding of the toxin to host-cell-surface glycoproteins, allowing the active A subunit to be transported into the cell by receptor-mediated endocytosis. The A subunit has adenylate cyclase activity and interferes with host-cell signal-transduction pathways by ADP-ribosylating γ-subunits of G-proteins, including G_i, G_o, and G_t. This has various effects, depending upon the tissue affected, including increased insulin production, sensitization to histamine, and attenuation of the immune system (Cascales and Christie, 2003; Krueger and Barbieri, 1995).

Although T4SSs are usually considered to engage in one-step protein secretion across the bacterial cell wall, the two-step secretion of PT is not entirely anomalous; even in the archetypal system, VirE2, VirD2, and VirF can be exported to the periplasm independently of the VirB apparatus, albeit in a Sec-independent fashion.

CagA: the Trojan horse

H. pylori is associated with peptic ulceration and chronic gastritis. Strains containing the *cag* pathogenicity island (PAI), encoding a type IVA secretion system, are linked with increased virulence and are associated with gastric carcinoma (Blaser *et al.*, 1995). So far, the only known substrate for this system is CagA, a powerful cytotoxin that interferes with host signal transduction (Nagai and Roy, 2003). The CagA effector contains up to six copies, dependent on strain, of an EPIYA (glutamine–proline–isoleucine–tyrosine–alanine) amino-acid repeat motif (Covacci *et al.*, 1993). CagA migrates to the plasma membrane after phosphorylation of tyrosine residues in the EPIYA repeats by an unidentified host tyrosine kinase. Activation results in cortical actin polymerization and cytoskeletal rearrangement, leading to elongation and spreading of eukaryotic cells, known as the "humming bird phenotype" (Backert *et al.*, 2001). This resembles closely the morphological changes presented following activation of the c-Met receptor by hepatocyte growth factor (HGF) and is characterized by dramatic elongation of the cell and production of filopodia and lamellipodia (Segal *et al.*, 1999).

CagA acts like a Trojan horse, taking control of the cell while limiting neutralizing reactions (Censini *et al.*, 2001). Cortical actin polymerization is a result of the direct association of CagA with N-WASP and Arp2/3, altering cytoskeletal arrangement. Several other cellular effects, including activation of Rho-GTPases and induction of NF-κB and activator protein 1 (AP-1) are CagA-independent but have been linked experimentally to the *cag* PAI, hinting at other undefined type IV effectors (Nagai and Roy, 2003).

Legionella pneumophila Dot/Icm type IVB system

L. pneumophila is the agent responsible for legionnaires' disease, a form of severe pneumonia in which bacteria replicate to high numbers within vacuoles in alveolar macrophages while evading endosomal degradation pathways (Lammertyn and Anne, 2004). This modulation of vesicular trafficking, reminiscent of the action of the Spi-2 system of *Salmonella*, is mediated by the Dot/Icm (defective for organelle trafficking/intracellular multiplication) type IVB secretion system.

The Dot/Icm system is encoded by genes from two different regions of the chromosome, which share homology with the plasmid-encoded CollbP9 plasmid conjugation system. The system is not specific to pathogenesis and also functions as a classic conjugal system to transfer DNA between organisms (Cascales and Christie, 2003). The Dot/Icm T4SS is essential for virulence by creating a nutrient-rich vacuole capable of evading the default lysosomal degradation pathway. These *Legionella*-containing phagosomes do not undergo endocytic maturation, preventing fusion with lysosomes. The phagosomes also recruit host secretory vesicles from the endoplasmic reticulum (ER), intercepting them before they reach the Golgi. This results in the remodeling of the phagosomal membrane, making it morphologically identical to the rough ER (Nagai and Roy, 2003). Other intracellular processes are also affected, including caspase-3-dependent apoptosis, increased phagocytosis, macropinocytosis, and pore-formation-mediated cytotoxicity (Lammertyn and Anne, 2004).

It is thought that the T4SS mediates formation of a pore in the host cell membrane and secretion of effector molecules. However, only a few proteins encoded by the Dot/Icm system have been identified and assayed for function. DotA, the best-characterized of the *Legionella* type IV effectors, is a polytopic inner-membrane protein containing eight hydrophobic domains (Roy and Isberg, 1997). DotA was assumed to be part of the T4SS apparatus. However, work by Nagai and Roy (2001) demonstrated secretion of DotA into culture media; thus, it appears that DotA acts as a translocator/effector. In support of this, other work has shown DotA to be able to form pores in the eukaryotic plasma membrane, and it is now widely believed to play this role during *L. pneumophila* infection (Kirby *et al.*, 1998).

LidA is secreted across the membrane of *L. pneumophila*-containing phagosomes in a Dot/Icm-dependent manner. LidA is exported early in the infection process and remains localized to the vicinity of the phagosome. Its functions are poorly understood – potential roles include vesicle recruitment to the replication vacuole, docking of vesicles to the phagosomal membrane, and the formation of a scaffold on the phagosome, allowing immobilization of other effectors at appropriate signaling sites (Conover *et al.*, 2003). RalF functions as a guanine nucleotide exchange factor for the ADP ribosylation factor (ARF) family of G-proteins. ARF1 acts as a key regulator of vesicle traffic from the ER to the Golgi, and there is evidence to suggest that RalF acts to recruit ARF1 to the phagosome surface, perhaps enhancing the efficiency of the replication vacuole (Nagai *et al.*, 2002).

Legionella infection ultimately triggers apoptosis, which, along with the extracellular activity of degradative bacterial enzymes, causes extensive damage of alveolar tissue. Caspase-3-mediated apoptosis in cultured macrophages has been shown to be dependent on the Dot/Icm system. Insertion of pores into the eukaryotic membrane causing osmotic lysis also allows release of intracellular bacteria. For a more detailed review of the mechanisms of *Legionella* infection, see Molmeret *et al.* (2004).

TYPE V SECRETION

In contrast to the type I-IV secretion systems, the term "type V secretion" describes three related and comparatively simple protein secretion systems:

- Va, the autotransporters (ATs);
- Vb, the two-partner system (TPS);
- Vc, the oligomeric coiled-coil (Oca) system, also termed AT-2 (Figure 3.3).

Each of these systems secretes extracellular moieties with a variety of functional activities, including adhesins, toxins, proteases, and mediators of

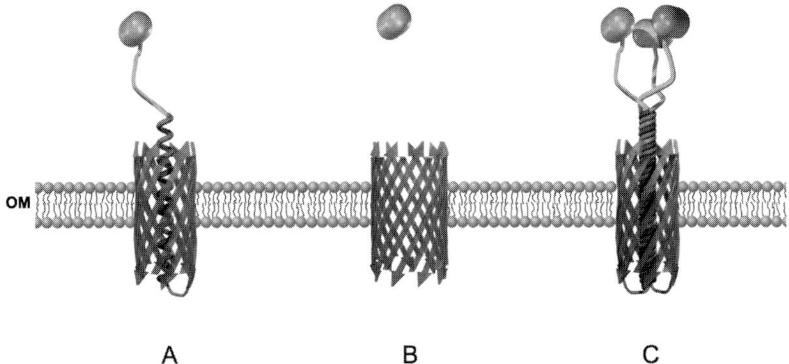

Figure 3.3 Schematic overview of the type V secretion systems. The secretion pathway across the outer membrane (OM) is depicted for the autotransporter (A), the two-partner (B), and the oligomeric coiled-coil system (C) secretion systems. After translocation through the inner membrane, the β-domains form β-barrel pore-like structures in the OM. After formation of the pore, the effector molecule inserts into the pore and is translocated to the bacterial cell surface, where it may or may not undergo further processing. The linker regions required for secretion of the autotransporters and oligomeric coiled-coil proteins are shaded in red and the autochaperone regions green. Equivalent regions appear to be missing in the two-partner secretion pathway, where the effector molecule and the β-barrel structure are encoded as separate proteins. See also Color plate 4.

motility (Henderson *et al.*, 2004). Some examples of the type V pathway are listed in Table 3.5.

The three systems have several unifying characteristics, including:

- The molecules targeted for secretion possess an N-terminal signal sequence, which mediates translocation across the inner membrane via the Sec translocon.
- In each case, a periplasmic intermediate is formed.
- An outer-membrane β-barrel pore permits translocation of the effector molecule to the cell surface.
- The effector molecules have been demonstrated, or predicted, to have β-helix structures.

However, despite these similarities, there remain distinct differences in each pathway. Notably, in the case of the AT and Oca systems, the N-terminal signal sequence mediating inner-membrane translocation, the effector molecule, and the C-terminal β-barrel mediating outer-membrane translocation of the effector molecule are encoded in a single polypeptide.

Table 3.5 *Examples of type V secretion systems*

Transporter Organism	Protein	Function
Va		
Bordetella pertussis	BrkA	Serum resistance
Chlamydia trachomatis	PmpD	Adherence
Escherichia coli	Antigen 43	Biofilm formation
Helicobacter pylori	VacA	Cytotoxin
Neisseria meningitidis	IgA1 protease	Degradation of immunoglobulin A1 (IgA1)
Pasteurella haemolytica	NanB	Sialidase
Salmonella enterica	ShdA	Adherence
Shigella flexneri	IcsA	Mediator of intra- and intercellular motility
Vb		
Bordetella pertussis	FhaB/C	Adhesin
Erwinia chrysanthemi	HecA/B	Adhesin
Serratia marcesens	ShlA/B	Hemolysin
Vc		
Halmophilus influenzae	Hia	Adhesin
Moraxella catarrhalis	UspA2	Serum resistance
Yersinia enterocolitica	YadA	Adhesin

In contrast, the TPS system effector molecule and β-barrel outer-membrane translocator are synthesized as two separate molecules (termed TpsA and TpsB, respectively), each with its own N-terminal signal sequence for inner-membrane translocation. Furthermore, in contrast to the AT and TPS systems, following translocation through the inner membrane, the Oca proteins must form a trimeric structure to direct translocation of the effector molecule across the outer membrane (Cotter *et al.*, 2005; Henderson *et al.*, 2004; Jacob-Dubuisson *et al.*, 2004).

As mentioned above, all proteins secreted via the type V pathway possess N-terminal signal sequences that mediate translocation across the inner membrane. In most cases, it appears that the molecules targeted for export possess typical signal sequences targeting the molecule to the Sec translocon and, at least in the case of *E. coli*, demonstrate a dependence on the SecB cytoplasmic chaperone. However, in some cases, proteins secreted via the type V pathway possess alternative signal sequences, including those consistent with lipidation of the mature protein and those that possess an N-terminal amino-acid extension termed the extended signal peptide region (ESPR) (Henderson *et al.*, 2004). In at least one case where the lipidation motif has been described, it has been shown that the mature effector molecule is a lipoprotein in which the N-terminus interacts with the outer membrane to tether the molecule to the bacterial cell surface; this interaction is required to allow the molecule to function correctly (Coutte *et al.*, 2003). In contrast to the signal sequences with the lipidation motif, the function of the signal sequences possessing the ESPR remains somewhat enigmatic.

Normal signal sequences directing Sec-dependent translocation across the cytoplasmic membrane possess:

- an N-domain of positively charged amino acids;
- an H-domain of hydrophobic amino acids; and
- a C-domain with a consensus signal peptidase recognition domain (Dalbey and Chen, 2004).

Proteins possessing the ESPR also contain N-, H-, and C-domains reminiscent of Sec-dependent translocation (Henderson *et al.*, 2004). The ESPR region is located N-proximal to these three domains and consists of a highly conserved charged domain termed N1 and a highly conserved hydrophobic domain termed H1. Such sequence conservation is highly unusual in signal sequences; thus, it was speculated that the ESPR might serve to recruit accessory proteins or to direct translocation across the inner membrane via an alternative pathway such as the Tat or the cotranslational signal-recognition particle (SRP) pathways (Henderson *et al.*, 1998). In support of this initial hypothesis, Sijbrandi *et al.* (2003) reported that a protein possessing the ESPR (the Hbp protein of *E. coli*) was translocated across the inner membrane via the SRP. However, this observation is somewhat controversial, as another protein (filamentous hemagglutinin) also possessing the ESPR was not found to use the SRP (Chevalier *et al.*, 2004). Szabady *et al.* (2005) have demonstrated that the presence of the ESPR in another protein (EspP) decreased the efficiency of translocation into the periplasm, purportedly transiently tethering the whole molecule to the inner membrane while the C-terminus of the AT folded and inserted into the outer membrane.

The process of secretion across the outer membrane distinguishes the type V subfamilies (Figure 3.3). In the case of the AT and Oca proteins, the C-termini adopt an outer-membrane β-barrel structure that forms a pore through which the effector molecule is translocated to the extracellular milieu (Henderson *et al.*, 2004). Although the exact structure of the AT β-barrel and the mode of translocation have been the subject of considerable debate, it appears from the crystal structure of the β-domain of the NalP AT that the original hypothesis of the effector polypeptide passing through the β-barrel pore was correct; the structure of NalP showed clearly a β-barrel with a polypeptide embedded within the pore (Oomen *et al.*, 2004). Mutagenesis studies revealed that the N-terminal region of ATs and Oca effector molecules were not required for secretion, suggesting that the signals for recognizing the β-barrel pore and effecting secretion were contained within the C-terminal region of the effector protein (Oliver, Huang, and Fernandez, 2003; Roggenkamp *et al.*, 2003). Interestingly, the polypeptide embedded in the NalP pore was revealed to adopt an α-helical conformation, and mutagenesis studies demonstrated that this α-helical region was an absolute requirement for secretion of other ATs (Oliver, Huang, and Fernandez, 2003a). In addition, immediately upstream of the α-helical region in the ATs is a relatively conserved domain termed the "autochaperone" (AC) domain (Oliver, Huang, and Fernandez, 2003; Oliver, Huang, Nodel *et al.*, 2003). It has been demonstrated that the AC domain is not required for secretion but is required for folding of the effector molecule on the bacterial cell surface. It has been proposed that the members of the Oca family also possess similar domains that are required for secretion and folding of the effector molecule (Roggenkamp *et al.*, 2003). Thus, it appears in the case of the ATs and Oca proteins that after insertion of the β-barrel into the outer membrane, a linear polypeptide forms a hairpin structure threaded through the pore of the β-barrel, locating the AC domain on the outside of the cell. Subsequently, as the effector molecule is extruded through the pore, the AC domain folds the molecule, such that folding occurs vectorially from the C- to the N-terminus (Henderson *et al.*, 2004). It is also worth mentioning that one investigation has demonstrated the necessity of an essential outer-membrane protein (Omp85) for the correct secretion of ATs, presumably by mediating the correct insertion of the β-barrel into the outer membrane (Voulhoux *et al.*, 2003).

Secretion of the TpsA effector molecules is somewhat different to that of the ATs and Oca functional moieties. As the TpsA and TpsB proteins are synthesized as separate molecules, a specific targeting signal must exist that directs the TpsA protein to the pore formed by TpsB. This motif has been localized to the extreme N-terminus of the effector molecule and is an absolute requirement for secretion (Jacob-Dubuisson *et al.*, 2004). In contrast to the ATs and Oca proteins, the C-terminal end of the effector molecule is not required for secretion or folding. Interestingly, the TpsB is related evolutionarily to the Omp85 proteins required for AT secretion (Surana *et al.*, 2004). In some cases, the TpsB β-barrel has also been demonstrated to possess additional function, where, in addition to mediating secretion, it is required for the correct maturation of the secreted TpsA effector molecule (Braun *et al.*, 1993; Walker *et al.*, 2004).

FUTURE DIRECTIONS

The advent of genome sequencing has resulted in an explosion in our knowledge of the distribution of secretion systems. Unfortunately, our ability to detect such systems has not been matched by progress in our fundamental understanding of the mechanisms of protein targeting and secretion. It is unlikely that we have exhaustively identified all bacterial protein-secretion systems, and undoubtedly a growing appreciation of the function of genes classified as "unknown function" will reveal new effector molecules and new secretion pathways. Furthermore, the presence of multiple similar secretion systems within a single genome raises several questions, the answers to which will be fundamental to our understanding of how bacteria behave in a given niche. Are effector molecules targeted to only one system? Is there redundancy? If not, how do the different systems differentiate between their cognate effector molecules? Is more than one system active at any given time? How is expression of the different systems coordinated? As most secretion systems often encode regulators controlling expression of the system, how is this regulation integrated into the bacterial cells global regulatory network? How much cross-talk exists between similar and different secretion systems? Certainly, one thing is apparent in the post-genome era: the abundance of unstudied secretion systems and associated questions will keep bacteriologists busy for years to come.

REFERENCES

Aizawa, S. I. (2001). Bacterial flagella and type III secretion systems. *FEMS Microbiol. Lett.* **202**, 157–164.

Aktories, K., Schmidt, G., and Just, I. (2000). Rho GTPases as targets of bacterial protein toxins. *Biol. Chem.* **381**, 421–426.

Alfano, J. R. and Collmer, A. (2004). Type III secretion system effector proteins: double agents in bacterial disease and plant defense. *Annu. Rev. Phytopathol.* **42**, 385–414.

Allaoui, A., Schulte, R., and Cornelis, G. R. (1995). Mutational analysis of the *Yersinia enterocolitica virC* operon: characterization of *yscE, F, G, I, J, K* required for Yop secretion and *yscH* encoding YopR. *Mol. Microbiol.* **18**, 343–355.

Andersen, C., Koronakis, E., Bokma, E., *et al.* (2002). Transition to the open state of the TolC periplasmic tunnel entrance. *Proc. Natl. Acad. Sci. U. S. A.* **99**, 11 103–11 108.

Backert, S., Moese, S., Selbach, M., Brinkmann, V., and Meyer, T. F. (2001). Phosphorylation of tyrosine 972 of the *Helicobacter pylori* CagA protein is essential for induction of a scattering phenotype in gastric epithelial cells. *Mol. Microbiol.* **42**, 631–644.

Barz, C., Abahji, T. N., Trulzsch, K., and Heesemann, J. (2000). The *Yersinia* Ser/Thr protein kinase YpkA/YopO directly interacts with the small GTPases RhoA and Rac-1. *FEBS Lett.* **482**, 139–143.

Baumann, U., Wu, S., Flaherty, K. M., and McKay, D. B. (1993). Three-dimensional structure of the alkaline protease of *Pseudomonas aeruginosa*: a two-domain protein with a calcium binding parallel beta roll motif. *EMBO J.* **12**, 3357–3364.

Benabdelhak, H., Kiontke, S., Horn, C., *et al.* (2003). A specific interaction between the NBD of the ABC-transporter HlyB and a C-terminal fragment of its transport substrate haemolysin. *Am. J. Mol. Biol.* **327**, 1169–1179.

Binet, R. and Wandersman, C. (1995). Protein secretion by hybrid bacterial ABC-transporters: specific functions of the membrane ATPase and the membrane fusion protein. *EMBO J.* **14**, 2298–2306.

Binet, R., Letoffe, S., Ghigo, J. M., Delepelaire, P., and Wandersman, C. (1997). Protein secretion by Gram-negative bacterial ABC exporters: a review. *Gene.* **192**, 7–11.

Bitter, W. (2003). Secretins of *Pseudomonas aeruginosa*: large holes in the outer membrane. *Arch. Microbiol.* **179**, 307–314.

Bitter, W., Koster, M., Latijnhouwers, M., de Cock, H., and Tommassen, J. (1998). Formation of oligomeric rings by XcpQ and PilQ, which are involved in protein transport across the outer membrane of *Pseudomonas aeruginosa*. *Mol. Microbiol.* **27**, 209–219.

Blaser, M. J., Perez-Perez, G. I., Kleanthous, H., *et al.* (1995). Infection with *Helicobacter pylori* strains possessing *cagA* is associated with an increased risk of developing adenocarcinoma of the stomach. *Cancer Res.* **55**, 2111–2115.

Bleves, S. and Cornelis, G. R. (2000). How to survive in the host: the *Yersinia* lesson. *Microbes Infect.* **2**, 1451–1460.

Bleves, S., Lazdunski, A., and Filloux, A. (1996). Membrane topology of three Xcp proteins involved in exoprotein transport by *Pseudomonas aeruginosa*. *J. Bacteriol.* **178**, 4297–4300.

Bleves, S., Voulhoux, R., Michel, G., *et al.* (1998). The secretion apparatus of *Pseudomonas aeruginosa*: identification of a fifth pseudopilin, XcpX (GspK family). *Mol. Microbiol.* **27**, 31–40.

Bleves, S., Gerard-Vincent, M., Lazdunski, A., and Filloux, A. (1999). Structure–function analysis of XcpP, a component involved in general secretory pathway-dependent protein secretion in *Pseudomonas aeruginosa*. *J. Bacteriol.* **181**, 4012–4019.

Bliska, J. B., Guan, K. L., Dixon, J. E., and Falkow, S. (1991). Tyrosine phosphate hydrolysis of host proteins by an essential *Yersinia* virulence determinant. *Proc. Natl. Acad. Sci. U. S. A.* **88**, 1187–1191.

Bolin, I., Portnoy, D. A., and Wolf-Watz, H. (1985). Expression of the temperature-inducible outer membrane proteins of *Yersiniae*. *Infect. Immun.* **48**, 234–240.

Bouley, J., Condemine, G., and Shevchik, V. E. (2001). The PDZ domain of OutC and the N-terminal region of OutD determine the secretion specificity of the type II out pathway of *Erwinia chrysanthemi*. *J. Mol. Biol.* **308**, 205–219.

Boyd, A. P., Lambermont, I., and Cornelis, G. R. (2000). Competition between the Yops of *Yersinia enterocolitica* for delivery into eukaryotic cells: role of the SycE chaperone binding domain of YopE. *J. Bacteriol.* **182**, 4811–4821.

Boyd, E. F. and Brussow, H. (2002). Common themes among bacteriophage-encoded virulence factors and diversity among the bacteriophages involved. *Trends Microbiol.* 10, 521–529.

Braun, V., Ondraczek, R., and Hobbie, S. (1993). Activation and secretion of *Serratia* hemolysin. *Zentralbl. Bakteriol.* **278**, 306–315.

Burghout, P., van Boxtel, R., Van Gelder, P., *et al.* (2004). Structure and electrophysiological properties of the YscC secretin from the type III secretion system of *Yersinia enterocolitica*. *J. Bacteriol.* **186**, 4645–4654.

Burns, D. L. (2003). Type IV transporters of pathogenic bacteria. *Curr. Opin. Microbiol.* 6, 29–34.

Buttner, D. and Bonas, U. (2002). Getting across: bacterial type III effector proteins on their way to the plant cell. *EMBO J.* **21**, 5313–5322.

Campellone, K. G. and Leong, J. M. (2003). Tails of two Tirs: actin pedestal formation by enteropathogenic *E. coli* and enterohemorrhagic *E. coli* O157:H7. *Curr. Opin. Microbiol.* **6**, 82–90.

Cascales, E. and Christie, P. J. (2003). The versatile bacterial type IV secretion systems. *Nat. Rev. Microbiol.* **1**, 137–149.

Censini, S., Stein, M., and Covacci, A. (2001). Cellular responses induced after contact with *Helicobacter pylori*. *Curr. Opin. Microbiol.* **4**, 41–46.

Chen, H. D. and Frankel, G. (2005). Enteropathogenic *Escherichia coli*: unravelling pathogenesis. *FEMS Microbiol. Rev.* **29**, 83–98.

Chen, L. Y., Chen, D. Y., Miaw, J., and Hu, N. T. (1996). XpsD, an outer membrane protein required for protein secretion by *Xanthomonas campestris* pv. *campestris*, forms a multimer. *J. Biol. Chem.* **271**, 2703–2708.

Chevalier, N., Moser, M., Koch, H. G., *et al.* (2004). Membrane targeting of a bacterial virulence factor harbouring an extended signal peptide. *J. Mol. Microbiol. Biotechnol.* **8**, 7–18.

Christie, P. J. and Vogel, J. P. (2000). Bacterial type IV secretion: conjugation systems adapted to deliver effector molecules to host cells. *Trends Microbiol.* **8**, 354–360.

Conover, G. M., Derre, I., Vogel, J. P., and Isberg, R. R. (2003). The *Legionella pneumophila* LidA protein: a translocated substrate of the Dot/Icm system associated with maintenance of bacterial integrity. *Mol. Microbiol.* **48**, 305–321.

Cordes, F. S., Komoriya, K., Larquet, E., *et al.* (2003). Helical structure of the needle of the type III secretion system of *Shigella flexneri*. *J. Biol. Chem.* **278**, 17 103–17 107.

Cotter, S. E., Surana, N. K., and St Geme, J. W., 3rd (2005). Trimeric autotransporters: a distinct subfamily of autotransporter proteins. *Trends Microbiol.* **13**, 199–205.

Coutte, L., Willery, E., Antoine, R., *et al.* (2003). Surface anchoring of bacterial subtilisin important for maturation function. *Mol. Microbiol.* **49**, 529–539.

Covacci, A., Censini, S., Bugnoli, M., *et al.* (1993). Molecular characterization of the 128-kDa immunodominant antigen of *Helicobacter pylori* associated with cytotoxicity and duodenal ulcer. *Proc. Natl. Acad. Sci. U. S. A.* **90**, 5791–5795.

Dalbey, R. E. and Chen, M. (2004). Sec-translocase mediated membrane protein biogenesis. *Biochim. Biophys. Acta.* **1694**, 37–53.

Daniell, S. J., Kocsis, E., Morris, E., *et al.* (2003). 3D structure of EspA filaments from enteropathogenic *Escherichia coli*. *Mol. Microbiol.* **49**, 301–308.

Darwin, K. H. and Miller, V. L. (2000). The putative invasion protein chaperone SicA acts together with InvF to activate the expression of *Salmonella typhimurium* virulence genes. *Mol. Microbiol.* **35**, 949–960.

Das, A. and Xie, Y. H. (1998). Construction of transposon Tn3phoA: its application in defining the membrane topology of the *Agrobacterium tumefaciens* DNA transfer proteins. *Mol. Microbiol.* **27**, 405–414.

Dean, P. and Kenny, B. (2004). Intestinal barrier dysfunction by enteropathogenic *Escherichia coli* is mediated by two effector molecules and a bacterial surface protein. *Mol. Microbiol.* **54**, 665–675.

De Groot, A., Filloux, A., and Tommassen, J. (1991). Conservation of *xcp* genes, involved in the two-step protein secretion process, in different *Pseudomonas* species and other gram-negative bacteria. *Mol. Gen. Genet.* **229**, 278–284.

Delepelaire, P. (2004). Type I secretion in gram-negative bacteria. *Biochim. Biophys. Acta* **1694**, 149–161.

Deleuil, F., Mogemark, L., Francis, M. S., Wolf-Watz, H., and Fallman, M. (2003). Interaction between the *Yersinia* protein tyrosine phosphatase YopH and eukaryotic Cas/Fyb is an important virulence mechanism. *Cell. Microbiol.* **5**, 53–64.

Denecker, G., Declercq, W., Geuijen, C. A., *et al.* (2001). *Yersinia enterocolitica* YopP-induced apoptosis of macrophages involves the apoptotic signaling cascade upstream of bid. *J. Biol. Chem.* **276**, 19 706–19 714.

Ding, Z., Atmakuri, K., and Christie, P. J. (2003). The outs and ins of bacterial type IV secretion substrates. *Trends Microbiol.* **11**, 527–535.

Eichelberg, K., Ginocchio, C. C., and Galan, J. E. (1994). Molecular and functional characterization of the *Salmonella typhimurium* invasion genes *invB* and *invC*: homology of InvC to the F0F1 ATPase family of proteins. *J. Bacteriol.* **176**, 4501–4510.

Espinosa, A. and Alfano, J. R. (2004). Disabling surveillance: bacterial type III secretion system effectors that suppress innate immunity. *Cell. Microbiol.* **6**, 1027–1040.

Feldman, M. F. and Cornelis, G. R. (2003). The multitalented type III chaperones: all you can do with 15 kDa. *FEMS Microbiol. Lett.* **219**, 151–158.

Filloux, A. (2004). The underlying mechanisms of type II protein secretion. *Biochim. Biophys. Acta* **1694**, 163–179.

Francis, M. S., Wolf-Watz, H., and Forsberg, A. (2002). Regulation of type III secretion systems. *Curr. Opin. Microbiol.* **5**, 166–172.

Galan, J. E. (2001). *Salmonella* interactions with host cells: type III secretion at work. *Annu. Rev. Cell. Dev. Biol.* **17**, 53–86.

Galan, J. E. and Collmer, A. (1999). Type III secretion machines: bacterial devices for protein delivery into host cells. *Science* **284**, 1322–1328.

Garcia-del Portillo, F., Zwick, M. B., Leung, K. Y., and Finlay, B. B. (1993). *Salmonella* induces the formation of filamentous structures containing lysosomal membrane glycoproteins in epithelial cells. *Proc. Natl. Acad. Sci. U. S. A.* **90**, 10 544–10 548.

Genin, S. and Boucher, C. (2004). Lessons learned from the genome analysis of ralstonia solanacearum. *Annu. Rev. Phytopathol.* **42**, 107–134.

Gentschev, I., Dietrich, G. and Goebel, W. (2002). The *E. coli* alpha-hemolysin secretion system and its use in vaccine development. *Trends Microbiol.* **10**, 39–45.

Gerard-Vincent, M., Robert, V., Ball, G., *et al.* (2002). Identification of XcpP domains that confer functionality and specificity to the *Pseudomonas aeruginosa* type II secretion apparatus. *Mol. Microbiol.* **44**, 1651–1665.

Ghigo, J. M., Letoffe, S., and Wandersman, C. (1997). A new type of hemophore-dependent heme acquisition system of *Serratia marcescens* reconstituted in *Escherichia coli*. *J. Bacteriol.* **179**, 3572–3579.

Ghosh, P. (2004). Process of protein transport by the type III secretion system. *Microbiol. Mol. Biol. Rev.* **68**, 771–795.

Gophna, U., Ron, E. Z., and Graur, D. (2003). Bacterial type III secretion systems are ancient and evolved by multiple horizontal-transfer events. *Gene* **312**, 151–163.

Gruenheid, S., Sekirov, I., Thomas, N. A., *et al.* (2004). Identification and characterization of NleA, a non-LEE-encoded type III translocated virulence factor of enterohaemorrhagic *Escherichia coli* O157:H7. *Mol. Microbiol.* **51**, 1233–1249.

Guzzo, J., Duong, F., Wandersman, C., Murgier, M., and Lazdunski, A. (1991). The secretion genes of *Pseudomonas aeruginosa* alkaline protease are functionally related to those of *Erwinia chrysanthemi* proteases and *Escherichia coli* alpha-haemolysin. *Mol. Microbiol.* **5**, 447–453.

Hakansson, S., Bergman, T., Vanooteghem, J. C., Cornelis, G., and Wolf-Watz, H. (1993). YopB and YopD constitute a novel class of *Yersinia* Yop proteins. *Infect. Immun.* **61**, 71–80.

Hardie, K. R., Lory, S., and Pugsley, A. P. (1996). Insertion of an outer membrane protein in *Escherichia coli* requires a chaperone-like protein. *EMBO J.* **15**, 978–988.

He, S. Y. (1998). Type III protein secretion systems in plant and animal pathogenic bacteria. *Annu. Rev. Phytopathol.* **36**, 363–392.

He, S. Y. and Jin, Q. (2003). The Hrp pilus: learning from flagella. *Curr. Opin. Microbiol.* **6**, 15–19.

He, S. Y., Nomura, K., and Whittam, T. S. (2004). Type III protein secretion mechanism in mammalian and plant pathogens. *Biochim. Biophys. Acta* **1694**, 181–206.

Henderson, I. R., Navarro-Garcia, F., and Nataro, J. P. (1998). The great escape: structure and function of the autotransporter proteins. *Trends Microbiol.* **6**, 370–378.

Henderson, I. R., Navarro-Garcia, F., Desvaux, M., Fernandez, R. C., and Ala'Aldeen, D. (2004). Type V protein secretion pathway: the autotransporter story. *Microbiol. Mol. Biol. Rev.* **68**, 692–744.

Hernandez, L. D., Pypaert, M., Flavell, R. A., and Galan, J. E. (2003). A *Salmonella* protein causes macrophage cell death by inducing autophagy. *J. Cell. Biol.* **163**, 1123–1131.

Hinsa, S. M., Espinosa-Urgel, M., Ramos, J. L., and O'Toole, G. A. (2003). Transition from reversible to irreversible attachment during biofilm formation by *Pseudomonas fluorescens* WCS365 requires an ABC transporter and a large secreted protein. *Mol. Microbiol.* **49**, 905–918.

Hobbie, S., Chen, L. M., Davis, R. J., and Galan, J. E. (1997). Involvement of mitogen-activated protein kinase pathways in the nuclear responses and cytokine production induced by *Salmonella typhimurium* in cultured intestinal epithelial cells. *J. Immunol.* **159**, 5550–5559.

Hoiczyk, E. and Blobel, G. (2001). Polymerization of a single protein of the pathogen *Yersinia enterocolitica* into needles punctures eukaryotic cells. *Proc. Natl. Acad. Sci. U. S. A.* **98**, 4669–4674.

Holland, I. B. (2004). Translocation of bacterial proteins: an overview. *Biochim. Biophys. Acta* **1694**, 5–16.

Hueck, C. J. (1998). Type III protein secretion systems in bacterial pathogens of animals and plants. *Microbiol. Mol. Biol. Rev.* **62**, 379–433.

Innes, R. (2003). New effects of type III effectors. *Mol. Microbiol.* **50**, 363–365.

Iriarte, M. and Cornelis, G. R. (1998). YopT, a new *Yersinia* Yop effector protein, affects the cytoskeleton of host cells. *Mol. Microbiol.* **29**, 915–929.

Jackson, M. W., Day, J. B., and Plano, G. V. (1998). YscB of *Yersinia pestis* functions as a specific chaperone for YopN. *J. Bacteriol.* **180**, 4912–4921.

Jacob-Dubuisson, F., Fernandez, R., and Coutte, L. (2004). Protein secretion through autotransporter and two-partner pathways. *Biochim. Biophys. Acta* **1694**, 235–257.

Jarchau, T., Chakraborty, T., Garcia, F., and Goebel, W. (1994). Selection for transport competence of C-terminal polypeptides derived from *Escherichia coli* hemolysin: the shortest peptide capable of autonomous HlyB/HlyD-dependent secretion comprises the C-terminal 62 amino acids of HlyA. *Mol. Gen. Genet.* **245**, 53–60.

Jin, Q. and He, S. Y. (2001). Role of the Hrp pilus in type III protein secretion in *Pseudomonas syringae. Science* **294**, 2556–2558.

Jin, Q., Thilmony, R., Zwiesler-Vollick, J., and He, S. Y. (2003). Type III protein secretion in *Pseudomonas syringae. Microbes Infect.* **5**, 301–310.

Kaniga, K., Uralil, J., Bliska, J. B., and Galan, J. E. (1996). A secreted protein tyrosine phosphatase with modular effector domains in the bacterial pathogen *Salmonella typhimurium. Mol. Microbiol.* **21**, 633–641.

Kenny, B. (2002). Mechanism of action of EPEC type III effector molecules. *Int. J. Med. Microbiol.* **291**, 469–477.

Kenny, B., DeVinney, R., Stein, M., *et al.* (1997). Enteropathogenic *E. coli.* (EPEC) transfers its receptor for intimate adherence into mammalian cells. *Cell* **91**, 511–520.

Kirby, J. E., Vogel, J. P., Andrews, H. L., and Isberg, R. R. (1998). Evidence for pore-forming ability by *Legionella pneumophila. Mol. Microbiol.* **27**, 323–336.

Kirn, T. J., Bose, N., and Taylor, R. K. (2003). Secretion of a soluble colonization factor by the TCP type 4 pilus biogenesis pathway in *Vibrio cholerae*. *Mol. Microbiol.* **49**, 81–92.

Knight, D. A. and Barbieri, J. T. (1997). Ecto-ADP-ribosyltransferase activity of *Pseudomonas aeruginosa* exoenzyme. *S. Infect. Immun.* **65**, 3304–3309.

Knodler, L. A., Celli, J., Hardt, W. D., *et al.* (2002). *Salmonella* effectors within a single pathogenicity island are differentially expressed and translocated by separate type III secretion systems. *Mol. Microbiol.* **43**, 1089–1103.

Knutton, S., Rosenshine, I., Pallen, M. J., *et al.* (1998). A novel EspA-associated surface organelle of enteropathogenic *Escherichia coli* involved in protein translocation into epithelial cells. *EMBO J.* **17**, 2166–2176.

Koronakis, V., Hughes, C. and Koronakis, E. (1991). Energetically distinct early and late stages of HlyB/HlyD-dependent secretion across both *Escherichia coli* membranes. *EMBO J.* **10**, 3263–3272.

Koronakis, E., Hughes, C., Milisav, I., and Koronakis, V. (1995). Protein exporter function and in vitro ATPase activity are correlated in ABC-domain mutants of HlyB. *Mol. Microbiol.* **16**, 87–96.

Koronakis, V., Sharff, A., Koronakis, E., Luisi, B. and Hughes, C. (2000). Crystal structure of the bacterial membrane protein TolC central to multidrug efflux and protein export. *Nature.* **405**, 914–919.

Krueger, K. M. and Barbieri, J. T. (1995). The family of bacterial ADP-ribosylating exotoxins. *Clin. Microbiol. Rev.* **8**, 34–47.

Kubori, T., Sukhan, A., Aizawa, S. I. and Galan, J. E. (2000). Molecular characterization and assembly of the needle complex of the *Salmonella typhimurium* type III protein secretion system. *Proc. Natl. Acad. Sci. U. S. A.* **97**, 10 225–10 230.

Kuhle, V., Jackel, D., and Hensel, M. (2004). Effector proteins encoded by *Salmonella* pathogenicity island 2 interfere with the microtubule cytoskeleton after translocation into host cells. *Traffic.* **5**, 356–370.

Lammertyn, E. and Anne, J. (2004). Protein secretion in *Legionella pneumophila* and its relation to virulence. *FEMS Microbiol. Lett.* **238**, 273–279.

Lee, A. H., Zareei, M. P., and Daefler, S. (2002). Identification of a NIPSNAP homologue as host cell target for *Salmonella* virulence protein SpiC. *Cell. Microbiol.* **4**, 739–750.

Letoffe, S., Ghigo, J. M., and Wandersman, C. (1994a). Iron acquisition from heme and hemoglobin by a *Serratia marcescens* extracellular protein. *Proc. Natl. Acad. Sci. U.S.A.* **91**, 9876–9880.

Letoffe, S., Ghigo, J. M., and Wandersman, C. (1994b). Secretion of the *Serratia marcescens* HasA protein by an ABC transporter. *J. Bacteriol.* **176**, 5372–5377.

Letoffe, S., Delepelaire, P., and Wandersman, C. (1996). Protein secretion in gram-negative bacteria: assembly of the three components of ABC protein-mediated exporters is ordered and promoted by substrate binding. *EMBO J.* **15**, 5804–5811.

Lindeberg, M., Salmond, G. P. and Collmer, A. (1996). Complementation of deletion mutations in a cloned functional cluster of *Erwinia chrysanthemi* out genes with *Erwinia carotovora out* homologues reveals OutC and OutD as candidate gatekeepers of species-specific secretion of proteins via the type II pathway. *Mol. Microbiol.* **20**, 175–190.

Liu, S., Yahr, T. L., Frank, D. W., and Barbieri, J. T. (1997). Biochemical relationships between the 53-kilodalton (Exo53) and 49-kilodalton (ExoS) forms of exoenzyme S of *Pseudomonas aeruginosa*. *J. Bacteriol.* **179**, 1609–1613.

Lu, L. and Walker, W. A. (2001). Pathologic and physiologic interactions of bacteria with the gastrointestinal epithelium. *Am. J. Clin. Nutr.* **73**, 1124S–1130S.

Luo, Y., Bertero, M. G., Frey, E. A., *et al.* (2001). Structural and biochemical characterization of the type III secretion chaperones CesT and SigE. *Nat. Struct. Biol.* **8**, 1031–1036.

Macnab, R. M. (2004). Type III flagellar protein export and flagellar assembly. *Biochim. Biophys. Acta* **1694**, 207–217.

Marenne, M. N., Journet, L., Mota, L. J., and Cornelis, G. R. (2003). Genetic analysis of the formation of the Ysc-Yop translocation pore in macrophages by *Yersinia enterocolitica*: role of LcrV, YscF and YopN. *Microb. Pathog.* **35**, 243–258.

Marlovits, T. C., Kubori, T., Sukhan, A., *et al.* (2004). Structural insights into the assembly of the type III secretion needle complex. *Science* **306**, 1040–1042.

Mattick, J. S. (2002). Type IV pili and twitching motility. *Annu. Rev. Microbiol.* **56**, 289–314.

McGhie, E. J., Hayward, R. D., and Koronakis, V. (2001). Cooperation between actin-binding proteins of invasive *Salmonella*: SipA potentiates SipC nucleation and bundling of actin. *EMBO J.* **20**, 2131–2139.

Miki, T., Okada, N., Shimada, Y., and Danbara, H. (2004). Characterization of *Salmonella* pathogenicity island 1 type III secretion-dependent hemolytic activity in *Salmonella enterica* serovar Typhimurium. *Microb. Pathog.* **37**, 65–72.

Molmeret, M., Bitar, D. M., Han, L., and Kwaik, Y. A. (2004). Cell biology of the intracellular infection by *Legionella pneumophila*. *Microbes Infect.* **6**, 129–139.

Mudgett, M. B. (2005). New insights to the function of phytopathogenic baterial type III effectors in plants. *Annu. Rev. Plant. Biol.* **56**, 509–531.

Nagai, H. and Roy, C. R. (2001). The DotA protein from *Legionella pneumophila* is secreted by a novel process that requires the Dot/Icm transporter. *EMBO J.* **20**, 5962–5970.

Nagai, H. and Roy, C. R. (2003). Show me the substrates: modulation of host cell function by type IV secretion systems. *Cell. Microbiol.* **5**, 373–383.

Nagai, H., Kagan, J. C., Zhu, X., Kahn, R. A. and Roy, C. R. (2002). A bacterial guanine nucleotide exchange factor activates ARF on *Legionella* phagosomes. *Science.* **295**, 679–682.

Navarro, L., Alto, N. M., and Dixon, J. E. (2005). Functions of the *Yersinia* effector proteins in inhibiting host immune responses. *Curr. Opin. Microbiol.* **8**, 21–27.

Neyt, C. and Cornelis, G. R. (1999). Role of SycD, the chaperone of the *Yersinia* Yop translocators YopB and YopD. *Mol. Microbiol.* **31**, 143–156.

Norris, F. A., Wilson, M. P., Wallis, T. S., Galyov, E. E., and Majerus, P. W. (1998). SopB, a protein required for virulence of *Salmonella dublin*, is an inositol phosphate phosphatase. *Proc. Natl. Acad. Sci. U. S. A.* **95**, 14 057–14 059.

Nouwen, N., Ranson, N., Saibil, H., *et al.* (1999). Secretin PulD: association with pilot PulS, structure, and ion-conducting channel formation. *Proc. Natl. Acad. Sci. U. S. A.* **96**, 8173–8177.

Nouwen, N., Stahlberg, H., Pugsley, A. P. and Engel, A. (2000). Domain structure of secretin PulD revealed by limited proteolysis and electron microscopy. *EMBO J.* **19**, 2229–2236.

Oliver, D. C., Huang, G., and Fernandez, R. C. (2003). Identification of secretion determinants of the *Bordetella pertussis* BrkA autotransporter. *J. Bacteriol.* **185**, 489–495.

Oliver, D. C., Huang, G., Nodel, E., Pleasance, S., and Fernandez, R. C. (2003). A conserved region within the *Bordetella pertussis* autotransporter BrkA is necessary for folding of its passenger domain. *Mol. Microbiol.* **47**, 1367–1383.

Oomen, C. J., van Ulsen, P., van Gelder, P., *et al.* (2004). Structure of the translocator domain of a bacterial autotransporter. *EMBO J.* **23**, 1257–1266.

Pallen, M. J. (2002). The ESAT-6/WXG100 superfamily – and a new Gram-positive secretion system? *Trends Microbiol.* **10**, 209–212.

Pallen, M. J. and Ponting, C. P. (1997). PDZ domains in bacterial proteins. *Mol. Microbiol.* **26**, 411–413.

Pallen, M. J., Francis, M. S., and Futterer, K. (2003). Tetratricopeptide-like repeats in type-III-secretion chaperones and regulators. *FEMS Microbiol. Lett.* **223**, 53–60.

Pallen, M. J., Beatson, S. A., and Bailey, C. M. (2005a). Bioinformatics, genomics and evolution of non-flagellar type-III secretion systems: a Darwinian perspective. *FEMS Microbiol. Rev.* **29**, 201–229.

Pallen, M. J., Beatson, S. A., and Bailey, C. M. (2005b). Bioinformatics analysis of the locus for enterocyte effacement provides novel insights into type-III secretion. *BMC Microbiol.* **5**, 9.

Pallen, M. J., Penn, C. W., and Chaudhuri, R. R. (2005). Bacterial flagellar diversity in the post-genomic era. *Trends Microbiol.* **13**, 143–149.

Parsot, C., Hamiaux, C., and Page, A. L. (2003). The various and varying roles of specific chaperones in type III secretion systems. *Curr. Opin. Microbiol.* **6**, 7–14.

Pepe, J. C. and Lory, S. (1998). Amino acid substitutions in PilD, a bifunctional enzyme of *Pseudomonas aeruginosa*: effect on leader peptidase and N-methyltransferase activities in vitro and in vivo. *J. Biol. Chem.* **273**, 19 120–19 129.

Pettersson, J., Holmstrom, A., Hill, J., *et al.* (1999). The V-antigen of *Yersinia* is surface exposed before target cell contact and involved in virulence protein translocation. *Mol. Microbiol.* **32**, 961–976.

Plano, G. V., Day, J. B., and Ferracci, F. (2001). Type III export: new uses for an old pathway. *Mol. Microbiol.* **40**, 284–293.

Possot, O. M., Vignon, G., Bomchil, N., Ebel, F., and Pugsley, A. P. (2000). Multiple interactions between pullulanase secreton components involved in stabilization and cytoplasmic membrane association of PulE. *J. Bacteriol.* **182**, 2142–2152.

Puhler, A., Arlat, M., Becker, A., *et al.* (2004). What can bacterial genome research teach us about bacteria–plant interactions?. *Curr. Opin. Plant Biol.* **7**, 137–147.

Py, B., Loiseau, L., and Barras, F. (2001). An inner membrane platform in the type II secretion machinery of Gram-negative bacteria. *EMBO Rep.* **2**, 244–248.

Ren, C. P., Chaudhuri, R. R., Fivian, A., *et al.* (2004). The ETT2 gene cluster, encoding a second type III secretion system from *Escherichia coli*, is present in the majority of strains but has undergone widespread mutational attrition. *J. Bacteriol.* **186**, 3547–3560.

Robinson, C. and Bolhuis, A. (2004). Tat-dependent protein targeting in prokaryotes and chloroplasts. *Biochim. Biophys. Acta* **1694**, 135–147.

Roggenkamp, A., Ackermann, N., Jacobi, C. A., *et al.* (2003). Molecular analysis of transport and oligomerization of the *Yersinia enterocolitica* adhesin YadA. *J. Bacteriol.* **185**, 3735–3744.

Roy, C. R. and Isberg, R. R. (1997). Topology of *Legionella pneumophila* DotA: an inner membrane protein required for replication in macrophages. *Infect. Immun.* **65**, 571–578.

Ruckdeschel, K., Harb, S., Roggenkamp, A., *et al.* (1998). *Yersinia enterocolitica* impairs activation of transcription factor NF-kappaB: involvement in the induction of programmed cell death and in the suppression of the macrophage tumor necrosis factor alpha production. *J. Exp. Med.* **187**, 1069–1079.

Sandkvist, M. (2001a). Type II secretion and pathogenesis. *Infect. Immun.* **69**, 3523–3535.

Sandkvist, M. (2001b). Biology of type II secretion. *Mol. Microbiol.* **40**, 271–283.

Sauvonnet, N., Vignon, G., Pugsley, A. P., and Gounon, P. (2000). Pilus formation and protein secretion by the same machinery in *Escherichia coli*. *EMBO J.* **19**, 2221–2228.

Schwartz, M. (2004). Rho signalling at a glance. *J. Cell. Sci.* **117**, 5457–5458.

Segal, E. D., Cha, J., Lo, J., Falkow, S., and Tompkins, L. S. (1999). Altered states: involvement of phosphorylated CagA in the induction of host cellular growth changes by *Helicobacter pylori*. *Proc. Natl. Acad. Sci. U. S. A.* **96**, 14 559–14 564.

Sharff, A., Fanutti, C., Shi, J., Calladine, C., and Luisi, B. (2001). The role of the TolC family in protein transport and multidrug efflux: from stereochemical certainty to mechanistic hypothesis. *Eur. J. Biochem.* **268**, 5011–5026.

Shea, J. E., Hensel, M., Gleeson, C., and Holden, D. W. (1996). Identification of a virulence locus encoding a second type III secretion system in *Salmonella typhimurium*. *Proc. Natl. Acad. Sci. U. S. A.* **93**, 2593–2597.

Shotland, Y., Kramer, H., and Groisman, E. A. (2003). The *Salmonella* SpiC protein targets the mammalian Hook3 protein function to alter cellular trafficking. *Mol. Microbiol.* **49**, 1565–1576.

Sijbrandi, R., Urbanus, M. L., ten Hagen-Jongman, C. M., *et al.* (2003). Signal recognition particle (SRP)-mediated targeting and Sec-dependent translocation of an extracellular *Escherichia coli* protein. *J. Biol. Chem.* **278**, 4654–4659.

Smith, C. L. and Hultgren, S. J. (2001). Bacteria thread the needle. *Nature.* **414**, 29–31.

Stebbins, C. E. and Galan, J. E. (2001). Maintenance of an unfolded polypeptide by a cognate chaperone in bacterial type III secretion. *Nature.* **414**, 77–81.

Stein, M. A., Leung, K. Y., Zwick, M., Garcia-del Portillo, F., and Finlay, B. B. (1996). Identification of a *Salmonella* virulence gene required for formation of filamentous structures containing lysosomal membrane glycoproteins within epithelial cells. *Mol. Microbiol.* **20**, 151–164.

Sukhan, A., Kubori, T., and Galan, J. E. (2003). Synthesis and localization of the *Salmonella* SPI-1 type III secretion needle complex proteins PrgI and PrgJ. *J. Bacteriol.* **185**, 3480–3483.

Surana, N. K., Grass, S., Hardy, G. G., *et al.* (2004). Evidence for conservation of architecture and physical properties of Omp85-like proteins throughout evolution. *Proc. Natl. Acad. Sci. U. S. A.* **101**, 14 497–14 502.

Szabady, R. L., Peterson, J. H., Skillman, K. M., and Bernstein, H. D. (2005). An unusual signal peptide facilitates late steps in the biogenesis of a bacterial autotransporter. *Proc. Natl. Acad. Sci. U. S. A.* **102**, 221–226.

Takizawa, N. and Murooka, Y. (1985). Cloning of the pullulanase gene and overproduction of pullulanase in *Escherichia coli* and *Klebsiella aerogenes*. *Appl. Environ. Microbiol.* **49**, 294–298.

Tamano, K., Aizawa, S., Katayama, E., *et al.* (2000). Supramolecular structure of the *Shigella* type III secretion machinery: the needle part is changeable in length and essential for delivery of effectors. *EMBO J.* **19**, 3876–3887.

Tamano, K., Katayama, E., Toyotome, T., and Sasakawa, C. (2002). *Shigella* Spa32 is an essential secretory protein for functional type III secretion machinery and uniformity of its needle length. *J. Bacteriol.* **184**, 1244–1252.

Tampakaki, A. P., Fadouloglou, V. E., Gazi, A. D., Panopoulos, N. J., and Kokkinidis, M. (2004). Conserved features of type III secretion. *Cell. Microbiol.* **6**, 805–816.

Tardy, F., Homble, F., Neyt, C., *et al.* (1999). *Yersinia enterocolitica* type III secretion-translocation system: channel formation by secreted Yops. *EMBO J.* **18**, 6793–6799.

Thanabalu, T., Koronakis, E., Hughes, C., and Koronakis, V. (1998). Substrate-induced assembly of a contiguous channel for protein export from *E. coli*: reversible bridging of an inner-membrane translocase to an outer membrane exit pore. *EMBO J.* **17**, 6487–6496.

Ton-That, H., Marraffini, L. A., and Schneewind, O. (2004). Protein sorting to the cell wall envelope of Gram-positive bacteria. *Biochim. Biophys. Acta* **1694**, 269–278.

Tu, X., Nisan, I., Yona, C., Hanski, E., and Rosenshine, I. (2003). EspH, a new cytoskeleton-modulating effector of enterohaemorrhagic and enteropathogenic *Escherichia coli*. *Mol. Microbiol.* **47**, 595–606.

Tucker, S. C. and Galan, J. E. (2000). Complex function for SicA, a *Salmonella enterica* serovar typhimurium type III secretion-associated chaperone. *J. Bacteriol.* **182**, 2262–2268.

Uchiya, K., Barbieri, M. A., Funato, K., *et al.* (1999). A Salmonella virulence protein that inhibits cellular trafficking. *EMBO J.* **18**, 3924–3933.

Van Montagu, M., Holsters, M., Zambryski, P., *et al.* (1980). The interaction of *Agrobacterium* Ti-plasmid DNA and plant cells. *Proc. R. Soc. Lond. B Biol. Sci.* **210**, 351–365.

Vazquez-Torres, A. and Fang, F. C. (2001a). Salmonella evasion of the NADPH phagocyte oxidase. *Microbes. Infect.* **3**, 1313–1320.

Vazquez-Torres, A. and Fang, F. C. (2001b). Oxygen-dependent anti-*Salmonella* activity of macrophages. *Trends Microbiol.* **9**, 29–33.

Voulhoux, R., Ball, G., Ize, B., *et al.* (2001). Involvement of the twin-arginine translocation system in protein secretion via the type II pathway. *EMBO J.* **20**, 6735–6741.

Voulhoux, R., Bos, M. P., Geurtsen, J., Mols, M., and Tommassen, J. (2003). Role of a highly conserved bacterial protein in outer membrane protein assembly. *Science* **299**, 262–265.

Wainwright, L. A. and Kaper, J. B. (1998). EspB and EspD require a specific chaperone for proper secretion from enteropathogenic *Escherichia coli*. *Mol. Microbiol.* **27**, 1247–1260.

Walker, G., Hertle, R., and Braun, V. (2004). Activation of *Serratia marcescens* hemolysin through a conformational change. *Infect. Immun.* **72**, 611–614.

Wandersman, C., Delepelaire, P., Letoffe, S., and Schwartz, M. (1987). Characterization of *Erwinia chrysanthemi* extracellular proteases: cloning and expression of the protease genes in *Escherichia coli*. *J. Bacteriol.* **169**, 5046–5053.

Waterman, S. R. and Holden, D. W. (2003). Functions and effectors of the *Salmonella* pathogenicity island 2 type III secretion system. *Cell. Microbiol.* **5**, 501–511.

Weber, E., Ojanen-Reuhs, T., Huguet, E., *et al.* (2005). The type III-dependent Hrp pilus is required for productive interaction of *Xanthomonas campestris* pv. *vesicatoria* with pepper host plants. *J. Bacteriol.* **187**, 2458–2468.

Woestyn, S., Allaoui, A., Wattiau, P., and Cornelis, G. R. (1994). YscN, the putative energizer of the *Yersinia* Yop secretion machinery. *J. Bacteriol.* **176**, 1561–1569.

Wong, K. R., McLean, D. M., and Buckley, J. T. (1990). Cloned aerolysin of *Aeromonas hydrophila* is exported by a wild-type marine *Vibrio* strain but remains periplasmic in pleiotropic export mutants. *J. Bacteriol.* **172**, 372–376.

Yahr, T. L., Vallis, A. J., Hancock, M. K., Barbieri, J. T., and Frank, D. W. (1998). ExoY, an adenylate cyclase secreted by the *Pseudomonas aeruginosa* type III system. *Proc. Natl. Acad. Sci. U. S. A.* **95**, 13 899–13 904.

Zhang, L., Chaudhuri, R. R., Constantinidou, C., *et al.* (2004). Regulators encoded in the *Escherichia coli* type III secretion system 2 gene cluster influence expression of genes within the locus for enterocyte effacement in enterohemorrhagic E. coli O157:H7. *Infect. Immun.* **72**, 7282–7293.

Zhao, Z., Sagulenko, E., Ding, Z., and Christie, P. J. (2001). Activities of virE1 and the VirE1 secretion chaperone in export of the multifunctional VirE2 effector via an *Agrobacterium* type IV secretion pathway. *J. Bacteriol.* **183**, 3855–3865.

Zhou, D. and Galan, J. (2001). *Salmonella* entry into host cells: the work in concert of type III secreted effector proteins. *Microbes. Infect.* **3**, 1293–1298.

Ziemienowicz, A. (2001). Odyssey of *Agrobacterium* T-DNA. *Acta Biochim. Pol.* **48**, 623–635.

CHAPTER 4

Microbial molecular patterns and host defense

Matam Vijay-Kumar and Andrew T. Gewirtz

INTRODUCTION

As an interface with the outside world, epithelial cells play an important role in host defense against the vast number of microbes within our midst. A central feature in such microbial–epithelial cell interactions is the use of host pattern-recognition receptors (PRR) to recognize microbes, thus enabling an appropriate response. This chapter reviews the PRRs that mediate epithelial responses, considers the mechanisms by which they function, and discusses the role of these receptors in host defense and homeostasis. The chapter is based primarily on the intestinal mucosa in light of the authors' knowledge in this area, but many of the concepts should be applicable to other mucosal surfaces. Due to the rapid pace of advancement in this area of research, PRRs are often described and partially characterized well ahead of full determinations of the tissues and cells that actually express these PRRs; thus, this chapter proceeds with a general review of PRRs and then discusses direct and indirect PRR activation of epithelial cells and the role of these activation events in host defense.

PATTERN-RECOGNITION RECEPTORS: MEDIATORS OF HOST–BACTERIAL INTERACTIONS

The immune system traditionally has been divided into innate and adaptive components, which are differentiated functionally on the basis of whether their antimicrobial action requires previous exposure to the microbe or its components. Innate immunity has long been thought to rely on some

Bacterial–Epithelial Cell Cross-Talk: Molecular Mechanisms in Pathogenesis, ed. Beth A. McCormick. Published by Cambridge University Press.

germline-encoded mechanisms to recognize foreign products. For example, mannose receptors, which play an important role in activating the complement cascade, have long been known to interact with a wide range of microorganisms or microbial products, including *Candida albicans, Pneumocystis carinii,* human immunodeficiency virus (gp120), Mycobacterial lipoarabinomannan, *Leishmania, Streptococcus pneumoniae,* and *Klebsiella pneumoniae* (Taylor *et al.*, 2005). The formyl-methionyl-leucyl-phenylyalanyl (f-MLP) receptor is a multispanning G-protein receptor that is expressed on phagocytic cells and plays an important role in chemotaxis during infection (Le *et al.*, 2001). Other receptors include C-type-lectin-like receptor, Dectin-1, the major receptor for fungal cell-wall β-glucans and demonstrated to play a significant role in mediating cytokine secretion by macrophages (Herre *et al.*, 2004). In addition, moesin (membrane-organizing extension spike protein), a 78-kDa protein, functions as a lipopolysaccharide receptor on human monocytes (Tohme *et al.*, 1999), whereas the C-reactive protein and other members of the pentraxin family have been demonstrated to bind microbes directly through a lectin-like interaction.

However, two important discoveries since 1996 have accelerated research in this field of study dramatically. First, in 1996, the Drosophila protein Toll, formerly known only for its role in ontogenesis was shown to be required for flies to mount an effective immune response to the fungus *Aspergillus fumigatus* (Lemaitre *et al.*, 1996). Shortly thereafter, a mammalian family of homologous proteins termed Toll-like receptors (TLRs) was identified. Second, in 1998, Toll-like receptor 4 (TLR-4) was identified via positional cloning as the lipopolysaccharide (LPS) receptor, encoded by the *Lps* locus and known to be required for mice to induce effective responses to Gram-negative bacteria in which LPS is a fundamental part of the outer cell membrane (Poltorak *et al.*, 1998). These findings revealed that the innate immune receptors of insects and mammals are evolutionarily conserved, suggesting a vital role for TLRs in the initial recognition of microbial pathogens ranging from protozoa to bacteria to fungi to viruses (Beutler, 2004).

Since 1998, our understanding of the innate immune system has been revolutionized by the identification of various classes of PRR that recognize specific molecular structures on pathogens. Such structures are sometimes referred to as pathogen-associated molecular patterns (PAMPs) (Kopp and Medzhitov, 1999); however, this term is somewhat a misnomer as these patterns are in general similar on pathogenic and commensal microbes alike. Thus, we would suggest that the term microbial patterns (MPs) might be more appropriate, at least in the context of the intestine. The encounter between PRRs and MPs can take place on the cell surface or intracellularly or in soluble form. TLRs are the best-described class of PRRs. Thirteen mammalian paralogs have now been identified (10 in humans, 12 in mice) (Beutler *et al.*, 2004; Tabeta *et al.*, 2004). Both forward and reverse genetic tools have been applied to determine the molecular specificity of most of the TLRs (Akira *et al.*, 2001). TLRs constitute type I, germline-encoded, evolutionarily conserved PRRs, which serve as central signal-transducing elements for the induction of immunoregulatory effector molecules that trigger innate immunity and instruct the development of adaptive immunity (Schnare *et al.*, 2001) via mechanisms discussed later in this chapter. TLRs contain N-terminal leucine-rich repeats (LRRs), which are located extracellularly for cell-surface PRRs. Among different PRRs, such LRRs contain sufficient diversity in order to provide a versatile structural framework for the formation of protein–protein or protein–carbohydrate/lipid interactions with the microbial ligands that PRRs recognize (Kobe and Kajava, 2001). For example, the LRRs of TLR-4 bind to LPS complexed to LPS binding protein (LBP), whereas flagellin interacts with the LRRs of TLR-5 (Tschopp *et al.*, 2003). The cytoplasmic portion of TLRs shows high similarity to that of the interleukin 1 (IL-1) receptor family, specifically known as the Toll/IL-1 receptor (TIR) domain, and shares similarity to plant resistance (R) proteins (O'Neill *et al.*, 2003).

Intracellular pattern-recognition receptors

While TLR-1, -2, -4, -5, -6, and -10 are expressed on the cell surface, TLR -7, -8, and -9 form an evolutionary cluster and are targeted to endosomes based on structural features residing in the cytoplasmic domain (Lee *et al.*, 2003; Nishiya and DeFranco, 2004; Wagner, 2004). The cellular distribution of TLR-3 appears to be cell-type-dependent, since this receptor is expressed on the surface of human fibroblasts as well as in intracellular organelles such as endosomes in monocyte-derived immature dendritic cells (DCs) and blood CD11c^{+} cells (Matsumoto *et al.*, 2003). TLR-3, -7, -8, and -9 together constitute nucleic-acid-recognizing PRRs and are thus strategically located intracellularly where replication of most of the viruses occurs, thus generating either single- or double-stranded RNA or DNA.

A second large class of intracellular PRRs is the family of nucleotide-binding oligomerization domain (NOD) proteins defined by a NOD domain that mediates oligomerization (Athman and Philpott, 2004; Inohara and Nunez, 2003). The NOD proteins, which are soluble and localized intracellularly, have an N-terminal signaling domain in addition to a C-terminal ligand-binding domain. Interestingly, for most NOD proteins the ligand-binding domain is an LRR domain and is similar to what has been observed for

TLRs, suggesting that these proteins are capable of serving as PRRs. NOD1 and NOD2 (also referred to as caspase-activating and recruitment domain 15 (CARD15) and CARD4, respectively) are well-studied PRRs that sense intracellular pathogens and activate nuclear factor kappa B (NF-κB). While NOD2 mediates intracellular recognition of muramyl dipeptide (MDP), a building block for bacterial cell walls (Girardin, Boneca, Viala *et al.*, 2003; Inohara *et al.*, 2003), NOD1 specifically detects a unique muropeptide fragment, diaminopimelate-containing N-acetylglucosamine-N-acetylmuramic acid (GlcNAc-MurNAc). This tripeptide motif is a breakdown product of Gram-negative bacterial peptidoglycan (Girardin, Boneca, Carneiro *et al.*, 2003). Therefore, NOD proteins are able to recognize breakdown products of peptidoglycan mediated by TLR-2, suggesting not only that NOD proteins influence the signaling downstream of TLR-2 but also that there is a tight interaction between extracellular and intracellular PRRs (Watanabe *et al.*, 2004). CLAN (caspase-associated recruitment domain, leucine-rich repeat, and NAIP CIIA HET-E, and TPI-containing protein) is an intracellular PRR. Although CLAN contains spans of LRRs, the characteristic nucleotide-binding domain appears to function independently of the TLRs and typically resides within the cytosol of specific cells, where it modulates caspase-1 activation and subsequent interleukin 1β (IL-1β) secretion in response to LPS, peptidoglycan, and pathogenic bacteria (Damiano *et al.*, 2004).

Another important intracellular PRR is the interferon (IFN)-inducible dsRNA-dependent protein kinase R (PKR), a serine/threonine kinase that has long been known to represent an intracellular receptor for dsRNA and that is believed to be a major viral signature. PKR also mediates the activation of signal-transduction pathways by pro-inflammatory stimuli, including LPS, tumor necrosis factor alpha (TNFα), and IL-1, in addition to its other physiological functions, especially during cellular stress (Barber, 2005; Williams, 2001). Following binding of dsRNA, PKR dimerizes and becomes activated via autophosphorylation. The main substrate for PKR is the eukaryotic initiation factor 2, explaining the ability of PKR to inhibit mRNA translation. Thus, activation of PKR downmodulates protein synthesis in virus-infected cells and, hence, inhibits viral replication. Furthermore, PKR is capable of activating other inflammatory signal-transduction pathways (Goh *et al.*, 2000; Kumar *et al.*, 1994; Kumar *et al.*, 2004; Uetani *et al.*, 2000); as well as playing a role in dsRNA and rotavirus-induced interleukin 8 (IL-8) expression (Vijay-Kumar *et al.*, 2005) in intestinal epithelial cells.

Research in this area continues to progress rapidly as other PRRs continue to be discovered. Most recently at the time of writing, RNA helicases have been identified as a third class of PRRs (Yoneyama *et al.*, 2004). Two cytoplasmic proteins, retinoic-acid-inducible gene I (RIG- I) and melanoma-differentiation-associated gene 5 (MDA5), each with two N-terminal signaling domains and a C-terminal helicase domain, were shown to be activated in response to virus infection and to transduce signals via the signaling domains. RIG-I and MDA5 are DExD/H box RNA helicases that contain a caspase-recruitment domain as an essential regulator for dsRNA-induced signaling and a helicase domain that possesses intact ATPase activity, which is responsible for the dsRNA-mediated signaling. The caspase-recruitment domain transmits downstream signals, resulting in the activation of NF-κB and interferon-responsive elements (Heim, 2005). The role of these and other PRRs in microbial pathogenesis remains under investigation.

Soluble pattern-recognition receptors

PRRs are likely to play a primary role against microbial defense not only as cell-surface receptors but also as systemic or humoral factors. Indeed, at least some PRRs are naturally found in serum in a soluble form, which may have contributory or regulatory effects on the function of cell-surface PRRs. A classic example is CD14, a 55-kDa glycosylphosphatidylinositol-anchored membrane protein that is devoid of a cytoplasmic domain (Haziot *et al.*, 1988; Pugin *et al.*, 1994) and that does not elicit intracellular signaling (Ulevitch and Tobias, 1995). Soluble TLRs have also been observed and may play an important role in host defense. Soluble PRRs such as CD14 are devoid of transmembrane and intracellular domains and do not initiate signaling; however, they can bind their respective ligands and thus help in controlling the MPs induced pro-inflammatory signaling. Soluble TLR-2 (sTLR-2) has been described in human plasma and milk. sTLR-2 levels were found to be lower in people with tuberculosis, and it appears that blocking TLR-2 signaling in response to bacterial lipopeptide is the main function of the sTLR-2 (LeBouder *et al.*, 2003). Presumably, low serum sTLR-2 levels in people with acquired immunodeficiency syndrome (AIDS) may be associated with increased cytokine responses to TLR-2-containing pathogens such as *Mycobacterium* species, Gram-positive bacteria, and fungi, all of which contribute to the progression of human immunodeficiency virus (HIV)-related cachexia (Heggelund *et al.*, 2004). Moreover, in vitro studies suggest that sTLR-2 partially attenuates the insoluble peptidoglycan derived from *Staphylococcus aureus*-induced NF-κB activation and IL-8 secretion (Iwaki *et al.*, 2002). An alternatively spliced murine TLR-4 mRNA has been described that results in the expression of the soluble form of TLR-4 (sTLR-4), which inhibits

LPS-induced NF-κB activation and TNFα release (Iwami *et al.*, 2000). Further, addition of sTLR-4 to the medium attenuates LPS-induced NF-κB activation and IL-8 secretion in wild-type TLR-4-expressing cells (Hyakushima *et al.*, 2004). Based on these observations, it may be possible to develop TLR-based treatment strategies and to use sTLRs to regulate the uncontrolled immune response in various disease states (Zuany-Amorim *et al.*, 2002).

Defining pattern-recognition receptor ligands

Although sequence homology allows straightforward identification of additional members to various classes of PRR, defining the various specific ligands that such PRRs recognize has been a more complex undertaking. In general, candidate ligands have emerged through a combination of guesses and biochemical purification of a complex bioactivity, while the verification of candidates has relied upon genetic gain or loss of function in in vitro or in vivo models. A major breakthrough, however, came from utilizing the C3H/HeJ mouse. This mouse contains a single point mutation in the BB loop of the TIR domain of TLR-4, a Pro712His that converted this mutant TLR-4 into a dominant-negative gene (Poltorak *et al.*, 1998) that is hyporesponsive to LPS. Table 4.1 describes the ligands for various PRRs. TLR-1 is involved in modulating the responses to TLR-2 and TLR-4, but at present no specific ligand has been identified. It has also been shown that TLR-1 is involved in the recognition of Mycobacterial lipoprotein and triacylated lipopetides (Takeuchi *et al.*, 2002). Furthermore, TLRs can form multimeric complexes (homodimers or heterodimers) to increase the spectrum of molecules that they recognize. For example, the cytoplasmic domain of TLR-2 can form functional pairs with TLR-6 and TLR-1, leading to signal transduction and cytokine expression after ligand activation (Takeuchi *et al.*, 1999). TLR-2 recognizes peptidoglycan and lipopeptides of Gram-positive bacterial origin. TLR-3 is the receptor for ds-RNA. TLR-4, the most extensively studied TLR, is a receptor for the LPS of Gram-negative bacteria. Interestingly, it has been proposed that the shape of the lipid A component determines the bioactivity of LPS. For instance, LPS with a conical shape (e.g. from *Escherichia coli*) induces cytokine production through TLR-4, whereas LPS with a more cylindrical form (e.g. from *Porphyromonas gingivalis*) induces expression of a different set of cytokines through TLR-2. Strictly cylindrical shaped LPS molecules (e.g. the lipid A precursor Ia or from *Rhodobacter sphaeroides*) have antagonistic properties at the level of TLRs (Netea *et al.*, 2002).

F-protein from respiratory syncitial virus is also a ligand for TLR-4. Moreover, TLR-4 interacts directly with NAD(P)H oxidase 4 isozyme (Nox4) and

Table 4.1 *Pattern-recognition receptor (PRR) ligands*

Receptor	Microbial product
TLR-1 (with TLR-2)	Mycobacterial lipoprotein
	Triacylated lipopeptides
TLR-2 (with TLR-1 or TLR-6)	Gram-positive bacteria
	Peptidoglycan, lipoteichoic acid
	Zymosan, liparabinomannan
	Bacterial glycolipids, yeast mannan
	GPI anchors of *Trypanosoma cruzi*
	LPS from *Leptospira interrogans*
	LPS from *Porphyromonas gingivalis* (more cylindrical)
TLR-3	Viral dsRNA, synthetic polyinosinic acid: cytidylic acid (poly (I:C))
TLR-4	Gram-negative bacteria
	LPS (conical shape), pneumolysin
	Lipid A (strictly cylindrical, antagonist)
	LPS from *Rhodobacter sphaeroides* (strictly cylindrical)
	Flavolipin from *Flavobacterium meningosepticum*
	Respiratory syncitial virus protein F
	Aspergillus fumigatus hyphae
	HSP 60 and 70, hyaluronan
	Fibronectin A domain, fibrinogen
	Necrotic cells, saturated fatty acids, Taxol*
TLR-5	Flagellin
TLR-6 (with TLR-2)	Mycoplasma lipoproteins, lipoteichoic acid, peptidoglycan
TLR-7 and TLR-8	Single-stranded RNA, imidazoquinalones
TLR-9	CpG DNA, hemozoin
TLR-10	Unknown
TLR-11	Uropathogenic bacteria
	Profilin-like protein molecules in *Toxoplasma gondii**
NOD-1	Muropeptide fragment diaminopimelate-containing GlcNAc-MurNAc
NOD-2	Muramyl dipeptide
Protein kinase R	dsRNA
RNA helicases	dsRNA
Dectin-I	β-Glucans
Mannose receptor	Liparabinomannan
f-MLP receptor	f-MLP
Moesin	LPS

*Only in mouse.

C_pG, see text; dsRNA, double-stranded RNA; f-MLP, formyl-methionyl-leucyl-phenylalanyl; GPI, glycophosphatidylinosital; HSP, heat-shock protein; LPS, lipopolysaccharide; NOD, nucleotide-binding oligomerization domain.

appears to be involved in LPS-mediated reactive oxygen generation and NF-κB activation (Park *et al.*, 2004). TLR-5 is a receptor for bacterial flagellin. TLR-6 recognizes mycoplasma lipopeptide. Murine TLR-7 and human TLR-8 mediate species-specific recognition of guanine and uracil (GU)-rich ssRNA (Heil *et al.*, 2004). Human TLR-7 also recognizes ssRNA molecules of non-viral origin (Diebold *et al.*, 2004). TLR-7 and TLR-8 recognize antiviral imidazoquinoline compounds (Bowie and Haga, 2005; Jurk *et al.*, 2002). TLR-9 is a receptor for bacterial DNA, which is enriched with consecutive unmethylated cytosine-guanosine residues, and referred to as CpG DNA. The ligand for TLR-10 has not been identified. TLR-11 has been discovered in mice and found to recognize uropathogenic bacteria, whereas in humans TLR-11 appears to be a pseudogene as it contains a series of stop codons and thus appears not to encode a functional protein (Zhang *et al.*, 2004). Additionally, it recognizes a novel profilin-like protein of the parasite *Toxoplasma gondii* and is essential for murine resistance to this pathogen (Yarovinsky *et al.*, 2005).

Toll-like receptor coreceptors

Some PRRs appear not to recognize bacterial ligands directly but instead to recognize microbial products complexed to coreceptors. The best-studied example is the interaction of TLR-4 with LPS. Lipopolysaccharide-binding protein (LBP), CD14, MD-2 and β2 integrins are known to act as coreceptors for TLR-4. LBP is produced in the liver and has the properties of an acute-phase reactant with opsonic activity. Its importance as a protein involved in the LPS response lies within its ability to accelerate the binding of LPS to CD14 (Fitzgerald *et al.*, 2004). For example, studies with LBP knockout mice have revealed that LBP enhances the sensitivity to LPS by approximately 300-fold (Wurfel and Wright, 1997). The sentinel function of LBP is also shared by CD14, which was originally discovered as the receptor for LBP-bound LPS. CD14 is central to mammalian responses to LPS and is present in two forms: a glycophosphatidylinositol (GPI)-linked form and a soluble form found in blood. Soluble CD14 functions to enhance LPS responses in cells that do not ordinarily express CD14. Thus, the presence of CD14 increases the sensitivity of the LPS response by 1000-fold (Fitzgerald *et al.*, 2004). The ability of β_2 integrins to enhance responsiveness to endotoxin is much greater in response to LPS encountered in an insoluble aggregate, such as a whole bacterium (Moore *et al.*, 2000). The discovery of MD-2 further aided in deciphering the central role of TLR-4 in LPS signaling (Shimazu *et al.*, 1999). MD-2 is an 18–25-kDa protein; it appears that only the monomeric form of MD-2 is biologically important for LPS signaling. Once LPS binds to TLR-4/MD-2, the mechanism by which the receptor becomes activated is unclear. Several studies, however, suggest that the aggregation of TLR-4, either artificially using monoclonal antibodies or as a result of LPS binding, is sufficient to activate signal transduction (Latz *et al.*, 2002). For TLR-5, it appears that gangliosides act as coreceptors for signaling induced by flagellin, thereby promoting hBD-2 expression via MAP kinase (Ogushi *et al.*, 2004). The extent of coreceptor usage by other PRRs has not been investigated.

Pattern-recognition receptor signaling

Although this chapter does not discuss signaling mechanisms of PRRs in depth, some mention of general PRR signaling paradigms is necessary in order to get an overall sense of how PRRs function. Each of the aforementioned classes of PRR uses distinct signaling adapters, and yet they activate substantially overlapping sets of signaling cascades, consistent with the fact that there is a substantial amount of overlap in the gene expression activated by different PRRs. Clearly some differences in effector usage by different PRRs have been demonstrated to result in differential gene expression – for example, TLR-4 activates interferons but TLR-2 does not (Toshchakov *et al.*, 2002) – but such examples at present are more the exception than the rule. Although this view may ultimately prove crude as research progresses in this area, at present it may be most generally helpful to view multiple PRRs as activating the same general sets of downstream signaling pathways.

TLR signaling is by far the most extensively studied PRR family. An essential component of nearly all TLR signaling is the effector protein MyD88, which is necessary for TLR activation of canonical pro-inflammatory signaling pathways, namely the transcription factor NF-κB and the family of mitogen-activated kinases, particularly p38 (Akira, 2001). TLRs also use other adapters to activate these same pathways, including a protein referred to as both Toll interleukin 1 receptor domain-containing adapter protein (TIRAP) and MyD88-adapter-like (MAL) and a host of downstream kinases, such as interleukin-1-receptor-associated kinase (IRAK) (Takeda and Akira, 2005). The N-terminal region of NOD proteins mediates the activation of other kinases (particularly those of the RICK family), resulting in activation of many of the same downstream effectors as TLRs, notably NF-κB and p38 (Inohara *et al.*, 2005). These signals are essential and generally rate-limiting regulators of an extensive panel of genes that are discussed below. The best-defined example of differential gene expression by different TLRs (as described above) is perhaps exhibited by the TLR signaling adapter TRIF (Toll/IL-1R domain-containing adapter-inducing interferon beta), which has the ability to activate interferon regulatory factor 3 and subsequently interferon expression (Yamamoto *et al.*, 2002). Nonetheless, this chapter focuses largely on the downstream results of PRR signaling without distinguishing one PRR-activated pathway from another, as the knowledge to do so is simply not yet available.

PATTERN-RECOGNITION RECEPTORS MEDIATE MULTIPLE MECHANISMS OF HOST DEFENSE

The existence of extensive evolutionarily conserved families of PRRs suggests an important role for these molecules. This section considers the experimental evidence to support and define this role and then broadly considers the important biologic events that are activated in PRRs, even though the precise roles of individual PRRs in host defense have only begun to be defined.

Global role and evidence of pattern-recognition receptor function in immunity

A substantial body of in vitro and in vivo evidence indicates a pre-eminent role for PRRs in this process. Although other mechanisms may also lead to activation of these signaling pathways – for example, some bacteria, particularly those with type III secretion systems, may have the ability to directly activate host-signaling pathways (Hardt *et al.*, 1998) – it is nonetheless likely that PRRs represent the major means of activating innate immunity. In vitro, epithelial cells will respond to a variety of microbial products/extracts with an extensive array of pro-inflammatory gene expression, including flagellin (Gewirtz, Simon *et al.*, 2001), MDP (Girardin, Boneca, Carneiro *et al.*, 2003), unmethylated CpG DNA (Platz *et al.*, 2004), dsRNA (Guillot *et al.*, 2005; Kumar, 2004), and, under some conditions, LPS and lipopeptide (Abreu *et al.*, 2002). Cells that are unresponsive to these products often can be made responsive by engineered expression of a particular PRR. Conversely, the responses of epithelial cells to whole bacteria can be substantially ablated by engineered expression of dominant-negative PRRs. All of these approaches have been used for characterizing epithelial pro-inflammatory gene expression in response to *Salmonella typhimurium* and indicate a dominant role for recognition of flagellin by TLR-5 (Gewirtz, Navas *et al.*, 2001; Hayashi *et al.*, 2001; Tallant *et al.*, 2004). Given the broad ability of TLR-5 to recognize an extensive array of flagellins from a variety of bacteria (to date shown to include

Salmonella, *E. coli*, *Pseudomonas*, and *Listeria*), it seems likely that TLR-5 will play similarly important roles for epithelial responses to a variety of other motile bacteria. Analogously, NOD1 plays a major role in recognizing the intracellular non-flagellated pathogen *Shigella flexneri*, and thus it seems reasonable to speculate a generally important role for this PRR against other non-motile invasive pathogens (Girardin *et al.*, 2001).

In vivo evidence for the role of particular classes of PRR in mediating innate immune responses has come largely from genetically modified mice. For PRRs in which the ligand is well defined, simple deletion of that particular gene will result in a complete loss of response to systemic application of that bacterial product. In vivo evidence for the role of PRRs in responding to whole bacteria has emerged more slowly and often with less clarity but nonetheless broadly supports the concept that PRRs are a major and essential component of innate immunity. Mice lacking TLR-4 show substantially elevated susceptibility to Gram-negative pathogens such as *Salmonella* (Weiss *et al.*, 2004), while mice lacking TLR-2 and TLR-4 are more susceptible to Gram-positive bacteria such as *Mycobacterium* (Reiling *et al.*, 2002). Interestingly, depending on the particular pathogen and the mode of challenge, the immune deficiency of mice lacking an individual TLR can be quite mild, as is somewhat the case in the above examples. In contrast, mice lacking TLR signaling, are in general, due to lack of the broadly used TLR signaling effector MyD88, highly susceptible to a variety of pathogens (Feng *et al.*, 2003; Scanga *et al.*, 2004; Villamon *et al.*, 2004). This indicates that although PRR recognition of individual bacterial products may lack redundancy, there is likely substantial redundancy in the ability of PRRs in general to recognize multiple components of most bacteria.

Pattern-recognition receptors regulate key aspects of host defense

PRRs in the intestinal mucosa play important roles in host defense by several mechanisms. Namely, ligation of PRRs by bacterial products can directly induce expression of soluble antimicrobial mediators, orchestrate recruitment and activation of phagocytes, and promote development of adaptive immunity. Additionally, studies also indicate a role for PRRs in regulation of mucosal barrier function, epithelial apoptosis, and intestinal homeostasis. Although at present the precise roles of individual PRRs in mediating specific elements of these various responses are not well defined, this section gives a general overview of the role of the downstream effects of PRR activation, particularly those mediated by epithelial cells.

Pattern-recognition receptors mediate expression of antimicrobial mediators

An important means by which the intestine controls the growth of and thus protects the host from the extensive microbial biomass that resides in the mucosa is by the production of a panel of antimicrobial peptides (Ganz, 2003). Some of these molecules are secreted by enterocytes (i.e. epithelial cells), while the expression of others is exclusive to or highly enriched in paneth cells residing in intestinal crypts. Epithelial antimicrobial secretions include antimicrobial lytic enzymes such as lysozyme, nutrient-binding proteins such as lactoferrin, and proteins that kill microbes via oxidative mechanisms, such as lactoperoxidase. Moreover, the epithelium also secretes an increasingly appreciated panel of smaller antimicrobial peptides with fewer than 100 amino acids, of which the major class is the defensins. In general, these peptides interact with the anionic part of the prokaryotic membrane due to their cationic charge and insert into microbial membranes, creating pores and resulting in cell lysis (Grandjean *et al.*, 1997). Defensins are a family of evolutionarily related arginine-rich small cationic peptides of 3.5–4.5 kDa molecular mass and with 29–35 amino-acid residues. They are abundant in both hematopoietic cells and non-hematopoietic cells and are presumed to be involved in host defense such as intestinal epithelial cells (IECs). Concentration of defensins in intestinal crypts can be quite high, reaching $> 10\ \mathrm{mg\ ml^{-1}}$ (Ayabe *et al.*, 2000). α-Defensins are expressed primarily by neutrophils and paneth cells, whereas β-defensins are expressed more widely in the intestine and are expressed particularly prominently in IEC. A dramatic example of the power of defensins to shape the outcome of host–pathogen interactions is that mice engineered to constitutively express a human defensin (HD-5) were markedly resistant to oral challenge with virulent *S. typhimurium* (Salzman, Ghosh *et al.*, 2003). Another antimicrobial mediator secreted in substantial amounts by both leukocytes and epithelial cells is the protein lipocalin 2, also referred to as neutrophil-gelatinase granule-associated lipocalin (NGAL). Lipocalin 2 is induced by ligands of TLRs and PKR (Vijay-Kumar *et al.*, 2005) and has been shown to play a crucial role in host defense against *E. coli* infections in mice (Flo *et al.*, 2004).

Although regulation of defensin expression has not yet been well defined, their expression is rapidly induced in response to bacteria and, with similar efficiency, to bacterial products, indicating an important role for PRRs in this process. Paneth cell defensins are stored in secretory granules, which are secreted rapidly in response to live or dead bacteria. Surface glycolipid appears to be one of the inducing components of this response, although the specific molecular determinants have not yet been defined and a role for TLR-4 has been excluded. β-Defensin expression, especially human beta-defensin (hBD), is particularly highly induced in the gut and by epithelial cells in vitro (O'Neil *et al.*, 2000). In vitro studies have demonstrated roles for TLR-2 and TLR-4 (Vora *et al.*, 2004). Other in vitro studies have reported bacterial flagellin to be an especially potent inducer of defensin expression by epithelial cells (Takahashi *et al.*, 2001).

Pattern-recognition receptors coordinate innate immunity

Although the above-described mechanisms play an important role in preventing bacterial overgrowth at mucosal surfaces and in protecting against large numbers of bacteria, in general they are insufficient to defend against a number of bacteria, especially pathogens. Thus, the epithelium has evolved the ability to coordinate immune responses by regulating the "professional" immune cells of both the innate and the adaptive immune systems. Specifically, via both secretion of soluble molecules and expression of surface molecules, the epithelium can direct movement and functional responsiveness of immune cells. These events appear to be regulated in large part by PRRs.

Upon activation of their PRRs, epithelial cells secrete an extensive panel of cytokines that regulate immune cells (Gewirtz *et al.*, 2002), a substantial portion of which appears to be regulated in large part via direct or indirect epithelial recognition of microbial patterns (Zeng *et al.*, 2003). The specific list has grown primarily due to development of large-scale analytical techniques such as cDNA microarray and thus has outpaced knowledge regarding functional significance; nonetheless, the list includes cytokines that direct movement and regulate function of innate and adaptive immune cells. Cytokine secretion by epithelial cells has been observed in a large number of epithelial cell lines and primary cells and in vivo by intracellular cytokine staining and in situ hybridization (Cromwell *et al.*, 1992; Jung *et al.*, 1995; Mazzucchelli *et al.*, 1994). Epithelial cells secrete a number of chemotactic cytokines (i.e. chemokines) that direct the movement, thereby controlling mucosal populations of both innate and adaptive immune cells. Epithelial secreted chemokines demonstrated to regulate neutrophil movement include IL-8 (McCormick *et al.*, 1993), epithelial neutrophil attractant-78 (ENA-78) (Keates *et al.*, 1997), Groα and Groβ (CXCL8, CXCL5, CXCL1, and CXCL2, respectively) (Dwinell *et al.*, 2003). Epithelial recruitment of monocytes appears to be driven primarily by monocyte chemotactic protein 1 (MCP-1), macrophage inflammatory protein 1α (MIP1α), and RANTES (CCL2, CCL3, and CCL5, respectively) (Lin *et al.*, 1998). MIP-1α also appears to play a particularly important role in recruitment of mucosal dendritic cells (Sierro *et al.*, 2001). Lastly, although less well studied in vitro or in vivo, epithelial cells also secrete a variety of chemokines that direct the movement of macrophages, mast cells, basophils, and eosinophils (Dwinell *et al.*, 2003).

Beyond orchestrating the movement of immune cells, the epithelium also secretes a number of immunoregulatory cytokines, with TNFα and IL-6 being the best examples (Cromwell *et al.*, 1992; Sitaraman *et al.*, 2001). Although epithelial secretion of these cytokines would certainly seem to affect the local milieu, they may also exhibit a substantial effect on the overall levels of serum cytokines. These cytokines can have profound effects on both epithelial cells and immune cells. TNFα strongly promotes expression of other pro-inflammatoy cytokines and chemokines, thus tending to amplify these responses. Further, TNFα will prime effector functions of innate immune cells, for example, causing neutrophils to make more superoxide and release more granule content when encountering bacterial stimuli. As these innate immune effector mechanisms can have potent effects on both microbes and host cells, the levels of these immunomodulating cytokines would seem to have a profound effect upon the outcome of host–microbial interactions that occur routinely at mucosal surfaces.

In addition to soluble factors, the epithelium also regulates a number of adhesion molecules that will affect the interaction of the epithelium with immune cells. Epithelial cells express basal levels of several ligands for intraepithelial lymphocytes (IEL), of which αεβ7 integrin plays a particularly prominent role (Shaw *et al.*, 1994), and upregulate several other T-cell ligands, e.g. intercellular cell-adhesion molecule 1 (ICAM-1), CD54, lymphocyte functional antigen 3 (LFA-3), and CD58 (Cruickshank *et al.*, 1998). Moreover, epithelial expression of neutrophil ligands is thought to play a major role in regulating both the adherence and the transepithelial migration of neutrophils. Particularly, important roles of epithelial CD47 and Signal regulatory protein 1 alpha (SIRP1α) in regulating the transmigration process have been reported (Liu *et al.*, 2002; Parkos *et al.*, 1996). ICAM-1 is markedly upregulated in inflammatory conditions and likely plays a role in the increased neutrophil–epithelial adherence associated with inflammatory bowel disease (Colgan *et al.*, 1993). Overall, regulation of epithelial expression of immune cell ligands and of chemokines likely work in concert in coordinating immune responses consistent with observations that many genes appear to be coregulated in these cells in inflammatory clusters (Zeng *et al.*, 2003). As we discuss below, there is strong in vitro and in vivo evidence to indicate a dominant role for PRRs in regulating expression of the above-described epithelial orchestration of immune cells.

Pattern-recognition receptors regulate adaptive immunity

In addition to orchestrating innate immune responses, growing evidence also suggests a role for epithelial cells in regulating adaptive immunity, with PRRs taking center stage as major regulators in this process. In addition to the above-described innate immune cell chemokines, the epithelium also secretes chemokines that direct movement of various populations of T-cells in the mucosa. Chemokines thought to be important for directing the movement of intraepithelial lymphocytes (IEL) include CXCL10 (i.e. interferon gamma-inducible protein 10, IP-10) and CXCL9 (i.e. monokine induced by interferon gamma, Mig) (Shibahara *et al.*, 2001). Consistent with the fact that IELs, unlike neutrophils, are normally present in the mucosa, these chemokines are constitutively expressed by epithelial cells; their expression can, nonetheless, be upregulated significantly by bacterial products (Shaw *et al.*, 1998; Shibahara *et al.*, 2001). These same chemokines also regulate movement of CD4 T-cells, although the fact that increased numbers of these cells are associated with inflammation and that other T-cell chemoattractants such as MIP1α, and liver and activation-regulated chemokine (LARC) are among the epithelial genes most upregulated by pro-inflammatory stimuli suggests a more prominent role for these latter chemokines in being more dynamic regulators of T-cell movement (Gewirtz *et al.*, 2002; Zeng *et al.*, 2003). In addition to controlling movement of lymphocytes, PRR-activated immunomodulating cytokines such as TNFα and IL-6 will also have a substantial influence on the extent and type of adaptive immune response. Lastly, an important role of intestinal epithelial cells in regulating adaptive immunity in general is in regulating antigen uptake through the mucosa. It has been demonstrated that TLR-2 regulates epithelial barrier function, via regulation of tight junctions, indicating that PRRs also play a role in regulating which antigens are seen by the adaptive immune system (Cario *et al.*, 2004).

Regulating pattern-recognition receptor function in IEC

The above sections of this chapter have attempted to give a broad overview of the range of PRR detection of microbes and the effector function that PRRs can activate. However, understanding how PRRs are used to appropriately regulate epithelial contributions to innate immunity is a considerably complex matter. Specifically, since epithelial cells, especially gut epithelial cells, are constantly in close proximity to large numbers of commensal bacteria whose molecular patterns are not easily distinguishable from pathogens, they must regulate their PRRs in such a manner so as to respond appropriately to pathogens but not commensal microbes. In considering the importance of such regulation, it is instructive to first consider the potential consequences if some of these functions were regulated inappropriately. Insufficient expression of basal mediators of epithelial immunity such as hBD-1 would seem to render the mucosa susceptible to bacterial overgrowth, while overexpression might be metabolically wasteful or might have a detrimental effect on the normal mucosal microflora, which may itself lead to metabolic problems or may render the host more susceptible to attack by pathogens. The consequences of improper regulation of inducible epithelial genes that regulate immune function seem more acute. For example, the inability to rapidly recruit immune cells in response to pathogens would make the host more susceptible to systemic infection. Conversely, recruitment of immune cells, especially neutrophils, in the absence of a perturbing pathogen is likely to be detrimental, as it can lead to substantial damage of host tissue, as is thought to occur in chronic inflammatory diseases. Thus, the epithelium exerts tight control over its immunomodulating genes, especially those associated with acute inflammation, such as the neutrophil chemoattractant IL-8. At present, it appears that some mechanisms by which the epithelium achieves appropriate control over PRRs rely on carefully controlling the spatial and temporal expression of TLRs and by tightly regulating the signaling pathways to which they have access.

Location, location, location

One epithelial strategy of reducing the potential of activating pro-inflammatory gene expression in response to bacterial products is to place PRRs in a locale where they would be activated by invasive pathogens but not luminal microflora. Perhaps the most dramatic example of this strategy is the case of TLR-5, the receptor for flagellin. Epithelial cell expression of TLR-5 is highly polarized to the basolateral surface, so that these cells exhibit a more than 1000-fold greater response to flagellin added basolaterally than to an equal amount added apically (Gewirtz, Navas *et al.*, 2001). Thus, on intestinal epithelial cells, TLR-5 will become activated only when the epithelium has been breached by either a flagellated microbe or its soluble flagellar component (i.e. flagellin). There has been some indication that this strategy may also be used by some other TLRs (Abreu *et al.*, 2003), although lack of quality of reliable anti-serum has greatly stymied in vivo localization of TLRs in general. Regardless, it is clear that expressing PRRs in locales to be reached only by invasive pathogens is a broadly used strategy in the intestinal mucosa. NODs and other intracellular PRRs will be activated only by invasive microbes and thus should not normally be prone to being activated by commensal microbes. Some microbial components, such as LPS, do not potently activate intestinal epithelial cells directly but rather are highly potent activators of immune cells in the lamina propria, i.e. monocytes and macrophages. As these immune cells will, in response to LPS, secrete cytokines that will activate the pro-inflammatory functions of epithelial cells, the expression of TLR-4 in the intestinal mucosa is in effect also spatially controlled, such that LPS only from invasive bacteria will lead to potent epithelial activation. This mechanism has been demonstrated in ex vivo studies of human skin (Liu *et al.*, 2003), but our unpublished results suggest that similar mechanisms may also be operative in the gut, and thus it may be a widely used mechanism. This concept is illustrated in Figure 4.1.

Not for ordinary use

Another means of reducing the possibility of unwarranted gut inflammation in response to commensal microbes while retaining the option to use an extensive array of PRRs in host defense is to limit PRR expression under ordinary circumstances and to express PRRs only when the risk of infection may be greatest. For example, under normal (i.e. non-inflammed) conditions, human intestinal epithelial cells appear to express little TLR-4 (Naik *et al.*, 2001) and most intestinal epithelial cell lines do not exhibit significant responses to LPS (Eaves-Pyles *et al.*, 1999; Gewirtz, Simon *et al.*, 2001; Eckmann *et al.*, 1993; Savkovic *et al.*, 1997). However, epithelial TLR-4 appears to be upregulated in inflamed mucosa consistent with observations that TLR-4 and its associated signaling apparatus are upregulated by IFNγ in cell culture (Abreu *et al.*, 2001, 2002). As several PRRs are also known to be induced by IFN, this mechanism also appears to have fairly broad use. As IFN expression is elevated in many states when there is known to be a risk of opportunistic infection, it seems reasonable to speculate that IFN-inducible epithelial PRR expression likely helps to reduce the risk of opportunistic infection. Thus, expressing some PRRs only under "high-alert" status may be one way of guarding against inappropriate epithelial expression of immunoactivating molecules.

Function to match environment

Although it is a relatively widely observed phenomenon that epithelial cells in general, and gut epithelial cells in particular, are not highly

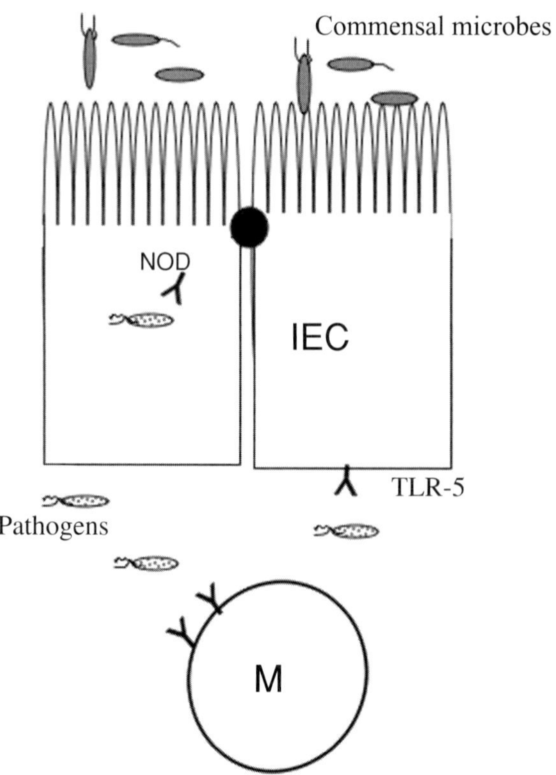

Figure 4.1 Pattern-recognition receptor (PRR) location can reduce the potential for activation of intestinal inflammation in response to lumenal commensal microbes. Both Toll-like receptors (TLRs) and nucleotide-binding oligomerization domain (NODs) proteins appear to play a key role in host defense. A means to prevent their activation by commensal bacteria is by placing those PRRs that induce pro-inflammatory signals in locations that should allow them to distinguish between lumenal and invasive pathogens. IEC, intestinal epithelial cell.

responsive by classic criteria (i.e. secretion of pro-inflammatory cytokines) to some bacterial components such as LPS and peptidoglycan, the reasons for this hyporesponsiveness are not entirely clear. Expression of TLR-2 and TLR-4 can be described as relatively low in non-inflammatory conditions, the best-available quantitative indices of PRR expression (using reverse transcriptase polymerase chain reaction, RT-PCR) generally indicate similar levels of TLR-2, -4, and -5 (Zarember and Godowski, 2002). Since all intestinal epithelial cell lines seem exquisitely responsive to flagellin but are relatively hyporesponsive to TLR-2 and TLR-4 ligands, simple regulation of TLR expression may be insufficient to explain relative responsiveness to PRR ligands. Indeed, for LPS at least one of the other critical components of TLR-4 signaling, coreceptor MD-2, is also highly regulated and made available only under pro-inflammatory conditions, as mimicked in vitro by interferons (Abreu *et al.*, 2002). However, more recent work suggests that some gut epithelial TLRs, particularly TLR-2, may be present and functional but have effector functions that are different from those classically described for PRRs. Specifically, ligation of gut epithelial TLR-2 does not result in robust chemokine secretion but rather mediates the rapid alteration of intestinal epithelial tight junctions, resulting in an increased epithelial barrier function (Cario *et al.*, 2004). Such a change in barrier function can be imagined to increase protection from microbial toxins, many of which must breach the epithelium or at least its apical surface in order to exert their effects (e.g. anthrax toxin acts on basolateral but not apical membrane of IECs). There is also in vivo evidence of a role for TLR signaling, particularly TLR-4 signaling in maintaining mucosal wellbeing in the absence of any pathogens. Consistent with such a non-traditional role for PRRs having a level of tonic activation and playing a role in gut homeostasis, the basal expression of TLR-2 and TLR-4 has, in contrast to TLR-5, largely been observed to be on the apical surface of the epithelium (Cario *et al.*, 2002). Although the signaling mechanisms by which TLR-2 ligands modulate epithelial junctions have not been defined in detail, it is clear that some of the signaling pathways used are distinct from those used centrally by TLR-mediated pro-inflammatory cascades. For example, TLR-mediated pro-inflammatory gene expression relies heavily on activation of the transcription factor NF-κB, but TLR-2-mediated junctional alterations relies on activation of protein kinase C (PKC) (α and δ). Whether gut epithelial TLR-2 uses different signaling effector proteins to mediate such PKC activation has not been defined.

POLYMORPHISM IN PATTERN-RECOGNITION RECEPTORS: VULNERABILITY TO INFECTIONS/DISEASES

Knockout mice are the most actively used resource in the efforts to define the roles of individual PRRs, but an obvious limitation is that these findings in mice may not always hold true in humans, particularly in the many cases where murine models do not resemble closely the corresponding human disease. Thus, it can be most informative to examine the effects of human PRR polymorphisms in particular disease states. Mounting data suggest that the ability of certain individuals to respond properly to TLR ligands may be impaired by single-nucleotide polymorphisms (SNPs) within TLR genes, resulting in an altered susceptibility to or course of infectious or inflammatory disease. Several studies have focused on two cosegregating SNPs, Asp299→Gly and Thr399→Ile, within the gene encoding TLR-4; the former SNP is associated with significantly blunted responses to LPS compared the latter SNP. These SNPs are present in approximately 10% of white individuals and are correlated positively with several infectious diseases. Replacement of Gly with Asp decreases α-helical content in the extracellular domain of the TLR-4, attenuates receptor signaling, and diminishes the inflammatory response to Gram-negative pathogens, increases Gram-negative septic shock, increases the risk of premature birth, and decreases the risk of atherosclerosis (Kiechl *et al.*, 2002; Lorenz *et al.*, 2002). A TLR-2 mutation was observed in lepromatous leprosy patients (Bochud *et al.*, 2003; Kang and Chae, 2001) and the TLR-2 polymorphism Arg753Gln is associated with increased Staphylococcal infection in humans (Lorenz *et al.*, 2000).

A common stop codon polymorphism has been identified in the ligand-binding domain of TLR-5 (TLR-$5^{392STOP}$). This is unable to mediate flagellin signaling, acts in a dominant fashion, and is associated with susceptibility to legionnaires' disease, pneumonia caused by the flagellated bacterium *Legionella pneumophila* (Hawn *et al.*, 2003). Interestingly, TLR-4 polymorphic individuals have been found to be resistant to legionnaires' disease (Hawn *et al.*, 2005). This protective association shows that a PRR can mediate either beneficial or deleterious inflammatory responses and that these outcomes vary with different pathogens. Another frame-shift mutation that increases individual susceptibility to Crohn's disease has been linked to the NOD family, in particular NOD2 (Hugot *et al.*, 2001; Ogura *et al.*, 2001). In addition to Crohn's disease, NOD2 is also involved in the susceptibility to a granulomatous disorder, Blau syndrome (Miceli-Richard *et al.*, 2001; van Duist *et al.*, 2005). These human phenotypes indicate the importance of PRRs to mammalian host defense.

Evasion of pattern-recognition receptors

Given the seemingly extensive efforts that mammalian hosts have made at developing an extensive array of PRRs to recognize a broad variety of microbes, it is perhaps not surprising that some microbes seem to have evolved strategies to evade PRR signaling. One such mechanism may be simply not making molecules that are recognized by PRR if they are not absolutely essential. For example, *Shigella* spp. seem to have the genes necessary to make flagella but apparently do not express these genes when colonizing mammalian hosts (Giron, 1995). Whether these genes are ever expressed by *Shigella* at some point in its lifecycle or are simply "evolutionary left-overs" remains unclear. Food-borne listeria are flagellated but rapidly shut off flagellar expression at 37 °C, which would presumably correspond with their infection of mammalian hosts (Peel *et al.*, 1988). Although it is possible that one explanation for shutting off flagellar expression is due to their high metabolic cost, it seems likely that evading detection by TLR-5 also figures centrally into this "decision". Another means of avoiding TLR-5 detection may be by cloaking flagella in a sheath, as is done by some *Helicobacter* and *Vibrio* species.

Another approach to avoiding detection by PRRs may be to modify the structures that PRRs are targeting. Microbes are fairly adept at mutating their way around road blocks, but evasion of PRRs appears generally difficult, as the targeted motifs are necessary for function. For example, most alterations in flagellin that eliminate recognition by TLR-5 also ablate the ability of flagellin to polymerize into a functional flagella (Smith *et al.*, 2003). However, at least one pathogen, *H. pylori* (the causative agent of gastric ulcers and other gastrointestinal pathologies), expresses functional flagella but does not produce a flagellin with potent ability to activate TLR-5 (Gewirtz *et al.*, 2004). Interestingly, *H. pylori* also produces an LPS that is a relatively poor ligand for TLR-4 (Perez-Perez *et al.*, 1995) and, thus, this microbe's evasion of PRRs in general may be an important factor in allowing it to persist in the gastrointestinal tract for the life of the host. *Salmonella* species use multiple variants of the strategies described above. For example, some results suggest that *Salmonella* may shut off flagellar expression once the intestinal mucosa has been breached, thus seeming to avoid the potent innate and adaptive immune responses directed towards flagellin (McSorley *et al.*, 2002). Further, *Salmonella* have a variety of complex ways of modifying their LPS structures so as to minimize detection by TLR-4 (Miller *et al.*, 2005). In addition, some *Salmonella* species have another way of reducing detection by PRRs, namely inducing downregulation of TLR expression in their hosts (Salzman, Chou *et al.*, 2003). Interestingly, non-inflammatory *Salmonella* have evolved a means to attenuate their activation of NF-κB-mediated pro-inflammatory gene expression by interfering with a key aspect of NF-κB signaling (Neish *et al.*, 2000). Although the precise consequences and underlying mechanisms of these observations are not yet fully understood, it seems likely that further study of PRR–microbe interactions will reveal a complex relationship of induced changes in host and microbe.

CONCLUSIONS

Extensive efforts in the first few years of the twenty-first century have begun to rapidly identify many of the key molecules of both microbes and hosts that play a central role in mediating the complex biological interactions between mammals and the extensive microbiota that inhabit their world. There is clearly a great deal still to be understood about how these molecules interact with each other in the complex interplay between microbe and host, but the great amount that has been learned thus far allows us to imagine that these complex interactions will become more elucidated in the coming years. Such understanding ought to enable the development of novel strategies to treat and prevent infectious diseases and to prevent inappropriate inflammation.

REFERENCES

Abreu, M. T., Vora, P., Faure, E., *et al.* (2001). Decreased expression of Toll-like receptor-4 and MD-2 correlates with intestinal epithelial cell protection against dysregulated proinflammatory gene expression in response to bacterial lipopolysaccharide. *J. Immunol.* **167**, 1609–1616.

Abreu, M. T., Arnold, E. T., Thomas, L. S., *et al.* (2002). TLR4 and MD-2 expression are regulated by immune-mediated signals in human intestinal epithelial cells. *J. Biol. Chem.* **277**, 20 431–20 437.

Abreu, M. T., Thomas, L. S., Arnold, E. T., *et al.* (2003). TLR signaling at the intestinal-epithelial interface. *J. Endotoxin Res.* **9**, 322–330.

Akira, S. (2001). [Toll-like receptors and innate immune system.] *Tanpakushitsu Kakusan Koso* **46**, 562–566.

Akira, S., Takeda, K., and Kaisho, T. (2001). Toll-like receptors: critical proteins linking innate and acquired immunity. *Nat. Immunol.* **2**, 675–680.

Athman, R. and Philpott, D. (2004). Innate immunity via Toll-like receptors and Nod proteins. *Curr. Opin. Microbiol.* **7**, 25–32.

Ayabe, T., Satchell, D. P., Wilson, C. L., *et al.* (2000). Secretion of microbicidal alpha-defensins by intestinal Paneth cells in response to bacteria. *Nat. Immunol.* **1**, 113–118.

Barber, G. N. (2005). The dsRNA-dependent protein kinase, PKR and cell death. *Cell Death Differ* **12**, 563–570.

Beutler, B. (2004). Inferences, questions and possibilities in Toll-like receptor signalling. *Nature* **430**, 257–263.

Beutler, B., Hoebe, K., and Shamel, L. (2004). Forward genetic dissection of afferent immunity: the role of TIR adapter proteins in innate and adaptive immune responses. *C. R. Biol.* **327**, 571–580.

Bochud, P. Y., Hawn, T. R., and Aderem, A. (2003). Cutting edge: a Toll-like receptor 2 polymorphism that is associated with lepromatous leprosy is unable to mediate mycobacterial signaling. *J. Immunol.* **170**, 3451–3454.

Bowie, A. G. and Haga, I. R. (2005). The role of Toll-like receptors in the host response to viruses. *Mol. Immunol.* **42**, 859–867.

Cario, E., Brown, D., McKee, M., *et al.* (2002). Commensal-associated molecular patterns induce selective toll-like receptor-trafficking from apical membrane to cytoplasmic compartments in polarized intestinal epithelium. *Am. J. Pathol.* **160**, 165–173.

Cario, E., Gerken, G., and Podolsky, D. K. (2004). Toll-like receptor 2 enhances ZO-1-associated intestinal epithelial barrier integrity via protein kinase C. *Gastroenterology* **127**, 224–238.

Colgan, S. P., Parkos, C. A., Delp, C., Arnaout, M. A., and Madara, J. L. (1993). Neutrophil migration across cultured intestinal epithelial monolayers is modulated by epithelial exposure to IFN-gamma in a highly polarized fashion. *J. Cell Biol.* **120**, 785–798.

Cromwell, O., Hamid, Q., Corrigan, C. J., *et al.* (1992). Expression and generation of interleukin-8, IL-6 and granulocyte–macrophage colony-stimulating factor by bronchial epithelial cells and enhancement by IL-1 beta and tumour necrosis factor-alpha. *Immunology* **77**, 330–337.

Cruickshank, S. M., Southgate, J., Selby, P. J., and Trejdosiewicz, L. K. (1998). Expression and cytokine regulation of immune recognition elements by normal human biliary epithelial and established liver cell lines in vitro. *J. Hepatol.* **29**, 550–558.

Damiano, J. S., Newman, R. M., and Reed, J. C. (2004). Multiple roles of CLAN (caspase-associated recruitment domain, leucine-rich repeat, and NAIP CIIA HET-E, and TP1-containing protein) in the mammalian innate immune response. *J. Immunol.* **173**, 6338–6345.

Diebold, S. S., Kaisho, T., Hemmi, H., Akira, S., and Reis e Sousa, C. (2004). Innate antiviral responses by means of TLR7-mediated recognition of single-stranded RNA. *Science* **303**, 1529–1531.

Dwinell, M. B., Johanesen, P. A., and Smith, J. M. (2003). Immunobiology of epithelial chemokines in the intestinal mucosa. *Surgery* **133**, 601–607.

Eaves-Pyles, T., Szabo, C., and Salzman, A. L. (1999). Bacterial invasion is not required for activation of NF-kappaB in enterocytes. *Infect. Immun.* **67**, 800–804.

Eckmann, L., Kagnoff, M., and Fierer, J. (1993). Epithelial cells secrete the chemokine interleukin-8 in response to bacterial entry. *Infect. Immun.* **61**, 4569–4574.

Feng, C. G., Scanga, C. A., Collazo-Custodio, C. M., *et al.* (2003). Mice lacking myeloid differentiation factor 88 display profound defects in host resistance and immune responses to *Mycobacterium avium* infection not exhibited by Toll-like receptor 2 (TLR2)- and TLR4-deficient animals. *J. Immunol.* **171**, 4758–4764.

Fitzgerald, K. A., Rowe, D. C., and Golenbock, D. T. (2004). Endotoxin recognition and signal transduction by the TLR4/MD2-complex. *Microbes Infect.* **6**, 1361–1367.

Flo, T. H., Smith, K. D., Sato, S., *et al.* (2004). Lipocalin 2 mediates an innate immune response to bacterial infection by sequestrating iron. *Nature* **432**, 917–921.

Ganz, T. (2003). Defensins: antimicrobial peptides of innate immunity. *Nat. Rev. Immunol.* **3**, 710–720.

Gewirtz, A. T., Navas, T. A., Lyons, S., Godowski, P. J., and Madara, J. L. (2001). Cutting edge: bacterial flagellin activates basolaterally expressed tlr5 to induce epithelial proinflammatory gene expression. *J. Immunol.* **167**, 1882–1885.

Gewirtz, A. T., Simon, P. O., Jr, Schmitt, C. K., *et al.* (2001). *Salmonella typhimurium* translocates flagellin across intestinal epithelia, inducing a proinflammatory response. *J. Clin. Invest.* **107**, 99–109.

Gewirtz, A. T., Collier-Hyams, L. S., Young, A. N., *et al.* (2002). Lipoxin a(4) analogs attenuate induction of intestinal epithelial proinflammatory gene expression and reduce the severity of dextran sodium sulfate-induced colitis. *J. Immunol.* **168**, 5260–5267.

Gewirtz, A. T., Yu, Y., Krishna, U. S., *et al.* (2004). *Helicobacter pylori* flagellin evades toll-like receptor 5-mediated innate immunity. *J. Infect. Dis.* **189**, 1914–1920.

Girardin, S. E., Tournebize, R., Mavris, M., *et al.* (2001). CARD4/Nod1 mediates NF-kappaB and JNK activation by invasive *Shigella flexneri*. *EMBO Rep.* **2**, 736–742.

Girardin, S. E., Boneca, I. G., Carneiro, L. A., *et al.* (2003). Nod1 detects a unique muropeptide from gram-negative bacterial peptidoglycan. *Science* **300**, 1584–1587.

Girardin, S. E., Boneca, I. G., Viala, J., *et al.* (2003). Nod2 is a general sensor of peptidoglycan through muramyl dipeptide (MDP) detection. *J. Biol. Chem.* **278**, 8869–8872.

Giron, J. A. (1995). Expression of flagella and motility by Shigella. *Mol. Microbiol.* **18**, 63–75.

Goh, K. C., deVeer, M. J., and Williams, B. R. (2000). The protein kinase PKR is required for p38 MAPK activation and the innate immune response to bacterial endotoxin. *EMBO J.* **19**, 4292–4297.

Grandjean, V., Vincent, S., Martin, L., Rassoulzadegan, M., and Cuzin, F. (1997). Antimicrobial protection of the mouse testis: synthesis of defensins of the cryptdin family. *Biol. Reprod.* **57**, 1115–1122.

Guillot, L., Le Goffic, R., Bloch, S., *et al.* (2005). Involvement of toll-like receptor 3 in the immune response of lung epithelial cells to double-stranded RNA and influenza A virus. *J. Biol. Chem.* **280**, 5571–5580.

Hardt, W.-D., Chen, L.-M., Schuebel, K. E., Bustelo, X. R., and Galan, J. E. (1998). *S. typhimurium* encodes an activator of rho GTPases that induces membrane ruffling and nuclear responses in host cells. *Cell* **93**, 815–826.

Hawn, T. R., Verbon, A., Lettinga, K. D., *et al.* (2003). A common dominant TLR5 stop codon polymorphism abolishes flagellin signaling and is associated with susceptibility to legionnaires' disease. *J. Exp. Med.* **198**, 1563–1572.

Hawn, T. R., Verbon, A., Janer, M., *et al.* (2005). Toll-like receptor 4 polymorphisms are associated with resistance to Legionnaires' disease. *Proc. Natl. Acad. Sci. U.S.A.* **102**, 2487–2489.

Hayashi, F., Smith, K. D., Ozinsky, A., *et al.* (2001). The innate immune response to bacterial flagellin is mediated by Toll- like receptor 5. *Nature* **410**, 1099–1103.

Haziot, A., Chen, S., Ferrero, E., *et al.* (1988). The monocyte differentiation antigen, CD14, is anchored to the cell membrane by a phosphatidylinositol linkage. *J. Immunol.* **141**, 547–552.

Heggelund, L., Flo, T., Berg, K., *et al.* (2004). Soluble toll-like receptor 2 in HIV infection: association with disease progression. *Aids* **18**, 2437–2439.

Heil, F., Hemmi, H., Hochrein, H., *et al.* (2004). Species-specific recognition of single-stranded RNA via toll-like receptor 7 and 8. *Science* **303**, 1526–1529.

Heim, M. H. (2005). RIG-I: an essential regulator of virus-induced interferon production. *J. Hepatol.* **42**, 431–433.

Herre, J., Willment, J. A., Gordon, S., and Brown, G. D. (2004). The role of Dectin-1 in antifungal immunity. *Crit. Rev. Immunol.* **24**, 193–203.

Hugot, J. P., Chamaillard, M., Zouali, H., *et al.* (2001). Association of NOD2 leucine-rich repeat variants with susceptibility to Crohn's disease. *Nature* **411**, 599–603.

Hyakushima, N., Mitsuzawa, H., Nishitani, C., *et al.* (2004). Interaction of soluble form of recombinant extracellular TLR4 domain with MD-2 enables lipopolysaccharide binding and attenuates TLR4-mediated signaling. *J. Immunol.* **173**, 6949–6954.

Inohara, N. and Nunez, G. (2003). NODs: intracellular proteins involved in inflammation and apoptosis. *Nat. Rev. Immunol.* **3**, 371–382.

Inohara, N., Ogura, Y., Fontalba, A., *et al.* (2003). Host recognition of bacterial muramyl dipeptide mediated through NOD2: implications for Crohn's disease. *J. Biol. Chem.* **278**, 5509–5512.

Inohara, N., Chamaillard, M., McDonald, C., and Nunez, G. (2005). NOD-LRR proteins: role in host–microbial interactions and inflammatory disease. *Annu. Rev. Biochem.* **74**, 355–383.

Iwaki, D., Mitsuzawa, H., Murakami, S., *et al.* (2002). The extracellular toll-like receptor 2 domain directly binds peptidoglycan derived from *Staphylococcus aureus*. *J. Biol. Chem.* **277**, 24 315–24 320.

Iwami, K. I., Matsuguchi, T., Masuda, A., *et al.* (2000). Cutting edge: naturally occurring soluble form of mouse Toll-like receptor 4 inhibits lipopolysaccharide signaling. *J. Immunol.* **165**, 6682–6686.

Jung, H. C., Eckmann, L., Yang, S.-K., *et al.* (1995). A distinct array of proinflammatory cytokines is expressed in human colon epithelial cells in response to bacterial invasion. *J. Clin. Invest.* **95**, 55–65.

Jurk, M., Heil, F., Vollmer, J., *et al.* (2002). Human TLR7 or TLR8 independently confer responsiveness to the antiviral compound R-848. *Nat. Immunol.* **3**, 499.

Kang, T. J. and Chae, G. T. (2001). Detection of Toll-like receptor 2 (TLR2) mutation in the lepromatous leprosy patients. *FEMS Immunol. Med. Microbiol.* **31**, 53–58.

Keates, S., Keates, A. C., Mizoguchi, E., Bhan, A., and Kelly, C. P. (1997). Enterocytes are the primary source of the chemokine ENA-78 in normal colon and ulcerative colitis. *Am. J. Physiol.* **273**, G75–82.

Kiechl, S., Lorenz, E., Reindl, M., *et al.* (2002). Toll-like receptor 4 polymorphisms and atherogenesis. *N. Engl. J. Med.* **347**, 185–192.

Kobe, B. and Kajava, A. V. (2001). The leucine-rich repeat as a protein recognition motif. *Curr. Opin. Struct. Biol.* **11**, 725–732.

Kopp, E. B. and Medzhitov, R. (1999). The Toll-receptor family and control of innate immunity. *Curr. Opin. Immunol.* **11**, 13–18.

Kumar, A., Haque, J., Lacoste, J., Hiscott, J., and Williams, B. R. (1994). Double-stranded RNA-dependent protein kinase activates transcription factor NF-kappa B by phosphorylating I kappa B. *Proc. Natl. Acad. Sci. U. S. A.* **91**, 6288–6292.

Kumar, M. V., Nagineni, C. N., Chin, M. S., Hooks, J. J., and Detrick, B. (2004). Innate immunity in the retina: Toll-like receptor (TLR) signaling in human retinal pigment epithelial cells. *J. Neuroimmunol.* **153**, 7–15.

Latz, E., Visintin, A., Lien, E., *et al.* (2002). Lipopolysaccharide rapidly traffics to and from the Golgi apparatus with the toll-like receptor 4-MD-2-CD14 complex in a process that is distinct from the initiation of signal transduction. *J. Biol. Chem.* **277**, 47 834–47 843.

Le, Y., Oppenheim, J. J., and Wang, J. M. (2001). Pleiotropic roles of formyl peptide receptors. *Cytokine Growth Factor Rev.* **12**, 91–105.

LeBouder, E., Rey-Nores, J. E., Rushmere, N. K., *et al.* (2003). Soluble forms of Toll-like receptor (TLR)2 capable of modulating TLR2 signaling are present in human plasma and breast milk. *J. Immunol.* **171**, 6680–6689.

Lee, J., Chuang, T. H., Redecke, V., *et al.* (2003). Molecular basis for the immunostimulatory activity of guanine nucleoside analogs: activation of Toll-like receptor 7. *Proc. Natl. Acad. Sci. U. S. A.* **100**, 6646–6651.

Lemaitre, B., Nicolas, E., Michaut, L., Reichhart, J. M., and Hoffmann, J. A. (1996). The dorsoventral regulatory gene cassette spatzle/Toll/cactus controls the potent antifungal response in Drosophila adults. *Cell* **86**, 973–983.

Lin, Y., Zhang, M., and Barnes, P. F. (1998). Chemokine production by a human alveolar epithelial cell line in response to *Mycobacterium tuberculosis. Infect. Immun.* **66**, 1121–1126.

Liu, Y., Buhring, H. J., Zen, K., *et al.* (2002). Signal regulatory protein (SIRPalpha), a cellular ligand for CD47, regulates neutrophil transmigration. *J. Biol. Chem.* **277**, 10 028–10 036.

Liu, L., Roberts, A. A., and Ganz, T. (2003). By IL-1 signaling, monocyte-derived cells dramatically enhance the epidermal antimicrobial response to lipopolysaccharide. *J. Immunol.* **170**, 575–580.

Lorenz, E., Mira, J. P., Cornish, K. L., Arbour, N. C., and Schwartz, D. A. (2000). A novel polymorphism in the toll-like receptor 2 gene and its potential association with staphylococcal infection. *Infect. Immun.* **68**, 6398–6401.

Lorenz, E., Hallman, M., Marttila, R., Haataja, R., and Schwartz, D. A. (2002). Association between the Asp299Gly polymorphisms in the Toll-like receptor 4 and premature births in the Finnish population. *Pediatr. Res.* **52**, 373–376.

Matsumoto, M., Funami, K., Tanabe, M., *et al.* (2003). Subcellular localization of Toll-like receptor 3 in human dendritic cells. *J. Immunol.* **171**, 3154–3162.

Mazzucchelli, L., Hauser, C., Zgraggen, K., *et al.* (1994). Expression of interleukin-8 gene in inflammatory bowel disease is related to the histological grade of active inflammation. *Am. J. Pathol.* **144**, 997–1007.

McCormick, B. A., Colgan, S. P., Archer, C. D., Miller, S. I., and Madara, J. L. (1993). *Salmonella typhimurium* attachment to human intestinal epithelial monolayers: transcellular signalling to subepithelial neutrophils. *J. Cell. Biol.* **123**, 895–907.

McSorley, S. J., Asch, S., Costalonga, M., Reinhardt, R. L., and Jenkins, M. K. (2002). Tracking salmonella-specific CD4 T cells in vivo reveals a local mucosal response to a disseminated infection. *Immunity* **16**, 365–377.

Miceli-Richard, C., Lesage, S., Rybojad, M., *et al.* (2001). CARD15 mutations in Blau syndrome. *Nat. Genet.* **29**, 19–20.

Miller, S. I., Ernst, R. K., and Bader, M. W. (2005). LPS, TLR4 and infectious disease diversity. *Nat. Rev. Microbiol.* **3**, 36–46.

Moore, K. J., Andersson, L. P., Ingalls, R. R., *et al.* (2000). Divergent response to LPS and bacteria in CD14-deficient murine macrophages. *J. Immunol.* **165**, 4272–4280.

Naik, S., Kelly, E. J., Meijer, L., Pettersson, S., and Sanderson, I. R. (2001). Absence of Toll-like receptor 4 explains endotoxin hyporesponsiveness in human intestinal epithelium. *J. Pediatr. Gastroenterol. Nutr.* **32**, 449–453.

Neish, A. S., Gewirtz, A. T., Zeng, H., *et al.* (2000). Prokaryotic regulation of epithelial responses by inhibition of IkappaB-alpha ubiquitination. *Science* **289**, 1560–1563.

Netea, M. G., van Deuren, M., Kullberg, B. J., Cavaillon, J. M., and Van der Meer, J. W. (2002). Does the shape of lipid A determine the interaction of LPS with Toll-like receptors? *Trends Immunol.* **23**, 135–139.

Nishiya, T. and DeFranco, A. L. (2004). Ligand-regulated chimeric receptor approach reveals distinctive subcellular localization and signaling properties of the Toll-like receptors. *J. Biol. Chem.* **279**, 19 008–19 017.

Ogura, Y., Bonen, D. K., Inohara, N., *et al.* (2001). A frameshift mutation in NOD2 associated with susceptibility to Crohn's disease. *Nature* **411**, 603–606.

Ogushi, K., Wada, A., Niidome, T., *et al.* (2004). Gangliosides act as co-receptors for *Salmonella enteritidis* FliC and promote FliC induction of human beta-defensin-2 expression in Caco-2 cells. *J. Biol. Chem.* **279**, 12 213–12 219.

O'Neil, D. A., Cole, S. P., Martin-Porter, E., *et al.* (2000). Regulation of human beta-defensins by gastric epithelial cells in response to infection with *Helicobacter pylori* or stimulation with interleukin-1. *Infect. Immun.* **68**, 5412–5415.

O'Neill, L. A., Fitzgerald, K. A., and Bowie, A. G. (2003). The Toll-IL-1 receptor adaptor family grows to five members. *Trends Immunol.* **24**, 286–290.

Park, H. S., Jung, H. Y., Park, E. Y., *et al.* (2004). Cutting edge: direct interaction of TLR4 with NAD(P)H oxidase 4 isozyme is essential for lipopolysaccharide-induced production of reactive oxygen species and activation of NF-kappa B. *J. Immunol.* **173**, 3589–3593.

Parkos, C. A., Colgan, S. P., Liang, T. W., *et al.* (1996). CD47 mediates post-adhesive events required for neutrophil migration across polarized intestinal epithelia. *J. Cell Biol.* **132**, 437–450.

Peel, M., Donachie, W., and Shaw, A. (1988). Temperature-dependent expression of flagella of *Listeria monocytogenes* studied by electron microscopy, SDS-PAGE and western blotting. *J. Gen. Microbiol.* **134**, 2171–2178.

Perez-Perez, G. I., Shepherd, V. L., Morrow, J. D., and Blaser, M. J. (1995). Activation of human THP-1 cells and rat bone marrow-derived macrophages by *Helicobacter pylori* lipopolysaccharide. *Infect. Immun.* **63**, 1183–1187.

Platz, J., Beisswenger, C., Dalpke, A., *et al.* (2004). Microbial DNA induces a host defense reaction of human respiratory epithelial cells. *J. Immunol.* **173**, 1219–1223.

Poltorak, A., He, X., Smirnova, I., *et al.* (1998). Defective LPS signaling in C3H/HeJ and C57BL/10ScCr mice: mutations in Tlr4 gene. *Science* **282**, 2085–2088.

Pugin, J., Heumann, I. D., Tomasz, A., *et al.* (1994). CD14 is a pattern recognition receptor. *Immunity* **1**, 509–516.

Reiling, N., Holscher, C., Fehrenbach, A., *et al.* (2002). Cutting edge: Toll-like receptor (TLR)2- and TLR4-mediated pathogen recognition in resistance to airborne infection with *Mycobacterium tuberculosis. J. Immunol.* **169**, 3480–3484.

Salzman, N. H., Chou, M. M., de Jong, H., *et al.* (2003). Enteric salmonella infection inhibits Paneth cell antimicrobial peptide expression. *Infect. Immun.* **71**, 1109–1115.

Salzman, N. H., Ghosh, D., Huttner, K. M., Paterson, Y., and Bevins, C. L. (2003). Protection against enteric salmonellosis in transgenic mice expressing a human intestinal defensin. *Nature* **422**, 522–526.

Savkovic, S. D., Koutsouris, A., and Hecht, G. (1997). Activation of NF-kappaB in intestinal epithelial cells by enteropathogenic *Escherichia coli. Am. J. Physiol.* **273**, C1160–1167.

Scanga, C. A., Bafica, A., Feng, C. G., *et al.* (2004). MyD88-deficient mice display a profound loss in resistance to *Mycobacterium tuberculosis* associated with partially impaired Th1 cytokine and nitric oxide synthase 2 expression. *Infect. Immun.* **72**, 2400–2404.

Schnare, M., Barton, G. M., Holt, A. C., *et al.* (2001). Toll-like receptors control activation of adaptive immune responses. *Nat. Immunol.* **2**, 947–950.

Shaw, S. K., Cepek, K. L., Murphy, E. A., *et al.* (1994). Molecular cloning of the human mucosal lymphocyte integrin alpha E subunit: unusual structure and restricted RNA distribution. *J. Biol. Chem.* **269**, 6016–6025.

Shaw, S. K., Hermanowski-Vosatka, A., Shibahara, T., *et al.* (1998). Migration of intestinal intraepithelial lymphocytes into a polarized epithelial monolayer. *Am. J. Physiol.* **275**, G584–591.

Shibahara, T., Wilcox, J. N., Couse, T., and Madara, J. L. (2001). Characterization of epithelial chemoattractants for human intestinal intraepithelial lymphocytes. *Gastroenterology* **120**, 60–70.

Shimazu, R., Akashi, S., Ogata, H., *et al.* (1999). MD-2, a molecule that confers lipopolysaccharide responsiveness on Toll-like receptor 4. *J. Exp. Med.* **189**, 1777–1782.

Sierro, F., Dubois, B., Coste, A., *et al.* (2001). Flagellin stimulation of intestinal epithelial cells triggers CCL20-mediated migration of dendritic cells. *Proc. Natl. Acad. Sci. U. S. A.* **98**, 13 722–13 727.

Sitaraman, S. V., Merlin, D., Wang, L., *et al.* (2001). Neutrophil-epithelial crosstalk at the intestinal lumenal surface mediated by reciprocal secretion of adenosine and IL-6. *J. Clin. Invest.* **107**, 861–869.

Smith, K. D., Andersen-Nissen, E., Hayashi, F., *et al.* (2003). Toll-like receptor 5 recognizes a conserved site on flagellin required for protofilament formation and bacterial motility. *Nat. Immunol.* **4**, 1247–1253.

Tabeta, K., Georgel, P., Janssen, E., *et al.* (2004). Toll-like receptors 9 and 3 as essential components of innate immune defense against mouse cytomegalovirus infection. *Proc. Natl. Acad. Sci. U. S. A.* **101**, 3516–3521.

Takahashi, A., Wada, A., Ogushi, K., *et al.* (2001). Production of beta-defensin-2 by human colonic epithelial cells induced by *Salmonella enteritidis* flagella filament structural protein. *FEBS Lett.* **508**, 484–488.

Takeda, K. and Akira, S. (2005). Toll-like receptors in innate immunity. *Int. Immunol.* **17**, 1–14.

Takeuchi, O., Kawai, T., Sanjo, H., *et al.* (1999). TLR6: a novel member of an expanding toll-like receptor family. *Gene* **231**, 59–65.

Takeuchi, O., Sato, S., Horiuchi, T., *et al.* (2002). Cutting edge: role of Toll-like receptor 1 in mediating immune response to microbial lipoproteins. *J. Immunol.* **169**, 10–14.

Tallant, T., Deb, A., Kar, N., *et al.* (2004). Flagellin acting via TLR5 is the major activator of key signaling pathways leading to NF-kappa B and proinflammatory gene program activation in intestinal epithelial cells. *BMC Microbiol.* **4**, 33.

Taylor, P. R., Gordon, S., and Martinez-Pomares, L. (2005). The mannose receptor: linking homeostasis and immunity through sugar recognition. *Trends Immunol.* **26**, 104–110.

Tohme, Z. N., Amar, S., and Van Dyke, T. E. (1999). Moesin functions as a lipopolysaccharide receptor on human monocytes. *Infect. Immun.* **67**, 3215–3220.

Toshchakov, V., Jones, B. W., Perera, P. Y., *et al.* (2002). TLR4, but not TLR2, mediates IFN-beta-induced STAT1alpha/beta-dependent gene expression in macrophages. *Nat. Immunol.* **3**, 392–398.

Tschopp, J., Martinon, F., and Burns, K. (2003). NALPs: a novel protein family involved in inflammation. *Nat. Rev. Mol. Cell. Biol.* **4**, 95–104.

Uetani, K., Der, S. D., Zamanian-Daryoush, M., *et al.* (2000). Central role of double-stranded RNA-activated protein kinase in microbial induction of nitric oxide synthase. *J. Immunol.* **165**, 988–996.

Ulevitch, R. J. and Tobias, P. S. (1995). Receptor-dependent mechanisms of cell stimulation by bacterial endotoxin. *Annu. Rev. Immunol.* **13**, 437–457.

Van Duist, M. M., Albrecht, M., Podswiadek, M., *et al.* (2005). A new CARD15 mutation in Blau syndrome. *Eur. J. Hum. Genet.* **13**, 742–747.

Vijay-Kumar, M., Gentsch, J. R., Kaiser, W. J., *et al.* (2005). Protein kinase R mediates intestinal epithelial gene remodeling in response to double stranded RNA and live rotavirus. *J. Immunol.* **174**, 6322–6331.

Villamon, E., Gozalbo, D., Roig, P., *et al.* (2004). Myeloid differentiation factor 88 (MyD88) is required for murine resistance to *Candida albicans* and is critically involved in Candida-induced production of cytokines. *Eur. Cytokine Netw.* **15**, 263–271.

Vora, P., Youdim, A., Thomas, L. S., *et al.* (2004). Beta-defensin-2 expression is regulated by TLR signaling in intestinal epithelial cells. *J. Immunol.* **173**, 5398–5405.

Wagner, H. (2004). The immunobiology of the TLR9 subfamily. *Trends Immunol.* **25**, 381–386.

Watanabe, T., Kitani, A., Murray, P. J., and Strober, W. (2004). NOD2 is a negative regulator of Toll-like receptor 2-mediated T helper type 1 responses. *Nat. Immunol.* **5**, 800–808.

Weiss, D. S., Raupach, B., Takeda, K., Akira, S., and Zychlinsky, A. (2004). Toll-like receptors are temporally involved in host defense. *J. Immunol.* **172**, 4463–4469.

Williams, B. R. (2001). Signal integration via PKR. *Sci. STKE* **2001**, RE2.

Wurfel, M. M. and Wright, S. D. (1997). Lipopolysaccharide-binding protein and soluble CD14 transfer lipopolysaccharide to phospholipid bilayers: preferential interaction with particular classes of lipid. *J. Immunol.* **158**, 3925–3934.

Yamamoto, M., Sato, S., Mori, K., *et al.* (2002). Cutting edge: a novel Toll/IL-1 receptor domain-containing adapter that preferentially activates the IFN-beta promoter in the Toll-like receptor signaling. *J. Immunol.* **169**, 6668–6672.

Yarovinsky, F., Zhang, D., Andersen, J. F., *et al.* (2005). TLR11 activation of dendritic cells by a protozoan profilin-like Protein. *Science* **308**, 1626–1629.

Yoneyama, M., Kikuchi, M., Natsukawa, T., *et al.* (2004). The RNA helicase RIG-I has an essential function in double-stranded RNA-induced innate antiviral responses. *Nat. Immunol.* **5**, 730–737.

Zarember, K. A. and Godowski, P. J. (2002). Tissue expression of human Toll-like receptors and differential regulation of Toll-like receptor mRNAs in leukocytes in response to microbes, their products, and cytokines. *J. Immunol.* **168**, 554–561.

Zeng, H., Carlson, A. Q., Guo, Y., *et al.* (2003). Flagellin is the major proinflammatory determinant of enteropathogenic *Salmonella*. *J. Immunol.* **171**, 3668–3674.

Zhang, D., Zhang, G., Hayden, M. S., *et al.* (2004). A toll-like receptor that prevents infection by uropathogenic bacteria. *Science* **303**, 1522–1526.

Zuany-Amorim, C., Hastewell, J., and Walker, C. (2002). Toll-like receptors as potential therapeutic targets for multiple diseases. *Nat. Rev. Drug Discov.* **1**, 797–807.

CHAPTER 5

Roles of flagella in pathogenic bacteria and bacterial–host interactions

Glenn M. Young

INTRODUCTION

The quintessential event that initiates an infection is defined by the pathogen encountering a targeted tissue of a host. Mucosal surfaces often serve as an entry point for pathogens, and the pathogens frequently access epithelial cells at these sites with the aid of flagellar-mediated motility. Flagella are common to a diversity of pathogenic bacteria, and their structure is well conserved (Harshey and Toguchi, 1996). The number of flagella produced by a bacterium is tightly regulated, ranging from one up to several dozen per cell depending on the species (Aldridge and Hughes, 2002). Likewise, these organelles may be localized to a single pole or both poles of a cell or may be in a peritrichous arrangement. In some cases, flagellar number and cellular position undergo a regulated change in response to specific environmental conditions. Directional control of motility, either towards a favorable situation or away from a stressful circumstance, is achieved by a chemosensory phosphorelay system, which integrates environmental signals and adjusts the frequency of motor reversals that cause reorientation of bacterial movement (Sournik, 2004; Wadhams and Armitage, 2004). The cumulative affect invoked by the chemosensory system in response to chemical, physical, and physiological cues offers the bacterium a behavioral activity to fine tune its progression through an environment and consequently enhances the ability of many bacteria to infect their hosts.

The primary structure of a flagellum is well conserved among the eubacteria, consisting of the cell-envelope-bound motor and basal body connected to the hook and filament extending out from the cell surface. Flagella are

Bacterial–Epithelial Cell Cross-Talk: Molecular Mechanisms in Pathogenesis, ed. Beth A. McCormick. Published by Cambridge University Press.

complex left-handed helical shaped organelles that can extend up to 15 μM from the surface of the bacterium (Macnab, 2003). Its high degree of structural conservation has marked the flagellum, and in particular the filament subunit flagellin, as a target recognized by the innate immune system through the activity of Toll-like receptor 5 (TLR-5) present on the surface of host cells (Ramos *et al.*, 2004). The flagellum is a hollow structure and a conduit through which flagellar proteins pass during assembly of the organelle (Macnab, 2004). This secretion pathway is functionally and structurally homologous to bacterial contact-dependent type III secretion (T3S) systems used to target virulence factors directly into host cells (Aizawa, 2001; Hueck, 1998). It has been recognized that flagella are bona fide T3S systems that can affect the outcome of an infection by providing the bacterium with a mechanism to transport non-flagellar proteins to extracellular sites in order to influence host–pathogen interactions (Journet *et al.*, 2005; Macnab, 2004).

FLAGELLA CONTRIBUTE TO BACTERIAL PATHOGENESIS

The mucous layer over the epithelium forms an elaborate physical and chemical barrier for an invading pathogen. Other involuntary cleansing activities, including peristalsis in the intestine and the constant outward flow of mucus in the respiratory system, add additional hurdles pathogens must face when establishing themselves in the host. Flagella provide the bacterium with the means to transit through the mucosa, and the chemosensory system ensures progress to an appropriate site (Figure 5.1). Flagella can also enhance adhesion to and invasion of the epithelium and serve as a conduit through which virulence factors may be secreted into the extrabacterial milieu. Several bacterial species have been shown to require motility in order to establish or maintain colonization in their human or animal hosts. Much of this information was gained through numerous efforts to understand the infectious cycles of a variety of bacteria. The most compelling evidence emanates from studies that compared the infectious behavior of wild-type bacteria with that of isogenic non-motile mutants. The following sections will highlight the roles of flagella in the initial and persistent stages of infections using different bacteria as examples.

A ROLE FOR FLAGELLA DURING THE INITIAL STAGES OF HOST COLONIZATION

Motility appears to play a role during the initial phases of infection for many pathogens, including enteropathogenic *Escherischia coli*, *Salmonella*,

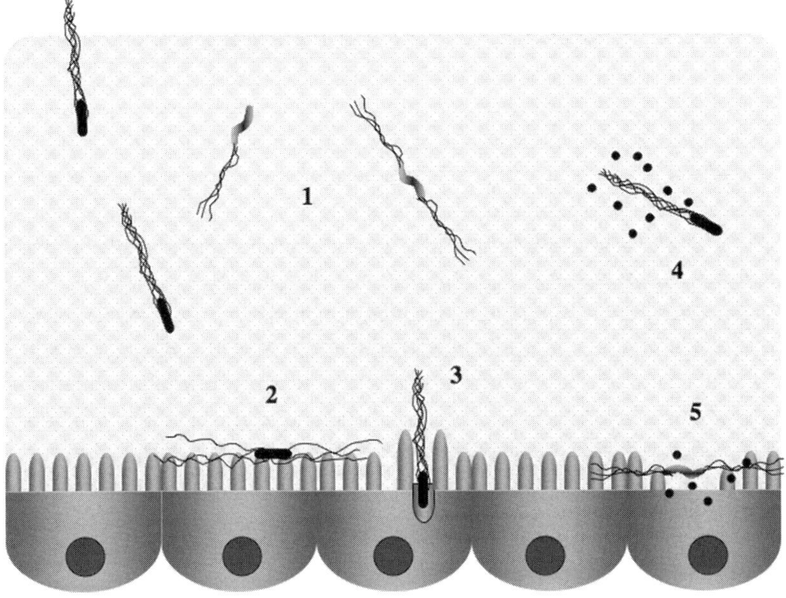

Figure 5.1 Functions of flagella that have been proposed to affect bacterial infections in mucosa. (1) Motility and chemotaxis provide many bacteria with the ability to penetrate mucus to access favorable sites (*Proteus mirabilis, Salmonella*, and many others) or to persist within mucus (*Helicobacter, Campylobacter*). (2) Flagella directly promote adherence to epithelial cells or mucus (enteropathogenic *Escherichia coli, Pseudomonas aeruginosa*). (3) Flagellar-mediated motility ensures bacterial-epithelial cell contact, which is needed to promote invasion (*Salmonella, Yersinia enterocolitica*). (4) Export of factors by the flagellar type III secretion (T3S) system enhances bacterial survival (*Y. enterocolitica*). (5) Specific targeting of virulence factors to the epithelium by the flagellar T3S system promotes bacterial adherence and invasion (*Campylobacter*).

and *Yersinia*, as well as the respiratory pathogens *Legionella pneumophila* and *Pseudomonas aeruginosa* (Deitrich *et al.*, 2001; Feldman *et al.*, 1998; Giron *et al.*, 2002; Marchetti *et al.*, 2004; Stecher *et al.*, 2004; Young *et al.*, 2000). Recent examinations of *Salmonella*, which has peritrichously arranged flagella, provides intriguing insight. A variety of *Salmonella* cause gastroenteritis in humans and, in the case of *Salmonella enterica* serovar Typhi, typhoid fever. Investigators most often utilize the inbred mouse model to study *Salmonella*–host interactions, a model that is best suited for studies of systemic typhoid-like illness and is not especially well suited to the investigation of gastroenteritis. Collectively, investigations with mice have not strongly supported a role for flagella as a virulence factor (Lockman and Curtis, 1990; Schmitt *et al.*, 2001). However, there is strong evidence for flagella as a virulence factor in other animals that are better suited to investigation of *Salmonella* infection of the intestinal mucosa. Colonization of the rabbit appendix by *S. enterica* serovar Typhimurium depends on flagella (Marchetti *et al.*, 2004). A comparison of wild-type *S.* Typhimurium with an isogenic flagellin mutant revealed that the non-motile bacterium was not able to adhere to M-cells, which serve as the predominant portal for invasion. Genetic complementation of the mutation restored motility to the mutant strain and restored the ability of the bacteria to adhere to the underlying M-cells. However, when similar experiments were conducted with appendix explants, flagella did not affect the ability of *S.* Typhimurium to adhere to M-cells. One consequence of these experimental conditions is that appendix explants lack the normal mucous layer that covers the surface of the appendix in vivo. This reveals the importance of flagella in allowing bacteria to transit across an intact mucosal barrier and highlights the challenges of establishing experimental models to study how flagella influence pathogen interactions with hosts. From this analysis, we can conclude for *Salmonella* that flagella do not mediate attachment per se to M-cells but likely promote efficient migration of the bacterium to the cell surface.

Although M-cells do actively take up and transport bacteria to the underlying tissue, it is important to consider the fact that *S.* Typhimurium maintains the ability to actively promote cellular invasion of epithelial cells by a mechanism that involves the *Salmonella* pathogenicity island 1 (SPI-1) T3S system (Suarez and Russmann, 1998). Furthermore, expression of flagella and the SPI-1 T3S system is coregulated by the product of the flagellar gene *fliZ* (Lucas *et al.*, 2000). This regulatory connection between flagellar-mediated motility and the SPI-1 invasion system points out the remarkable coordination of pathogenic activities by *Salmonella*. Although fewer comprehensive studies have been completed, studies of bacterial invasion of cultured epithelial cells similarly point to a requirement for flagella to initiate cell contact in other bacteria, including the respiratory pathogen *L. pneumophila* and the enteropathogen *Yersinia enterocolitica* (Deitrich *et al.*, 2001; Young *et al.*, 2000).

Colonization of the respiratory system by the opportunistic human pathogen *P. aeruginosa*, possessing a single polar flagellum, also points to other roles for flagella during initial bacterial–host interaction. *P. aeruginosa* mutants lacking the ability to produce flagellin, the filament subunit, are known to display a reduced ability to cause mortality in the newborn mouse model of acute pneumonia (Feldman *et al.*, 1998). More specifically, however, flagella provide a means for *P. aeruginosa* attachment to human respiratory mucin, which may limit the host's ability to clear the pathogen (Ramphal and

Arora, 2001). Disruption of *fliD*, the gene encoding the flagellar cap, reduced adhesion, suggesting that this protein localized to the tip of the flagellum acts as a specific adhesin (Arora *et al.*, 1998). Studies aimed at defining the carbohydrate moieties to which FliD binds indicate that FliD from strain PAO1, but not from strain PAK, has an affinity for glycoconjugates bearing Lewis^x or sialyl-Lewis^x structures (Scharfman *et al.*, 2001). It appears, therefore, that there is some variability in the affinity of FliD from different *P. aeruginosa* isolates. Consistent with the importance of flagella having a role only during initial colonization of the respiratory system, clinical isolates of *P. aeruginosa* from chronically colonized people with cystic fibrosis have a propensity to be non-motile (Mahenthiralinggam *et al.*, 1994). *P. aeruginosa* also causes acute opportunistic infections of the eye and wreaks havoc on people suffering from burn-induced wounds. Animal and cellular models of infection implicate flagella as virulence factors in these situations (Arora *et al.*, 2005; Drake and Monte, 1988; Fleizig *et al.*, 2001). Roles for flagella in adherence to mucous or epithelial cells have been proposed for other distantly related bacteria. Crude preparations of flagella from the Gram-positive opportunistic enteropathogen *Clostridium difficile* bind to the cecal mucus of germ-free mice, which correlates with adherence of this bacterium to cecal tissue (Tasteyre *et al.*, 2001). For some pathogenic strains of *E. coli*, flagella appear to promote efficient attachment to intestinal mucosa and to cultured epithelial cells (Giron *et al.*, 2002). This highlights the possibility that flagellar-mediated binding to mucosal components is a trait of various bacteria that may have been independently selected on numerous occasions.

THE CONTRIBUTION OF FLAGELLA TO PERSISTENT COLONIZATION OF HOSTS

It is quite clear for some bacteria that flagella play a dominant role in persistent long-term colonization of hosts. Two intriguing examples are *Campylobacter jejuni*, which is one of the greatest incidental causes of bacterial diarrhea, and *Helicobacter pylori*, which has a high worldwide infection rate among humans and is associated with gastritis, peptic ulcers, and gastric cancer. For each of these organisms, flagella are needed for the bacterium to reach a favorable host niche within the gastrointestinal tract and to avoid removal by the flow of intestinal contents. In part, each of these species survives by establishing quarters within the mucosal layer covering the epithelial surface.

Campylobacter spp. are curved or spiral-shaped cells with a single polar flagellum at one or both ends. *C. jejuni*, although a causative agent for diarrhea in humans, is a commensal intestinal resident of the avian intestinal tract, where its population can reach 10^9 cfu per 1/g of feces. They are highly adapted for motility in viscous solutions that restrict the movement of other types of bacteria, such as *E. coli* and *Salmonella* (Ferrero and Lee, 1988; Szymanski *et al.*, 1995). Using specific pathogen-free mice, the bacterium was observed to preferentially colonize the cecal portion of the large intestine (Lee *et al.*, 1986). In particular, observations of intestinal scrapings by phase-contrast microscopy revealed that the mucus-filled crypts were heavily colonized by *C. jejuni* and that the bacteria were actively motile. More direct evidence for a role for the flagellum came from examination of a library of motility mutants derived by chemical mutagenesis, which demonstrated that strains lacking the capacity to produce a flagellum, producing a paralyzed flagellum, or affected for chemotaxis were impaired in their ability to colonize the intestine of suckling mice (Morooka *et al.*, 1985). Additional support for the idea that the flagellum is necessary for host colonization comes from studies with non-motile phase-variant strains. *C. jejuni* undergoes a low-frequency reversible shift from motile to non-motile phases (Caldwell *et al.*, 1985). The frequency from the non-motile to motile phase is approximately 1×10^6 per generation. However, there is a near-complete shift to the motile phase among bacteria recovered following an infection. This shift is seen when a pure culture of a non-motile variant of *C. jejuni* or with mixtures of motile and non-motile strains of bacteria are used to infect rabbits, hamsters, and humans volunteers (Aguero-Rosenfeld *et al.*, 1990; Black *et al.*, 1988; Caldwell *et al.*, 1985). At a minimum, we can conclude that conditions in vivo strongly select for the motile phenotype. The advent of techniques for the construction of defined mutations in *C. jejuni* has led to a more direct assessment of the role for flagella during an infection. *C. jejuni* has two genes, *flaA* and *flaB*, that encode the flagellin subunits of the filament. FlaA is the predominant protein subunit of the flagellum, and FlaB appears to be synthesized in response to specific specialized growth conditions. Two independent studies confirmed that FlaA mutants of *C. jejuni* are defective for colonization of 3-day-old chicks (Nachamkim *et al.*, 1993; Wassenaar *et al.*, 1993). Consistent with these results, a study showed that flagellar and chemotaxis mutants formed the predominant class of chick colonization-defective mutants identified by signature-tagged mutagenesis (Hendrixson and DiRita, 2004). Follow-up analysis demonstrated that these strains were specifically affected for competitive colonization of the chick intestine. Colonization of the intestine by *C. jejuni* undoubtedly requires multiple factors, but flagella clearly play a primary role in both commensal and pathogenic interactions.

H. pylori is strikingly well adapted for survival in the extreme environment of the stomach. *H. pylori* are spiral-shaped bacteria with multiple membrane-sheathed flagella clustered at a single pole that, much like *C. jejuni*, retains motility in highly viscous solutions (Worku *et al.*, 1999). A primary role for motility in gastric colonization was proposed soon after its discovery, when examination of biopsy specimens revealed that a small population of *H. pylori* adhere to the epithelium but the vast majority of bacteria display active motility and reside within the gastric mucus (Hazell *et al.*, 1986). One of the first experimental tests of the role played by flagella in host colonization was completed using spontaneously non-motile phase variants and, at that time, the newly developed gnotobiotic piglet model (Eaton *et al.*, 1992). Motile *H. pylori* colonized gnotobiotic piglets at a frequency of 90%, compared with only 25% for the non-motile variants. The motile form survived for 21 days in the infected piglets, but survival was reduced to 6 days for the non-motile variants. Subsequently, an isogenic set of flagellar mutants was tested that carried single or double mutations that inactivated *flaA* and *flaB*, genes that encode major and minor components of the filament. Mutation of either flagellin gene caused a reduction in motility and decreased the efficiency of colonization by *H. pylori* (Eaton *et al.*, 1996). Later studies readdressed the contribution of motility to infection through the infection of FVB/N mice with a *motB* mutant, which retains a complete surface-exposed flagellar structure but lacks a critical component of the rotary motor (Ottemann and Lowenthal, 2002). The *motB* mutant failed to efficiently colonize the mice, providing strong evidence that motility itself, rather than the flagellar structure, was necessary for infectivity. Consistent with this conclusion, mutations that interfere with the chemosensory system controlling chemotaxis and the rotational direction of the flagellar motor compromise *H. pylori* colonization in mice, gerbils, and piglets (Foynes *et al.*, 2000; Guo and Mekalanos, 2002; McGee *et al.*, 2005; Terry *et al.*, 2005). Clearly, the preponderance of evidence points towards a significant role for flagella affecting *H. pylori* interactions with its hosts.

REV UP THE ENGINES: SWARMING IS AN AGGRESSIVE FORM OF MOTILITY

Swarming is a widely distributed form of bacterial differentiation that leads to cells with the ability to rapidly migrate over surfaces (Harshey, 1994). Bacteria that enter into swarming states are characteristically elongated and multinucleated, with prolific production of flagella on their surface. Movement by swarming bacteria involves cooperation between individual cells that form groups or rafts, which together transit rapidly over surfaces. This social behavior is displayed prominently by the urinary-tract pathogen *Proteus mirabilis* (Coker *et al.*, 2000). Infections by this bacterium frequently occur in hospitalized patients with urinary catheters. The bacteria gain access by migrating against the flow of urine, to initially colonize the urethra. If untreated, bacteria progressively move up the urinary tract to the bladder, the ureter, and, finally, the kidney. Advancement of the infection to the kidney results in pyelonephritis and renal-stone formation. During the course of an infection, *P. mirabilis* grows within the urine and invades the epithelium. One test of the hypothesis that flagella contribute to ascending urinary-tract infection was conducted by assessment of the effect that a *flaD* mutation would have on bacterial colonization in the mouse model of ascending urinary-tract infection and invasion of cultured human proximal renal epithelial cells (Mobley *et al.*, 1996). FlaD (called FliD in *Salmonella*) forms the cap at the tip of the flagellar filament responsible for polymerization of flagellin subunits. Consequently, FlaD mutants synthesize and secrete flagellin subunits but lack full-length flagella on their surface. In agreement with a role for flagella in the infection process, the *flaD* mutant displayed reduced colonization of CBA mice following transurethral inoculation. After 1 week of infection, the *flaD* mutant was recovered from urine, bladder, and kidney samples at levels 100-fold less than fully motile bacteria (Mobley *et al.*, 1996). Furthering the central role for flagella in *P. mirabilis* virulence, it turns out that this bacterium has placed regulation of other virulence determinants under the control of *flhDC*, the flagellar master regulatory operon, including a hemolysin and the secreted ZapA protease (Fraser *et al.*, 2002). It is assumed that swarming contributes to the infectious process, but this inference is actually supported by only a limited number of studies. Swarmer cells have been observed in fixed specimens from infected mice, but the majority of *P. mirabilis* seemed to be swimmer cells (Zhao *et al.*, 1998). Genetic analysis of swarming has led to the isolation of strains that are locked into the swimmer-cell state, providing the opportunity to tackle the hypothesis more directly. Swarming mutants clearly display a reduced ability to form biofilms that block catheters, supporting the idea that swarming predisposes patients to infection (Jones et al., 2004). The only direct assessment, however, was completed by injection of bacteria into the bladders of mice with motile but swarming-defective mutants, which demonstrated that swarming was necessary for colonization of the bladder and spread of the infection to the kidneys (Allison *et al.*, 1994). A true assessment of the contribution made by swarming to bacterial infections ascending from a catheter remains to be completed, but flagella clearly have an important role during infection by *P. mirabilis*.

THE FLAGELLIN SURVEILLANCE SYSTEM OF THE HOST

The interplay between pathogens and their hosts is a symphony of events. Hosts are equipped to restrict pathogen colonization through numerous means, such as renewal of epithelial cells, maintenance of tight junctions between cells, and production of antimicrobial peptides. As these barriers are compromised, an inducible innate immune response, or pro-inflammatory response, is stimulated. Epithelial cells, reinforced by macrophages, monocytes, and dendritic cells that infiltrate the epithelium, are all equipped with pattern-recognition receptors (PRRs) that serve as the forward sentries that signal a breach of the barrier by detection of pathogen-associated molecular patterns (PAMPs) (Akira and Takeda, 2004). Surface-localized PRRs of mammalian cells belong to the Toll-like receptor (TLR) family of proteins. Each of the TLRs has a central role in the recognition of a particular PAMP, and each PAMP is a conserved molecule among diverse bacterial pathogens, some examples of which are lipopolysaccharide, lipopeptides, peptidoglycan, and unmethylated CpG DNA. Flagellin is a PAMP recognized by TLR-5, which, like other TLRs, stimulates activation of nuclear factor-κB (NF-κB) and mitogen-activated protein kinase (MAPK) pathways that regulate transcription of genes encoding immune system mediators that affect the short- and long-term response of the host (Hayashi *et al.*, 2001). TLR-5 is expressed by epithelial cells of the intestine and the lung (Gerwitz *et al.*, 2001; Hawn *et al.*, 2003). The monomeric form of flagellin serves as the ligand, with binding dependent upon residues of the highly conserved amino and carboxyl termini (Eaves-Pyles *et al.*, 2001; Smith *et al.*, 2003). The mechanism by which flagellin is released into the mucosa by bacteria during colonization of the epithelium is crucial to understanding how the innate immune system is activated. Flagellin could be passively released or actively secreted by the bacterium. We can imagine that flagella could be released from the surface of bacteria by extreme changes in pH, as is encountered during passage through the stomach, by proteases, surfactants, and bile salts. Flagella may be ejected in response to specific environmental cues, as occurs for the free-living bacterium *Calobacter crecentus* (Shapiro and Maizel, 1973). Free flagellin monomers can also be directly secreted to extracellular surroundings by the flagellar T3S system, as is commonly observed when bacteria are cultured in vitro (Young *et al.*, 1998). Regardless of the mechanism, the

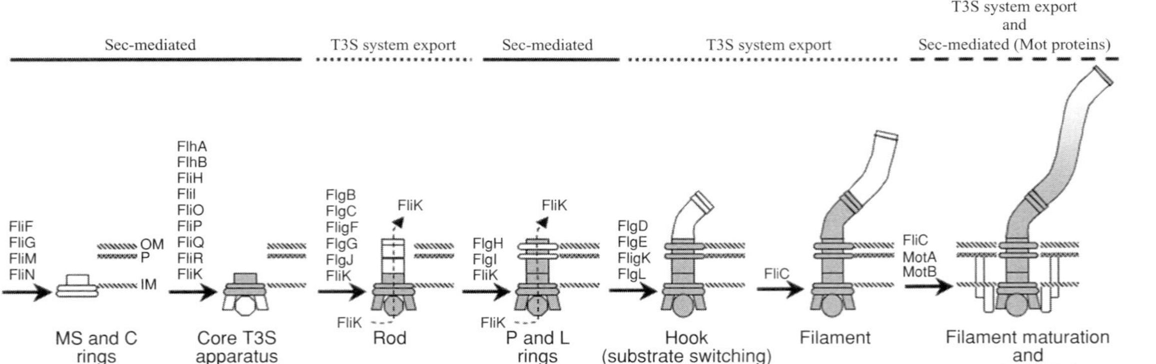

Figure 5.2 The flagellar biogenesis pathway of *Salmonella enterica* serovar Typhimurium. At each major step in the pathway, newly added domains of the flagellum are indicated in white, with completed regions shaded in gray. A subset of flagellar components is targeted to the correct location either by the Sec-mediated pathway or involving the flagellar type III secretion (T3S) pathway, which is an integral part of the flagellar structure. During the early steps, the flagellar T3S system recognizes basal body and hook components. Completion of the hook causes the flagellar T3S system to change its specificity to filament components. For some bacteria, the flagellar T3S system can also recognize and export non-flagellar proteins to extracellular sites.

result of flagellin binding by TLR-5 promotes activation of host immune responses.

STRUCTURE AND ASSEMBLY OF FLAGELLA

The process of flagellar biogenesis is best understood for bacteria belonging to the family Enterobacteriaceae. The sequence of events and the mechanisms of flagellar assembly have been studied extensively in *Salmonella*, but the same or similar processes occur in other bacteria (Aizawa, 1996; Macnab, 1996) (Figure 5.2). Flagellar biogenesis is a highly ordered dynamic process with a series of stages that requires the coordination of the Sec pathway and a pathway that is an integral part of the flagellum. The first stage constitutes assembly of the flagellar secretion apparatus, a bona fide T3S system containing ten different proteins that share a high degree of amino-acid sequence similarity with components that form the secretion machinery of contact-dependent T3S systems (Figure 5.2). This apparatus is the platform on which the rest of the flagellum is built and provides a pathway needed for export of structural subunits. The identity between proteins of the flagellar and contact-dependent T3S systems ranges from 20% to 35%, and their similarity reaches 50%. The relatedness of these proteins reflects structural and functional conservation, but it also suggests an evolutionary relationship between the secretion machinery of flagella and contact-dependent T3S systems. Consistent with this notion, microscopic procedures have revealed remarkably similar images of the *Salmonella* flagellar basal body (Aizawa *et al.*, 1985), the *Salmonella* SPI-1 TTS system (Kubori *et al.*, 1998), and the *Shigella flexneri mxi/spa*-encoded TTS system (Blocker *et al.*, 1999). The wealth of information about how the flagellum is assembled has provided significant insight on how contact-dependent T3S systems are assembled and are used by many pathogens in order to deliver toxic virulence factors during an infection.

The first stages of flagellar assembly are marked by insertion of the integral membrane protein FliF, probably by the Sec pathway, into the cytoplasmic membrane, which leads to formation of the MS ring (Kubori *et al.*, 1992; Lyons *et al.*, 2004). This allows FliG, FliM, and FliN to associate peripherally with FliF, forming the C-ring (Francis *et al.*, 1994). Eight additional proteins (FlhA, FlhB, FliH, FliI, FliO, FliP, FliQ, FliR) complete the core flagellar T3S apparatus (Minamino and Macnab, 1999). The rod proteins (FlgB, FlgC, FlgF, FlgG) are transported by and through a central channel of the T3S apparatus to the periplasm, where they self-assemble (Homma *et al.*, 1990). Completion of the rod is followed by transport of FlgI and FlgH across the cytoplasmic membrane by the Sec pathway, with concomitant cleavage of their signal peptides, where they then proceed to form the periplasmic P-ring and the outer-membrane L-ring (Homma *et al.*, 1987; Jones *et al.*, 1990). Completion of this structural intermediate, called the basal body, marks the point at which further flagellar segments extend from the surface of the cell and where subsequent localization of all other subunits depends solely on the flagellar T3S system. The hook-capping protein FlgD is targeted to the growing organelle and ensures that secreted FlgE polymerizes to form the hook (Ohnishi *et al.*, 1994). The hook-filament junction proteins FlgK and FlgL, along with the filament-capping protein FliD, are exported to the tip of the hook (Ikeda *et al.*, 1989). Control of substrate specificity by the flagellar T3S system during rod and hook assembly involves FliK, a protein that is secreted and released (Minamino *et al.*, 1999). By an ill-defined mechanism, completion of the hook leads to retention of FliK, which interacts with the T3S apparatus protein FlhB and serves as a stimulus for substrate switching by the flagellar T3S apparatus (Minamino *et al.*, 2004). This substrate switching ensures that the T3S system specifically recognizes the final set of flagellar subunits such as the FliC flagellin. During the final stages of organelle assembly, the T3S apparatus exports flagellin to the distal end of the growing filament, where polymerization is mediated by FliD (Yonekura *et al.*, 2000). Simultaneously, the motor, consisting of MotA and MotB, along with the chemosensory system are completed, providing the cell with the capacity for directional control of organelle rotation (Amosti and Chamberlin, 1989; Blair and Berg, 1988; Kutsukake *et al.*, 1990). Completion of the flagellum and chemosensory system is required for bacteria to respond behaviorally to their surroundings and to seek out favorable environmental niches.

REGULATION OF FLAGELLAR GENE EXPRESSION

The production of flagella by bacteria is controlled in response to a number of environmental cues, such as temperature, pH, and nutrient availability, and in response to a variety of environmental stresses. For most bacteria, the mechanisms that control regulation of flagellar gene expression in response to these external stimuli have not been determined precisely. However, the accumulating evidence indicates that most or all regulatory cues affecting flagellar gene expression appear to be integrated through precise modulation of transcription factors. Although certainly not representative of all bacteria, the flagellar regulatory cascade of *Salmonella* and *E. coli* serve as the paradigm (Fernandez and Berenguer, 2000; Hughes and Aldridge, 2002).

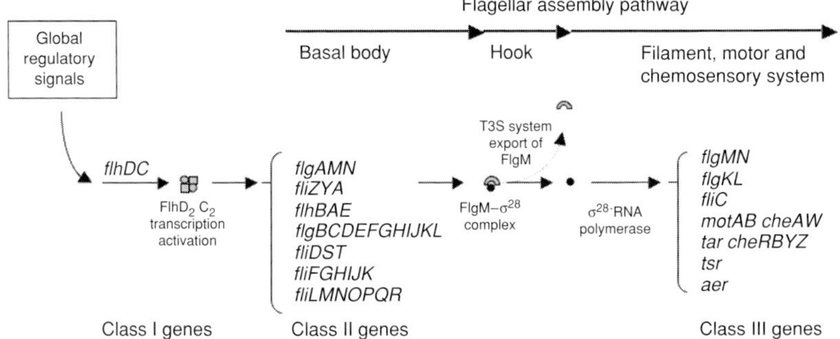

Figure 5.3 Coupling of the flagellar transcriptional cascade to flagellar assembly in *Salmonella enterica* serovar Typhimurium. More than 50 genes belong to the flagellar regulon that encodes components of the flagellar organelle and the chemosensory system that controls the direction of flagellar rotation. The genes are transcribed in operons expressed during the early (class I), middle (class II), and late (class III) stages of flagellar assembly. A few genes contain both class II and III promoters, driving their transcription through the middle and late stages of flagellar assembly. The class I genes form a single operon, *flhDC*, encoding proteins that form a heterotetrameric transcription activator that induces σ^{70}-RNA polymerase to initiate transcription of genes with class II promoters. The class II genes encode structural and accessory proteins required for assembly of intermediate proteins of the organelle, including the flagellar type III secretion (T3S) system, basal body, and hook. In addition, the FlgM and σ^{28} regulatory proteins are expressed at this stage. The FlgM-σ^{28} complex remains inactive until the hook is completed, at which time FlgM is exported from the cell by the flagellar T3S system. Free σ^{28} then becomes available to associate with RNA polymerase, thereby redirecting its promoter specificity. σ^{28}-RNA polymerase initiates transcription at class III promoters located upstream of genes needed for late stages of flagellar assembly. The class III genes encode components of the flagellar filament, motor, and chemosensory system.

Flagellar gene expression corresponds with the major stages of morphological development of the flagellar structure, which, as would be suggested by its complexity, involves the coordination of about 50 genes (Kalir *et al.*, 2001; Kutsukake *et al.*, 1990). The hierarchical regulatory cascade contains three major flagellar gene classes (Figure 5.3). Class I genes, consisting of *flhD* and *flhC*, form a single operon that is expressed at the top of the hierarchy and is required for the expression of all other flagellar genes. Mutations that inactivate these genes result in a transcriptional block of the entire flagellar cascade. In *S. enterica* serovar Typhimurium, the *flhDC* operon is preceded by six σ^{70}-dependent transcriptional start sites, reflecting the sophisticated level of signal integration imparting regulatory effects on expression (Yanagihara *et al.*, 1999). Binding sites for the transcription factor CRP have been identified in the *flhDC* promoter regions for both *E. coli* and *S.* Typhimurium (Soutourina *et al.*, 1999; Yanagihara *et al.*, 1999). Sites of binding have also been mapped for several other transcription factors, including H-NS, LrhA, OmpR, and RcsB (Francez-Chalot *et al.*, 2003; Lehnen *et al.*, 2002; Shin and Park, 1995; Soutourina *et al.*, 1999). Less well characterized are mechanisms by which other transcription factors exert control, such as HdfR and SirA (Ko and Park, 2000; Teplitski *et al.*, 2003). Each of these regulators affects a variety of genes in response to a diverse set of signals, thereby fine tuning decisions to maintain motility with the physiological state of the cell. Adding another mechanism of modulation, the stability of *flhDC* mRNA is affected by the global regulatory RNA encoded by CsrA (Wei *et al.*, 2001). Control of these class I genes is clearly the fulcrum in the flagellar regulatory cascade.

The FlhD and FlhC proteins form a heterotetrameric transcription activator that is required for the expression of σ^{70}-dependent class II genes (Liu and Matsumura, 1994) (Figure 5.3). Most class II genes encode structural and accessory proteins required for assembly of the basal body and hook components of the flagellum, but two genes, *flgM* and *fliA*, encode the FlgM and σ^{28} regulatory proteins that form a regulatory checkpoint (Gillen and Hughes, 1991; Ohnishi *et al.*, 1990). Expression of class II genes coincides with completion of the flagellar basal body and hook, at which time the flagellar T3S system exports FlgM (Hughes *et al.*, 1993). While it is retained within the cell, FlgM complexes with σ^{28}, preventing it from activating class III genes. Upon export of FlgM, σ^{28} associates with RNA polymerase activating class III genes, each of which contains σ^{28}-dependent promoters (Kutsukake *et al.*, 1994). Class III genes encode proteins required for maturation of the flagellar filament, motor, and chemosensory system. The flagellar T3S machinery exports some of these proteins, such as flagellin, to the outer surface of the cell. Consistent with this regulatory model, mutations in *fliA* render the cell with a truncated paralyzed organelle. Conversely, a mutation in *flgM* results in unregulated expression of σ^{28}-dependent genes, leading to overproduction and secretion of flagellin subunits.

CONTROL OF DIRECTIONAL MOVEMENT OF BACTERIA BY CHEMOTAXIS

Propulsion for motility provided by flagella must be controlled in order to achieve bacterial progress towards, and maintenance within, favorable host niches. Bacteria move towards higher concentrations of attractants and avoid high concentrations of repellents by sensing local fluctuations in the concentration of chemoeffectors (Sournik, 2004; Wadhams and Armitage, 2004). Chemotactic motility characteristically consists of periods of smooth forward swimming (runs) interrupted by momentary reorientations (tumbles). Runs are generated by the counterclockwise (CCW) rotation of flagella, and tumbles take place when the rotary motor is temporarily reversed to the clockwise (CW) direction. The change in swimming direction is random, with favorable progress of the bacterium accomplished through suppression of tumbles leading to longer runs. The fundamental mechanisms of chemotaxis are shared among diverse bacteria, with the *E. coli* system being the best characterized (Figure 5.4). An extensive assortment of sensory cues are processed by the chemosensory system, allowing precise responses to the concentration of a variety of small molecules, such as amino acids, sugars, and short peptides, as well as responses to changes in pH, temperature, and redox state.

Specific receptors, many of which reside on the cytoplasmic membrane, modulate the activity of a phosphotransfer relay system in response to ligand binding (Figure 5.4). The central enzyme in this relay is the histidine kinase CheA (Borkovich *et al.*, 1989). Its activity is modulated by the receptor in a complex that also includes CheW. The net effect is that occupation of the receptor by an attractant inhibits CheA autophosphorylation, while occupation of a receptor by a repellent stimulates CheA autophosphorylation (CheA-P). CheA-P transfers phosphate to the response regulator CheY. Phosphorylated CheY (CheY-P) diffuses to flagellar motors, switching the rotational direction from CCW to CW, and thus promoting tumbles. CheY-P is rapidly dephosphorylated by CheZ phosphatase, ensuring that the cell recovers quickly (Hess *et al.*, 1988). Thus, increases in CheA-P correlate with increased frequencies of tumbling.

Further control of the chemosensory system is achieved through controlled methylation of four glutamyl residues on receptors (Figure 5.4). Often called sensory adaptation, this allows the organism to desensitize its signal transduction system to the original stimulus and to respond to higher concentrations of the same chemoeffectors or other types of chemoeffectors. For receptors that detect an attractant or detect a repellent, the unregulated CheR methyltransferase modifies the receptor, which overides structural changes caused by ligand binding (Springer and Koshland, 1977). This returns the receptor to the baseline state and nullifies the effect of the receptors on CheA activity. Demethylation of receptors by the CheB methylesterase keeps the system in balance (Yonekawa *et al.*, 1983). The activity of CheB is modulated

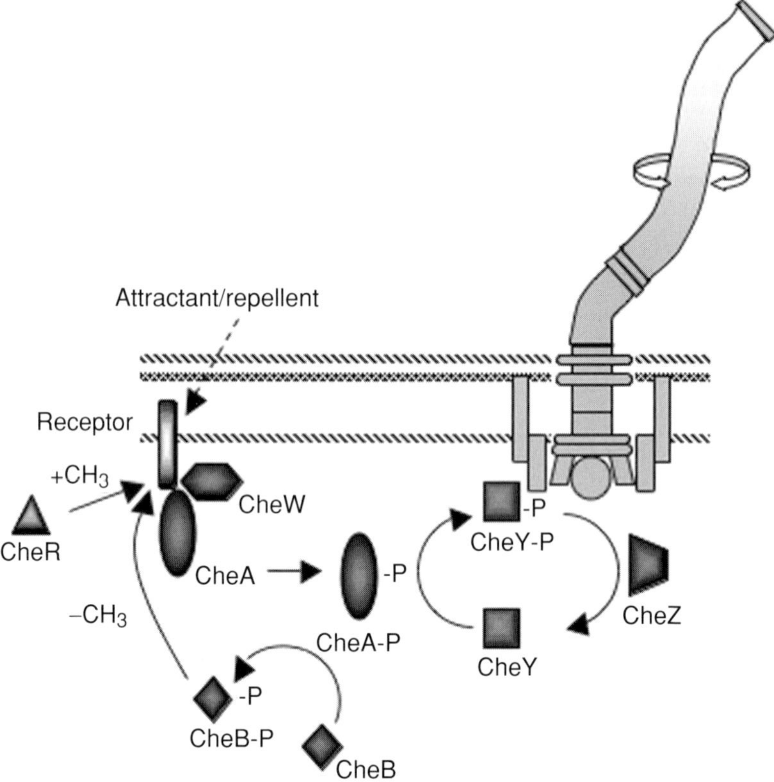

Figure 5.4 Operation of the chemosensory system. Changes in the concentration of an attractant or repellent are recognized by specific receptors that are in a complex with the adaptor proteins CheW and the histidine kinase CheA. Autophosporylation of CheA is stimulated by repellents and inhibited by attractants. Phosphorylated CheA (CheA-P) efficiently transfers phosphate to the response regulator CheY, generating CheY-P, which in turn diffuses to the flagellar motor and switches its direction from counterclockwise to clockwise, promoting tumbles. CheY-P is rapidly reduced to CheY by the phosphatase CheZ. Two enzymes, CheR and CheB, mediate adaptation during chemotaxis. The methyl transferase CheR slowly adds methyl groups to the receptor, desensitizing it to its ligand and increaseing CheA autophosphorylation. Receptor sensitivity is restored by a feedback loop through activation of CheB methylesterase activity by CheA-P.

through phosphorylation by the CheA kinase, directly linking adaptation to the decision to switch the direction of the flagellar motor (Ninfa *et al.*, 1991). Different bacterial pathogens likely monitor a variety of different chemoattractants during infection through a plethora of ligand receptors, but the well-integrated Che-type phosphorelay network, and close variations of it, serve as a powerful signaling network to control bacterial behavior.

NEW ROLES FOR FLAGELLA IN BACTERIAL VIRULENCE: EXPORT OF NON-FLAGELLAR PROTEINS BY THE FLAGELLAR TYPE III SECRETION SYSTEM

The integral protein secretion system of the flagellum was described many years ago and was then considered to be rather unique (Macnab, 1999). It was the subsequent revelation that many Gram-negative bacteria utilize a related secretion pathway, eventually called T3S, to deliver virulence factors into targeted host cells that spawned the new view that T3S is quite prevalent among the eubacteria (Hueck, 1998). All T3S apparatus contain a related core set of components and appear to recognize their substrates by a similar mechanism. The extensive similarity between the core components of the flagellar and contact-dependent T3S systems indicates that these two types of apparatus are evolutionarily related (Hueck, 1998). Delivery of a protein by a T3S system is completed without modification of the peptide, takes places without the accumulation of a transport intermediate, and is driven by the hydrolysis of adenosine triphosphate (ATP). The flagellar T3S apparatus originally was considered to have an ancillary role in organelle biogenesis, but more recent studies have revealed a more expansive role that includes the secretion of non-flagellar proteins (Young *et al.*, 1999). *C. jejuni* uses the flagellar T3S system to secrete the *Campylobacter* invasion antigen (Cia) proteins and the FlaC protein (Konkel *et al.*, 2004; Song *et al.*, 2004). The Cia proteins are necessary for adherence and invasion of eukaryotic cells by *C. jejuni*. Export of these proteins by the flagellar T3S system implicates the flagellum in this process, even if it only provides the conduit though which the Cia proteins pass. However, YplA of *Y. enterocolitica*, which was the first documented example of a secreted non-flagellar protein, remains the most extensively studied non-flagellar protein secreted by a flagellar T3S system (Schmiel *et al.*, 1998). This protein is one of a group of secreted proteins called flagellar outer proteins (Fops), whose secretion depends on the flagellar T3S system (Young *et al.*, 1999). YplA is a phospholipase implicated in virulence because it is required for survival of *Y. enterocolitica* in experimentally infected mice (Schmiel *et al.*, 1998). YplA also induces lecithin-dependent hemolysis of rabbit erythrocytes and cytotoxicity of cultured HeLa epithelial cells (Tsubokura *et al.*, 1979). YplA secretion by the flagellar system was first observed during a study of the *Y. enterocolitica* flagellar master regulatory operon *flhDC* (Young *et al.*, 1998). Strikingly, a *flhDC* mutant not only was blocked for production of the flagellin filament proteins but also lacked the ability to secrete several extracellular proteins (Young *et al.*, 1998, 1999). Further analysis revealed that all mutants incapable of producing flagella were deficient for Fop secretion, implicating the flagellar T3S system as the responsible export apparatus (Young *et al.*, 1999). Furthermore, a mutation in *flgM*, which encodes a negative regulator of flagellar genes, caused overproduction of flagellin and increased secretion of the Fops (Young *et al.*, 1999). Complementation of the *flgM* mutation reduced extracellular protein production to wild-type levels. These results indicate that the expression of flagella and the secretion of Fops are coupled.

The provocative idea that a flagellum could serve as a bona fide pathway for protein secretion was not accepted immediately. The proof of principle for establishing the flagellar T3S system as an export apparatus of non-flagellar proteins came through further study of *Y. enterocolitica* YplA (Young and Young, 2002; Young *et al.*, 1999). However, the situation was complicated by the fact that *yplA* expression is coupled to regulation of the flagellar TTS system as a class III flagellar gene (Schmiel *et al.*, 2000). To separate regulatory effects from secretion effects, *yplA* expression was placed under control of the P*tac* or P*cat* promoter, ensuring that transcription of *yplA* occurred independently of factors that control flagellar gene expression (Young and Young, 2002; Young *et al.*, 1999). This provided the opportunity to examine directly how specific flagellar mutations affect YplA secretion. The results from this analysis demonstrated that the flagellar T3S system was necessary for YplA export. Mutations that blocked production of the core components of the flagellar T3S system prevented YplA secretion. This included strains carrying mutations in the flagellar master regulator operon, *flhDC*, or in genes encoding T3S apparatus components (*flgA*, *flgB*, *flgF*, *flhA*, *flhB*, *fliE*). Interestingly, YplA secretion was not affected by mutations in *fliA*, *flgM*, *fleA*, or *motA*, indicating that the flagellar filament and the torque generating motor complex are not necessarily required for YplA secretion. These experiments, therefore, defined the basal-body–hook complex as the minimal structure capable of acting as a T3S system for non-flagellar proteins (Young and Young, 2002; Young *et al.*, 1999).

Unequivocal documentation supporting the hypothesis that YplA is exported by a type III mechanism came about when it was demonstrated that YplA could serve as a substrate for two other T3S systems in *Y. enterocolitica* (Young and Young, 2002). The Ysa and Ysc T3S systems are contact-dependent systems that have the capacity to deliver toxic effector proteins into targeted host cells. Both the Ysa and Ysc T3S systems can also be induced to export proteins under specific laboratory conditions in the absence of host cells. When YplA was produced in wild-type *Y. enterocolitica* under conditions that induced the Ysc T3S system, it was found that YplA was secreted along with the Yop proteins (Young and Young, 2002). Similarly, YplA was secreted by the Ysa T3S system (Young and Young, 2002). Further investigations have determined that YplA has an amino-terminal secretion signal that is essential for its recognition by the Ysa, Ysc, and flagellar T3S systems (Warren and Young, 2005). These data have helped advance the concept that flagellar and contact-dependent T3S systems are mechanistically very similar and form a unified family of bacterial protein targeting pathways. Further investigations of other bacterial flagellar T3S systems will likely reveal other non-flagellar proteins that are secreted and that affect the outcome of bacterial–host interactions.

REFERENCES

Aguero-Rosenfeld, M. E., Yang, X. H., and Nachamkim, I. (1990). Infection of Syrian hamsters with flagellar variants of *Campylobacter jejuni*. *Infect. Immun.* **58**, 2214–2219.

Aizawa, S. I. (1996). Flagellar assembly in *Salmonella typhimurium*. *Mol. Microbiol.* **19**, 1–5.

Aizawa, S. I. (2001). Bacterial flagella and type III secretion systems. *FEMS Microbiol. Lett.* **202**, 157–164.

Aizawa, S. I., Dean, G. E., Jones, C. J., Macnab, R. M., and Yamaguchi, S. (1985). Purification and characterization of the flagellar hook–basal body complex of *Salmonella typhimurium*. *J. Bacteriol.* **161**, 836–849.

Akira, S. and Takeda, K. (2004). Toll-like receptor signalling. *Nat. Rev. Immunol.* **4**, 499–511.

Aldridge, P. and Hughes, K. T. (2002). Regulation of flagellar assembly. *Curr. Opin. Microbiol.* **5**, 160–165.

Allison, C., Emody, L., Coleman, N., and Hughes, C. (1994). The role of swarm differentiation and multicellular migration in the uropathogenicity of *Proteus mirabilis*. *J. Infect. Dis.* **169**, 1155–1158.

Amosti, D. N. and Chamberlin, M. J. (1989). Secondary sigma factor controls transcription of flagellar and chemotaxis genes in *Escherichia coli*. *Proc. Natl. Acad. Sci. U. S. A.* **86**, 830–834.

Arora, K., Ritchings, B. W., Almira, E. C., Lory, S., and Ramphal, R. (1998). The *Pseudomonas aeruginosa* flagellar cap protein, FliD, is responsible for mucin binding. *Infect. Immun.* **66**, 1474–1479.

Arora, S. K., Neely, A. N., Blair, B., Lory, S., and Ramphal, R. (2005). Role of motility and flagellin glycosylation in the pathogenesis of *Pseudomonas aeruginosa* burn wound infections. *Infect. Immun.* **73**, 4395–4398.

Black, R. E., Levinne, M. M., Clements, M. L., and Blaser, M. L. (1988). Experimental *Campylobacter jejuni* infection in humans. *J. Infect. Dis.* **157**, 472–479.

Blair, D. F. and Berg, H. C. (1988). Restoration of torque in defective flagellar motors. *Science* **242**, 1678–1681.

Blocker, A., Gounon, P., Larquet, E., *et al.* (1999). The tripartite type III secreton of *Shigella flexneri* inserts IpaB and IpaC into host membranes. *J. Cell Biol.* **147**, 683–693.

Borkovich, K. A., Kaplan, N., Hess, J. F., and Simon, M. I. (1989). Transmembrane signal transduction in bacterial chemotaxis involves ligand-dependent activation of phosphate group transfer. *Proc. Natl. Acad. Sci. U. S. A.* **86**, 1208–1212.

Caldwell, M. B., Guerry, P., Lee, J. P., and Walker, R. I. (1985). Reversible expression of flagella in *Campylobacter jejuni*. *Infect. Immun.* **50**, 941–943.

Coker, C., Poore, C. A., Li, X., and Mobley, H. T. (2000). Pathogenesis of *Proteus mirabilis* urinary tract infection. *Microbes Infect.* **2**, 1479–1505.

Deitrich, C., Heuner, K., Brand, B. C., Hacker, J., and Steinert, M. (2001). Flagellum of *Legionella pneumophila* positively affects the early phase of infection of eukaryotic host cells. *Infect. Immun.* **69**, 2116–2122.

Drake, D. and Monte, T. C. (1988). Flagella, motility and invasive virulence of *Pseudomonas aeruginosa*. *J. Gen. Microbiol.* **134**, 43–52.

Eaton, K. A., Morgan, D. R., and Krakowka, S. (1992). Motility as a factor in the colonization of gnotobiotic piglets by *Helicobacter pylori*. *J. Med. Microbiol.* **37**, 123–127.

Eaton, K. A., Suerbaum, S., Josenhans, C., and Krakowka, S. (1996). Colonization of gnotobiotic piglets by *Helicobacter pylori* deficient in two flagellin genes. *Infect. Immun.* **64**, 2445–2448.

Eaves-Pyles, T. D., Wong, H. R., Odoms, K., and Pyles, R. B. (2001). Salmonella flagellin-dependent proinflammatory responses are localized to the conserved amino and carboxyl regions of the protein. *J. Immunol.* **167**, 7009–7016.

Feldman, M., Bryan, R., Rajan, S., *et al.* (1998). Role of flagella in pathogenesis of *Pseudomonas aeruginosa* pulmonary infection. *Infect. Immun.* **66**, 43–51.

Fernandez, L. A. and Berenguer, J. (2000). Secretion and assembly of regular surface structures in Gram-negative bacteria. *FEMS Microbiol. Rev.* **24**, 21–44.

Ferrero, R. L. and Lee, A. (1988). Motility of *Campylobacter jejuni* in a viscous environment: comparison with conventional rod-shaped bacteria. *J. Gen. Microbiol.* **134**, 53–59.

Fleizig, S. M., Arora, S. K., Van, R., and Ramphal, R. (2001). FlhA, a component of the flagellum assembly apparatus of *Pseudomonas aeruginosa*, plays a role in internalization by corneal epithelial cells. *Infect. Immun.* **69**, 4931–4937.

Foynes, S., Dorrell, N., Ward, S. J., *et al.* (2000). *Helicobacter pylori* possesses two CheY response regulators and a histidine kinase sensor, CheA, which are essential for chemotaxis and colonization of the gastric mucosa. *Infect. Immun.* **68**, 2016–2023.

Francez-Chalot, A., Laugel, B., Van Gemert, A., *et al.* (2003). RcsCDB His-Asp phosphorelay system negatively regulates the *flhDC* operon in *Escherichia coli*. *Mol. Microbiol.* **49**, 823–832.

Francis, N. R., Sosinsky, G. E., Thomas, D., and DeRosier, D. J. (1994). Isolation, characterization and structure of the bacterial flagellar motors containing the switch complex. *J. Mol. Biol.* **235**, 1261–1270.

Fraser, G. M., Claret, L., Furness, R., Gupta, S., and Hughes, C. (2002). Swarming-coupled expression of the *Proteus mirabilis hpmBA* hemolysin operon. *Microbiology* **148**, 2191–2201.

Gerwitz, A. T., Navas, T. A., Lyons, S., Godowski, P. J., and Madara, J. L. (2001). Cutting edge: bacterial flagellin activates basolaterally expressed TLR5 to induce epithelial proinflammatory gene expression. *J. Immunol.* **167**, 1882–1885.

Gillen, K. L. and Hughes, K. T. (1991). Molecular characterization of *flgM*, a gene encoding a negative regulator of flagellin synthesis in *Salmonella typhimurium*. *J. Bacteriol.* **173**, 6453–6459.

Giron, J. A., Torres, A. G., Freer, E., and Kaper, J. B. (2002). The flagella of enteropathogenic *Escherichia coli* mediate adherence to epithelial cells. *Mol. Microbiol.* **44**, 361–379.

Guo, B. P. and Mekalanos, J. J. (2002). Rapid genetic analysis of *Helicobacter pylori* gastric mucosal colonization in suckling mice. *Proc. Natl. Acad. Sci. U. S. A.* **99**, 8354–8359.

Harshey, R. M. (1994). Bees aren't the only ones: swarming in Gram-negative bacteria. *Mol. Microbiol.* **13**, 389–394.

Harshey, R. M. and Toguchi, A. (1996). Spinning tails: homologies among bacterial flagellar systems. *Trends Microbiol.* **4**, 226–231.

Hawn, T. R., Verbon, A., Lettinga, K. D., *et al.* (2003). A common TLR5 stop codon polymorphism abolishes flagellin signaling and is associated with susceptibility to Legionnaires' disease. *J. Exp. Med.*, 1563–1572.

Hayashi, F., Smith, K. D., Ozinsky, A., *et al.* (2001). The innate immune response of bacterial flagellin is mediated by Toll-like receptor 5. *Nature* **410**, 1099–1103.

Hazell, S. L., Lee, A., Brady, L., and Hennessy, W. (1986). *Campylobacter pylori* and gastritis: association with intercellular spaces and adaptation to an environment of mucus as important factors in colonization of the gastric epithelium. *J. Infect. Dis.* **153**, 658–663.

Hendrixson, D. R. and DiRita, V. J. (2004). Identification of *Campylobacter jejuni* genes involved in commensal colonization of the chick gastrointestinal tract. *Mol. Microbiol.* **52**, 471–484.

Hess, J. F., Oosawa, K., Kaplan, N., and Simon, M. I. (1988). Phosphorylation of three proteins in the signaling pathway of bacterial chemotaxis. *Cell* **53**, 79–87.

Homma, M., Komeda, Y., Iino, T., and Macnab, R. M. (1987). The *flaFIX* gene product of *Salmonella typhimurium* is a flagellar basal body component with a signal peptide. *J. Bacteriol.* **169**, 1493–1498.

Homma, M., Kutsukake, K., Hasebe, M., Iino, T., and Macnab, R. M. (1990). FlgB, FlgC, FlgD and FlgG. A family of structurally related proteins in the flagellar basal body of *Salmonella typhimurium. J. Mol. Biol.* **211**, 465–477.

Hueck, C. J. (1998). Type III secretion systems in bacterial pathogens of animals and plants. *Microbiol. Mol. Biol. Rev.* **62**, 379–433.

Hughes, K. T. and Aldridge, P. (2002). Regulation of flagellar assembly. *Curr. Opin. Microbiol.* **5**, 160–165.

Hughes, K. T., Gillen, K. L., Semon, M. J., and Karlinsey, J. E. (1993). Sensing structural intermediates in bacterial flagellar assembly by export of a negative regulator. *Science* **262**, 1277–1280.

Ikeda, T., Asakura, S., and Kamiya, R. (1989). Total reconstitution of *Salmonella* flagellar filaments from hook and purified flagellin and hook-associated proteins *in vitro. J. Mol. Biol.* **209**, 109–114.

Jones, B. V., Young, R., Mahenthiralingam, E., and Stickler, D. J. (2004). Ultrastructure of *Proteus mirabilis* swarmer cell rafts and role of swarming in catheter-associate urinary tract infection. *Infect. Immun.* **72**, 3941–3950.

Jones, C. J., Macnab, R. M., Okino, H., and Aizawa, S. I. (1990). Stoichiometric analysis of the flagellar hook–(basal body) complex of *Salmonella typhimurium. J. Mol. Biol.* **212**, 377–387.

Journet, L., Hughes, K. T., and Cornelis, G. R. (2005). Type III secretion: a secretory pathway serving both motility and virulence. *Mol. Membr. Biol.* **22**, 41–50.

Kalir, S., McClure, J., Pabbaraju, K., *et al.* (2001). Ordering genes in the flagellar pathway by analysis of expression from living bacteria. *Science* **292**, 2080–2083.

Ko, M. and Park, C. (2000). H-NS-dependent regulation of flagellar synthesis is mediated by a LysR family protein. *J. Bacteriol.* **182**, 4670–4672.

Konkel, M. E., Klena, J. D., Rivera-Amill, V., *et al.* (2004). Secretion of virulence proteins from *Campylobacter jejuni* is dependent on a functional flagellar export apparatus. *J. Bacteriol.* **186**, 3296–3303.

Kubori, T., Shimamoto, N., Yamaguchi, S., Namba, K., and Aizawa, T. (1992). Morphologic pathway of flagellar assembly in *Salmonella typhimurium*. *J. Mol. Biol.* **226**, 433–446.

Kubori, T., Matsushima, Y., Nakamura, D., *et al.* (1998). Supramolecular structure of the *Salmonella typhimurium* type III protein secretion system. *Science* **280**, 602–605.

Kutsukake, K., Ohya, Y., and Iino, T. (1990). Transcriptional analysis of the flagellar regulon of *Salmonella typhimurium*. *J. Bacteriol.* **172**, 741–747.

Kutsukake, K., Iyoda, S., Onishi, K., and Iiono, T. (1994). Genetic and molecular analyses of the interaction between the flagellum-specific sigma and anti-sigma factors in *Salmonella typhimurium*. *EMBO J.* **13**, 4568–4576.

Lee, A., O'Rourke, J. L., Barrington, P. J., and Trust, T. J. (1986). Mucus colonization as determinant of pathogenicity in intestinal infection by *Campylobacter jejuni*: a mouse cecal model. *Infect. Immun.* **51**, 536–546.

Lehnen, D., Blumer, C., Polen, T., *et al.* (2002). LlhA as a new transcription key regulator of flagella, motility and chemotaxis genes in *Escherichia coli*. *Mol. Microbiol.* **45**, 521–532.

Liu, X. and Matsumura, P. (1994). The FlhD/FlhC complex, a transcriptional activator of *Escherichia coli* flagellar class II operons. *J. Bacteriol.* **176**, 7345–7351.

Lockman, H. A. and Curtis, R. III (1990). *Salmonella typhimurium* mutants lacking flagella and motility remain virulent in BALB/c mice. *Infect. Immun.* **58**, 137–143.

Lucas, R. L., Lostroh, C. P., DiRusso, C. C., *et al.* (2000). Multiple factors independently regulate *hilA* and invasion gene expression in *Salmonella enterica* Serovar *Typhimurium*. *J. Bacteriol.* **182**, 1872–1882.

Lyons, S., Wang, L., Casanova, J. E., *et al.* (2004). *Salmonella typhimurium* translocates flagellin via an SPI2-mediated vesicular transport pathway. *J. Cell Sci.* **117**, 5771–5780.

Macnab, R. M. (1996). Flagella and motility. In Escherichia coli *and Salmonella typhimurium: cellular and molecular biology*, ed. F. C. Neidhardt. Washington, DC: ASM Press, pp. 123–145.

Macnab, R. M. (1999). The bacterial flagellum: reversible rotary propellor and type III export apparatus. *J. Bacteriol.* **181**, 7149–7153.

Macnab, R. M. (2003). How bacteria assemble flagella. *Annu. Rev. Microbiol.* **57**, 77–100.

Macnab, R. M. (2004). Type III flagellar protein export and flagellar assembly. *Biochim. Biophys. Acta* **1694**, 207–217.

Mahenthiralinggam, E., Campbell, M. E., and Speert, D. P. (1994). Nonmotility and phagocytic resistance of *Pseudomonas aeruginosa* isolates from chronically infected patients with cystic fibrosis. *Infect. Immun.* **62**, 596–605.

Marchetti, M., Sirard, J. C., Sansonetti, P., Pringault, E., and Kerneis, S. (2004). Interaction of pathogenic bacteria with rabbit appendix M cells: bacterial motility is a key feature in vivo. *Microbes Infect.* **6**, 521–528.

McGee, D. J., Langford, M. L., Watson, E. L., *et al.* (2005). Colonization and inflammation deficiencies in Mongolian gerbils infected by *Helicobacter pylori* chemotaxis mutants. *Infect. Immun.* **73**, 1820–1827.

Minamino, T. and Macnab, R. M. (1999). Components of the *Salmonella* flagellar export apparatus and classification of export substrates. *J. Bacteriol.* **181**, 1388–1394.

Minamino, T., Gonzalez-Pedrajo, B., Yamaguchi, K., Aizawa, S. I., and Macnab, R. M. (1999). FliK, the protein responsible for flagellar hook length control in *Salmonella*, is exported during hook assembly. *Mol. Microbiol.* **34**, 295–304.

Minamino, T., Saijo-Hamano, Y., Furukawa, Y., *et al.* (2004). Domain organization and function of the *Salmonella* FliK, a flagellar hook-length control protein. *J. Mol. Biol.* **341**, 491–502.

Mobley, H. L., Belas, R., Lockatell, V., *et al.* (1996). Construction of a flagellum-negative mutant of *Proteus mirabilis*: effect on internalization by human renal epithelial cells and virulence in a mouse model of ascending urinary tract infection. *Infect. Immun.* **64**, 5332–5340.

Morooka, T., Umeda, A., and Amako, K. (1985). Motility as an intestinal colonization factor for *Campylobacter jejuni*. *J. Gen. Microbiol.* **131**, 1973–1980.

Nachamkim, I., Yang, X. H., and Stern, N. J. (1993). Role of *Campylobacter jejuni* flagella as colonization factors for three-day-old chicks: analysis with flagellar mutants. *Appl. Envir. Microbiol.* **59**, 1269–1273.

Ninfa, E. G., Stock, A., Mowbray, S., and Stock, J. (1991). Reconstitution of the bacterial chemotaxis signal transduction system from purified components. *J. Biol. Chem.* **266**, 9764–9770.

Ohnishi, K., Kutsukake, K., Suzuki, H., and Iino, T. (1990). Gene *fliA* encodes an alternative sigma factor specific for flagellar operons in *Salmonella typhimurium*. *Mol. Gen. Genet.* **221**, 139–147.

Ohnishi, K., Ohto, Y., Aizawa, S.-I., Macnab, R. M., and Iiono, T. (1994). FlgD is a scaffolding protein needed for flagellar hook assembly in *Salmonella typhimurium*. *J. Bacteriol.* **176**, 2272–2281.

Ottemann, K. M. and Lowenthal, A. C. (2002). *Helicobacter pylori* uses motility for initial colonization and to attain robust infection. *Infect. Immun.* **70**, 1984–1990.

Ramos, H. C., Rumbo, M., and Sirard, J.-C. (2004). Bacterial flagellins: mediators of pathogenicity and host immune responses in mucosa. *Trends Microbiol.* **12**, 509–517.

Ramphal, R. and Arora, S. K. (2001). Recognition of mucin components by *Pseudomonas aeruginosa. Glycoconj. J.* **18**, 709–713.

Scharfman, A., Arora, S. K., Delmotee, P., *et al.* (2001). Recognition of Lewis^{x} derivatives present on mucins by flagellar components of *Pseudomonas aeruginosa. Infect. Immun.* **69**, 5243–5248.

Schmiel, D. H., Wagar, E., Karamanou, L., Weeks, D., and Miller, V. L. (1998). Phospholipase A of *Yersinia enterocolitica* contributes to pathogenesis in a mouse model. *Infect. Immun.* **66**, 3941–3951.

Schmiel, D. S., Young, G. M., and Miller, V. L. (2000). The *Yersinia enterocolitica* phospholipase gene *yplA* is part of the flagellar regulon. *J. Bacteriol.* **182**, 2314–2320.

Schmitt, C. K., Ikeda, J. S., Darnell, S. C., *et al.* (2001). Absence of all components of the flagellar export and synthesis machinery differentially alters virulence of *Salmonella enterica* serovar Typhimurium in models of typhoid fever, survival in macrophages, tissue culture invasiveness, and calf enterocolitis. *Infect. Immun.* **69**, 5619–5625.

Shapiro, L. and Maizel, J. (1973). Synthesis and structure of *Calobacter crescentus* flagella. *J. Bacteriol.* **113**, 478–485.

Shin, S. and Park, C. (1995). Modulation of flagellar expression in *Escherichia coli* by acetyl phosphate and the osmoregulator OmpR. *J. Bacteriol.* **177**, 4696–4702.

Smith, K. D., Andersen-Nissen, E., Hayashi, F., *et al.* (2003). Toll-like receptor 5 recognizes a conserved site on flagellin required for protofilament formation and bacterial motility. *Nat. Immunol.* **5**, 1159–1160.

Song, Y. C., Jin, S., Louie, H., *et al.* (2004). FlaC, a protein of *Campylobacter jejuni* TGH9011 (ATCC43431) secreted through the flagellar apparatus, binds epithelial cells and influences cell invasion. *Mol. Microbiol.* **53**, 541–553.

Sournik, V. (2004). Receptor clustering and signal processing in *E. coli* chemotaxis. *Trends Microbiol.* **12**, 569–576.

Soutourina, O., Kolb, A., Krin, E., *et al.* (1999). Multiple control of flagellum biosynthesis in *Escherichia coli*: role of H-NS protein and the cyclic AMP-catabolite activator protein complex in transcription of the *flhDC* master operon. *J. Bacteriol.* **181**, 7500–7508.

Springer, W. R. and Koshland, D. E. J. (1977). Identification of a protein methyltransferase as the *cheR* gene product in the bacterial sensing system. *Proc. Natl. Acad. Sci. U. S. A.* **74**, 533–537.

Stecher, B., Hapfelmeier, S., Muller, C., *et al.* (2004). Flagella and chemotaxis are required for efficient induction of *Salmonella enterica* serovar Typhimurium colitis in streptomycin-treated mice. *Infect. Immun.* **72**, 4138–4150.

Suarez, M. and Russmann, H. (1998). Molecular mechanisms of *Salmonella* invasion: the type III secretion system of pathogenicity island 1. *Int. Microbiol.* **1**, 197–204.

Szymanski, C. M., King, M., Haardt, M., and Armstrong, G. D. (1995). *Campylobacter jejuni* motility and invasion of CaCo-2 cells. *Infect. Immun.* **63**, 4295–4300.

Tasteyre, A., Barc, M. C., Collingnon, A., Boureau, H., and Karajalainen, T. (2001). Role of FliC and FliD flagellar proteins of *Clostridium difficile* in adherence and gut colonization. *Infect. Immun.* **69**, 7937–7940.

Teplitski, M., Goodier, R. I., and Ahmer, B. M. (2003). Pathways leading from BarA/SirA to motility and virulence gene expression in *Salmonella. J. Bacteriol.* **185**, 7257–7265.

Terry, K., Williams, S. M., Connolly, L., and Ottemann, K. M. (2005). Chemotaxis plays multiple roles during *Helicobacter pylori* animal infection. *Infect. Immun.* **73**, 803–811.

Tsubokura, M., Otsuki, K., Shimohira, I., and Yamamoto, H. (1979). Production of indirect hemolysin by *Yersinia enterocolitica* and its properties. *Infect. Immun.* **25**, 939–942.

Wadhams, G. H. and Armitage, J. P. (2004). Making sense of it all: bacterial chemotaxis. *Nat. Rev. Mol. Cell. Biol.* **5**, 1024–1037.

Warren, S. M. and Young, G. M. (2005). An amino-terminal secretion signal is required for YplA export by the Ysa, Ysc and flagellar type III secretion systems of *Yersinia enterocolitica* Biovar 1B. *J. Bacteriol.* **187**, 6075–6083.

Wassenaar, T. M., van der Zeijst, B. A., Ayling, R., and Newell, D. G. (1993). Colonization of chicks by motility mutants of *Campylobacter jejuni* demonstrates the importance of flagellin A expression. *J. Gen. Microbiol.* **139**, 1171–1175.

Wei, B. L., Brun-Zinkernagel, A. M., Simecka, J. W., *et al.* (2001). Positive regulation of motility and *flhDC* expression by the RNA-binding protein CsrA of *Escherichia coli. Mol. Microbiol.* **40**, 245–256.

Worku, M. L., Sidebotham, R. L., Baron, J. H., *et al.* (1999). Motility of *Helicobacter pylori* in a viscous environment. *Eur. J. Gastroenterol. Hepatol.* **11**, 1143–1150.

Yanagihara, S., Iyoda, S., Ohnishi, K., and Kutsukake, K. (1999). Structure and transcriptional control of the flagellar master operon of *Salmonella typhimurium. Genes Genet. Syst.* **74**, 105–111.

Yonekawa, H., Hayashi, H., and Parkinson, J. S. (1983). Requirement of the *cheB* function for sensory adaptation in *Escherichia coli. J. Bacteriol.* **156**, 1228–1235.

Yonekura, K., Maki, S., Morgan, D. G., *et al.* (2000). The bacterial flagellar cap as the rotary promoter of flagellin self-assembly. *Science* **290**, 2148–2152.

Young, B. M. and Young, G. M. (2002). YplA is exported by the Ysc, Ysa and flagellar type III secretion systems of *Yersinia enterocolitica*. *J. Bacteriol.* **184**, 1324–1334.

Young, G. M., Smith, M., Minnich, S. A., and Miller, V. L. (1998). The *Yersinia enterocolitica* motility master regulatory operon, *flhDC*, is required for flagellin production, swimming motility and swarming motility. *J. Bacteriol.* **181**, 2823–2833.

Young, G. M., Badger, J. L., and Miller, V. L. (2000). Motility is required to initiate host cell invasion by *Yersinia enterocolitica*. *Infect. Immun.* **68**, 4323–4326.

Young, G. M., Schmiel, D. H., and Miller, V. L. (1999). A new pathway for the secretion of virulence factors by bacteria: the flagellar export apparatus functions as a protein-secretion system. *Proc. Natl. Acad. Sci. U. S. A.* **96**, 6456–6461.

Zhao, H., Thompson, R. B., Lockatell, V., Johnson, D. E., and Mobley, H. L. (1998). Use of green fluorescent protein to assess urease gene expression by uropathogenic *Proteus mirabilis* during experimental ascending urinary tract infection. *Infect. Immun.* **66**, 330–335.

CHAPTER 6

The role of bacterial adhesion to epithelial cells in pathogenesis

Christof R. Hauck

INTRODUCTION

Colonizing host epithelia represents a formidable challenge to bacterial pathogens. To a large extent, epithelial surfaces are designed to shield the multicellular organism from the environment and to protect the body interior from potentially harmful microbes. Where epithelial surfaces permit exchange of molecular components with the external world, a multitude of innate and acquired host defence mechanisms keep microorganisms in check. In addition, invading pathogens either have to compete successfully with resident commensal bacteria for space and nutrients or have to reach and establish themselves in otherwise sterile parts of the body. As we will discuss in this chapter, the specific interaction with host surface components and the tight adhesion to epithelial cells form one of the common ways in which bacterial pathogens have evolved to successfully accomplish the colonization of their respective host organism. It is important to emphasize that in many cases, this initial host–microbe encounter at the epithelial barrier is not only a critical determinant of pathogen–host specificity and range but also, to a large extent, a decisive point for the infection process as a whole.

Adhesion of pathogenic bacteria to host cells had been observed at the beginning of the twentieth century in the early days of investigations into the microbiological origin of infectious diseases (Guyot, 1908). However, the concept that bacterial adhesion to host surfaces often represents an essential step in the development of infection matured only several decades later (Beachey, 1981; Duguid, 1959; Eden *et al.*, 1976; McNeish *et al.*, 1975; Punsalang and Sawyer, 1973). Mechanical cleansing processes, ranging from eye-blinking to

Bacterial–Epithelial Cell Cross-Talk: Molecular Mechanisms in Pathogenesis, ed. Beth A. McCormick. Published by Cambridge University Press.

the flushing of the urethra and the constant turnover of the mucous layer will remove any non-attached particles from the epithelium. Contacting the host surface and establishing a dependable adhesive connection is, therefore, key to successful colonization. The elucidation of adhesion components and the ability to genetically manipulate these components on both the pathogen and the host side have underscored the critical importance of bacterial adhesion in infectious diseases. In several models of bacterial infection, the molecular nature of the adhesin and the respective host receptors is now well documented. Although other bacterial surface structures, such as lipopolysaccharide, wall teichoic acid, and lipoteichoic acid, can contribute to cell adhesion (Edwards *et al.*, 2000; Ofek *et al.*, 1975; Paradis *et al.*, 1999; Weidenmaier *et al.*, 2004), the main bacterial adhesins characterized so far are proteins.

FIMBRIAL ADHESINS

In a broad sense, adhesive bacterial proteins can be classified into fimbrial and non-fimbrial (or afimbrial) adhesins. As a rule, fimbriae are long ($> 1\ \mu m$) and usually thin (2–8 nm) protrusions of the bacterial surface that are sometimes referred to as pili. The term "pilus" was originally used to describe bacterial organelles involved in the conjugative transfer of DNA, but it is now also applied regularly to adhesive surface structures. Fimbrial protrusions are made up of major (with more than 100 copies per pilus) and minor protein subunits that often, but not always, carry out the structural and the adhesive functions, respectively, of these microbial organelles. Classical examples of fimbrial adhesins are found in the Pap and Fim systems of uropathogenic *Escherichia coli* (UPEC) and the type IV pili of *Pseudomonas aeruginosa*, enteropathogenic *E. coli*, and pathogenic *Neisseriae*; the molecular components and the roles of these fimbriae for epithelial cell adhesion have been summarized in several reviews (Craig *et al.*, 2004; Jonson *et al.*, 2005; Kau *et al.*, 2005; Sauer *et al.*, 2004).

It appears that the long surface protrusions are an evolutionarily optimized tool to establish the initial bacterial contact with the epithelial barrier. Indeed, the thin and bendable fimbriae are ideally suited for the task of long-range contact, as they help to circumvent the electrostatic repulsion forces between the two negatively charged surfaces of the eukaryotic cell and the microbe. Moreover, at least type IV pili also seem to work in a manner similar to grappling hooks, as the pilus fiber can be retracted by the bacterium and thereby exert force (Maier *et al.*, 2002). This fascinating dynamic nature of the pilus not only can confer motility to single suspended bacteria (Merz

Table 6.1 *Fimbrial adhesins and recognized host cell structures*

Species	Adhesin	Recognized structure
Escherichia coli	FimH	Mannose
	FaeG	Gaβ1-3Gal
	GafD	N-acetyl-D-glucosamine
	PapG	Galβ1-4Gal
	SfaS	NeuAcβ2-3Galβ1-3GalNac
	CfaB	NeuAc-GM2
	G-Fimbrien	GlcNAc
Neisseria gonorrhoeae	TypIV pili	CD46 (?)
Vibrio cholerae	TypIV pili	L-Fucose

et al., 2000) but also endows the microbes with a means to actively narrow the gap between the piliated bacterium and the tissue surface. The result of pilus retraction in cell-associated bacteria, therefore, is a close apposition between the two membranes, allowing additional short-reach interactions to take place. An overview of fimbrial adhesins from different pathogenic microorganisms and their host receptors is given in Table 6.1.

As can be seen in Table 6.1, fimbrial adhesins most often function as lectins and recognize carbohydrate moieties found on membrane-embedded glycoproteins and glycolipids. Accordingly, low-complexity carbohydrates have the ability to interfere with the adhesion of bacterial pathogens, a phenomenon recognized well before the molecular characterization of the respective adhesins (Old, 1972). However, so far it has not been possible to translate these findings into novel clinical treatment options that prohibit bacterial attachment to epithelial cells and thereby block bacterial infections right from the start (Bouckaert *et al.*, 2005).

AFIMBRIAL ADHESINS

In addition to fimbrial adhesins, a large variety of non-fimbrial adhesins has been characterized. As the name implies, these adhesins are either embedded into or attached to the outer surface of the microbe and are usually not localized on surface protrusions. As we will see in the following examples, this type of adhesin often engages in direct protein–protein interactions with host components. Non-fimbrial adhesins are found in a variety of Gram-negative bacteria, but they are also common in Gram-positive bacteria, where few fimbrial adhesins have been described so far (Ton-That and Schneewind, 2004). Examples of non-fimbrial adhesins from Gram-positive microbes include internalin A (InlA) and InlB from *Listeria monocytogenes*. Similar to many other adhesins, these two proteins not only mediate tight binding of the bacteria to host cells but also trigger, via their recognized cellular receptors, actin cytoskeleton rearrangements, leading to bacterial internalization (Pizarro-Cerda *et al.*, 2004).

The initial InlA-mediated interaction of *L. monocytogenes* with intestinal epithelial cells is a prominent example of the essential role that adhesion to epithelia often plays in bacterial pathogenesis. Guinea pig and human intestinal epithelial cells allow InlA-dependent attachment of *L. monocytogenes*; this is due to the presence of a highly homologous E-cadherin, the cellular receptor for InlA, in these two mammalian species (Mengaud *et al.*, 1996). In contrast to humans and guinea pigs, mice are very resilient to oral infection, the normal entry site of food contaminating *L. monocytogenes*; this resistance could be attributed by a series of elegant experiments to a single amino-acid difference in mouse E-cadherin (glutamine at position 16) compared with human or guinea pig E-cadherin (proline at position 16) (Lecuit *et al.*, 1999). Moreover, when mice are engineered to express human E-cadherin in their intestines, these animals now become highly susceptible to oral infection by *L. monocytogenes* (Lecuit *et al.*, 2001).

The atomic structure of the adhesin–receptor complex has further illuminated the intricate handshake-like protein–protein interaction between InlA and E-cadherin (Schubert *et al.*, 2002). This submolecular view has also underpinned the critical position of Pro-16 in the center of the InlA–E-cadherin interface (Schubert *et al.*, 2002). The elucidation of InlA binding specificity and the molecular details of InlA–E-cadherin interaction have highlighted the essential role of this interaction in the establishment of the disease. Together, these investigations provide us with a marvelous example of how host specificity can be determined by bacterial adhesin–host receptor pairs.

INTEGRIN-BINDING AFIMBRIAL ADHESINS OF STAPHYLOCOCCI

Other Gram-positive bacteria, such as *Staphylococcus aureus* and *Streptococcus pyogenes*, have evolved a number of surface proteins that bind to serum and matrix components of their host (Foster and Hook, 1998; Patti *et al.*, 1994). These types of afimbrial adhesin include fibronectin-binding protein A and B (FnBP-A and -B) of *S. aureus* and streptococcal fibronectin-binding protein 1 (Sfb1) (also termed F1) of *S. pyogenes*. Both, FnBPs and Sfb1 attach the plasma component and extracellular matrix protein fibronectin to the surface of the bacteria by an intriguing tandem beta-zipper mechanism (Schwarz-Linek *et al.*, 2003).

It is thought that binding to extracellular matrix proteins such as fibronectin allows the bacteria to colonize matrix-coated surfaces such as implanted medical devices. Furthermore, the bacterial adhesin does not only mediate direct binding to fibronectin-covered surfaces but also utilizes fibronectin as a molecular bridge, indirectly linking the bacterial surface with the principal host fibronectin receptor, the integrin $\alpha_5\beta_1$ (Ozeri *et al.*, 1998; Sinha *et al.*, 1999). Upon bacterial adhesion, integrin ligation can lead to efficient internalization of the bacteria into eukaryotic cells in vitro and in vivo (Agerer *et al.*, 2003, 2005; Brouillette *et al.*, 2003; Ozeri *et al.*, 1998; Sinha *et al.*, 1999). Interference with fibronectin binding to the bacteria by either genetic deletion of FnBP proteins or administration of a recombinant fibronectin binding domain of FnBP as a competitive inhibitor during infections can attenuate staphylococcal virulence in several disease models, such as abscess formation, mastitis, and endocarditis (Brouillette *et al.*, 2003; Kuypers and Proctor, 1989; Menzies *et al.*, 2002).

However, the role of FnBPs in staphylococcal infections is not clear-cut, and results from experimental models have been reported where FnBPs did not contribute to the virulence of this pathogen (for review, see (Menzies, 2003). As staphylococci are associated with a large variety of clinical manifestations, often involving secreted toxins, it is very likely that cell adhesion and, in particular, FnBPs will act as bona fide virulence factors only in some of these situations. It is also conceivable that such FnBP-mediated adhesion and invasion processes are not only found in acute disease settings but also are of importance for the persistence of staphylococci in their host. Together, current experimental and epidemiologic data support the view that FnBPs contribute in some settings to the virulence of *S. aureus* and that cellular invasion via integrins represents one of the functional properties conferred by FnBP expression.

STAPHYLOCOCCAL INTERNALIZATION VIA MATRIX-BINDING INTEGRINS

Interestingly, most integrins usually operate in the context of immobilized matrix proteins and are not considered to mediate endocytosis of attached ligands. In particular, integrin $\alpha_5\beta_1$, the fibronectin receptor, has been implied in organizing extracellular fibronectin into a fibrillar network by exerting force on the immobilized matrix protein (Schwarzbauer and Sechler, 1999). It is interesting to speculate that in contrast to immobilized fibronectin, the integrin-attached bacteria are pulled into the cell by the same cellular force-generating machinery that is used to remodel the fibronectin matrix under physiologic conditions.

In general, extracellular matrix contact and integrin ligation induce the formation of protein complexes at the cytoplasmic aspect of cell adhesion sites. These protein complexes have been termed focal adhesions or focal contacts, as they occur at discrete focal spots along the matrix-facing surface of the adherent eukaryotic cell. Focal complexes have important structural and signaling functions, as they dynamically link the clustered and ligand-bound integrins to the intracellular actin network that, together with myosin, is responsible for force generation (Zaidel-Bar *et al.*, 2004). Due to their morphologic appearance and their functional connection to fibronectin fibril assembly, integrin $\alpha_5\beta_1$-initiated protein complexes have been termed fibrillar adhesions (Zamir and Geiger, 2001). A characteristic component of such adhesive structures is tensin, an actin-binding adaptor molecule (Zamir *et al.*, 2000). In addition, signaling enzymes, such as the protein tyrosine kinases Src and focal adhesion kinase (FAK), have been implied in the generation of fibrillar adhesions and the integrin $\alpha_5\beta_1$-mediated assembly of a fibrillar fibronectin network (Ilic *et al.*, 2004; Volberg *et al.*, 2001).

Work has now started to address the role of these integrin-associated host cell factors for the FnBP-mediated attachment and invasion of *S. aureus*. Engagement of integrin $\alpha_5\beta_1$ by fibronectin-binding staphylococci indeed induces the formation of fibrillar adhesion-like protein complexes at the site of bacterial attachment, as characterized by the recruitment of tensin, FAK, zyxin, and vinculin (Agerer *et al.*, 2005). Furthermore, interference with Src or FAK function abrogates the internalization of the bacteria via integrin $\alpha_5\beta_1$ and suppresses the increased tyrosine phosphorylation observed at bacterial attachment sites (Agerer *et al.*, 2003, 2005). One of the effectors of activated FAK and Src kinases during integrin-mediated internalization has been identified as cortactin, an actin-binding protein that can associate with the Arp2/3 complex and promote actin polymerization but that can also bind to dynamin-2, a regulator of membrane endocytosis (McNiven *et al.*, 2000; Selbach and Backert, 2005). Together, these investigations support the view that fibronectin-coated staphylococci induce fibrillar adhesion-like contact sites that are regulated by protein tyrosine kinase signaling and link the bacteria-occupied integrins with the intracellular actin cytoskeleton. It is tempting to speculate that in this case bacterial internalization is promoted not only by increased actin polymerization generating membrane protrusions that enclose the pathogen but also, at least partially, by the contraction forces generated by the integrin-connected intracellular actin-myosin network that, under physiological conditions, promotes fibronectin fibrillogenesis.

INTEGRIN ENGAGEMENT BY ENTEROPATHOGENIC *YERSINIAE*

In contrast to the indirect way in which staphylococci and streptococci exploit host integrin $\alpha_5\beta_1$, the enteropathogenic *Yersinia* species, *Y. enterocolitica* and *Y. pseudotuberculosis*, bind directly to β_1-containing integrins and thereby trigger uptake by host cells. Because of its invasive property, the bacterial adhesin responsible for integrin binding and cellular internalization has been coined invasin; it serves as the prototype for this class of proteins (Isberg and Leong, 1990). Latex beads coated with invasin or non-pathogenic *E. coli* expressing invasin are taken up efficiently by different cell types, demonstrating that this protein is sufficient to confer invasiveness (Dersch and Isberg, 1999; Isberg and Leong, 1990). Since its discovery, several seminal contributions have illuminated the structural determinants of invasin required for integrin binding as well as the host factors required to allow invasin-initiated uptake (see review in Isberg *et al.*, 2000). It is interesting to note that the C-terminal domains of this bacterial surface protein seem to mimic fibronectin type III repeats 9 and 10 that are involved in integrin binding. Indeed, invasin can competitively inhibit fibronectin association with integrin $\alpha_5\beta_1$ demonstrating that these two proteins bind to the same site in integrins (Isberg *et al.*, 2000).

Invasin expression seems to allow the orally ingested *Yersinia* to overcome the intestinal barrier. Indeed, invasin-deficient bacteria are impaired in their ability to transverse the intestinal lining (Marra and Isberg, 1997). In this case, it is thought that invasin-expressing *Yersinia* do not attach to the regular enterocytes, polarized epithelial cells that do not expose integrins on their apical surface. However, a specialized subset of intestinal epithelial cells, the so-called microfold or M-cells, does expose integrins to the gut lumen (Clark *et al.*, 1998). In this context, it is also worth mentioning that a close relative of these pathogens, the plague bacterium *Yersinia pestis*, does not express invasin. The respective coding sequence is found in the genome, and yet it is inactivated by the insertion of an insertion sequence (IS) element (Simonet *et al.*, 1996). As *Y. pestis* is transmitted by an arthropod vector directly into the bloodstream of the host, these bacteria do not have to overcome an epithelial barrier and, therefore, invasin expression might not be required any more.

Table 6.2 *Pathogens targeting extracellular matrix (ECM)-binding integrins*

Species	ECM protein/integrin subunit
Borrelia burgdorferi	FN/β1 integrins
Mycobacterium leprae	FN, LN/β1, and β4 integrins
Mycobacterium bovis BCG	FN/β1 integrins
Neisseria gonorrhoeae and *N. meningitidis*	FN, VN/β1, and β3 integrins
Porphyromonas gingivalis	β1 integrins
Shigella flexneri	β1 integrins
Staphylococcus aureus	FN, LN, Col/β1 integrins
Streptococcus pyogenes and *S. dysgalactiae*	FN/β1 integrins
Yersinia pseudotuberculosis and *Y. enterocolitica*	βx1 integrins

Col, collagen; FN, fibronectin; LN, laminin; VN, vitronectin.

Direct or indirect engagement of matrix-binding integrins is also observed for other pathogenic bacteria (listed in Table 6.2). Although integrins are exploited by multiple bacterial and viral pathogens as well as parasites, it is not straightforward to conceive the role of this interaction in the context of an intact epithelium. In particular, integrins are distributed on the basolateral side of polarized epithelial cells, which would be inaccessible for bacteria colonizing the throat, intestine, or urogenital tract. Therefore, it has been speculated that these types of adhesin come into play only after the initial contact of the microbes with the tissue surface. As shown in several examples, pathogenic bacteria can influence the integrity of the epithelial barrier either directly by secreted toxins or indirectly by inducing granulocyte influx and tissue destruction (McCormick, 2003). Damage to the epithelial lining might then allow microbes to gain access to basolateral components such as integrins or cadherins.

COOPERATION BETWEEN FIMBRIAL AND AFIMBRIAL ADHESINS: THE PARADIGM OF PATHOGENIC *NEISSERIAE*

Fimbrial and afimbrial adhesins are not mutually exclusive, but often they are expressed simultaneously on the same microorganism. The pathogenic *Neisseriae* are one of the well-characterized examples, where both types of adhesin seem to act in concert and together coordinate a multistep and complex process of cell attachment. The remainder of this chapter concentrates on this paradigm in order to highlight some of the features of bacterial adhesins and receptor targeting. Furthermore, in light of advances, this overview is extended beyond the direct adhesin–receptor interaction to cover some of the specific responses of the host epithelial cell following bacterial binding.

Both pathogenic neisserial species, *N. gonorrhoeae* and *N. meningitidis* (the gonococcus and the meningococcus, respectively), are highly specialized colonizers of the human mucosa. Gonococci and meningococci are known to express fimbriae; the presence of these adhesive surface structures is subject to phase variation. Fimbriated *Neisseriae* have a high tendency to stick together in microcolonies, as the pilus supports bacteria–bacteria adhesion. Moreover, it is generally accepted that type IV pilus-mediated interactions are the initial event allowing both unencapsulated *N. gonorrhoeae* and encapsulated meningococci to colonize the epithelial surface of the urogenital tract or the nasopharynx, respectively (Kellogg *et al.*, 1963; Punsalang and Sawyer, 1973), although direct in vivo proof for this concept is still lacking (Cohen and Cannon, 1999). In vitro, piliated *Neisseria* adhere avidly to polarized human epithelial cells and form microcolonies on the apical surface (Pujol *et al.*, 1997). These microcolonies are due to bacterial replication but also seem to arise from pilus-initiated bacterial aggregates that attach to the cell surface. The PilX protein has been identified as a factor promoting bacterial aggregation in the presence of pili (Helaine *et al.*, 2005). Deletion of the PilX encoding gene not only disrupts bacterial aggregates but also dramatically reduces attachment to host cells (Helaine *et al.*, 2005). These data support the view that pilus-mediated cell attachment requires prior bacteria–bacteria binding and suggests that bacterial aggregates allow bacterial adhesion even in the case of low-affinity interaction between the bacterial adhesin and the cellular receptor by increasing avidity. This would also be in line with the obvious difficulty in unequivocally identifying a cellular receptor for the neisserial pilus, as such a low-affinity interaction might be hard to detect by standard biochemical approaches.

The main subunit of the gonococcal type IV pilus, pilin encoded by the *pilE* gene, is a textbook paradigm for its astounding antigenic variability. Antigenic variation of pilin is generated by a recombination-based exchange of coding sequences between promoter-less silent gene loci and actively expressed PilE loci (Meyer *et al.*, 1990). Although pilin can agglutinate erythrocytes, it has always been questioned whether the variable PilE can encode the principal adhesin of the neisserial type IV pilus. Work from several laboratories has suggested that a minor pilus subunit, the PilC protein encoded by two alleles in the genome, is the adhesive factor (Morand et al., 2001; Rudel et al., 1995). Indeed, purified gonococcal PilC binds to human cells and inhibits adhesion of piliated *N. gonorrhoeae* and *N. meningitidis* (Scheuerpflug *et al.*, 1999). PilC-deficient bacteria are impaired in their adhesion to human epithelial cells but also seem to be compromised in their capacity for natural transformation and motility, both pilus-dependent traits (Morand *et al.*, 2004; Ryll *et al.*, 1997), suggesting that PilC could have multiple functions. At least in meningococci it appears that only one of the two *pilC* alleles, *pilC1*, entails adhesive properties (Morand *et al.*, 2001). So far, no binding partner for the PilC adhesin on human cells has been described (Kirchner and Meyer, 2005). However, work from the group of Jonsson has suggested that human CD46 serves as a cellular receptor for pilitated *Neisseria* on epithelial cell lines (Kallstrom *et al.*, 1997). In other situations, a strict correlation between CD46 expression and pilus-mediated adhesion of *Neisseria* has not been detected (Tobiason and Seifert, 2001), and CD46-mediated binding of piliated gonococci has not been observed in other studies (Kirchner *et al.*, 2005). Interestingly, although mice transgenic for human CD46 show a higher mortality rate after intraperitoneal injection of meningococci, the virulence of the bacteria in this in vivo model is independent of a piliated phenotype (Johansson *et al.*, 2003).

Despite the still debated role of CD46 as a cellular receptor for pilus-dependent interactions, it is clear that pilus-mediated attachment to epithelial cells evokes a number of host responses, ranging from changes in cytosolic Ca^{2+} levels to tyrosine phosphorylation of host proteins (Ayala *et al.*, 2001; Hoffmann *et al.*, 2001; Kallstrom *et al.*, 1998; Lee *et al.*, 2002). These cellular events are observed within minutes to a few hours after infection, coinciding with the initial local adherence of the microbes, when bacterial microcolonies are found on top of the cells. However, several hours after the pilus-mediated contact, pilus and capsule expression are downregulated in the case of meningococci (Deghmane *et al.*, 2002), suggesting that the bacteria sense the presence of, or the attachment to, the eukaryotic cells by a currently unknown mechanism. As pilus expression ceases, the pattern of adhesion is altered, indicating that another type of adhesive contact is established. By scanning electron microscopy, it has been observed that the bacterial microcolonies are resolved, and singly attached bacteria are found distributed over the epithelial cell surface (Pujol *et al.*, 1997). This form of attachment has been termed "diffuse adherence" and seems to represent a further step in the microbe–host cell interaction.

It is thought that at this point a second group of neisserial adhesins, the colony opacity-associated (Opa) proteins, come into play. In both pathogenic neisserial species, multiple functional copies of *opa* genes are distributed over the entire genome. Transcription of Opa proteins is constitutive; however, expression undergoes a high rate of phase variation by a translation-based mechanism (Stern and Meyer, 1987; Stern *et al.*, 1986). Therefore, a natural population of these pathogens will comprise both non-opaque and opaque organisms, with some having a single Opa protein but others expressing multiple Opas at a given time. As there are up to 12 copies of *opa* genes in the gonococcal genome, the possibilities for variation are enormous. It is interesting to note that in human-volunteer challenge experiments, an initial inoculum of non-opaque gonococci can establish disease at infectious doses comparable with opaque phenotypes. However, most of the bacteria that are reisolated from the infected volunteers have switched to an opaque phenotype (Jerse *et al.*, 1994; Swanson *et al.*, 1988). These results suggest that Opa proteins, although not essential for the initial contact with the host, play a beneficial role for the pathogen during colonization and multiplication in vivo.

In contrast to the bacterial type IV pilus, Opa proteins are embedded within the outer bacterial membrane and therefore belong to the group of afimbrial adhesins. Secondary structure predictions suggest that mature Opa proteins possess eight membrane-spanning domains arranged as anti-parallel β-strands, giving rise to a membrane-embedded β-barrel with four extracellular loops (Bhat *et al.*, 1991; de Jonge *et al.*, 2002; Malorny *et al.*, 1998). Although the crystal structure of Opa proteins is currently unknown, the structure of the related NspA protein from *N. meningitidis* has been solved and suggests that the extracellular loops constitute a conformational binding interface (Vandeputte-Rutten *et al.*, 2003). On the basis of sequence comparison of multiple gonococcal and meningococcal Opa proteins, the amino-acid sequence of the central two extracellular loops has been found to be hypervariable, and they have been termed hypervariable domain 1 (HV-1) and 2 (HV-2) (Bhat *et al.*, 1991). New Opa protein variants constantly emerge, not only due to point mutations within HV-1 and HV-2 but also due to modular exchange of domains between different Opa proteins (Hobbs *et al.*, 1994). Interestingly, the interaction of Opa proteins with different receptors on human cells has also been pinned down to the HV-1 and HV-2 regions (see below), posing the puzzling question of specific receptor recognition in the context of these sequence alterations.

Work since 1995 has identified the cellular receptors targeted by various Opa proteins. Historically, the first cellular receptor characterized was found to belong to the family of heparansulfate proteoglycans (HSPGs), highly glycosylated proteins that occur in transmembrane (syndecans) or glycosylphosphatidylinositol-anchored (glypicans) forms (David, 1993).

Initially, it was observed that heparin addition or heparinase treatment abolishes gonococcal adhesion to different epithelial cells and that hamster cell lines deficient in heparansulfate biosynthesis were poorly recognized by the respective opaque gonococcal variants (Chen *et al.*, 1995; van Putten and Paul, 1995). Additional investigations have revealed that both syndecan-1 and syndecan-4 can support not only the binding but also the internalization of Opa_{HSPG}-expressing gonococci (Freissler *et al.*, 2000). So far, only a limited set of Opa proteins with specificity for HSPGs (Opa_{HSPG} protein) have been identified, including OpaA/Opa_{30} of gonococcal strain MS11 and $Opa_{27.5}$ of strain VP1 (Kupsch *et al.*, 1993; van Putten and Paul, 1995). Furthermore, Opa proteins with this type of specificity have not been described in meningococci.

On the bacterial side, HSPG binding has been mapped to the hypervariable extracellular loops of Opa proteins. From mutagenesis studies, it appears that a series of positively charged amino acids in HV-1 and HV-2 is critical in order to mediate the association with the HSPGs that have a high negative charge (Bos *et al.*, 2002). Interestingly, HSPG-specific Opa proteins have additional binding capabilities. Depending on the cell line used, an increased Opa_{HSPG} protein-triggered invasion has been observed in the presence of serum (Gomez-Duarte *et al.*, 1997). Further analysis has suggested that Opa_{HSPG} proteins also bind to the serum proteins vitronectin and fibronectin and, in a manner similar to the above described staphylococcal FnBPs, can therefore mediate an indirect engagement of host cell integrins (Dehio *et al.*, 1998; van Putten *et al.*, 1998). Again, indirect binding and clustering of integrins allow enhanced internalization by human non-professional phagocytes. As integrins and also syndecans are found ubiquitously on most human cells, Opa_{HSPG} protein-mediated attachment could connect these bacteria with numerous cell types during the infection. Whereas a few Opa proteins with binding specificity for HSPGs have been characterized, Opa_{HSPG} proteins represent only a minor fraction of the total Opa protein repertoire of most characterized strains.

In contrast, the majority of the currently characterized meningococcal and gonococcal Opa proteins display binding specificity for human surface receptors of the carcinoembryonic antigen (CEA)-related cell-adhesion molecule (CEACAM) family (Opa_{CEA} proteins). Prototypes of this group of Opa proteins comprise, for example, Opa_{52} of gonococcal strain MS11 and Opa_{132} of meningococcal strain C751 (for Opa protein nomenclature, refer to Malorny *et al.*, 1998). As is true for Opa_{HSPG} proteins, the binding sites for CEACAMs reside within the hypervariable loops of CEACAM-specific Opa proteins. So far, no consensus motif has been elucidated that would predict the capabilities of a given Opa protein to bind to CEACAMs

(Bos *et al.*, 2002). What is more, the combination of hypervariable loops from two Opa_{CEA} proteins in one chimeric molecule does not result in a CEACAM-binding protein (Bos *et al.*, 2002). These results are in line with the idea that CEACAM recognition is based on a three-dimensional structure created by the proper combination of two complementary hypervariable loops of Opa proteins. It is interesting to note that shuffling of hypervariable loops derived from CEACAM-binding Opa proteins in some cases created chimeric Opa proteins with a novel specificity for HSPGs (Bos *et al.*, 2002). These striking observations suggest that HSPG recognition by Opa proteins might be a side product of the ongoing evolution and optimization of Opa_{CEA} molecules.

On the host side, all Opa_{CEA} proteins bind to the N-terminal immunoglobulin variable-like (Ig_V) domain characteristic for CEACAMs. Although this N-terminal domain is highly conserved among CEACAMs, opaque *Neisseria* have been found to recognize only four of the seven CEACAMs expressed by their human host, namely CEACAM1, CEACAM3, CEA (the product of the *ceacam*5 gene), and CEACAM6. Common to these CEACAM family members is the presence of the CD66 epitope in their N-terminal domain; therefore, these proteins have formerly been designated CD66a (CEACAM1), CD66c (CEACAM6), CD66d (CEACAM3), and CD66e (CEA). In contrast, the additional family members CEACAM4, CEACAM7, and CEACAM8 (CD66b) are not bound by any Opa_{CEA} protein analyzed so far (Popp *et al.*, 1999). On the basis of Opa_{CEA}-binding and non-binding CEACAM N-terminal domains, a number of receptor chimeras and mutants have been constructed to delineate the Opa_{CEA} protein binding site on the non-glycosylated C′CFG-face of the immunoglobulin domain fold (reviewed in Billker *et al.*, 2000). The crystal structure of murine CEACAM1, the only member of this glycoprotein family found in mice, has revealed a characteristic surface-exposed loop coordinated by Tyr-34 within the C′CFG face (Tan *et al.*, 2002). It is assumed that a similar prominent surface extension helps to anchor Opa_{CEA} proteins to human CEACAM N-terminal domains, as mutagenesis of the corresponding Tyr-34 residue in human CEACAM1 abolishes binding of Opa_{CEA}-expressing meningococci (Virji *et al.*, 1999). It is interesting to point out that in mice, CEACAM1 serves as the cellular receptor for mouse hepatitis virus (MHV), and MHV binding also takes place at the N-terminal Ig_V-like domain (Dveksler *et al.*, 1993). Importantly, the genetic ablation of the murine CEACAM1 N-terminal domain from the mouse genome has resulted in animals resistant to MHV infection, pointing towards the essential role that the pathogen–CEACAM interaction plays in this system (Blau *et al.*, 2001).

Although the physiologic function of CEACAMs in vivo is not understood completely, some family members are known to mediate cell–cell adhesion in vitro via both homotypic (CEACAM1, CEA, CEACAM6) and/or heterotypic (CEA–CEACAM6 and CEACAM6–CEACAM8) interactions (Benchimol *et al.*, 1989; Oikawa *et al.*, 1991). CEACAM1 and CEACAM6 on neutrophils are also involved in the adherence of activated neutrophils to cytokine-activated endothelial cells, both directly through their ability to present the sialylated Lewis^x antigen to E-selectin and indirectly by the CEACAM6-stimulated activation of CD18 integrins (Kuijpers *et al.*, 1992). Interestingly, CEACAM1 is also implicated in hepatic uptake of insulin, demonstrating that CEACAMs could participate in internalization processes under physiological conditions (Poy, Ruch *et al.*, 2002; Poy, Yang *et al.*, 2002).

An important aspect of CEACAM biology is the fact that several CEACAM family members can be expressed by epithelial cells, where they are usually located at the apical membrane of the polarized epithelium (Hammarstrom, 1999). This has important implications with respect to their role as bacterial receptors, as they are prominently exposed on mucosal surfaces and, therefore, accessible for incoming microbes. Furthermore, CEACAMs have been linked to signal transduction into the cell, and some isoforms seem to be connected to the intracellular cytoskeleton (Obrink, 1997).

Most work in this regard has focused on CEACAM1, the most widely expressed CEACAM family member. CEACAM1 is not only abundantly expressed on epithelia (including stomach, colon, kidney, gall bladder, liver, urinary bladder, prostate, cervix, and endometrium), sweat and sebaceous gland cells, and endothelia but also found on leukocytes such as granulocytes and B- and T-cells (Hammarstrom, 1999). In addition, CEACAM1 homologs are found in non-primate species such as mouse and rat, enabling better experimental access. In epithelial cells, CEACAM1 localizes to cell–cell contacts and is associated with the actin cytoskeleton under the control of Rho-family GTPases (Sadekova *et al.*, 2000). Moreover, the isolated cytoplasmic domain of CEACAM1 binds to actin and tropomyosin in vitro (Schumann *et al.*, 2001), supporting the view that CEACAM1 is connected directly to the cytoskeleton and plays a role in maintaining tissue integrity. Other groups report a direct association between the cytoplasmic domain of CEACAM1 and β3 integrins, indicating that there could also be an indirect linkage, via integrins, between CEACAMs and the actin cytoskeleton (Brummer *et al.*, 2001). However, pharmacological inhibitors of the actin cytoskeleton do not seem to influence CEACAM1- and CEACAM6-mediated uptake of opaque bacteria (Billker *et al.*, 2002; McCaw *et al.*, 2004).

Binding of Opa_{CEA} proteins to members of the CEACAM family is sufficient to induce the internalization of the bacteria into several cell types in vitro. Several CEACAMs are expressed on human granulocytes, where in particular CEACAM3 promotes an efficient opsonin-independent uptake of CEACAM-binding bacteria (Schmitter *et al.*, 2004). With regard to epithelial cells, CEACAM1, CEA, and CEACAM6 are often found to be coexpressed. In polarized T84 epithelial cell monolayers, CEA, CEACAM1, and CEACAM6 are transported apically, where they mediate invasion and subsequent transcytosis of Opa_{CEA}-expressing gonococci by an intracellular route (Wang *et al.*, 1998). In primary human umbilical vein endothelial cells (HUVECs) and in certain epithelial cells such as cells derived from human ovary, CEACAM1 expression is low in resting cells. However, CEACAM1 expression can be dramatically induced by infection with pathogenic Neisseriae, leading in turn to increased adherence and internalization of Opa-positive variants (Muenzner *et al.*, 2001, 2002).

It is conceivable, therefore, that Opa-mediated binding to CEACAMs is an important mechanism that allows the pathogens to successfully colonize human mucosal surfaces. According to this hypothesis, Opa expression by the bacteria and presence or upregulation of CEACAMs on epithelia or endothelia upon contact with the microorganisms would act as a switch to facilitate bacterial colonization and potentially also to enhance the passage of pathogenic Neisseriae through epithelial (and endothelial) barriers. It is important to point out that several other Gram-negative human-specific pathogens that share the same ecological niche and cause a similar spectrum of diseases as gonococci and meningococci have been found to possess CEACAM-binding adhesins. In an example of convergent evolution, typeable and non-typeable *Haemophilus influenzae, H. influenzae* biogroup *aegyptius*, and *Moraxella catarrhalis* have elaborated diverse surface antigens to engage CEACAMs (Hill and Virji, 2003; Virji *et al.*, 2000). For example, *H. influenzae* contacts CEACAM family members by the outer-membrane protein P5, whereas *M. catarrhalis* employs the UspA1 antigen (Hill and Virji, 2003; Hill *et al.*, 2001). These findings imply that there must be some major advantage for bacteria colonizing the human mucosa to specifically target members of the CEACAM family.

Novel insight suggests that in addition to providing a tight molecular anchor on the apical side of human epithelia, CEACAM recognition might serve an even more elaborate function in support of mucosal colonization. More specifically, CEACAM engagement by human pathogens might be a means to blunt an innate defense mechanism of stratified and squamous epithelial tissues, namely the exfoliation of superficial cells.

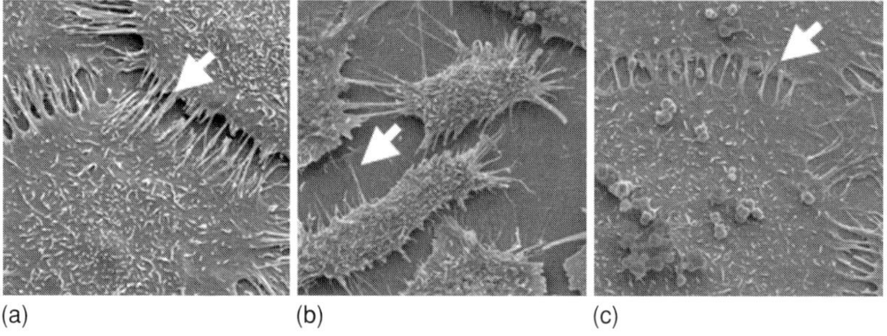

Figure 6.1 Epithelial cells infected with carcinoembryonic antigen-related cell-adhesion molecule (CEACAM)-binding *Neisseria gonorrhoeae* do not detach after prolonged infection. Confluent monolayers of human cervix epithelial cell line (ME-180) grown on collagen-coated surfaces were left (a) uninfected, or (b) infected for 14 h with piliated non-opaque *N. gonorrhoeae* (NgoP+), or (c) infected with non-piliated Opa_{CEA}-expressing *N. gonorrhoeae* (Ngo Opa_{CEA}). Cultures were fixed in situ and analyzed by scanning electron microscopy. Whereas uninfected monolayers display well-spread epithelial cells with numerous cell–cell contacts (a; arrow), ME-180 cells infected with piliated gonococci lose cell–cell contacts (B; arrow), round up and detach from the extracellular matrix surface, consistent with bacteria-induced exfoliation. In contrast, epithelial cells infected with CEACAM-binding gonococci (an OpaCEA-expressing strain) stay attached to the matrix, although cell–cell contacts are diminished (c; arrow). Figures courtesy of M. Rohda, GBF, Braunschweig, Germany.

Importantly, different non-opaque gonococcal variants cause detachment of epithelial cells after prolonged infection in culture, and such an exfoliation of urethral cells has been reported to occur during gonorrhea in vivo (Apicella *et al.*, 1996; Evans, 1977; Melly *et al.*, 1981; Mosleh *et al.*, 1997; Ward *et al.*, 1974). In striking contrast, prolonged infection with CEACAM-binding gonococci does not result in epithelial exfoliation (Muenzner *et al.*, 2005). This process can be documented by scanning electron microscopy, whereupon infection of a confluent monolayer of a human cervix epithelial cell line with piliated non-opaque gonococci, a reduction in cell–cell contacts, and the rounding and detachment of infected epithelial cells can be observed clearly (Figure 6.1). However, when these cervix epithelial cells that endogenously express CEACAM family members are challenged with Opa_{CEA}-expressing gonococci, they show loss of cell–cell contacts but still remain attached to the underlying extracellular matrix (Figure 6.1). The lack of detachment was attributed to a dramatically increased adhesive property of cells infected with opaque *Neisseria* (Muenzner *et al.*, 2005).

By microarray-based gene expression analysis and further functional studies, the enhanced adhesive properties of the infected cells could be pinned down to the de novo expression of CD105 following CEACAM engagement. Importantly, upregulation of CD105 occurs in response to a number of CEACAM-binding human pathogens, including *N. gonorrhoeae, N. meningitidis, H. influenzae,* and *M. catarrhalis,* and is sufficient to promote increased cell adhesion (Muenzner *et al.*, 2005). Therefore, it is conceivable that pathogen-initiated CEACAM stimulation, the stimulated expression of CD105, and the ensuing enhanced cell–matrix adhesion are central events that counteract the exfoliation of infected epithelial cells in vivo and, therefore, facilitate the colonization of the human mucosa by CEACAM-binding bacteria. In light of these results, it is tempting to speculate that the prevention of epithelial detachment attained through CEACAM binding is the major evolutionary driving force behind the appearance of distinct CEACAM-directed adhesins in several bacterial species colonizing the human mucosa.

CONCLUSIONS

Pathogen–epithelial cell recognition and the tight attachment of microorganisms to this host cell type are of fundamental importance in major infectious diseases. Since the initial discoveries of the first fimbrial and afimbrial adhesins, work by numerous research groups has led to the identification of numerous adhesin–receptor pairs, and these molecular investigations still yield exciting and often surprising insight. Fimbrial adhesins have long been recognized as critical bacterial surface structures mediating the initial contact between the prokaryotic and eukaryotic worlds. In addition, a large variety of afimbrial adhesins provides pathogens with an additional arsenal to interact intimately with target cells and to trigger specific responses upon receptor engagement. The more we learn about the intricate molecular communication taking place at the bacterial–epithelial cell interface, the more it is becoming evident that receptor targeting by bacterial adhesins has effects beyond the attachment to the host cell surface. The coming years will witness an increased appreciation and understanding of such post-adhesion events that modulate and shape the infection process.

ACKNOWLEDGMENTS

I thank the members of my laboratory for stimulating discussions during the writing process and the Bundesministerium für Bildung und Forschung as well as the Deutsche Forschungsgemeinschaft for support of our research in this area.

REFERENCES

Agerer, F., Michel, A., Ohlsen, K., and Hauck, C. R. (2003). Integrin-mediated invasion of *Staphylococcus aureus* into human cells requires Src family protein tyrosine kinases. *J. Biol. Chem.* **278**, 42 524–42 531.

Agerer, F., Lux, S., Michel, A., *et al.* (2005). Cellular invasion by *Staphylococcus aureus* reveals a functional link between focal adhesion kinase and cortactin in integrin-mediated internalisation. *J. Cell Sci.* **118**, 2189–2200.

Apicella, M. A., Ketterer, M., Lee, F. K. N., *et al.* (1996). The pathogenesis of gonococcal urethritis in men: confocal and immunoelectron microscopic analysis of urethral exsudates from men infected with *Neisseria gonorrhoeae. J. Infect. Dis.* **173**, 636–646.

Ayala, B. P., Vasquez, B., Clary, S., *et al.* (2001). The pilus-induced Ca2+ flux triggers lysosome exocytosis and increases the amount of Lamp1 accessible to *Neisseria* IgA1 protease. *Cell. Microbiol.* **3**, 265–275.

Beachey, E. H. (1981). Bacterial adherence: adhesin–receptor interactions mediating the attachment of bacteria to mucosal surface. *J. Infect. Dis.* **143**, 325–345.

Benchimol, S., Fuks, A., Jothy, S., *et al.* (1989). Carcinoembryonic antigen, a human tumor marker, functions as an intercellular adhesion molecule. *Cell* **57**, 327–334.

Bhat, K. S., Gibbs, C. P., Barrera, O., *et al.* (1991). The opacity proteins of *Neisseria gonorrhoeae* strain MS11 are encoded by a family of 11 complete genes. *Mol. Microbiol.* **5**, 1889–1901. [Published erratum appears in *Mol. Microbiol.* 1992, **6**, 1073–1076.]

Billker, O., Popp, A., Gray-Owen, S. D., and Meyer, T. F. (2000). The structural basis of CEACAM-receptor targeting by neisserial Opa proteins. *Trends Microbiol.* **8**, 258–260.

Billker, O., Popp, A., Brinkmann, V., *et al.* (2002). Distinct mechanisms of internalization of *Neisseria gonorrhoeae* by members of the CEACAM receptor family involving Rac1- and Cdc42- dependent and -independent pathways. *EMBO J.* **21**, 560–571.

Blau, D. M., Turbide, C., Tremblay, M., *et al.* (2001). Targeted disruption of the Ceacam1 (MHVR) gene leads to reduced susceptibility of mice to mouse hepatitis virus infection. *J. Virol.* **75**, 8173–8186.

Bos, M. P., Kao, D., Hogan, D. M., Grant, C. C., and Belland, R. J. (2002) Carcinoembryonic antigen family receptor recognition by gonococcal Opa proteins requires distinct combinations of hypervariable Opa protein domains. *Infect. Immun.* **70**, 1715–1723.

Bouckaert, J., Berglund, J., Schembri, M., *et al.* (2005). Receptor binding studies disclose a novel class of high-affinity inhibitors of the *Escherichia coli* FimH adhesin. *Mol. Microbiol.* **55**, 441–455.

Brouillette, E., Grondin, G., Shkreta, L., Lacasse, P., and Talbot, B. G. (2003). In vivo and in vitro demonstration that *Staphylococcus aureus* is an intracellular pathogen in the presence or absence of fibronectin-binding proteins. *Microb. Pathog.* **35**, 159–168.

Brummer, J., Ebrahimnejad, A., Flayeh, R., *et al.* (2001). cis Interaction of the cell adhesion molecule CEACAM1 with integrin beta(3). *Am. J. Pathol.* **159**, 537–546.

Chen, T., Belland, R. J., Wilson, J., and Swanson, J. (1995). Adherence of pilus-Opa+ gonococci to epithelial cells in vitro involves heparan sulfate. *J. Exp. Med.* **182**, 511–517.

Clark, M. A., Hirst, B. H., and Jepson, M. A. (1998). M-cell surface beta1 integrin expression and invasin-mediated targeting of *Yersinia pseudotuberculosis* to mouse Peyer's patch M cells. *Infect. Immun.* **66**, 1237–1243.

Cohen, M. S. and Cannon, J. G. (1999). Human experimentation with Neisseria gonorrhoeae: progress and goals. *J. Infect. Dis.* **179** (**Suppl 2**), S375–379.

Craig, L., Pique, M. E., and Tainer, J. A. (2004). Type IV pilus structure and bacterial pathogenicity. *Nat. Rev. Microbiol.* **2**, 363–378.

David, G. (1993). Integral membrane heparan sulfate proteoglycans. *FASEB J.* **7**, 1023–1030.

Deghmane, A. E., Giorgini, D., Larribe, M., Alonso, J. M., and Taha, M. K. (2002). Down-regulation of pili and capsule of *Neisseria meningitidis* upon contact with epithelial cells is mediated by CrgA regulatory protein. *Mol. Microbiol.* **43**, 1555–1564.

Dehio, M., Gomez-Duarte, O. G., Dehio, C., and Meyer, T. F. (1998). Vitronectin-dependent invasion of epithelial cells by *Neisseria gonorrhoeae* involves alpha(v) integrin receptors. *FEBS Lett.* **424**, 84–88.

De Jonge, M. I., Bos, M. P., Hamstra, H. J., *et al.* (2002). Conformational analysis of opacity proteins from *Neisseria meningitidis*. *Eur. J. Biochem.* **269**, 5215–5223.

Dersch, P. and Isberg, R. R. (1999). A region of the *Yersinia pseudotuberculosis* invasin protein enhances integrin-mediated uptake into mammalian cells and promotes self-association. *EMBO J.* **18**, 1199–1213.

Duguid, J. P. (1959). Fimbriae and adhesive properties in *Klebsiella* strains. *J. Gen. Microbiol.* **21**, 271–286.

Dveksler, G. S., Dieffenbach, C. W., Cardellichio, C. B., *et al.* (1993). Several members of the mouse carcinoembryonic antigen-related glycoprotein family are functional receptors for the coronavirus mouse hepatitis virus-A59. *J. Virol.* **67**, 1–8.

Eden, C. S., Hanson, L. A., Jodal, U., Lindberg, U., and Akerlund, A. S. (1976). Variable adherence to normal human urinary-tract epithelial cells of *Escherichia coli* strains associated with various forms of urinary-tract infection. *Lancet* **1**, 490–492.

Edwards, N. J., Monteiro, M. A., Faller, G., *et al.* (2000). Lewis X structures in the O antigen side-chain promote adhesion of *Helicobacter pylori* to the gastric epithelium. *Mol. Microbiol.* **35**, 1530–1539.

Evans, B. A. (1977). Ultrastructural study of cervical gonorrhea. *J. Infect. Dis.* **136**, 248–255.

Foster, T. J. and Hook, M. (1998). Surface protein adhesins of *Staphylococcus aureus*. *Trends Microbiol.* **6**, 484–488.

Freissler, E., Meyer auf der Heyde, A., David, G., Meyer, T. F., and Dehio, C. (2000). Syndecan-1 and syndecan-4 can mediate the invasion of Opa_{HSPG}-expressing *Neisseria gonorrhoeae* into epithelial cells. *Cell. Microbiol.* **2**, 69–82.

Gomez-Duarte, O. G., Dehio, M., Guzman, C. A., *et al.* (1997). Binding of vitronectin to Opa-expressing *Neisseria gonorrhoeae* mediates invasion of HeLa cells. *Infect. Immun.* **65**, 3857–3866.

Guyot, G. (1908). Über die bakterielle Adhäsion. *Zentralbl. Bakteriol.* **46**, 640–653.

Hammarstrom, S. (1999). The carcinoembryonic antigen (CEA) family: structures, suggested functions and expression in normal and malignant tissues. *Semin. Cancer Biol.* **9**, 67–81.

Helaine, S., Carbonnelle, E., Prouvensier, L., *et al.* (2005). PilX, a pilus-associated protein essential for bacterial aggregation, is a key to pilus-facilitated attachment of *Neisseria meningitidis* to human cells. *Mol. Microbiol.* **55**, 65–77.

Hill, D. J. and Virji, M. (2003). A novel cell-binding mechanism of *Moraxella catarrhalis* ubiquitous surface protein UspA: specific targeting of the N-domain of carcinoembryonic antigen-related cell adhesion molecules by UspA1. *Mol. Microbiol.* **48**, 117–129.

Hill, D. J., Toleman, M. A., Evans, D. J., *et al.* (2001). The variable P5 proteins of typeable and non-typeable *Haemophilus influenzae* target human CEACAM1. *Mol. Microbiol.* **39**, 850–862.

Hobbs, M. M., Seiler, A., Achtmann, M., and Cannon, J. G. (1994). Microevolution within a clonal population of pathogenic bacteria: recombination, gene duplication and horizontal genetic exchange in the *opa* gene family of *Neisseria meningitidis*. *Mol. Microbiol.* **12**, 171–180.

Hoffmann, I., Eugene, E., Nassif, X., Couraud, P. O., and Bourdoulous, S. (2001). Activation of ErbB2 receptor tyrosine kinase supports invasion of endothelial cells by *Neisseria meningitidis*. *J. Cell Biol.* **155**, 133–143.

Ilic, D., Kovacic, B., Johkura, K., *et al.* (2004). FAK promotes organization of fibronectin matrix and fibrillar adhesions. *J. Cell Sci.* **117**, 177–187.

Isberg, R. R. and Leong, J. M. (1990). Multiple b_1 chain integrins are receptors for invasin, a protein that promotes bacterial penetration into mammalian cells. *Cell* **60**, 861–871.

Isberg, R. R., Hamburger, Z., and Dersch, P. (2000). Signaling and invasin-promoted uptake via integrin receptors. *Microbes Infect.* **2**, 793–801.

Jerse, A. E., Cohen, M. S., Drown, P. M., *et al.* (1994). Multiple gonococcal opacity proteins are expressed during experimental urethral infection in the male. *J. Exp. Med.* **179**, 911–920.

Johansson, L., Rytkonen, A., Bergman, P., *et al.* (2003). CD46 in meningococcal disease. *Science* **301**, 373–375.

Jonson, A. B., Normark, S., and Rhen, M. (2005). Fimbriae, pili, flagella and bacterial virulence. *Contrib. Microbiol.* **12**, 67–89.

Kallstrom, H., Liszewski, M. K., Atkinson, J. P., and Jonsson, A. B. (1997). Membrane cofactor protein (MCP or CD46) is a cellular pilus receptor for pathogenic *Neisseria*. *Mol. Microbiol.* **25**, 639–647.

Kallstrom, H., Islam, M. S., Berggren, P.-O., and Jonsson, A.-B. (1998). Cell signaling by the type IV pili of pathogenic *Neisseria*. *J. Biol. Chem.* **273**, 21 777–21 782.

Kau, A. L., Hunstad, D. A., and Hultgren, S. J. (2005). Interaction of uropathogenic *Escherichia coli* with host uroepithelium. *Curr. Opin. Microbiol.* **8**, 54–59.

Kellogg, D. S., Peacock, W. L., Deacon, W. E., Brown, L., and Pirkle, C. I. (1963). *Neisseria gonorrhoeae*: I. Virulence linked to clonal variation. *J. Bacteriol.* **85**, 1274–1279.

Kirchner, M. and Meyer, T. F. (2005). The PilC adhesin of the Neisseria type IV pilus: binding specificities and new insights into the nature of the host cell receptor. *Mol. Microbiol.* **56**, 945–957.

Kirchner, M., Heuer, D., and Meyer, T. F. (2005). CD46-independent binding of neisserial type IV pili and the major pilus adhesin, PilC, to human epithelial cells. *Infect. Immun.* **73**, 3072–3082.

Kuijpers, T. W., Hoogerwerf, M., van der Laan, L. J., *et al.* (1992). CD66 nonspecific cross-reacting antigens are involved in neutrophil adherence to cytokine-activated endothelial cells. *J. Cell Biol.* **118**, 457–466.

Kupsch, E. M., Knepper, B., Kuroki, T., Heuer, I., and Meyer, T. F. (1993). Variable opacity (Opa) outer membrane proteins account for the cell tropisms displayed by *Neisseria gonorrhoeae* for human leukocytes and epithelial cells. *EMBO J.* **12**, 641–650.

Kuypers, J. M. and Proctor, R. A. (1989). Reduced adherence to traumatized rat heart valves by a low-fibronectin-binding mutant of *Staphylococcus aureus. Infect. Immun.* **57**, 2306–2312.

Lecuit, M., Dramsi, S., Gottardi, C., *et al.* (1999). A single amino acid in E-cadherin responsible for host specificity towards the human pathogen *Listeria monocytogenes. EMBO J.* **18**, 3956–3963.

Lecuit, M., Vandormael-Pournin, S., Lefort, J., *et al.* (2001). A transgenic model for listeriosis: role of internalin in crossing the intestinal barrier. *Science* **292**, 1722–1725.

Lee, S. W., Bonnah, R. A., Higashi, D. L., *et al.* (2002). CD46 is phosphorylated at tyrosine 354 upon infection of epithelial cells by *Neisseria gonorrhoeae. J. Cell Biol.* **156**, 951–957.

Maier, B., Potter, L., So, M., *et al.* (2002). Single pilus motor forces exceed 100 pN. *Proc. Natl. Acad. Sci. U. S. A.* **99**, 16 012–16 017.

Malorny, B., Morelli, G., Kusecek, B., Kolberg, J., and Achtman, M. (1998). Sequence diversity, predicted two-dimensional protein structure, and epitope mapping of neisserial Opa proteins. *J. Bacteriol.* **180**, 1323–1330.

Marra, A. and Isberg, R. R. (1997). Invasin-dependent and invasin-independent pathways for translocation of *Yersinia pseudotuberculosis* across the Peyer's patch intestinal epithelium. *Infect. Immun.* **65**, 3412–3421.

McCaw, S. E., Liao, E. H., and Gray-Owen, S. D. (2004). Engulfment of *Neisseria gonorrhoeae*: revealing distinct processes of bacterial entry by individual carcinoembryonic antigen-related cellular adhesion molecule family receptors. *Infect. Immun.* **72**, 2742–2752.

McCormick, B. A. (2003). The use of transepithelial models to examine host–pathogen interactions. *Curr. Opin. Microbiol.* **6**, 77–81.

McNeish, A. S., Turner, P., Fleming, J., and Evans, N. (1975). Mucosal adherence of human enteropathogenic *Escherichia coli. Lancet* **2**, 946–948.

McNiven, M. A., Kim, L., Krueger, E. W., *et al.* (2000). Regulated interactions between dynamin and the actin-binding protein cortactin modulate cell shape. *J. Cell Biol.* **151**, 187–198.

Melly, M. A., Gregg, C. R., and McGee, Z. A. (1981). Studies of toxicity of *Neisseria gonorrhoeae* for human fallopian tube mucosa. *J. Infect. Dis.* **143**, 423–431.

Mengaud, J., Ohayon, H., Gounon, P., Mege, R. M., and Cossart, P. (1996). E-cadherin is the receptor for internalin, a surface protein required for entry of *L. monocytogenes* into epithelial cells. *Cell* **84**, 923–932.

Menzies, B. E. (2003). The role of fibronectin binding proteins in the pathogenesis of *Staphylococcus aureus* infections. *Curr. Opin. Infect. Dis.* **16**, 225–229.

Menzies, B. E., Kourteva, Y., Kaiser, A. B., and Kernodle, D. S. (2002). Inhibition of staphylococcal wound infection and potentiation of antibiotic prophylaxis by a recombinant fragment of the fibronectin-binding protein of *Staphylococcus aureus*. *J. Infect. Dis.* **185**, 937–943.

Merz, A. J., So, M., and Sheetz, M. P. (2000). Pilus retraction powers bacterial twitching motility. *Nature* **407**, 98–102.

Meyer, T. F., Gibbs, C. P., and Haas, R. (1990). Variation and control of protein expression in *Neisseria*. *Annu. Rev. Microbiol.* **44**, 451–477.

Morand, P. C., Tattevin, P., Eugene, E., Beretti, J.-L., and Nassif, X. (2001). The adhesive property of the type IV pilus-associated component PilC1 of pathogenic *Neisseria* is supported by the conformational structure of the N-terminal part of the molecule. *Mol. Microbiol.* **40**, 846–856.

Morand, P. C., Bille, E., Morelle, S., *et al.* (2004). Type IV pilus retraction in pathogenic *Neisseria* is regulated by the PilC proteins. *EMBO J.* **23**, 2009–2017.

Mosleh, I. M., Boxberger, H. J., Sessler, M. J., and Meyer, T. F. (1997). Experimental infection of native human ureteral tissue with *Neisseria gonorrhoeae*: adhesion, invasion, intracellular fate, exocytosis, and passage through a stratified epithelium. *Infect. Immun.* **65**, 3391–3398.

Muenzner, P., Naumann, M., Meyer, T. F., and Gray-Owen, S. D. (2001). Pathogenic *Neisseria* trigger expression of their carcinoembryonic antigen-related cellular adhesion molecule 1 (CEACAM1; previously CD66a) receptor on primary endothelial cells by activating the immediate early response transcription factor, nuclear factor-kappa B. *J. Biol. Chem.* **276**, 24 331–24 340.

Muenzner, P., Billker, O., Meyer, T. F., and Naumann, M. (2002). Nuclear factor-kB directs CEACAM1 receptor expression in *Neisseria gonorrhoeae*-infected epithelial cells. *J. Biol. Chem.* **277**, 7438–7446.

Muenzner, P., Rohde, M., Kneitz, S., and Hauck, C. R. (2005). CEACAM engagement by human pathogens enhances cell adhesion and counteracts bacteria-induced detachment of epithelial cells. *J. Cell Biol.* **170**, 825–836.

Obrink, B. (1997). CEA adhesion molecules: multifunctional proteins with signal-regulatory properties. *Curr. Opin. Cell Biol.* **9**, 616–626.

Ofek, I., Beachey, E. H., Jefferson, W., and Campbell, G. L. (1975). Cell membrane-binding properties of group A streptococcal lipoteichoic acid. *J. Exp. Med.* **141**, 990–1003.

Oikawa, S., Inuzuka, C., Kuroki, M., *et al.* (1991). A specific heterotypic cell adhesion activity between members of carcinoembryonic antigen family, W272 and NCA, is mediated by N-domains. *J. Biol. Chem.* **266**, 7995–8001.

Old, D. C. (1972). Inhibition of the interaction between fimbrial haemagglutinins and erythrocytes by D-mannose and other carbohydrates. *J. Gen. Microbiol.* **71**, 149–157.

Ozeri, V., Rosenshine, I., Mosher, D. F., Fassler, R., and Hanski, E. (1998). Roles of integrins and fibronectin in the entry of *Streptococcus pyogenes* into cells via protein F1. *Mol. Microbiol.* **30**, 625–637.

Paradis, S. E., Dubreuil, J. D., Gottschalk, M., Archambault, M., and Jacques, M. (1999). Inhibition of adherence of *Actinobacillus pleuropneumoniae* to porcine respiratory tract cells by monoclonal antibodies directed against LPS and partial characterization of the LPS receptors. *Curr. Microbiol.* **39**, 313–320.

Patti, J. M., Allen, B. L., McGavin, M. J., and Hook, M. (1994). MSCRAMM-mediated adherence of microorganisms to host tissues. *Annu. Rev. Microbiol.* **48**, 585–617.

Pizarro-Cerda, J., Sousa, S., and Cossart, P. (2004). Exploitation of host cell cytoskeleton and signalling during *Listeria monocytogenes* entry into mammalian cells. *C. R. Biol.* **327**, 115–123.

Popp, A., Dehio, C., Grunert, F., Meyer, T. F., and Gray-Owen, S. D. (1999). Molecular analysis of neisserial Opa protein interactions with the CEA family of receptors: identification of determinants contributing to the differential specificities of binding. *Cell. Microbiol.* **1**, 169–181.

Poy, M. N., Ruch, R. J., Fernstrom, M. A., Okabayashi, Y., and Najjar, S. M. (2002). Shc and CEACAM1 interact to regulate the mitogenic action of insulin. *J. Biol. Chem.* **277**, 1076–1084.

Poy, M. N., Yang, Y., Rezaei, K., *et al.* (2002). CEACAM1 regulates insulin clearance in liver. *Nat. Genet.* **19**, 19.

Pujol, C., Eugene, E., de Saint Martin, L., and Nassif, X. (1997). Interaction of *Neisseria meningitidis* with a polarized monolayer of epithelial cells. *Infect. Immun.* **65**, 4836–4842.

Punsalang, A. P. Jr and Sawyer, W. D. (1973). Role of pili in the virulence of *Neisseria gonorrhoeae*. *Infect. Immun.* **8**, 255–263.

Rudel, T., Scheuerpflug, I., and Meyer, T. F. (1995). *Neisseria* PilC protein identified as type-4 pilus-tip located adhesin. *Nature* **373**, 357–359.

Ryll, R. R., Rudel, T., Scheuerpflug, I., Barten, R., and Meyer, T. F. (1997). PilC of *Neisseria meningitidis* is involved in class II pilus formation and restores pilus assembly, natural transformation competence and adherence to epithelial cells in PilC-deficient gonococci. *Mol. Microbiol.* **23**, 879–892.

Sadekova, S., Lamarche-Vane, N., Li, X., and Beauchemin, N. (2000). The CEACAM1-L glycoprotein associates with the actin cytoskeleton and localizes to cell–cell contact through activation of Rho-like GTPases. *Mol. Biol. Cell* **11**, 65–77.

Sauer, F. G., Remaut, H., Hultgren, S. J., and Waksman, G. (2004). Fiber assembly by the chaperone-usher pathway. *Biochim. Biophys. Acta* **1694**, 259–267.

Scheuerpflug, I., Rudel, T., Ryll, R., Pandit, J., and Meyer, T. F. (1999). Roles of PilC and PilE proteins in pilus-mediated adherence of *Neisseria gonorrhoeae* and *Neisseria meningitidis* to human erythrocytes and endothelial and epithelial cells. *Infect. Immun.* **67**, 834–843.

Schmitter, T., Agerer, F., Peterson, L., Muenzner, P., and Hauck, C. R. (2004). Granulocyte CEACAM3 is a phagocytic receptor of the innate immune system that mediates recognition and elimination of human-specific pathogens. *J. Exp. Med.* **199**, 35–46.

Schubert, W.-D., Urbanke, C., Ziehm, T., *et al.* (2002). Structure of internalin, a major invasion protein of *Listeria monocytogenes*, in complex with its human receptor E-cadherin. *Cell* **111**, 825.

Schumann, D., Chen, C. J., Kaplan, B., and Shively, J. E. (2001). Carcinoembryonic antigen cell adhesion molecule 1 directly associates with cytoskeleton proteins actin and tropomyosin. *J. Biol. Chem.* **276**, 47 421–47 433.

Schwarzbauer, J. E. and Sechler, J. L. (1999). Fibronectin fibrillogenesis: a paradigm for extracellular matrix assembly. *Curr. Opin. Cell Biol.* **11**, 622–627.

Schwarz-Linek, U., Werner, J. M., Pickford, A. R., *et al.* (2003). Pathogenic bacteria attach to human fibronectin through a tandem beta-zipper. *Nature* **423**, 177–181.

Selbach, M. and Backert, S. (2005). Cortactin: an Achilles' heel of the actin cytoskeleton targeted by pathogens. *Trends Microbiol.* **13**, 181–189.

Simonet, M., Riot, B., Fortineau, N., and Berche, P. (1996). Invasin production by *Yersinia pestis* is abolished by insertion of an IS200-like element within the inv gene. *Infect. Immun.* **64**, 375–379.

Sinha, B., Francois, P. P., Nusse, O., *et al.* (1999). Fibronectin-binding protein acts as *Staphylococcus aureus* invasin via fibronectin bridging to integrin alpha5beta1. *Cell. Microbiol.* **1**, 101–117.

Stern, A. and Meyer, T. F. (1987). Common mechanism controlling phase and antigenic variation in pathogenic neisseriae. *Mol. Microbiol.* **1**, 5–12.

Stern, A., Brown, M., Nickel, P., and Meyer, T. F. (1986). Opacity genes in *Neisseria gonorrhoeae*: control of phase and antigenic variation. *Cell* **47**, 61–71.

Swanson, J., Barrera, O., Sola, J., and Boslego, J. (1988). Expression of outer membrane protein II by gonococci in experimental gonorrhea. *J. Exp. Med.* **168**, 2121–2129.

Tan, K., Zelus, B. D., Meijers, R., *et al.* (2002). Crystal structure of murine sCEACAM1a[1,4]: a coronavirus receptor in the CEA family. *EMBO J.* **21**, 2076–2086.

Tobiason, D. M. and Seifert, H. S. (2001). Inverse relationship between pilus-mediated gonococcal adherence and surface expression of the pilus receptor, CD46. *Microbiology* **147**, 2333–2340.

Ton-That, H. and Schneewind, O. (2004). Assembly of pili in Gram-positive bacteria. *Trends Microbiol.* **12**, 228–234.

Vandeputte-Rutten, L., Bos, M. P., Tommassen, J., and Gros, P. (2003). Crystal structure of neisserial surface protein A (NspA), a conserved outer membrane protein with vaccine potential. *J. Biol. Chem.* **278**, 24 825–24 830.

Van Putten, J. P. and Paul, S. M. (1995). Binding of syndecan-like cell surface proteoglycan receptors is required for *Neisseria gonorrhoeae* entry into human mucosal cells. *EMBO J.* **14**, 2144–2154.

Van Putten, J. P., Duensing, T. D., and Cole, R. L. (1998). Entry of OpaA+ gonococci into HEp-2 cells requires concerted action of glycosaminoglycans, fibronectin and integrin receptors. *Mol. Microbiol.* **29**, 369–379.

Virji, M., Evans, D., Hadfield, A., *et al.* (1999). Critical determinants of host receptor targeting by *Neisseria meningitidis* and *Neisseria gonorrhoeae*: identification of Opa adhesiotopes on the N-domain of CD66 molecules. *Mol. Microbiol.* **34**, 538–551.

Virji, M., Evans, D., Griffith, J., *et al.* (2000). Carcinoembryonic antigens are targeted by diverse strains of typable and non-typable *Haemophilus influenzae*. *Mol. Microbiol.* **36**, 784–795.

Volberg, T., Romer, L., Zamir, E., and Geiger, B. (2001). pp60(c-src) and related tyrosine kinases: a role in the assembly and reorganization of matrix adhesions. *J. Cell Sci.* **114**, 2279–2289.

Wang, J., Gray-Owen, S. D., Knorre, A., Meyer, T. F., and Dehio, C. (1998). Opa binding to cellular CD66 receptors mediates the transcellular traversal of *Neisseria gonorrhoeae* across polarized T84 epithelial cell monolayers. *Mol. Microbiol.* **30**, 657–671.

Ward, M. E., Watt, P. J., and Robertson, J. N. (1974). The human fallopian tube: a laboratory model for gonococcal infection. *J. Infect. Dis.* **129**, 650–659.

Weidenmaier, C., Kokai-Kun, J. F., Kristian, S. A., *et al.* (2004). Role of teichoic acids in *Staphylococcus aureus* nasal colonization, a major risk factor in nosocomial infections. *Nat. Med.* **10**, 243–245.

Zaidel-Bar, R., Cohen, M., Addadi, L., and Geiger, B. (2004). Hierarchical assembly of cell-matrix adhesion complexes. *Biochem. Soc. Trans.* **32**, 416–420.

Zamir, E. and Geiger, B. (2001). Molecular complexity and dynamics of cell-matrix adhesions. *J. Cell Sci.* **114**, 3583–3590.

Zamir, E., Katz, M., Posen, Y., *et al.* (2000). Dynamics and segregation of cell-matrix adhesions in cultured fibroblasts. *Nat. Cell Biol.* **2**, 191–196.

CHAPTER 7

Bacterial toxins that modify the epithelial cell barrier

Joseph T. Barbieri

INTRODUCTION

Bacterial pathogens utilize invasive pathways and/or toxins to subvert the innate and acquired immune systems in order to damage the host epithelium. The infectious process requires that the pathogen can adhere and proliferate in the host. The capacity to colonize and cause disease varies among bacterial pathogens. For example, *Clostridium tetani* has only a limited ability to bind and proliferate within the host but is pathogenic due to the production of a potent neurotoxin, but the streptococcus and staphylococcus have strong adhesion factors that allow efficient colonization, with virulence due to the production of a multitude of virulence factors, including superantigens that simultaneously bind the major histocompatibility complex (MHC) and T-cell receptor of immune cells to stimulate production of antigen-independent cytokines.

The basic distinction between a member of our normal flora and a pathogen lies in the capacity to damage the host. However, this distinction is grayed by the immune status of the host, where host compromise converts commensal bacteria or even saprophytic bacteria into potent opportunistic pathogens. *Pseudomonas aeruginosa* is an opportunistic pathogen in many clinical situations but does not elicit disease in healthy individuals despite its ability to produce both a classical exotoxin and type III cytotoxins. *Clostridium difficile* can cause pseudomembrane colitis in patients undergoing antibiotic therapy. *C. difficile* pathogenesis is related to the ability to produce the exotoxins, toxin A and toxin B. *Escherichia coli*, a component of our normal gut flora, becomes a pathogen upon the acquisition of accessory genes that

Bacterial–Epithelial Cell Cross-Talk: Molecular Mechanisms in Pathogenesis, ed. Beth A. McCormick. Published by Cambridge University Press.

can encode several classes of toxins. These opportunistic pathogens contrast *Shigella dysenteriae*, which is highly infectious for healthy humans and is toxic through the action of a potent toxin that inhibits protein synthesis through the inactivation of rRNA. Host–pathogen interactions play a major role in the capacity of a bacterium to cause disease.

There are several sites of interaction between the bacterial pathogen and the epithelium, including the oral cavity, lung, stomach, colon, and vagina. Each pathogen damages the host through a unique molecular mechanism. Damage to the epithelium can be direct, as observed through the modification of actin by C2 toxin of the *Clostridia*, or indirect, through the modification of Rho GTPases that regulate the polymerization state of actin as catalyzed by the cytotoxic necrotizing factor of *E. coli*. Toxins modulate the epithelial cell barrier through covalent and non-covalent mechanisms. Covalent modifications include ADP-ribosylation, glucosylation, deamidation, and proteolysis, where the substrates modified include actin, G-proteins, adaptor and regulatory proteins of G-protein signaling, and RNA. Other toxins modulate host cell physiology though non-covalent mechanisms, including toxins that modulate G-protein signaling as mimics of host proteins that regulate the nucleotide state of the Rho GTPases. These covalent and non-covalent modulations to host physiology can produce subtle changes to cell function or even cell death. This chapter describes the general features of bacterial toxins that damage the epithelium.

MOLECULAR ORGANIZATION OF THE HOST TARGET OF BACTERIAL TOXINS

Bacterial toxins target several signaling pathways and protein complexes to effectively modulate the physiology of the host epithelium (Table 7.1). Host targets include heterotrimeric and monomeric G-proteins and actin (Figure 7.1) and proteins of the extracellular tight junctions.

Signal transduction by heterotrimeric G-proteins

Heterotrimeric G-proteins are non-covalently associated trimeric-protein complexes composed of an α-subunit that undergoes nucleotide exchange (GDP–GTP) and a $\beta\gamma$-subunit complex that is associated with a GDP–α-subunit (Gilman, 1995). Inactive GDP–G-protein is bound to a heptahelical receptor at the plasma membrane. Upon ligand binding to the extracellular surface of the heptahelical receptor, intracellular loops of receptor undergo

Table 7.1 *Bacterial toxins that modulate epithelial cells*

Toxin	Example	Structure–function organization	Mode of action	Substrate
Glucosylating toxins	Toxin A, Toxin B	AB	Glucosylation of Thr	Rho GTPase
Deamidating toxins	CNF1 397	AB	Deamidation of Gln	Rho A
Modulate nucleotide state (activate)	SopE	Type III	GEF 396	Rho GTPase
Modulate nucleotide state (inactivate)	ExoS	Type III	GAP 397	Rho GTPase
ADP-ribosylating toxin	C2	A–B	ADP-ribosylation of Arg	Actin
ADP-ribosylating toxin	ExoS	Type III	ADP-ribosylation of Arg	ERM proteins 394
ADP-ribosylating toxin	ExoT	Type III	ADP-ribosylation of Arg	Crk proteins 394
ADP-ribosylating toxin	Cholera toxin	AB5	ADP-ribosylation of Arg	α-Subunit of GTPase
Phosphatase toxins	YopH	Type III	Phosphatase of Pi-Y	Focal adhesion proteins
De-adenylation toxins	ST	AB5	De-adenylation	rRNA
Actin polymerizing toxin	RTX	?	Actin multimerization	Actin
Protease	YopT	Type III	Cleave C-terminal cysteine	Rho A
Protease	BFT	A	Cleavage	E cadherin
Physical association	CPE	AB	Association to focal adhesion	Claudins, occludin

CNF, cytotoxic necrotizing factor; CRK, CT10 regulator of kinase; ERM, ezrin, radixin, and moesin; GAP, GTPase-activating protein; GEF, guanine nucleotide exchange factor.

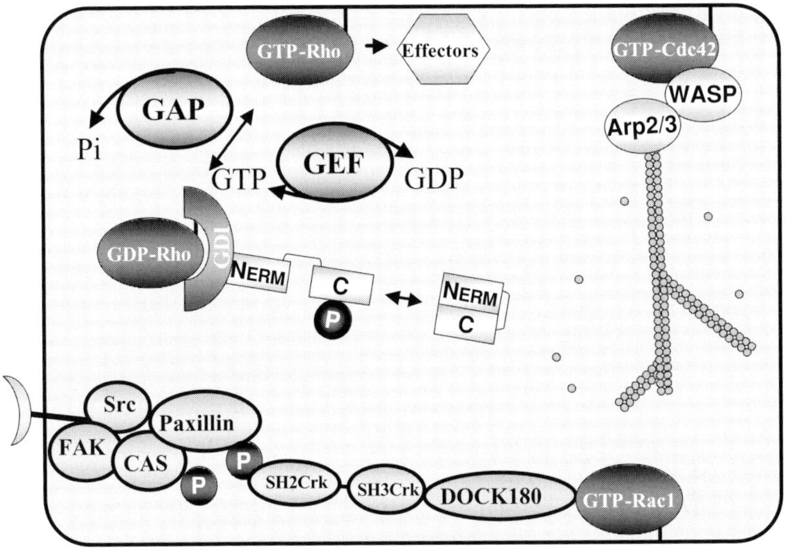

Figure 7.1 Components of the Rho GTPase cycle in mammalian cells. Rho proteins are molecular switches. GDP-Rho is inactive and bound to guanine nucleotide dissociation inhibitors (GDI). GDP-Rho is dissociated from GDI upon binding of activated ezrin, radixin, and moesin (ERM) proteins. ERM proteins are activated to an open conformation by phosphorylation. GDP/GTP exchange of Rho is stimulated by guanine nucleotide exchange factors (GEFs). Activated Rho GTPases, such as GTP-Cdc42, can initiate the site of actin polymerization by binding downstream effectors such as the Wiskott–Aldrich syndrome protein (WASP) family of proteins. The Rac GEF, DOCK180, is delivered to the cell membrane by a CT10 regulator of kinase (Crk) adaptor protein following activation by the intergin signaling pathway and stimulates cellular phagocytosis by GTP-Rac 1. GTP-Rho binds several downstream effectors, including kinases, lipases, and adaptor proteins, which are isoprenylated at their C-terminus, allowing membrane binding. GTP-Rho inactivation is stimulated by GTPase-activating proteins (GAPs).
FAK, focal adhesion kinase.

conformational change to stimulate nucleotide exchange (GDP $\rightarrow$ GTP) of the α-subunit of the G-protein, which dissociates the $\beta\gamma$-subunit. Both the α-subunit and $\beta\gamma$-subunit modulate the activity of downstream effectors. The α-subunit possesses intrinsic GTPase activity alone, or with regulators of G-protein stimulation, it hydrolyzes GTP $\rightarrow$ GDP, which allows reassociation of the GDP–α-subunit and the $\beta\gamma$-subunits. The α-subunits of the heterotrimer G-protein are targeted by several bacterial toxins for covalent modification. Cholera toxin ADP-ribosylates the α-subunit of G_s to reduce intrinsic GTPase activity, resulting in ligand-independent stimulation signal transduction and diarrhea; pertussis toxin ADP-ribosylates the α-subunit of G_i to reduce interactions with the G-protein heptahelical receptors, uncoupling signal transduction, which results in complex modulation of host cell physiology.

Dynamics of actin polymerization

Actin is a major target of bacterial pathogens. Toxins alter the actin cytoskeleton through direct covalent modification of actin or indirectly through modification of actin regulatory proteins, such as the Rho-GTPases, focal adhesion proteins, CT10 regulator of kinase (Crk), or the ezrin, radixin, and moesin (ERM) proteins. Toxins can stimulate or inhibit actin polymerization and can have irreversible or reversible effects on the polymerization state of actin.

The actin cytoskeleton is a dynamic structure involved in numerous cellular functions, including motility, phagocytosis, and proliferation. Actin polymerization is a polar process, whereby the intracellular pools of actin fluctuate between soluble monomeric G-actin and polymerized F-actin (Castellano *et al.*, 2001; Fujiwara *et al.*, 2002). F-actin contains slow-growing ends and fast-growing ends, which differ in their ability to sequester G-actin to facilitate polymerization. Several actin-binding proteins participate in actin nucleation and polymerization. A large protein complex that includes the Arp2/3 complex is the nucleation site of actin polymerization. The Arp2/3 complex binds ATP and is activated by Wiskott–Aldrich syndrome protein (WASP) family of proteins, which possess an Arp2/3 binding domain and a Cdc42-interactive binding domain that binds Cdc42 to WASP proteins. Thus, Cdc42 can directly activate actin nucleation. The Arp2/3 complex assembles and recruits profilin–ATP–actin. Profilin is a nucleotide exchange factor that binds ADP–actin to stimulate nucleotide exchange (ADP → ATP). Monomeric ATP–G-actin binds to the fast-growing ends of actin; this polymerization is responsible for actin protrusions such as filopodia. Over time, ATP–actin within the growing polymer is converted to ADP–actin, which can be released from the slow-growing end of actin unless capping proteins stabilize actin. Rapid actin depolymerization is stimulated by cofilin and other actin-binding proteins.

Cell signaling by Rho GTPases

There are more than 20 mammalian Rho GTPases, with RhoA, Rac1, and Cdc42 being the best characterized (Etienne-Manneville and Hall, 2002).

RhoA, Rac1, and Cdc42 have unique functions in the regulation of the actin cytoskeleton. RhoA regulates the formation of actin stress fibers. Rac1 modulates the formation of membrane ruffling or lamellipodia. Cdc42 modulates the formation of microspikes or filopodia (Bar-Sagi and Hall, 2000). Rho GTPases are monomeric G-proteins that act as molecular switches of the actin cytoskeleton. GDP–Rho is cytosolic and bound to guanine nucleotide dissociation inhibitors (GDI). Intracellular signals, through the direct binding of ERM proteins or phosphatidylinositol-4,5-biphosphate (PIP2), stimulate release of GDP–Rho from the Rho–GDI complex and allow movement of Rho to the cell membrane, where guanine nucleotide exchange factors (GEFs) stimulate nucleotide exchange (GDP → GTP). GTP–Rho binds effector proteins that modulate the polymerization of the actin cytoskeleton. These effectors include protein and lipid kinases, phospholipases, and adaptor proteins. GTP–Rho is inactivated through the hydrolysis of the γ-phosphate of GTP, which is stimulated by GTPase-activating proteins (GAPs). GDP–Rho is then extracted from the plasma membrane by GDI, which completes the signaling cycle. Bacterial toxins modulate Rho function through both covalent and non-covalent mechanisms. Examples are the large clostridial cytotoxins that inactivate Rho via glucosylation and the family of cytotoxic necrotizing factor (CNF) toxins that activate Rho through deamidation.

Ezrin, radixin, and moesin proteins modulate the actin cytoskeleton

ERM proteins regulate cell shape, motility, cell adhesion, and the ability of cells to phagocytose foreign objects. ERM proteins contain two functional domains. The N-terminus of the ERM proteins contains Rho–GDI and diffuse B-cell lymphoma (Dbl) (a Rho GEF), where binding to Rho–GDI stimulates dissociation of the Rho GTPase–Rho–GDI complex and subsequent binding of the GEF to GDP–Rho (Pearson *et al.*, 2000). The C-terminus of the ERM proteins binds actin and can serve as a nucleation site for actin polymerization. ERM proteins exist in an inactive closed form, where the N-terminus blocks F-actin binding to the C-terminus, and an active open form, where the C-terminus can bind F-actin (Mangeat *et al.*, 1999). ERM activation occurs through the phosphorylation of Thr558 by Rho kinase or protein kinase C (PKC) (Matsui *et al.*, 1998), which maintains ERM proteins in the open conformation. ERM proteins are adaptors for cell-surface receptors and for the direct binding to the actin cytoskeleton. ERM proteins have been identified as targets for ADP-ribosylation by *Pseudomonas* ExoS and for *Helicobacter* CagA-dependent dephosphorylation.

Crk proteins regulate Rac1 signal transduction

Crk I and II are adaptor proteins involved in eukaryotic signaling transduction (Matsuda *et al.*, 1992). Crk I contains an N-terminal SH2 domain, which recognizes phosphotyrosine, and a C-terminal SH3 domain, which binds proteins with polyproline motifs, such as DOCK180, a guanine exchange factor for Rac1. Crk II has an additional C-terminal SH3 domain. Cell mobility and phagocytosis are regulated by the integrin-activated Crk signaling pathway (Albert *et al.*, 2000). Integrins comprise a large family of transmembrane adhesion receptors. Upon activation, integrin receptors recruit focal adhesion protein to the plasma membrane, including focal adhesion kinase (FAK) and Src family kinase, as well as the scaffolding proteins paxillin and p130Cas. FAK phosphorylates paxillin (Boyd *et al.*, 2000), while Src kinase phosphorylates p130Cas at multiple tyrosines. Phosphorylation creates docking sites for Crk proteins (Goldberg *et al.*, 2003). Crk binding brings SH3 domain-associated proteins such as DOCK180 to the cell membrane in order to activate Rac 1 (Kiyokawa *et al.*, 1998) and stimulate cell migration and phagocytosis. Crk signaling is modulated by *Yersinia* YopH, *Pseudomonas* ExoT, and *Shigella flexneri*.

The polarized epithelium

The polarized epithelium is maintained by tight junctions, which are stabilized by actin through a series of transmembrane proteins, including occludin, claudins, and other adaptor proteins (Furuse *et al.*, 1994). Cell polarity and tight junctions are regulated by Cdc42. Tight junctions are targeted directly by a protease of *Bacteriodes fragillis*, which targets tight junction proteins, and an enterotoxin of *Clostridium perfringens*, which directly binds the epithelial tight junction.

PHYSICAL ORGANIZATION OF BACTERIAL TOXINS

Bacterial toxins often possess AB domains that define structure–function properties. The A domain is the catalytic domain where each toxin possesses a unique mode of action, e.g. by acting as a hydrolase or by catalyzing a post-translational modification of a specific amino acid on host proteins. The B domain has two functional subunits: the receptor-binding (R) domain, which binds to receptors on the surface of eukaryotic cells, and the translocation (T) domain, which translocates the A domain across the cell membrane into the cytosol of eukaryotic cells (Figure 7.2). The site of translocation can be

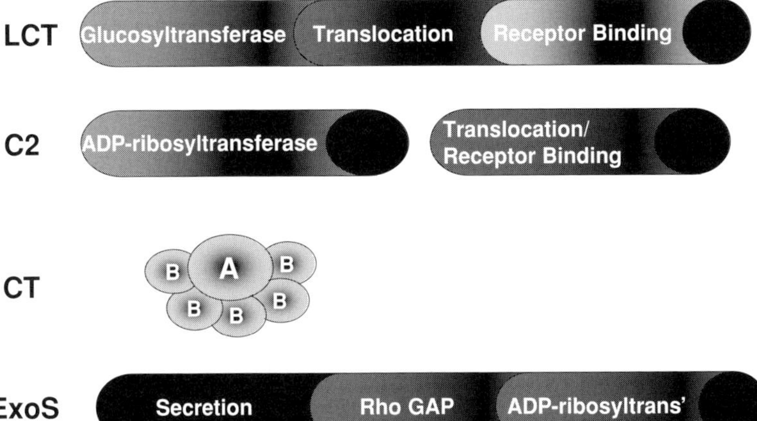

Figure 7.2 Structure–function organization of bacterial toxins. Bacterial toxins possess specific AB structure–function relationships. A is the catalytic domain; each toxin possesses a unique catalytic activity. B has two functional domains: the receptor-binding domain, which binds to the cell surface, and the translocation domain, which delivers the A domain across the cell membrane into the cytosol of eukaryotic cells. In bacterial toxins, the AB domains can be contained within a single protein, as observed for the large clostridial toxins (LCT); can be encoded on separate proteins, as observed with C2 toxin or; can be non-covalently associated, as observed for cholera toxin (CT). Alternatively, the A domain can be delivered by the type III secretion system, as outlined by ExoS.

the plasma membrane or an intracellular vesicle membrane. Each domain can be isolated and can perform a specific function independent of the other domains.

A–B domains may be associated covalently within a single protein, termed AB, as observed with the large clostridial toxins (LCTs) (Figure 7.2). LCTs have three functional domains that are physically organized in sequential linear order (Schirmer and Aktories, 2004). The N-terminus is the glucosyltransferase A domain, the internal domain is the translocation T domain, and the C-terminus is the receptor-binding R domain.

A–B domains may be associated non-covalently within a complex composed of a single A polypeptide and five B polypeptides, termed AB5, as observed with cholera toxin (Figure 7.3), the heat-labile enterotoxin of *Escherichia coli*, pertussis toxin, and Shiga toxin (Sixma *et al.*, 1993). In AB5 toxins, the A polypeptide is organized as a ~190-amino-acid N-terminal catalytic A1 domain that is contiguous with a ~25-amino-acid C-terminal A2 domain. The **A2** domain non-covalently links A1 to the B5 subunit. B5 comprises five small proteins, which may be identical, as in cholera toxin, or

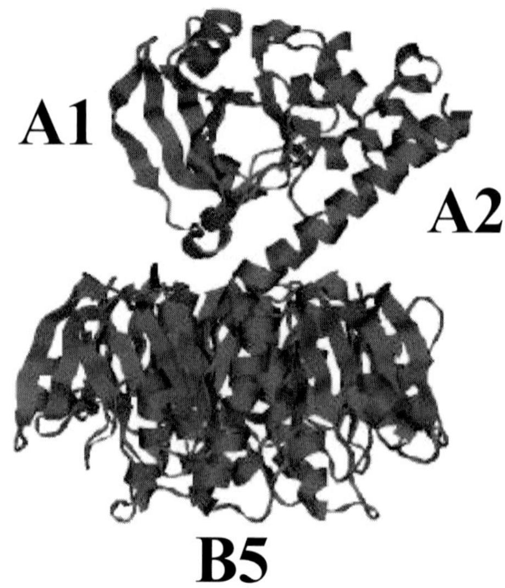

Figure 7.3 AB organization of bacterial toxins. The physical organization of an AB toxin can be represented by the AB5 organization of cholera toxin (CT). CT comprises an A1 domain, which encodes a catalytic ~190-amino-acid peptide that is the ADP-ribosyltransferase domain, and a B domain, which are linked non-covalently by the A2 peptide. The A2 linker may contain a KDEL sequence at the C terminus to facilitate intracellular trafficking in the mammalian cell. The structure 1LTS was downloaded from the Protein Data Base.

different, as in pertussis toxin. B5 is the receptor-binding domain and contributes to the transport and delivery of A1 into eukaryotic cells. The C-terminus of the A2 domain of several AB5 toxins possesses KDEL-like sequences that contribute to intracellular trafficking within the eukaryotic cells.

Some bacterial toxins are organized with the A–B domains synthesized as separate proteins that are not associated in solution but that associate on the surface of the eukaryotic cell, termed binary toxins (Barth *et al.*, 2000). Anthrax toxin of *Bacillus anthracis* and C2 toxin of *Clostridium botulinum* (Figure 7.2) are binary toxins. B monomers bind to cell-surface receptors, are cleaved by a eukaryotic protease, and are polymerized to a heptamer that forms a platform for A to bind. This AB complex then enters cells via receptor-mediated endocytosis, where the B domain forms a channel to translocate the A domain across the early acidified endosomal membrane and into the cytosol.

Certain bacterial toxins lack a B domain. Type III cytotoxins comprise N-terminal secretion sequences linked to an A domain that encodes a catalytic activity. Type III secretion systems of Gram-negative bacteria are similar and comprise two components, a secretion apparatus and a translocation pore. The prototropic *Yersinia* type III apparatus is composed of about 30 *ysc* (*Yersinia* secretion) genes (Cornelis and Van Gijsegem, 2000) and is derived from gene duplication of the flagellum operon. Studies with high-resolution microscopy have generated a structure for the secretion apparatus of *Salmonella* (Marlovits *et al.*, 2004). The functions of individual components of the type III apparatus are, for the most part, unknown, with few exceptions. YscC is essential for the secretion of Yops across the outer membrane of *Yersinia* and is a member of the family of outer-membrane secretin proteins that function in macromolecular transport. YscC forms an oligomeric complex in the outer membrane of *Yersinia* with an apparent internal pore (Burghout *et al.*, 2004). Secretion of type III cytotoxins also requires YscN, which possesses an ATP-binding motif (Woestyn *et al.*, 1994). Mutations within this ATP-binding motif interfere with Yop secretion, implicating a need for energy in the secretion process. Translocation of type III cytotoxins across the mammalian plasma membrane requires three bacterial proteins, YopB and Yop D, which function in pore formation, and LcrV, which plays a pleotrophic but essential role in translocation. The secretion and translocation signals are located within the N-terminal 100 amino acids of type III cytotoxins. In addition to the secretion and translocation apparatus, efficient type III secretion requires chaperone-like proteins that interact directly with the preformed type III cytotoxin in the cytoplasm of the bacterium to direct secretion through the type III apparatus. A novel but controversial model implicates a role for an RNA intermediate of some type III cytotoxin in the secretion process (Cheng and Schneewind, 2000).

Thus, bacterial toxins possess specific A–B organization and modes of action. Toxins have evolved to modify eukaryotic proteins that control cellular function. At equilibrium, the covalent modification of host proteins by bacterial toxins is an irreversible reaction. This contrasts the ability of some bacterial toxins to mimic host proteins in order to reversibly modulate the activity of mammalian physiology.

BACTERIAL TOXINS THAT COVALENTLY MODIFY ACTIN

There are three classes of bacterial toxins that covalently modify actin: the C2 family of toxins of clostridia and bacillus, *Salmonella* cytotoxins, and the actin polymerizing toxin of *Vibrio cholerae*.

C2 toxins ADP-ribosylate actin

C2 toxins catalyze the transfer of ADP-ribose from nicotinamide adenine dinucleotide (NAD) to the guanidine group of Arg-177 of actin. This depolymerizes actin and disrupts the actin cytoskeleton.

$$NAD^{+} + \text{actin} \rightarrow \text{actin(Arg-177)} - ADP\text{-}ribose + nicotinamide + \text{H}^{+}$$

ADP-ribosylation of actin correlates with the reorganization of the actin cytoskeleton, showing that actin is the primary target for C2 toxin (Aktories *et al.*, 1986; Ohishi and Tsuyama, 1986). C2-like toxins include the C2 toxin of *C. botulinum*, iota toxin of *Clostridium perfringens*, *Clostridium spiroforme* toxin, *C. difficile* ADP-ribosyltransferase, and vegetative insecticidal protein (VIP) of *Bacillus cereus*. C2 toxins are binary toxins and consist of an A protein, which possesses ADP-ribosyltransferase activity, and a B protein, which includes the receptor-binding and translocation domains. C2 toxins possess different capacities to ADP-ribosylate actin. C2 toxins ADP-ribosylate all actin isoforms, except C2 toxin, which ADP-ribosylates β/γ-actin but not α-actin. ADP-ribosylated actin does not polymerize into F-actin (Aktories *et al.*, 1986) and acts as a dominant negative protein to inhibit actin polymerization (Wegner and Aktories, 1988).

Cytotoxins that enhance *Salmonella* pathogenesis

SpvB is a type III cytotoxin (Lesnick *et al.*, 2001) of *Salmonella* that ADP-ribosylates β/γ-actin (Lesnick *et al.*, 2001; Tezcan-Merdol *et al.*, 2001). SpvB enhances *Salmonella* invasion of epithelial cells and macrophages. ADP-ribosylation causes actin to depolymerize, which prevents conversion of G-actin into F-actin. SpvB appears to complement the action of SptP, another type III effector of *Salmonella* (Fu and Galan, 1999). SipC, another type III cytotoxin of *Salmonella*, has similarities with the pore-forming protein of IpaC of *Shigella* and stimulates F-actin nucleation via an undefined mechanism (Chang *et al.*, 2005; Hayward and Koronakis, 1999).

Actin-polymerizing toxin of *Vibrio cholerae*

RtxA is the actin-polymerizing toxin of *Vibrio cholerae*. RtxA was originally identified within a gene cluster linked to the gene encoding cholera toxin in the El Tor strain of *V. cholerae* (Fullner and Mekalanos, 2000). RtxA has limited homology with the repeats in the toxin (Rtx) protein family and was originally named as a member of this protein family. However, RtxA does not contain the pore-forming domain that is a component of the RTX toxins and does not possess hemolytic activity. RtxA is predicted to comprise 4546 amino acids and elicits two phenotypes in clustered cells, cell rounding and the covalent cross-linking of actin. Actin depolymerization and actin cross-linking may be located on separate domains of Rtx, but the molecular basis for either phenotype is not known. RtxA has been implicated as an inducer of a local intestinal inflammatory response during *V. cholerae* pathogenesis.

TOXINS THAT INACTIVATE RHO GTPASES

Bacterial toxins utilize several strategies to inactivate the Rho GPTPases, including covalent modifications (glucosylation) by LCTs, proteolytic cleavage by *Yersinia* YopT, and as molecular mimics of the mammalian Rho GAPs by a family of type III cytotoxins.

Large clostridial toxins glucosylate Rho

Toxins A and B of *C. difficile* are single polypeptide AB toxins that are the best-characterized members of the family of glucosylating toxins. Toxin A is ~308 kDa and toxin B is ~270 kDa. A and B are organized as single-chain AB toxins that have ~50% homology at the primary amino-acid level. The N-terminus encodes the glucosyltransferase domain (A), while the C-terminus comprises the translocation/receptor-binding domains (B), but cell-surface receptors for toxin A and toxin B have not been identified. Toxin A and toxin B appear to enter cells through a pH-dependent vesicle pathway. The glucosylating toxins catalyze the mono-O-glucosylation of Rho GTPases (Chaves-Olarte *et al.*, 1996; Just *et al.*, 1995).

$$\textit{Glucose} + \text{Rho A} \rightarrow \text{Rho A (Thr37)-}\textit{glucose} + \text{H}^+$$

C. difficile is responsible for antibiotic-associated pseudomembrane colitis, and toxins A and B have been implicated in the development of inflammation of the colon. Toxin A was originally described as an enterotoxin that stimulated fluid secretion and inflammation in an experimental animal model, while toxin B gene was reported to be cytotoxic for cultured cells. More recent studies suggest that toxin A and toxin B have similar actions on cells and do not possess unique enterotoxic or cytotoxic activities.

Toxins A and B glucosylate Rho A at Thr37 and glucosylate Rac 1 and Cdc 42 at Thr35. Glucosylation of Rho changes cell morphology and results in cell rounding and subsequent cell release from the matrix. Glucosylation has multiple effects on Rho action, including inhibition of downstream effector protein binding, GEF-mediated nucleotide exchange, and GTPase activity. In addition, Rho-GDI does not extract glucosylated Rho from the cell membrane. Although the exact role of toxins A and B in inducing diarrhea and pseudomembranous colitis are not understood, F-actin depolymerization can modulate tight-junction function. In addition to eliciting a cytotoxic response in mammalian cells, toxin A and toxin B generate an inflammatory response in white blood cells, which leads to the release of numerous cytokines; this may also contribute to colonic inflammation.

YopT cleaves at the C-terminus of Rho

YopT is a 322-amino-acid type III cytotoxin of *Yersinia* that is a metalloprotease (Shao and Dixon, 2003). YopT cleaves the C-terminal cysteine of RhoA to disrupt the actin cytoskeleton. The C-terminal cysteine of the Rho GTPAses is part of the CAAX box that is isoprenylated in eukaryotic cells to anchor Rho GTPase to the cell membrane. During the native maturation of Rho, the cysteine is isoprenylated, followed by cleavage of AAX and carboxymethylation of the isoprenylated cysteine. Thus, YopT cleaves Rho proteins to release carboxymethylated isoprenylated cysteine, which disrupts binding of GTP–Rho to the cell membrane and interferes with interactions of GDP–Rho with Rho GDI. In vitro, YopT targets RhoA, Rac1, and Cdc42, but RhoA appears to be the preferred target in cells (Shao *et al.*, 2003). Cleavage of the post-translated cysteine by YopT contributes to the anti-internalization activity of *Yersinia*.

TOXINS THAT ACTIVATE RHO FUNCTION

Rho activation by bacterial toxins is caused by the deamidation of a catalytic glutamine on Rho by a family of cytotoxic necrotizing factor (CNF) toxins and transglutamidation by dermonecrotizing toxin of *Bordetella*.

Cytotoxic necrotizing factor deamidates Rho

CNFs are single-chain ~110-kDa AB proteins, where the N-terminus comprises the receptor-binding/translocation domain and the C-terminus comprises the deamidase domain. CNF was initially isolated from *E. coli*, but CNF homologs have also been isolated in *Yersinia*. CNF stimulates multinucleation, actin polymerization, and phagocytosis in cultured cells.

$$\text{Rho(Gln63)} + H_2O \rightarrow \text{Rho(Glu63)} + NH_2$$

CNF stimulates the deamidation of Gln63 of RhoA (Popoff *et al.*, 1996; Schmidt *et al.*, 1997). Gln63 coordinates a hydrolytic H_2O that is involved in intrinsic and RhoGAP-stimulated hydrolysis of the γ-phosphate of GTP. In this reaction, the amide nitrogen of Gln63 stabilizes a transition state for GTP hydrolysis, and deamidation of Gln63 gives Rho a dominant active phenotype. CNF contributes to colonization and tissue damage in mouse models and to the invasive properties of *E. coli*. In addition to Rac activation, an unexpected property of CFN is the stimulation of proteasome-mediated degradation of Rac1. The physiological significance for the modulation of the steady state of Rac1 by CFN is not yet apparent. CNF1 also increases intestinal permeability by decreasing transepithelial electrical resistance, with a decrease in the function of tight junctions. Thus, CFN can modulate barrier function of epithelial cells.

Dermonecrotizing toxin transglutamidates Rho

Bordetella dermonecrotizing toxin (DNT) is a ~150-kDa AB protein that also deamidates Gln63 of Rho to stimulate actin polymerization. DNT has an N-terminal receptor-binding/translocation domain and a C-terminal deamidase domain (Kashimoto *et al.*, 1999; Schmidt *et al.*, 1999). In addition to catalyzing the deamidase reaction, DNT also catalyzes a transglutamidase reaction, where polyamines, such as putrescine, are exchanged for the amide of Gln63.

$$\text{Rho(Gln63)} + \text{putrescine} \rightarrow \text{Rho(Glu63)-putrescine}$$

Transglutamidation blocks hydrolysis of the γ-phosphate from GTP–Rho and allows GDP–Rho to associate with Rho kinase and stimulate Rho signal transduction.

TOXINS THAT MODIFY RHO THROUGH NON-COVALENT MECHANISMS

The modulation of actin cytoskeleton by non-covalent modification provides a bacterial pathogen temporal regulation of the activation state of the Rho GTPases. Bacterial toxins can either activate or inactivate Rho proteins as molecular mimics of GEFs and GAPs, respectively.

Activation of Rho GTPases by guanine nucleotide exchange factor mimics

SopE is a type III cytotoxin of *Salmonella* that mimics eukaryotic GEFs through a mechanism that is similar to mammalian GEFs (Hardt *et al.*, 1998). SopE stimulates release of GTP from Rac1 and Cdc42. Expression of SopE stimulates the actin cytoskeleton in cultured cells and enhances the invasive properties of *Salmonella*. This was the first observation that a bacterial toxin could modulate host physiology through a non-covalent modification of host cell function. More recent studies have reported the presence of a SopE2 protein that also functions as a GEF.

Inactivation of Rho GTPases by RhoGTPase-activating protein mimics

Salmonella SptP, *Pseudomonas* ExoS and ExoT, and *Yersinia* YopE are type III cytotoxins that mimic the action of mammalian Rho GAPs (Fu and Galan, 1999). Bacterial Rho GAP domains comprise a small ~140-amino-acid domain that includes nine α-helices (Wurtele *et al.*, 2001). Stimulation of the inactivation of Rho GTPase by the bacterial Rho GAPs is similar but not identical to mammalian Rho GAP. Rho GAPs enhance the intrinsic GTPase activity of Rho GTPases by stabilizing a transition state of the catalytic Gln of the GTPase. Like mammalian Rho GAPs, bacterial Rho GAPs utilize an Arg to stimulate the hydrolytic activity of the Rho GTPases, but the catalytic Arg of bacterial Rho GAPs is a component of a α-helix rather than a flexible loop, as observed for mammalian Rho GAPs. Bacterial Rho GAPs utilize a main-chain carbonyl on a glycine to stabilize the active site Gln. The crystal structure of bacterial Rho GAPs has little homology with mammalian Rho GAPs, apart from both families of proteins being composed of α-helices, implicating convergent pathways in their evolution.

SptP is a bifunctional toxin of *Salmonella*, where the N-terminus comprises the Rho GAP domain that targets Rac and Cdc42 and a C-terminus comprises a tyrosine phosphatase domain (Fu and Galan, 1999). *Pseudomonas* ExoS and ExoT are also bifunctional toxins that comprise an N-terminal Rho GAP domain for Rho, Rac, and Cdc42 and a C-terminal ADP-ribosyltransferase domain (Barbieri and Sun, 2004). *Yersinia* YopE possesses only a Rho GAP domain (Von Pawel-Rammingen *et al.*, 2000). Although YopE, ExoS, and ExoT are anti-internalization factors, SptP works in concert with SopE to contribute to the invasive properties of *Salmonella*. YopE has also been reported to interfere with the caspase-1-mediated maturation of interleukin 1β through the modulation of Rac1, where Rac1 contributed to autoactivation of caspase-1 through activation of a host cofilin kinase (Yang *et al.*, 1998). This suggests a new function of Rho GTPases in the regulation of innate immunity.

TOXINS THAT MODULATE THE CRK SIGNALING PATHWAY

Yersinia YopH, *Pseudomonas* ExoT, and a *Shigella flexneri* type III related protein modulate the Crk signalling pathway.

YopH is a phosphatase of focal adhesion proteins

YopH is a 468-amino-acid type III cytotoxin of *Yersinia* that possesses tyrosine phosphatase activity (Black *et al.*, 2000). *Yersinia* invasion protein binds to the integrin receptor to stimulate recruitment of Src kinase and FAK to the focal adhesion complex. These kinases phosphorylate the focal adhesion complex protein, paxillin, and p130CAS. YopH dephosphorylates these focal adhesion proteins, which blocks interactions between the focal adhesion protein and Crk proteins (Phan *et al.*, 2003). The crystal structure of the YopH phosphatase domain complexed with a non-hydrolyzable substrate suggests that the active site of YopH acts through a similar mechanism as mammalian phosphatases. Although single *yop* mutant strains of *Yersinia* were not affected for colonization or persistence in a mouse model of infection (Logsdon and Mecsas, 2003), a strain of *Yersinia* that was mutated for *yop*H and *yop*E did not colonize intestinal tissues. This suggests that YopH and YopE have redundant functions. Studies suggest that YopH may also block early steps in T-cell signaling and inhibit development of an immune response to infection (Alonso *et al.*, 2004).

ExoT ADP-ribosylates Crk

Pseudomonas ExoT is a 457-amino-acid bifunctional type III cytotoxin that contains an N-terminal Rho GAP domain and a C-terminal ADP-ribosylation domain (Barbieri and Sun, 2004). The function of the ExoT ADP-ribosyltransferase domain has been resolved. Early studies observed that relative to ExoS, ExoT possessed only limited ADP-ribosyltransferase activity for Ras and SBTI. Therefore, ExoT was proposed to represent a defective

ADP-ribosyltransferase. Subsequent studies showed that a Rho GAP defective ExoT retained some capacity to reorganize the actin cytoskeleton and possessed antiphagocytic activity (Garrity-Ryan *et al.*, 2004). Furthermore, ExoT elicited a cytotoxic response independent of Rho GAP activity without ADP-ribosylating Ras (Sundin *et al.*, 2001). This suggested that ExoT might target host proteins distinct from ExoS. ExoT was subsequently shown to ADP-ribosylate Crk-I and Crk-II with a rate of ADP-ribosylation that is comparable to that of ExoS for SBTI. This observation linked the antiphagocytic activity of ExoT to a mechanism that was shown to be distinct from transient modulation of Rho GTPases by Rho GAP activity via Rac1-mediated phagocytotic signalling.

Shigella activates Crk signaling pathway

Shigella is an intracellular pathogen that stimulates a Crk-mediated pathway to facilitate uptake of the bacterium into mammalian cells (Bougneres *et al.*, 2004). The Abl family of tyrosine kinases is required for the internalization of *Shigella*, and Abl and Arg accumulate at the site of *Shigella* entry. These kinases phosphorylate Crk-II, which activates Rac-1, facilitating *Shigella* internalization. The bacterial protein responsible for the activation of the Abl family of tyrosine kinases has not been identified.

TOXINS THAT MODULATE EZRIN, RADIXIN, AND MOESIN PROTEINS

The ERM proteins are regulators of the actin cytoskeleton and function as foci for actin nucleation and activators of Rho signal transduction. *Pseudomonas* ExoS and *Helicobacter* CagA target ERM proteins.

ExoS ADP-ribosylates ezrin, radixin, and eosin proteins

Pseudomonas ExoS is a 453-amino-acid bifunctional type III cytotoxin that contains an N-terminal Rho GAP domain and a C-terminal ADP-ribosyltransferase domain (Barbieri and Sun, 2004). ExoS was initially identified as an ADP-ribosylating protein with broad substrate specificity (Coburn and Gill, 1991; Iglewski *et al.*, 1978). Subsequent studies showed that ADP-ribosylation disrupted Ras interactions with its GEF (Ganesan *et al.*, 1999). Although in vitro studies showed that ExoS ADP-ribosylated numerous substrates, only a few host proteins were early targets for ADP-ribosylation. One of the earliest targets for ADP-ribosylation were the ERM proteins (Maresso *et al.*, 2004).

CagA stimulates the dephosphorylation of ezrin

Helicobacter cytotoxin-associated gene A (CagA) is a type-IV secreted toxin. *H. pylori* infects many humans, leading to gastritis, gastric ulceration, and gastric cancer, although molecular association between *H. pylori* infection and cancer remains to be resolved. Following type IV delivery, CagA is phosphorylated and stimulates reorganization of the actin cytoskeleton. One of the outcomes of CagA internalization is the inhibition of Src kinases. This inhibition causes the steady-state dephosphorylation of several host cell proteins, including ezrin (Selbach *et al.*, 2004). The role of ERM protein inactivation in *H. pylori* pathogenesis is still under investigation.

TOXINS THAT DISRUPT TIGHT JUNCTIONS

Although many toxins elicit indirect effects on the tight junctions of polarized epithelial cells, the metalloprotease of *Bacteroides fragilis* has a direct disruptive effect on tight junction function.

Bacteroides fragilis toxin is a protease of E-cadherin

B. fragilis toxin (BFT) is a ~20-kDa protein that is associated with diarrhea in animals and humans (Sears, 2001). BFT stimulates reorganization of the actin cytoskeleton in cultured cells and reduces barrier function in polarized epithelial cells. BFT is an extracellular protease that cleaves E cadherin. Several distinct BFTs have been identified, which have between 87% and 96% identity but possess similar biochemical and biological activities (Wu *et al.*, 2002). The significance of allelic forms of BFT remains to be determined.

Clostridium perfringens enterotoxin associates with tight junctions

Clostridium perfringens enterotoxin (CPE) protein is a 35-kDa protein that directly binds epithelial tight junctions, including claudins and occludin. This association affects tight junction structure and function, changing the permeability and contributing to diarrhea.

TOXINS THAT MODULATE INTRACELLULAR CAMP IN MAMMALIAN CELLS

Cyclic adenosine monophosphate (cAMP) is a secondary messenger in mammalian cell signaling. Cholera toxin (CT) elevates intracellular cAMP through the ADP-ribosylation of G-proteins, while *Pseudomonas* ExoY mimics the action of the host adenylate cyclases to elevate intracellular cAMP to supraphysiological levels.

Cholera toxin ADP-ribosylates the α-subunit of G_S

CT is an AB5 protein that is produced by *V. cholerae* and is responsible for pandemic cholera (Mekalanos *et al.*, 1983). *E. coli* produces a closely related protein toxin, the heat-labile enterotoxin (LT) that causes traveler's diarrhea (Moseley and Falkow, 1980). The A domain of CT/LT is a ~22-kDa protein that ADP-ribosylates the α-subunit of the heterotrimeric G_S protein. ADP-ribosylation locks the α-subunit of G_S in a GTP-bound and constitutively active form. ADP-ribosylated $G_S\alpha$ activates mammalian adenlyate cyclase, leading to elevated cAMP and the secretion of electrolytes and H_2O from epithelial cells. The B5 domain comprises five identical peptides (~11 kDa) that assemble into a stable ring. The B5 domain of CT binds the ganglioside G_{M1} on the apical surface of polarized epithelial cells (Heyningen, 1974). CT enters epithelial cells through an endocytic pathway that includes retrograde trafficking to Golgi/endoplasmic reticulum (ER) (Badizadegan *et al.*, 2004). In the ER, the A domain unfolds and is translocated from the lumen of the ER into the cytosol. The C terminus of the A domain possesses an ER-retention (KDEL motif) sequence that enhances intracellular trafficking of CT but is not required for trafficking. Lipid rafts appear to contribute to the intracellular trafficking of the CT–G_{M1} complex (Fujinaga *et al.*, 2003).

ExoY is a mimic of host adenylate cyclase

Pseudomonas ExoY is a type III cytotoxin (Yahr *et al.*, 1998). ExoY has primary amino-acid homology with the active sites of the adenylate cyclase toxin of *B. pertussis* and the edema factor of *B. anthracis*. Type-III-delivered ExoY elevates cAMP levels about 2000-fold in cultured cells. This has pleotrophic effects on host cell metabolism, including the reorganization of the actin cytoskeleton. In vitro ExoY catalyzes only a basal adenylate cyclase and is stimulated by the addition of a host cell lysate. The identity of the mammalian activator of ExoY remains to be determined.

TOXINS THAT INHIBIT PROTEIN SYNTHESIS THROUGH DISRUPTION OF RNA FUNCTION

Shiga toxin (ST) and the Shiga-like toxins disrupt the host epithelium through a novel de-adenylation that uncouples the function of rRNA, thereby inhibiting protein synthesis.

Shiga toxin deadenylates rRNA

ST is an AB5 toxin produced by *Shigella dysenteriae* (O'Brien *et al.*, 1980). The A domain catalyzes the deadenylation of RNA, while the B5 domain binds the glycolipid receptor (Gb3). ST enters mammalian cells though clathrin-coated pits and clathrin-independent endocytosis (Sandvig and van Deurs, 2005). Preferred trafficking appears to vary among different cell lines. Transport to the Golgi involves a Rab-11-dependent pathway but is independent of Rab-9, suggesting that ST does not traffic through late endosomes (Iversen *et al.*, 2001). Transport of ST from endosomes to the Golgi is dependent on dynamin (Lauvrak *et al.*, 2004). In the ER, the A domain is translocated from the lumen of the ER to the cytosol. The A domain is an *N*-glycosidase that cleaves a specific adenine base from the 28-S rRNA of the 60-S ribosomal subunit (Endo *et al.*, 1988). De-adenylation inactivates a loop within rRNA that fails to bind elongation factor. This yields a potent inhibition of protein synthesis and is responsible for the pathology associated with ST-mediated disease.

Shiga-toxin-producing *Escherichia coli* and the hemolytic uremic syndrome

Several strains of *E. coli* produce AB5 toxins (Stx), which are related to the ST of *S. dysenteriae*. *E. coli* that produce Stx are termed Stx-producing *E. coli* (STEC). *E. coli* O157:H7 is responsible for many outbreaks of hemorrhagic colitis and diarrhea-associated hemolytic uremic syndrome (HUS) (O'Brien *et al.*, 1980, 1983). STEC produce either Stx1 or Stx2; the A domain in Stx1 differs from ST by only a single amino acid, and Stx1 and Stx2 have 55% identity. Like ST, the A domains of Stx1 and Stx2 are *N*-glycosidases that cleave 28-S rRNA and inhibit protein synthesis. The B5 domain of Stx binds glycolipid receptors for entry into cells. Several differences between Stx1 and Stx2 have been found, including the C-terminus of Stx2 A2 domain forming a short α-helix while the C-terminus of Stx1 A-subunit is disordered (Fraser *et al.*, 2004). This may be one explanation for the differential toxicities of Stx1 and Stx2.

CONCLUSION

Bacteria utilize numerous strategies to subvert the host cell physiology and damage the epithelium. Advances in our understanding of how bacterial toxins disrupt host cell functions parallel advances in our understanding of the molecular and cell biology of mammalian cells and studies on the physical and biochemical properties of these toxins. Future studies will utilize high-resolution microscopy, biophysical analysis, and functional biology to better correlate how the host responds to the insult of toxin action and will continue to unravel the molecular steps in the intoxication of the specialized cells of the epithelium.

ACKNOWLEDGMENTS

Research efforts of JTB are supported by NIH-NIAID-AI-30162.

REFERENCES

Aktories, K., Ankenbauer, T., Schering, B., and Jakobs, K. H. (1986). ADP-ribosylation of platelet actin by botulinum C2 toxin. *Eur. J. Biochem.* **161**, 155–162.

Albert, M. L., Kim, J. I., and Birge, R. B. (2000). alphavbeta5 integrin recruits the CrkII-Dock180-rac1 complex for phagocytosis of apoptotic cells. *Nat. Cell Biol.* **2**, 899–905.

Alonso, A., Bottini, N., Bruckner, S., *et al.* (2004). Lck dephosphorylation at Tyr-394 and inhibition of T cell antigen receptor signaling by *Yersinia* phosphatase YopH. *J. Biol. Chem.* **279**, 4922–4928.

Badizadegan, K., Wheeler, H. E., Fujinaga, Y., and Lencer, W. I. (2004). Trafficking of cholera toxin-ganglioside GM1 complex into Golgi and induction of toxicity depend on actin cytoskeleton. *Am. J. Physiol. Cell Physiol.* **287**, C1453–1462.

Barbieri, J. T. and Sun, J. (2004). *Pseudomonas aeruginosa* ExoS and ExoT. *Rev. Physiol. Biochem. Pharmacol.* **152**, 79–92.

Bar-Sagi, D. and Hall, A. (2000). Ras and Rho GTPases: a family reunion. *Cell* **103**, 227–238.

Barth, H., Blocker, D., Behlke, J., *et al.* (2000). Cellular uptake of *Clostridium botulinum* C2 toxin requires oligomerization and acidification. *J. Biol. Chem.* **275**, 18 704–18 711.

Black, D. S., Marie-Cardine, A., Schraven, B., and Bliska, J. B. (2000). The *Yersinia* tyrosine phosphatase YopH targets a novel adhesion-regulated signalling complex in macrophages. *Cell. Microbiol.* **2**, 401–414.

Bougneres, L., Girardin, S. E., Weed, S. A., *et al.* (2004). Cortactin and Crk cooperate to trigger actin polymerization during *Shigella* invasion of epithelial cells. *J. Cell. Biol.* **166**, 225–235.

Boyd, A. P., Lambermont, I., and Cornelis, G. R. (2000). Competition between the Yops of *Yersinia enterocolitica* for delivery into eukaryotic cells: role of the SycE chaperone binding domain of YopE. *J. Bacteriol.* **182**, 4811–4821.

Burghout, P., van Boxtel, R., Van Gelder, P., *et al.* (2004). Structure and electrophysiological properties of the YscC secretin from the type III secretion system of *Yersinia enterocolitica. J. Bacteriol.* **186**, 4645–4654.

Castellano, F., Chavrier, P., and Caron, E. (2001). Actin dynamics during phagocytosis. *Semin. Immunol.* **13**, 347–355.

Chang, J., Chen, J., and Zhou, D. (2005). Delineation and characterization of the actin nucleation and effector translocation activities of *Salmonella* SipC. *Mol. Microbiol.* **55**, 1379–1389.

Chaves-Olarte, E., Florin, I., Boquet, P., *et al.* (1996). UDP-glucose deficiency in a mutant cell line protects against glucosyltransferase toxins from *Clostridium difficile* and *Clostridium sordellii. J. Biol. Chem.* **271**, 6925–6932.

Cheng, L. W. and Schneewind, O. (2000). Type III machines of Gram-negative bacteria: delivering the goods. *Trends Microbiol.* **8**, 214–220.

Coburn, J. and Gill, D. M. (1991). ADP-ribosylation of p21ras and related proteins by *Pseudomonas aeruginosa* exoenzyme S. *Infect. Immun.* **59**, 4259–4262.

Cornelis, G. R. and Van Gijsegem, F. (2000). Assembly and function of type III secretory systems. *Annu. Rev. Microbiol.* **54**, 735–774.

Endo, Y., Tsurugi, K., Yutsudo, T., *et al.* (1988). Site of action of a Vero toxin (VT2) from *Escherichia coli* O157:H7 and of Shiga toxin on eukaryotic ribosomes. RNA *N*-glycosidase activity of the toxins. *Eur. J. Biochem.* **171**, 45–50.

Etienne-Manneville, S. and Hall, A. (2002). Rho GTPases in cell biology. *Nature* **420**, 629–635.

Fraser, M. E., Fujinaga, M., Cherney, M. M., *et al.* (2004). Structure of shiga toxin type 2 (Stx2) from *Escherichia coli* O157:H7. *J. Biol. Chem.* **279**, 27 511–27 517.

Fu, Y. and Galan, J. E. (1999). A salmonella protein antagonizes Rac-1 and Cdc42 to mediate host-cell recovery after bacterial invasion. *Nature* **401**, 293–297.

Fujinaga, Y., Wolf, A. A., Rodighiero, C., *et al.* (2003). Gangliosides that associate with lipid rafts mediate transport of cholera and related toxins from the plasma membrane to endoplasmic reticulum. *Mol. Biol. Cell.* **14**, 4783–4793.

Fujiwara, I., Takahashi, S., Tadakuma, H., Funatsu, T., and Ishiwata, S. (2002). Microscopic analysis of polymerization dynamics with individual actin filaments. *Nat. Cell. Biol.* **4**, 666–673.

Fullner, K. J. and Mekalanos, J. J. (2000). In vivo covalent cross-linking of cellular actin by the *Vibrio cholerae* RTX toxin. *EMBO J.* **19**, 5315–5323.

Furuse, M., Itoh, M., Hirase, T., *et al.* (1994). Direct association of occludin with ZO-1 and its possible involvement in the localization of occludin at tight junctions. *J. Cell Biol.* **127**, 1617–1626.

Ganesan, A. K., Vincent, T. S., Olson, J. C., and Barbieri, J. T. (1999). *Pseudomonas aeruginosa* exoenzyme S disrupts Ras-mediated signal transduction by inhibiting guanine nucleotide exchange factor-catalyzed nucleotide exchange. *J. Biol. Chem.* **274**, 21 823–21 829.

Garrity-Ryan, L., Shafikhani, S., Balachandran, P., *et al.* (2004). The ADP ribosyltransferase domain of *Pseudomonas aeruginosa* ExoT contributes to its biological activities. *Infect. Immun.* **72**, 546–558.

Gilman, A. G. (1995). Nobel Lecture. G proteins and regulation of adenylyl cyclase. *Biosci. Rep.* **15**, 65–97.

Goldberg, G. S., Alexander, D. B., Pellicena, P., *et al.* (2003). Src phosphorylates Cas on tyrosine 253 to promote migration of transformed cells. *J. Biol. Chem.* **278**, 46 533–46 540.

Hardt, W. D., Chen, L. M., Schuebel, K. E., Bustelo, X. R., and Galan, J. E. (1998). *S. typhimurium* encodes an activator of Rho GTPases that induces membrane ruffling and nuclear responses in host cells. *Cell* **93**, 815–826.

Hayward, R. D. and Koronakis, V. (1999). Direct nucleation and bundling of actin by the SipC protein of invasive *Salmonella*. *EMBO J.* **18**, 4926–4934.

Heyningen, S. V. (1974). Cholera toxin: interaction of subunits with ganglioside GM1. *Science* **183**, 656–657.

Iglewski, B. H., Sadoff, J., Bjorn, M. J., and Maxwell, E. S. (1978). *Pseudomonas aeruginosa* exoenzyme S: an adenosine diphosphate ribosyltransferase distinct from toxin A. *Proc. Natl. Acad. Sci. U. S. A.* **75**, 3211–3215.

Iversen, T. G., Skretting, G., Llorente, A., *et al.* (2001). Endosome to Golgi transport of ricin is independent of clathrin and of the Rab9- and Rab11-GTPases. *Mol. Biol. Cell* **12**, 2099–2107.

Just, I., Selzer, J., Wilm, M., *et al.* (1995). Glucosylation of Rho proteins by *Clostridium difficile* toxin B. *Nature* **375**, 500–503.

Kashimoto, T., Katahira, J., Cornejo, W. R., *et al.* (1999). Identification of functional domains of *Bordetella* dermonecrotizing toxin. *Infect. Immun.* **67**, 3727–3732.

Kiyokawa, E., Hashimoto, Y., Kobayashi, S., *et al.* (1998). Activation of Rac1 by a Crk SH3-binding protein, DOCK180. *Genes Dev.* **12**, 3331–3336.

Lauvrak, S. U., Torgersen, M. L., and Sandvig, K. (2004). Efficient endosome-to-Golgi transport of Shiga toxin is dependent on dynamin and clathrin. *J. Cell Sci.* **117**, 2321–2331.

Lesnick, M. L., Reiner, N. E., Fierer, J., and Guiney, D. G. (2001). The *Salmonella* spvB virulence gene encodes an enzyme that ADP-ribosylates actin and destabilizes the cytoskeleton of eukaryotic cells. *Mol. Microbiol.* **39**, 1464–1470.

Logsdon, L. K. and Mecsas, J. (2003). Requirement of the *Yersinia pseudotuberculosis* effectors YopH and YopE in colonization and persistence in intestinal and lymph tissues. *Infect. Immun.* **71**, 4595–4607.

Mangeat, P., Roy, C., and Martin, M. (1999). ERM proteins in cell adhesion and membrane dynamics. *Trends Cell Biol.* **9**, 187–192.

Maresso, A. W., Baldwin, M. R., and Barbieri, J. T. (2004). Ezrin/radixin/moesin proteins are high affinity targets for ADP-ribosylation by *Pseudomonas aeruginosa* ExoS. *J. Biol. Chem.* **279**, 38 402–38 408.

Marlovits, T. C., Kubori, T., Sukhan, A., *et al.* (2004). Structural insights into the assembly of the type III secretion needle complex. *Science* **306**, 1040–1042.

Matsuda, M., Reichman, C. T., and Hanafusa, H. (1992). Biological and biochemical activity of v-Crk chimeras containing the SH2/SH3 regions of phosphatidylinositol-specific phospholipase C-gamma and Src. *J. Virol.* **66**, 115–121.

Matsui, T., Maeda, M., Doi, Y., *et al.* (1998). Rho-kinase phosphorylates COOH-terminal threonines of ezrin/radixin/moesin (ERM) proteins and regulates their head-to-tail association. *J Cell Biol.* **140**, 647–657.

Mekalanos, J. J., Swartz, D. J., Pearson, G. D., *et al.* (1983). Cholera toxin genes: nucleotide sequence, deletion analysis and vaccine development. *Nature* **306**, 551–557.

Moseley, S. L. and Falkow, S. (1980). Nucleotide sequence homology between the heat-labile enterotoxin gene of *Escherichia coli* and *Vibrio cholerae* deoxyribonucleic acid. *J. Bacteriol.* **144**, 444–446.

O'Brien, A. D., LaVeck, G. D., Griffin, D. E., and Thompson, M. R. (1980). Characterization of *Shigella dysenteriae* 1 (Shiga) toxin purified by anti-Shiga toxin affinity chromatography. *Infect. Immun.* **30**, 170–179.

O'Brien, A. O., Lively, T. A., Chen, M. E., Rothman, S. W., and Formal, S. B. (1983). *Escherichia coli* O157:H7 strains associated with haemorrhagic colitis in the United States produce a Shigella dysenteriae 1 (SHIGA) like cytotoxin. *Lancet* **1**, 702.

Ohishi, I. and Tsuyama, S. (1986). ADP-ribosylation of nonmuscle actin with component I of C2 toxin. *Biochem. Biophys. Res. Commun.* **136**, 802–806.

Pearson, M. A., Reczek, D., Bretscher, A., and Karplus, P. A. (2000). Structure of the ERM protein moesin reveals the FERM domain fold masked by an extended actin binding tail domain. *Cell* **101**, 259–270.

Phan, J., Lee, K., Cherry, S., *et al.* (2003). High-resolution structure of the *Yersinia pestis* protein tyrosine phosphatase YopH in complex with a phosphotyrosyl mimetic-containing hexapeptide. *Biochemistry* **42**, 13 113–13 121.

Popoff, M. R., Chaves-Olarte, E., Lemichez, E., *et al.* (1996). Ras, Rap, and Rac small GTP-binding proteins are targets for *Clostridium sordellii* lethal toxin glucosylation. *J. Biol. Chem.* **271**, 10 217–10 224.

Sandvig, K. and van Deurs, B. (2005). Delivery into cells: lessons learned from plant and bacterial toxins. *Gene Ther.* **12**, 865–872.

Schirmer, J. and Aktories, K. (2004). Large clostridial cytotoxins: cellular biology of Rho/Ras-glucosylating toxins. *Biochim. Biophys. Acta* **1673**, 66–74.

Schmidt, G., Sehr, P., Wilm, M., *et al.* (1997). Gln 63 of Rho is deamidated by *Escherichia coli* cytotoxic necrotizing factor-1. *Nature* **387**, 725–729.

Schmidt, G., Goehring, U. M., Schirmer, J., Lerm, M., and Aktories, K. (1999). Identification of the C-terminal part of *Bordetella* dermonecrotic toxin as a transglutaminase for rho GTPases. *J. Biol. Chem.* **274**, 31 875–31 881.

Sears, C. L. (2001). The toxins of *Bacteroides fragilis. Toxicon* **39**, 1737–1746.

Selbach, M., Moese, S., Backert, S., Jungblut, P. R., and Meyer, T. F. (2004). The *Helicobacter pylori* CagA protein induces tyrosine dephosphorylation of ezrin. *Proteomics* **4**, 2961–2968.

Shao, F. and Dixon, J. E. (2003). YopT is a cysteine protease cleaving Rho family GTPases. *Adv. Exp. Med. Biol.* **529**, 79–84.

Shao, F., Vacratsis, P. O., Bao, Z., *et al.* (2003). Biochemical characterization of the *Yersinia* YopT protease: cleavage site and recognition elements in Rho GTPases. *Proc. Natl. Acad. Sci. U. S. A.* **100**, 904–909.

Sixma, T. K., Kalk, K. H., van Zanten, B. A., *et al.* (1993). Refined structure of *Escherichia coli* heat-labile enterotoxin, a close relative of cholera toxin. *J. Mol. Biol.* **230**, 890–918.

Sundin, C., Henriksson, M. L., Hallberg, B., Forsberg, A., and Frithz-Lindsten, E. (2001). Exoenzyme T of *Pseudomonas aeruginosa* elicits cytotoxicity without interfering with Ras signal transduction. *Cell. Microbiol.* **3**, 237–246.

Tezcan-Merdol, D., Nyman, T., Lindberg, U., *et al.* (2001). Actin is ADP-ribosylated by the *Salmonella enterica* virulence-associated protein SpvB. *Mol. Microbiol.* **39**, 606–619.

Von Pawel-Rammingen, U., Telepnev, M. V., Schmidt, G., *et al.* (2000). GAP activity of the *Yersinia* YopE cytotoxin specifically targets the Rho pathway: a mechanism for disruption of actin microfilament structure. *Mol. Microbiol.* **36**, 737–748.

Wegner, A. and Aktories, K. (1988). ADP-ribosylated actin caps the barbed ends of actin filaments. *J. Biol. Chem.* **263**, 13 739–13 742.

Woestyn, S., Allaoui, A., Wattiau, P., and Cornelis, G. R. (1994). YscN, the putative energizer of the *Yersinia* Yop secretion machinery. *J. Bacteriol.* **176**, 1561–1569.

Wu, S., Dreyfus, L. A., Tzianabos, A. O., Hayashi, C., and Sears, C. L. (2002). Diversity of the metalloprotease toxin produced by enterotoxigenic *Bacteroides fragilis. Infect. Immun.* **70**, 2463–2471.

Wurtele, M., Wolf, E., Pederson, K. J., *et al.* (2001). How the *Pseudomonas aeruginosa* ExoS toxin downregulates Rac. *Nat. Struct. Biol.* **8**, 23–26.

Yahr, T. L., Vallis, A. J., Hancock, M. K., Barbieri, J. T., and Frank, D. W. (1998). ExoY, an adenylate cyclase secreted by the *Pseudomonas aeruginosa* type III system. *Proc. Natl. Acad. Sci. U. S. A.* **95**, 13 899–13 904.

Yang, N., Higuchi, O., Ohashi, K., *et al.* (1998). Cofilin phosphorylation by LIM-kinase 1 and its role in Rac-mediated actin reorganization. *Nature* **393**, 809–812.

Part III Host cell signaling by bacteria

CHAPTER 8

Host-mediated invasion: the *Salmonella* Typhimurium trigger

Brit Winnen and Wolf-Dietrich Hardt

INTRODUCTION

The epithelial layer of mucosal surfaces in the gastrointestinal tract is a busy surface for communication between the host and both commensal and pathogenic bacteria. In the case of invasive pathogens, the epithelium represents the major barrier that has to be overcome. Important virulence determinants include those mediating adhesion to and invasion of the epithelium. Different strategies are used to induce the uptake of invasive bacteria (Figure 8.1). For example, *Yersinia* spp. and *Listeria monocytogenes* express adhesins that bind tightly to host cell-surface proteins. This binding interaction initiates invasion by triggering signaling pathways normally involved in the regulation of cell adhesion. This invasion mechanism has been termed "zipper" because the host cell membrane is wrapped tightly around the bacteria (Mengaud *et al.*, 1996). For more details on the zipper mechanism; see Alonso and Garcia-del Portillo (2004). More recently, a variant of this mechanism, termed the "tandem β-zipper," has been identified. As an example of this entry mechanism, pathogenic bacteria like *Staphylococcus aureus* express surface proteins that bind to human fibronectin. This fibronectin coat facilitates binding to host cell-surface receptors and mediates invasion (Figure 8.1b) (Schwarz-Linek *et al.*, 2004). A third strategy, termed "trigger mechanism," is employed by *Salmonella* and *Shigella* spp. This involves specialized protein-transport systems called type III secretion systems (TTSS) to inject virulence factors (effector proteins) directly into the cytosol of the host cells. The translocated effector proteins trigger host signaling cascades that mediate a variety of responses, including pathogen uptake (Hueck, 1998). Host

Bacterial–Epithelial Cell Cross-Talk: Molecular Mechanisms in Pathogenesis, ed. Beth A. McCormick. Published by Cambridge University Press.

Table 8.1 *Type III effectors involved in host cell invasion or inhibition of phagocytosis*

Pathogen	Effector	Enzymatic activity and function	Reference
Salmonella spp.	SopB	Inositol phosphatase; promotes indirect Rho GTPase activation and macropinocytosis by generating PtdIns(3)P and Ins(1,4,5,6)P_4; anti-apoptotic activity in epithelial cells	Hernandez *et al.* (2004); Knodler *et al.* (2005); Marcus *et al.* (2001); Terebiznik *et al.* (2002); Zhou *et al.* (2001)
	SopE/E2	GDP–GTP exchange factor (GEF) for Cdc42 and Rac1; stimulates actin reorganization to promote bacterial entry	Bakshi *et al.* (2000); Stender *et al.* (2000); Hardt *et al.* (1998); Wood *et al.* (1996)
	SipA	Binds actin; diminishes its critical concentration; induces actin bundling; enhances SipC effect; stabilizes F-actin; inhibits actin depolymerization; enhances actin polymerization and macropinocytosis	Lilic *et al.* (2003); McGhie *et al.* (2004); Zhou *et al.* (1999 a,b)
	SptP	Amino terminus: GTPase-activating protein (Cdc42 and Rac1); carboxy terminus: tyrosine phosphatase activity; reverses cellular changes induced by other effector proteins	Fu and Galan (1999); Kaniga *et al.* (1996); Stebbins and Galan (2000)
	SipC	Actin nucleation and bundling; TTS translocator	Collazo and Galan (1997); Hayward and Koronakis (1999)
Shigella spp.	IpgD	Inositol 4-phosphatase, dephosphorylates PtdIns(4,5)P_2 into PtdIns(5)P; uncouples cytoplasma membrane from actin cytoskeleton (homologous to SopB)	Allaoui *et al.* (1993); Niebuhr *et al.* (2002)

Table 8.1 *(cont.)*

	IpaA	Targets vinculin; initiates formation of focal adhesion-like structures required for efficient invasion (homologous to SipA)	Bourdet-Sicard *et al.* (1999); Tran Van Nhieu *et al.* (1997)
	IpaB	Interacts with CD44 receptor; induces apoptosis via caspase-1; binds cholesterol with high affinity (homologous to SipB)	Hayward *et al.* (2005); Hilbi *et al.* (1998); Skoudy *et al.* (2000)
	IpaC	Direct nucleation and bundling of actin; activates Cdc42 to promote bacterial entry (homologous to SipC)	Kueltzo *et al.* (2003); Picking *et al.* (2001); Tran Van Nhieu *et al.* (1999)
	VirA	Interacts with tubulin and destabilizes microtubules; important for bacterial entry by activation of Rac1	Uchiya *et al.* (1995); Yoshida *et al.* (2002)
Yersiniae spp.	YopE	GTPase-activating protein; Cdc42 and Rac1; disrupts actin filaments to inhibit phagocytosis	Black and Bliska (2000); Von Pawel-Rammingen *et al.* (2000)
	YopH	Tyrosine phosphatase; disrupts peripheral focal complexes; blocks phagocytosis	Bliska *et al.* (1991); Guan and Dixon (1990)
	YpkA (YopO)	Serine/threonine kinase; intracellular target unknown; activated by cellular actin; C-terminus interacts with RhoA and Rac1	Galyov *et al.* (1993); Juris *et al.* (2000); Navarro *et al.* (2005)
	YopT	Cysteine protease; releases Rho GTPases from cell membrane by cleaving C-terminus; irreversibly inhibits Rho family signaling; disrupts actin filaments to inhibit phagocytosis	Aepfelbacher *et al.* (2003); Navarro *et al.* (2005); Shao *et al.* (2002, 2003)

Ins (1,4,5,6)P_4, inositol 1,4,5,6-tetrakisphosphate; PtdIns(3)P, phosphatidylinositol-3-phosphate; PtdIns(5)P, phosphatidylinositol-5-phosphate; PtdIns(4,5)P_2, phosphatidylinositol-4,5-diphosphate; TTS, type 3 secretion.

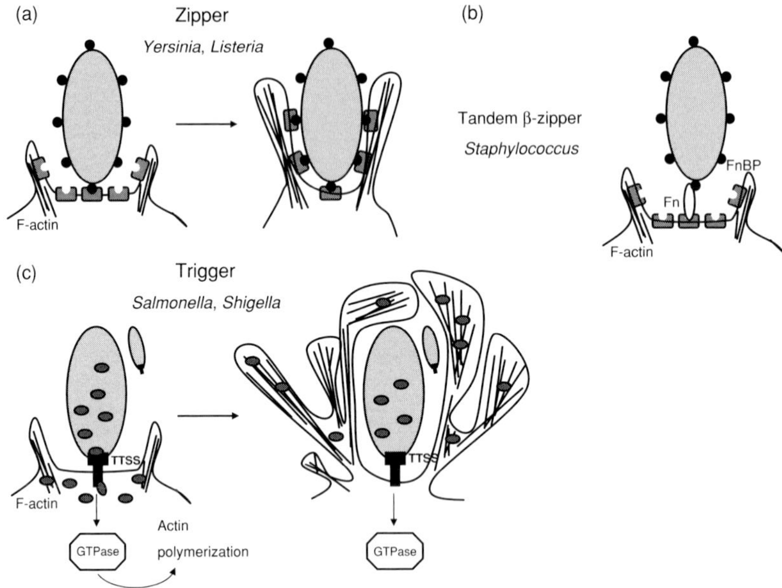

Figure 8.1 Bacterial pathogens can induce phagocytosis by different mechanisms. (a) The zipper invasion mechanism. Uptake of *Listeria monocytogenes* or *Yersinia* spp. into epithelial cells is directed by tight ligand–receptor interaction, resulting in uptake of one bacterium at a time. (b) Several other pathogenic bacteria, including *Staphylococcus aureus,* use a tandem β-zipper mechanism to attach to human fibronection (Fn). Fn connects a bacterial surface protein (Fn-binding protein, FnBP) and the host cell receptor (i.e. $\alpha_5\beta_1$-integrins) (Schwarz-Linek *et al.*, 2004). (c) The trigger invasion mechanism. *Salmonella* and *Shigella* spp. inject a cocktail of effector proteins via a type III secretion system (TTSS) into the host cell cytosol. These effector proteins induce membrane ruffles, which engulf fluid phase markers, the pathogen, as well as "bystander bacteria," leading to their uptake.

cell invasion of *Salmonella* Typhimurium via the trigger mechanism is the focus of this chapter.

The basic components of the TTSS and the mechanism of protein translocation itself are conserved among different Gram-negative bacteria. In contrast, the effector proteins differ dramatically (see Table 8.1). This likely reflects the coevolution of bacteria with their specific host and, as a result, there is no common pathway used to circumvent the host's defense mechanisms. Instead, each species has evolved its own unique tactic to establish infection, thereby reflecting the great diversity of pathogens encountered by humans. This is illustrated by the invasion of the intestinal mucosa by different pathogenic bacteria harboring TTSS (Figure 8.2). Bacteria can traverse the

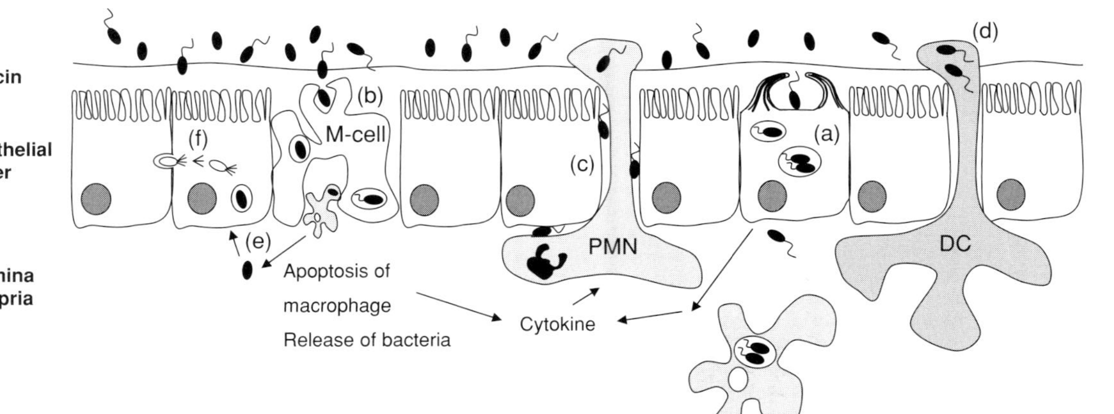

Figure 8.2 Strategies of enteroinvasive bacteria to breach the intestinal epithelium. (a) Bacteria (black; +/− flagella) on the luminal surface of intestinal epithelium inject type III secretion system (TTSS) effector proteins into the host cell cytoplasm. This induces focal actin polymerization and bacterial entry. (b) Bacteria enter M-cells, which lack microvilli and are located in the Peyer's patches. (c) Paracellular entry via disruption of tight junctions. (d) Uptake at the luminal side by dendritic cells (DCs) (Rescigno *et al.*, 2001). Once the bacteria have breached the epithelial barrier, they can replicate within macrophages (*Salmonella* requires the *Salmonella* pathogenicity island 2, SPI-2) or remain extracellular by avoiding phagocytosis and inducing host cell apoptosis. (e) Some bacteria are capable of invading polarized epithelial cells from the basolateral side, e.g *Shigella* (Mounier *et al.*, 1992), which is followed by intracellular motility and cell-to-cell spread (f). Other bacteria e.g. *Salmonella*, are phagocytosed by adjacent subepithelial macrophages and the bacteria can spread to other organs (liver, spleen, lymph nodes) and cause systemic infections. All pathways that have been discussed for *Salmonella* Typhimurium are marked by black bacteria.

DC, dendritic cell; IL, interleukin; PMN, polymonophonuclear cell.

intestinal epithelial barrier in several different ways. *Salmonella* spp. inject TTSS effector proteins into and thereby invade enterocytes (Figure 8.2a). Other bacteria (*Salmonella, Shigella, Yersinia*) enter via M-cells (Figure 8.2b) and/or are transported passively across the epithelial barrier by dendritic cells (Figure 8.2d) (Rescigno *et al.*, 2001). Once bacteria reach the lamina propria, they encounter phagocytic cells such as macrophages, polymorphonuclear cells (PMN), and dendritic cells.

Again, this encounter with phagocytes can involve TTSS function, which can have quite different consequences. Some bacteria (*Yersinia* spp.) remain strictly extracellular and inhibit macrophage phagocytosis via TTSS effector proteins (Table 8.1). Other bacteria are taken up but induce rapid macrophage apoptosis (*Salmonella* spp.) Table 8.1 In yet other cases, the bacteria can survive and even replicate inside macrophages and dendritic cells; this process is thought to allow dissemination within the host. In contrast to *Salmonella* spp., *Shigella* spp. "hide" from phagocytic cells by invading epithelial cells from the basolateral side and spreading between epithelial cells using actin-based motility (Figure 8.2e, f). In general, the bacterial insult leads to the release of pro-inflammatory cytokines and recruitment of phagocytes (macrophages, PMN) to the site of infection. As a side effect of this inflammatory response, one frequently observes tissue disruption, which may facilitate tissue entry of further pathogens (Figure 8.2c).

The molecular interaction of enteropathogenic bacteria with the gut epithelium has been studied extensively in vitro. For practical reasons, most studies on host cell invasion have been performed with non-polarized cells such as fibroblasts and HeLa cells. This has proven useful because the cellular targets of the invasion factors/effector proteins are present in these cells and the signaling pathways controlling the cytoskeletal architecture are well understood. However, polarized epithelial cells (mostly T84 and Caco2 cells) have the obvious advantage that they can mimic more closely the situation at the intestinal epithelium. In this respect, a tissue culture assay involving the transmigration of phagocytic cells (PMN) in response to bacterial infection of a polarized epithelial monolayer has been of special interest (Hurley and McCormick, 2003). Nevertheless, most of the available data that are discussed in this chapter stem from work on non-polarized cells. Future work will have to establish how far these mechanisms extend to polarized epithelia. Furthermore, animal models are required to learn which mechanisms operate in vivo when the polarized epithelium is just the outer layer of a complex and quite dynamic tissue.

In this chapter, we focus on the molecular mechanisms used by the enteropathogen *Salmonella* Typhimurium to invade host cells. For comparison, the entry mechanisms of *Shigella flexneri* and *Yersinia* spp. are also discussed.

MIMICRY OF HOST CELLULAR PROTEINS

Many pathogenic bacteria employ factors that manipulate signal-transduction cascades within the host cell. Generally, these virulence factors have been acquired by horizontal gene transfer, some from the host (e.g. RalF from *Legionella*) (Nagai *et al.*, 2002) and others from other bacteria. (For more details on horizontal gene transfer, see Brussow *et al.* (2004) and Lawrence (2005)). The latter have often evolved by "convergent evolution." These virulence factors often have identical functions but do not resemble their eukaryotic counterparts in sequence or structure. This has emerged as a common theme and is referred to as "molecular mimicry" (Stebbins and Galan, 2001). No matter how they were acquired, these virulence factors allow the pathogens to actuate signaling cascades that are hard-wired in the host cells that they infect. Therefore, it is not surprising that the bacteria-induced responses often resemble a certain facet of the cells normal response repertoire. For example, *Salmonella* and *Shigella* spp. invade host cells by triggering actin rearrangements. These resemble membrane ruffles induced by growth factors, including the recruitment of specific cytoskeletal associated proteins to the site of bacterial entry (Finlay *et al.*, 1991; Francis *et al.*, 1993; Ginocchio *et al.*, 1992).

During early steps of invasion, the *Salmonella* protein SopE functions as an exchange factor for Rho GTPases (Table 8.1), thereby mimicking host cellular guanine nucleotide exchange factors, despite the lack of sequence similarities. Thus, SopE provides an excellent example for molecular mimicry and has probably emerged in some unknown bacterial species by convergent evolution (Buchwald *et al.*, 2002; Hardt *et al.*, 1998; Schlumberger *et al.*, 2003). Similarly, the N-terminal domain of the *S.* Typhimurium TTSS effector protein SptP has similar function but no sequence similarity to eukaryotic GTPase-activating proteins (Stebbins and Galan, 2001). In contrast, the C-terminal domain of SptP shows significant sequence similarity to eukaryotic tyrosine phosphatases. Thus, this domain of SptP has probably been acquired by horizontal gene transfer from some unidentified eukaryote (Kaniga *et al.*, 1996; Stebbins and Galan, 2000). To understand the pathophysiological functions of the bacterial effector proteins triggering invasion, it is important to understand the signaling cascades of the host cell. Significant progress has been made during the past several years.

MODULATION OF THE HOST ACTIN CYTOSKELETON: THE CYTOSKELETON IS A MAJOR TARGET IN MAMMALIAN CELLS

Salmonella and *Shigella* spp. invade normally non-phagocytic host cells via the trigger mechanism. A number of TTSS effectors act in concert to induce actin rearrangements and engulfment of the bacteria (Garcia-del Portillo and Finlay, 1994). Inhibitors of microfilament formation (e.g. cytochalasin D) or inhibition of the appropriate signaling cascades prevents cytoskeletal rearrangements and bacterial entry (L. M. Chen *et al.*, 1996; Clerc *et al.*, 1987; Finlay *et al.*, 1991). Thus, it is helpful to provide a brief overview of the regulation of actin architecture and dynamics.

Small guanosine-5′-triphosphatases (GTPases) of the Rho subfamily are central regulators of the eukaryotic actin cytoskeleton. They are molecular switches that have a GTP-bound "on" state and a GDP-bound "off" state. In the eukaryotic host cell, these states are controlled by several families of regulatory proteins. The GTPase-activating proteins (GAPs) accelerate the intrinsic rate of GTP hydrolysis, thereby switching off the GTPase. In contrast, the guanine nucleotide exchange factors (GEFs) switch on the GTPase by increasing the rate of GDP-to-GTP exchange (Figure 8.3) (Etienne-Manneville and Hall, 2002; Molendijk *et al.*, 2004).

The active Rho GTPases, Rac1 and Cdc42, bind to and activate nucleation-promoting factors, i.e. Wiskott–Aldrich syndrome protein (WASP), neuronal Wiskott–Aldrich syndrome protein (N-WASP), and suppressor of cAMP receptor (Scar)/WASP-family verprolin-homologous protein (WAVE) (Machesky *et al.*, 1999; Prehoda *et al.*, 2000; Takenawa and Miki, 2001). Once targeted to the correct subcellular destination, these proteins bind and activate the Arp2/3 complex (Higgs and Pollard, 2001; Suetsugu *et al.*, 2002). The Arp2/3 complex then facilitates the rapid growth of actin-filaments during dynamic cellular events such as phagocytosis and chemotaxis (Welch *et al.*, 1997). Other actin-polymerizing proteins can also be involved, i.e. formins and vasodilator-stimulated phosphoprotein (VASP) (Sechi and Wehland, 2004; Zigmond, 2004).

Tight junctions are another important feature of the architecture of the intestinal epithelium. They are linked tightly to the cytoskeleton, are essential for polarization of the epithelial cell architecture, and prevent the passage of molecules and ions through the space between cells. Thereby, tight junctions regulate the permeability through the paracellular pathway. Moreover, manipulation of tight junctions can provide enteroinvasive pathogens with an alternative route to enter the gut tissue. However, so far little is known about manipulation of tight or adherence junctions by *Salmonella* spp. (Jepson

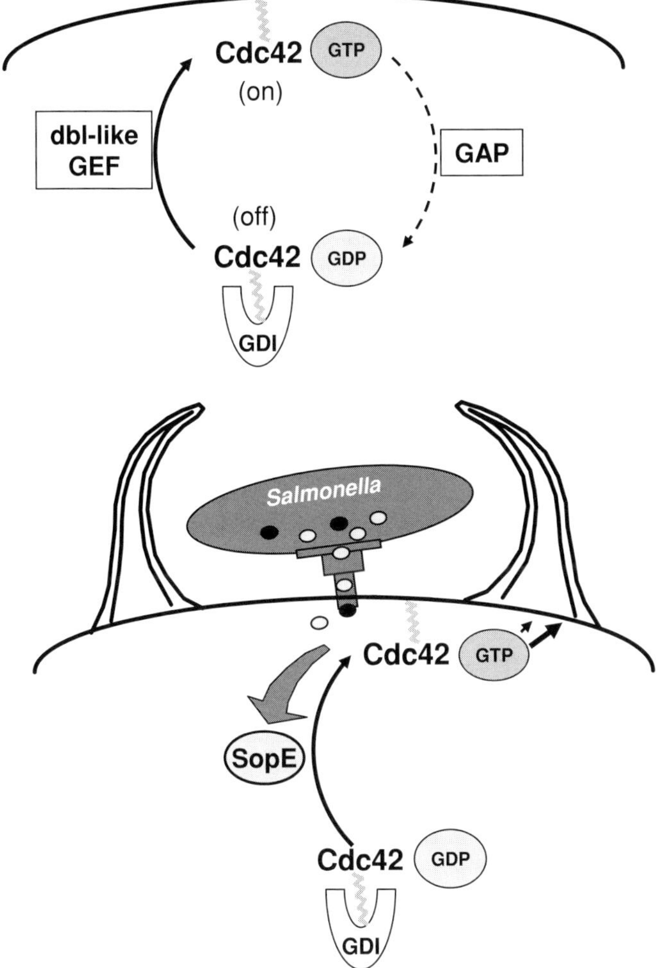

Figure 8.3 Rho GTPase signaling triggered by *Salmonella* Typhimurium. The Rho GTPases Rac1 and Cdc42 can be activated directly by host cell G-nucleotide exchange factors (GEF) (dbl-family) or by the *S.* Typhimurium effector proteins SopE and SopE2. GAP, GTPase-activating protein; GDI, guanine nucleotide dissociation inhibitor.

et al., 1995; McCormick *et al.*, 1993). Using human epithelial cells, it has been shown that *Shigella* spp. can modulate tight-junction-associated proteins and specifically remove claudin, dephosphorylate occludin, and downregulate zonula occludens 1 (ZO-1). This leads to disruption of the intestinal barrier function and allows paracellular passage of the intestinal epithelial barrier (Sakaguchi *et al.*, 2002).

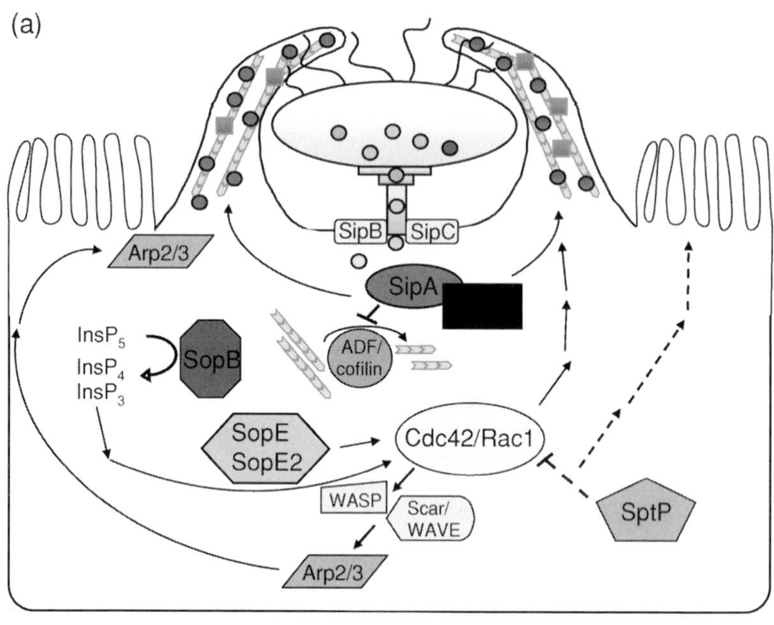
(a)
SipB
SipC
Arp2/3
SipA
InsP5
InsP4
InsP3
SopB
ADF/
cofilin
SopE
SopE2
Cdc42/Rac1
WASP
Scar/
WAVE
SptP
Arp2/3

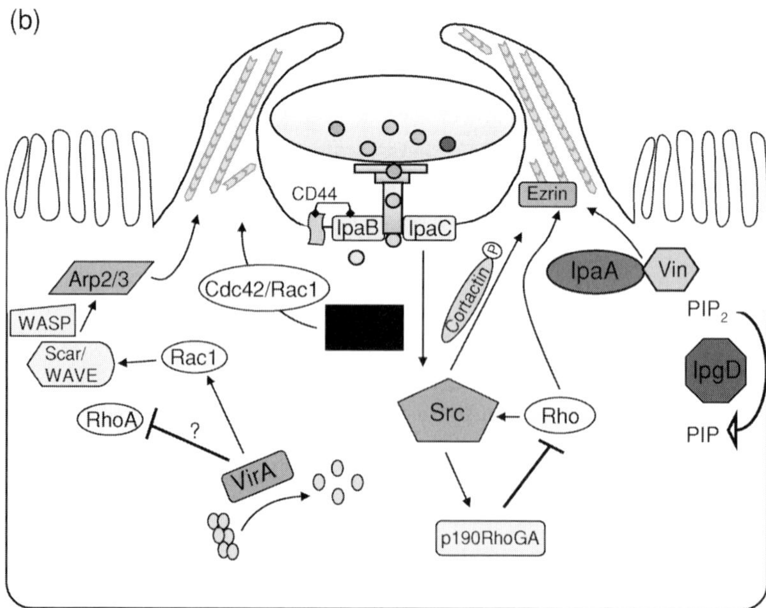
(b)
CD44
IpaB
IpaC
Ezrin
Arp2/3
Cdc42/Rac1
Cortactin
IpaA
Vin
PIP2
WASP
Scar/
WAVE
Rac1
IpgD
Src
Rho
RhoA
?
PIP
VirA
p190RhoGA

THE *SALMONELLA* PARADIGM FOR THE TRIGGER MECHANISM OF HOST CELL INVASION

Mutation analyses have revealed that at least five *S.* Typhimurium effector proteins (SopE, SopE2, SipA, SopB, SipC) translocated via the *Salmonella* pathogenicity island 1 (SPI-1) TTSS participate in triggering invasion of non-phagocytic cells. SopE and its homolog SopE2 are involved in host cell invasion (Bakshi *et al.*, 2000; Hardt *et al.*, 1998; Stender *et al.*, 2000; Wood *et al.*, 1996) and also in inducing intestinal inflammation in calves and streptomycin-pretreated mice (Hapfelmeier *et al.*, 2004; Wallis and Galyov, 2000; Zhang *et al.*, 2002). Both effectors act as GEFs for Cdc42 and Rac1 (Figures 8.3 and 8.4a; Table 8.1) (Hardt *et al.*, 1998; Stender *et al.*, 2000). SopE mimics the dbl-like GEFs for Rho GTPases but lacks any sequence and structural similarity to this eukaryotic protein family (Buchwald *et al.*, 2002; Schlumberger *et al.*, 2003). Specificity for certain Rho GTPases was observed, suggesting that SopE activates Cdc42 and Rac1 whereas SopE2 interacts preferentially with Cdc42 but not Rac1 (Friebel *et al.*, 2001; Hardt *et al.*, 1998; Stender *et al.*, 2000). However, it is still a matter of speculation how specificity for certain Rho GTPases may offer an advantage.

SopB is an inositol phosphatase and involved in stimulation of chloride secretion and enterocolitis in the *Salmonella* calf model (Norris *et al.*, 1998). Inositol metabolism seems to affect many aspects of the bacteria–host interaction. The inositol phosphatase activity of SopB is essential to promote actin rearrangements, and this activity depends on Cdc42 function (Figure 8.4a) (Zhou *et al.*, 2001). Purified SopB has a broad substrate spectrum. However, in vitro and in vivo SopB acts as a specific 3-phosphatase towards inositol 1,3,4,5,6-pentakisphosphate ($InsP_5$), thereby depleting the

←

Figure 8.4 Host cell invasion by *Salmonella* Typhimurium and *Shigella flexneri*. (a) *S.* Typhimurium host cell invasion. Cdc42/Rac1 are activated directly by SopE and SopE2 and indirectly by SopB. This triggers cytoskeletal rearrangements in the host cell. Actin remodeling is enhanced further by the actin-binding proteins SipC and SipA (see text for details). (b) *Shigella* invasion. A complex formed by invasion plasmid antigens (Ipa) B/C binds to beta-1 integrins and CD44 on the host cell surface. Subsequent injection of effector proteins triggers the formation of the entry structure. Central are the activities of virulence protein A (VirA), IpaA, invasion plasmid gene D (IpgD), and Src activation (see text for details).

ADF, cofilin; $InsP_3$, inositol triphosphate; $InsP_4$, inositol tetrakisphosphate; $InsP_5$, inositol pentakisphosphate; PIP, phosphatidylinositol-5-monophosphate; PIP2, phosphatidylinositol-4,5-bisphosphate.

cellular pool of $InsP_5$ and leading to an accumulation of inositol 1,4,5,6-tetrakisphosphate ($InsP_4$) (Zhou *et al.*, 2001). Evidence indicates that SopB also affects phosphatidylinositol-4,5-bisphosphate (PtdIns(4,5)P_2) and that hydrolysis of this phosphoinositide occurs at the base of membrane ruffles (Terebiznik *et al.*, 2002). This might be important for efficient formation of the *Salmonella*-containing vacuole (SCV) (Terebiznik *et al.*, 2002).

There seems to be significant functional redundancy between the effector proteins SopE, SopE2, and SopB: SopE and SopE2 can activate Cdc42 directly, while SopB triggers Cdc42-activating signaling pathways (Zhou *et al.*, 2001). Furthermore, *Salmonella* strains lacking SopE, or SopE2, or SopB can invade host cells with only a minor reduction in invasion efficiency, whereas a *Salmonella* strain lacking all three effector proteins is no longer able to remodel actin and shows an invasion-defective phenotype (Mirold *et al.*, 2001; Zhou *et al.*, 2001). This functional redundancy of SopE/E2/B may help to ensure that the bacteria are able to invade host cells even though the exact wiring of the actin regulatory pathways may differ between different cell types or hosts. This hypothesis awaits further experimentation.

Downstream of the Rho GTPases Rac1 and Cdc42, the Arp2/3 complex exerts its function in signaling and actin polymerization. The Arp2/3 complex localizes to ruffles in *Salmonella*-infected non-polarized and polarized cells (Criss and Casanova, 2003; Stender *et al.*, 2000). Interestingly, SopE seems to be involved in both apical and basolateral invasion of polarized cells, while only Rac1 was required for apical invasion, and neither Rac1 nor Cdc42 was required for basolateral invasion (Criss and Casanova, 2003). The implications of this observation for the role of SopE (acts efficiently on Cdc42 and Rac1) and SopE2 (acts efficiently on Cdc42 but not on Rac1) will be an interesting topic for further investigation.

ACTIN-BINDING PROTEINS ENHANCE *SALMONELLA* ENTRY

As well as the signaling cascades regulating de novo actin polymerization of eukaryotic actin, binding proteins provide a second way to regulate the highly dynamic actin cytoskeleton (Chen *et al.*, 2000; Winder and Ayscough, 2005). Accordingly, actin remodeling triggered by *S.* Typhimurium effector proteins SopE, SopE2, and SopB is enhanced further by the two actin-binding SPI-1 TTSS effector proteins SipA and SipC (Figure 8.4a) (Hayward and Koronakis, 1999; Zhou *et al.*, 1999a,b).

SipA contains an F-actin-binding domain at its C-terminus. Binding of SipA to actin reduces the critical actin concentration required for polymerization at the site of *Salmonella* entry and inhibits filament depolymerization in vitro (McGhie *et al.*, 2001; Zhou *et al.*, 1999a). By modulating the actin-bundling activity of the cellular actin-binding protein (ABP) fimbrin, SipA increases the stability of F-actin that drives and supports the growth of membrane ruffles (Jepson *et al.*, 2001; Zhou *et al.*, 1999a). Furthermore, SipA arrests cellular actin turnover by inhibiting ADF/cofilin- and gelsolin-directed actin depolymerization (McGhie *et al.*, 2004). Taken together, these activities ultimately result in increased efficiency of bacterial invasion. In vitro, SipA potentiates SipC nucleation and bundling of actin (McGhie *et al.*, 2001). For detailed information on actin binding proteins, see Hayward and Koronakis (2002).

SipC alone is insufficient to mediate host cell invasion, but it is essential for the invasion process (Collazo and Galan, 1997; Kaniga *et al.*, 1995; Mirold *et al.*, 2001; Zhou *et al.*, 2001). This phenomenon can be explained by the fact that SipC serves two functions in parallel: actin nucleation and translocation of all effectors (Hayward and Koronakis, 1999; Kaniga *et al.*, 1995). The latter function effectively shuts down SPI-1 TTSS function and has hindered the detailed analysis of SipC-mediated actin remodeling. Actin nucleation and effector translocation functions of SipC have been genetically separated (Chang *et al.*, 2005). The central region (residues 201–220) of SipC is essential for this actin nucleation activity, whereas the C-terminal region (residues 315–409) is essential for effector protein translocation (Chang *et al.*, 2005; Scherer *et al.*, 2000). In agreement with previous data (Hayward and Koronakis, 1999; Scherer *et al.*, 2000), SipC actin nucleation activity contributed to robust *Salmonella*-induced actin cytoskeleton rearrangements and promoted efficient bacterial invasion. These observations provide the basis for further elucidation of the SPI-1 TTSS effector protein functions and how they trigger *S.* Typhimurium host cell invasion.

THE *SHIGELLA* PARADIGM

Shigella flexneri is a Gram-negative pathogen related closely to *Salmonella* spp. *S. flevneri* invades the colonic mucosa and tissue culture cells that are normally non-phagocytic. This process resembles many aspects of *Salmonella* host cell invasion. However, the interaction with the polarized epithelia of the gut differs significantly. *S.* Typhimurium enters intestinal epithelia cells directly from the luminal side, but *Shigella* spp. are thought to breach the epithelial barrier via M-cells or tight-junction modifications and enter intestinal epithelial cells only from the basolateral side (Figure 8.2) (Mounier *et al.*, 1992). Further differences arise from the fact that *Shigella* spp. lyse the phagosome, move inside the host cell cytosol, and spread within epithelial monolayers direcly from one host cell to the next (Figure 8.2f) (Bourdet-Sicard *et al.*, 2000).

Most of the genes necessary for bacterial entry (including the TTSS encoding *mxi-spa* operon) are encoded on a large virulence plasmid. The TTSS secretion apparatus, the effector proteins, and the membrane ruffles induced by the TTSS effector proteins are quite similar to the *S.* Typhimurium counterparts. For example, the *S.* Typhimurium TTSS translocation complex proteins SipB, SipC, and SipD are similar to the *Shigella* proteins IpaB, IpaC, and IpaD, and they are also required for transport of effector proteins into host cells (Menard *et al.*, 1993; Sasakawa *et al.*, 1989). The *Shigella* effector protein IpgD is homologous to the *Salmonella* effector SopB, has phosphatidylinositol (PI)-phosphatase activity, and enhances S*higella* invasion (Figure 8.4b) (Allaoui *et al.*, 1993; Niebuhr *et al.*, 2002). IpgD acts as a inositol-4-phosphatase that specifically dephosphorylates phosphatidylinositol-4,5-bisphosphate [PtdIns(4,5)P_2] into phosphatidylinositol-5-monophosphate [PtdIns(5)P] (Niebuhr *et al.*, 2002). In line with these observations, *S.* Typhimurium SipB and *Shigella* IpaB are capable of binding cholesterol with high affinity, and cholesterol is required for efficient translocation of other effector proteins (Hayward *et al.*, 2005). The starting signal for effector protein delivery into the host cell may also be similar between bacterial species. For *Shigella,* it has been found that host-cell plasma-membrane domains called lipid rafts and proteins located in these membrane domains (e.g. CD44) (Skoudy *et al.*, 2000) are engaged for efficient infection and may provide the signal for injection of effector proteins into host cells (Garner *et al.*, 2002; Lafont *et al.*, 2002; van der Goot *et al.*, 2004).

The invasion plasmid antigen C (IpaC) also serves as an essential effector molecule for epithelial cell invasion. IpaC is homologous to the *Salmonella* effector protein SipC, and both proteins possess similar actin-modulating activities (Hayward and Koronakis, 1999; Tran Van Nhieu *et al.*, 1999). Complementation analysis suggests that the in vivo functions might also be similar. The *ipaC* gene was able to complement a *Salmonella* SipC strain. However, *sipC* failed to complement an *ipaC* null mutant *Shigella* strain. Thus, SipC and IpaC may not have 100% identical functions, and SipC may lack an activity possessed by IpaC (Osiecki *et al.*, 2001).

GTPases of the Rho family, Cdc42, Rac, and Rho, and tyrosine kinases are essential for *Shigella* invasion (Mounier *et al.*, 1999; Nhieu and Sansonetti, 1999). The *Shigella* TTSS effector IpaC plays a key role in this process and triggers actin polymerization and the Cdc42-dependent formation of filopodia (Adam *et al.*, 1995; Hall, 1998; Tran Van Nhieu *et al.*, 1999). *Shigella* entry also involves RhoA signaling, leading to ezrin recruitment, stress-fiber formation, and promotion of a structure facilitating bacterial entry (Bourdet-Sicard *et al.*, 2000; Dumenil *et al.*, 2000; Mounier *et al.*, 1999). In contrast to *S.* Typhimurium, *Shigella* spp. seems to activate Rho GTPases in an indirect way through the activation of the Src tyrosine kinase (Dehio *et al.*, 1995). Thereby, Src regulates Rho function via tyrosyl-phosphorylation of p190RhoGAP (Dumenil *et al.*, 2000). Several other host cell proteins, including cortactin, ezrin, and α-actinin, are also phosphorylated (Dumenil *et al.*, 1998). After bacterial invasion is completed, actin polymerization seems to be downregulated via a feedback mechanism (Figure 8.4b) (Dumenil *et al.*, 1998).

Overall, these mechanisms lead to formation of an entry structure, where actin-associated (e.g. plastin, α-actinin, cortactin) and focal adhesion components (e.g. vinculin, talin) accumulate (Dumenil *et al.*, 1998; Jockusch *et al.*, 1995; Nobes and Hall, 1995; Tran Van Nhieu *et al.*, 1997). *Salmonella* also triggers marked rearrangements in various cytoskeletal components, including recruitment of α-actinin, tropomyosin, talin, plastin (interacts with SipA), and tubulin. However, little accumulation has been observed for vinculin, and suppression of cortactin levels by RNA interference has no effect on *Salmonella* invasion or intracellular actin assembly (Finlay *et al.*, 1991; Unsworth *et al.*, 2004; Zhou *et al.*, 1999b).

Another difference relates to the TTSS effector protein IpaA. IpaA associates with vinculin during bacterial invasion, and this IpaA–vinculin interaction initiates the formation of focal adhesion-like structures required for efficient invasion (Tran Van Nhieu *et al.*, 1997). As mentioned above, the *Salmonella* effector protein SipA is involved in a similar phenotype, although it appears to exert its function by binding directly to actin (McGhie *et al.*, 2001; Zhou *et al.*, 1999a). This mechanism is completely independent of vinculin. Nevertheless, both effector proteins are responsible for the spatial restriction of the cytoskeletal rearrangements that result from the activation of Rho GTPases during infection.

Despite these subtle differences, the basic mechanisms of *Shigella* and *Salmonella* entry share many similarities. The effector proteins induce pronounced membrane ruffling, and the key effector proteins involved show significant sequence similarity. It will be interesting to find out whether any of the differences in the host-cell manipulation via the SPI-1 and the Ipa TTSS may explain the subtle differences observed between the diseases caused by these two organisms.

LATER STAGES OF *SALMONELLA*–HOST CELL INTERACTION

One to three hours post-invasion, the actin cytoskeletal rearrangements induced by *S.* Typhimurium entry are reversed. The cells regain their normal architecture despite the large number of intracellular bacteria. The *Salmonella* SPI-1 TTSS protein SptP is required to mediate this host cell recovery by antagonizing Cdc42 and Rac1 activity (Fu and Galan, 1999). SptP exerts its function by acting as a GTPase-activating protein for Rac1 and Cdc42 (Figure 8.4a) (Fu and Galan, 1999). This may prevent damage to the host cell once the bacteria are internalized. Two discrete domains are found in SptP: the N-terminal region is required for GAP activity and the C-terminal region encodes a tyrosine phosphatase. The tyrosine phosphatase activity of SptP is involved in reversing the MAP kinase activation. MAP kinase signaling is thought to mediate nuclear responses and cytokine production (Hobbie *et al.*, 1997; Murli *et al.*, 2001). Although GAP activity has been analysed in detail, the in vivo targets of the tyrosine phosphatase remain unknown. When SopE and SptP are microinjected simultaneously, no actin remodeling has been observed, showing that SopE and SptP can antagonize each other's effect on the actin cytoskeleton (Fu and Galan, 1999). Since SopE and SptP are translocated via the SPI-1 TTSS apparatus at the same time (or possibly shortly after one another), how can they achieve sequential formation and reversion of membrane ruffles? This question has been answered by Kubori and Galan (2003): SopE and SptP are temporally regulated by proteasome-dependent protein degradation. Both effector proteins differ in the length of their half-lives: SopE is degraded with a half-life of 30 min while SptP is more stable, with a half-life of 3 h. This temporal regulation enables the bacteria to induce transient actin rearrangements (Kubori and Galan, 2003). In summary, the SPI-1 TTSS allows *S.* Typhimurium to enter host cells and also sets the stage for later successful replication inside the host cell.

SUCCESSFUL ENTRY: WHAT NEXT?

As a consequence of host-cell invasion, pathogenic bacteria such as *Salmonella* Typhimurium, *Mycobacterium tuberculosis, Legionella pneumophila,* and *Brucella abortus* are found within discrete vacuoles. *Mycobacterium* and *Salmonella* block and redirect the maturation of their vacuole at different stages of the endocytic pathway. *Legionella* and *Brucella* are found in a multimembranous vacuole called an autophagosome and replicate within an endoplasmic reticulum (ER)-like compartment (Comerci *et al.*, 2001; Delrue *et al.*, 2001; Horwitz, 1983; Joshi *et al.*, 2001). In contrast, *Shigella* spp., *L. monocytogenes, Rickettsia* spp., *Mycobacterium marinum* (Stamm *et al.*, 2003), and *Burkholderia pseudomallei* (Kespichayawattana *et al.*, 2000) all lyse the vacuolar membrane after entering host cells, thus gaining access to the cell cytosol. Once inside the cytosol, the bacteria are often propelled by Arp2/3-mediated actin polymerization (Cossart, 2000). This also facilitates direct spreading from cell to cell, allowing the bacteria to avoid host immune responses (Cossart, 2000; Cossart and Sansonetti, 2004; Goldberg, 2001; Gouin *et al.*, 2005).

S. Typhimurium remains within a vacuolar compartment. Right after bacterial entry, the SPI-1 TTSS seems to modulate vesicle maturation and sets the stage for formation of the proper *Salmonella*-containing vacuole (SCV) (Steele-Mortimer *et al.*, 2002). Specifically, the SPI-1 TTSS effector protein SopB seems to be involved in this process by decorating the SCV membrane with phosphatidyinositol-3-phosphate (PtdIns(3)P), which contributes to the enlargement of the SCV by stimulating aggregation with other enveloped bacteria (Hernandez *et al.*, 2004). Once inside the host cell, expression of the SPI-1 TTSS is switched off and the SPI-2 TTSS is expressed. The SPI-2 TTSS effector proteins are required for modifying the membrane-bound compartment (SCV) in order to establish a replicative niche. Specifically, *S.* Typhimurium avoids the late endosome/pre-lysosome. Nevertheless, several lysosomal membrane proteins (Rab7, LAMP1) but no lysosomal hydrolases are acquired (Brumell and Grinstein, 2004; Holden, 2002; Steele-Mortimer *et al.*, 1999). This is essential for replication inside epithelial cells and also for survival and replication in phagocytic cells during later stages of the systemic infection. Furthermore, the manipulation of vesicular trafficking leads to formation of spectacular vacuolar structures called *Salmonella*-induced filaments (SIF), which are enriched in lysosomal glycoproteins (Garcia-del Portillo *et al.*, 1993a,b). In conclusion, the fate of the bacteria after host-cell invasion is determined by two different TTSS.

SALMONELLA PLASMID VIRULENCE

A number of non-typhoid *Salmonella* strains, including *S.* Typhimurium, carry the highly conserved *spv* (*salmonella* plasmid virulence) gene cluster encoded on either a plasmid or a chromosome (Boyd and Hartl, 1998; Guiney *et al.*, 1995; Gulig and Doyle, 1993). A protein encoded by *spv*, the intracellular toxin SpvB, is essential for *Salmonella* virulence in mice and is associated with severe systemic infections (Fierer *et al.*, 1992). This protein functions as a mono (ADP-ribosyl) transferase and transfers an ADP-ribose moiety from nicotinamide adenine dinucleotide (NAD) to G-actin monomers, thereby altering actin physiology during infection, and causes disruption of the host cell cytoskeleton (Browne *et al.*, 2002; Lesnick *et al.*, 2001; Tezcan-Merdol *et al.*, 2001). SpvB induces actin degradation, such that infected cells eventually become completely depleted of F-actin filaments (Browne *et al.*, 2002; Lesnick *et al.*, 2001; Tezcan-Merdol *et al.*, 2001).

Transfection of a SpvB-expressing vector into mammalian host cells led to the complete loss of the cytoskeleton (Lesnick *et al.*, 2001). The amino-terminal domain of SpvB is homologous to the insecticidal toxin TcaC from the bacterium *Photorhabdus luminescens* (Bowen *et al.*, 1998). The ADP-ribosylating activity is located in the carboxy-terminal domain (Lesnick *et al.*, 2001; Otto *et al.*, 2000). Interestingly, *spvB* mutant strains are not compromised in invasion of epithelial cells. However, it is still a matter of discussion how SpvB affects survival in macrophages. Nevertheless, expression is upregulated once the bacteria are intracellular, and it has been shown that SpvB is important for proliferation in macrophages during the systemic extraintestinal phase of the disease (Guiney and Lesnick, 2005; Gulig *et al.*, 1998). Ultimately, the cytotoxicity leads to apoptosis (C. Y. Chen *et al.*, 1996; Kurita *et al.*, 2003; Libby *et al.*, 2000), which could promote cell-to-cell spread of *Salmonella* by phagocytosis of infected apoptotic cells by macrophages. This model would also explain why aminoglycosides such as gentamicin are not feasible for therapy: the bacteria are never exposed to the extracellular surrounding during their infection cycle (Fierer *et al.*, 1990; Guiney and Lesnick, 2005).

THE *YERSINIA* PARADIGM

Yersinia enterocolitica and *Yersinia pseudotuberculosis* are Gram-negative enteropathogens that also breach the gut epithelial barrier to reach their niche. For this purpose, *Y. enterocolitica* and *Y. pseudotuberculosis* take advantage of M-cells (Autenrieth and Firsching, 1996; Clark *et al.*, 1998; Marra and Isberg, 1997) that are found above organized mucosal lymphoid follicles and deliver microorganisms by transepithelial transport from the lumen to organized lymphoid tissues within the small and large intestines; for review, see Neutra *et al.* (1996, 1999). The adhesion protein invasin plays a critical role in *Yersinia* invasion of M-cells via the zipper mechanism. Invasin binds β-1 integrins with high affinity, induces intracellular signaling activating Rac1 (Alrutz *et al.*, 2001), and finally mediates internalization.

After the bacteria reach their replicative niche below the M-cells, no further localization of the bacteria within host cells is observed (Heesemann *et al.*, 1993). Invasin expression is switched off. *Yersinia* replicate extracellulary and block phagocytosis using four principal TTSS effector proteins, YopH, YopE, YopT, and YopO/YpkA, which inhibit actin cytoskeleton dynamics. These proteins along with other virulence factors are encoded on the Yop virulon of the *Yersinia* virulence plasmid and are translocated via a TTSS. YopH is homologous to the C-terminal portion of SptP, exploiting the tyrosine phosphatase activity (Bliska *et al.*, 1991; Galan, 2001; Kaniga *et al.*, 1996). In vitro,

YopH preferentially dephosphorylates three proteins from the focal adhesion, p130^{Cas}, Fyb, and SKAP-HOM (Black and Bliska, 1997; Hamid *et al.*, 1999). This leads to the reorganization of the cytoskeleton (Black and Bliska, 1997; Persson *et al.*, 1997). YopE is homologous to the N-terminal domain of SptP and acts as a GAP, switching RhoA, Rac1, and Cdc42 to the inactive state (Black and Bliska, 2000; Rosqvist *et al.*, 1990; Von Pawel-Rammingen *et al.*, 2000). YopE might, therefore, exert the same negative action as SptP on membrane ruffling. YopT is a cysteine protease that modifies and inactivates Rho, Rac, and Cdc42 (Iriarte and Cornelis, 1998) through cleavage near their carboxyl termini, releasing them from the membrane. This leads to the disruption of the host cellular actin cytoskeleton (Shao *et al.*, 2002). Finally, YpkA (for *Yersinia* protein kinase A), the fourth effector important for phagocytosis inhibition is an autophosphorylating serine-threonine kinase (Galyov *et al.*, 1993) with sequence and structural similarity to RhoA-binding kinases (Dukuzumuremyi *et al.*, 2000). Furthermore, YpkA is activated by actin binding (Juris *et al.*, 2000). However, the protein target and the exact mode of action of YpkA remain unknown (Wong and Isberg, 2005). In conclusion, *Yersinia* spp. first employ a surface adhesin to gain access into the host cell followed by the deployment of an impressive number of TTSS effector proteins, which inhibit uptake by phagocytic host cells and inflammatory signaling once they have entered their niche inside the host.

CONCLUSIONS

In the past few years, there has been remarkable progress in the understanding of the interaction of pathogenic bacteria and their hosts. Pathogens have evolved a variety of strategies to manipulate host cell functions for the benefit of the bacterium. However, there is much more to be learned. Most insights originated from studies on simplified model systems. For instance, *Salmonella* virulence factors have often been studied using non-polarized cell lines. Therefore, many questions remain pertaining to their function in polarized cells and in vivo during the course of the real infection. These questions are pressing because a variety of mechanisms for breaching the epithelium can be envisaged in the settings of a real intestinal infection (Figure 8.2). Clearly, the years are going to provide new exciting insights into microbe–host interaction.

ACKNOWLEDGMENTS

We apologize to the many scientists whose work has not been cited or discussed properly because of space limitations. B. W. is supported by a fellowship from the Boehringer Ingelheim Foundation.

REFERENCES

Adam, T., Arpin, M., Prevost, M. C., Gounon, P., and Sansonetti, P. J. (1995). Cytoskeletal rearrangements and the functional role of T-plastin during entry of *Shigella flexneri* into HeLa cells. *J. Cell Biol.* **129**, 367–381.

Aepfelbacher, M., Trasak, C., Wilharm, G., *et al.* (2003). Characterization of YopT effects on Rho GTPases in *Yersinia enterocolitica*-infected cells. *J Biol Chem.* **278**, 33 217–33 223.

Allaoui, A., Menard, R., Sansonetti, P. J., and Parsot, C. (1993). Characterization of the *Shigella flexneri* ipgD and ipgF genes, which are located in the proximal part of the mxi locus. *Infect. Immun.* **61**, 1707–1714.

Alonso, A. and Garcia-del Portillo, F. (2004). Hijacking of eukaryotic functions by intracellular bacterial pathogens. *Int. Microbiol.* **7**, 181–191.

Alrutz, M. A., Srivastava, A., Wong, K. W., *et al.* (2001). Efficient uptake of *Yersinia pseudotuberculosis* via integrin receptors involves a Rac1-Arp 2/3 pathway that bypasses N-WASP function. *Mol. Microbiol.* **42**, 689–703.

Autenrieth, I. B. and Firsching, R. (1996). Penetration of M cells and destruction of Peyer's patches by *Yersinia enterocolitica*: an ultrastructural and histological study. *J. Med. Microbiol.* **44**, 285–294.

Bakshi, C. S., Singh, V. P., Wood, M. W., *et al.* (2000). Identification of SopE2, a *Salmonella* secreted protein which is highly homologous to SopE and involved in bacterial invasion of epithelial cells. *J. Bacteriol.* **182**, 2341–2344.

Black, D. S. and Bliska, J. B. (1997). Identification of p130C as a substrate of *Yersinia* YopH (Yop 51), a bacterial protein tyrosine phosphatasc that translocates into mammalian cells targets focal adhesions. *EMBO J.* **16**, 2730–2744.

Bliska, J. B., Guan, K. L., Dixon, J. E., and Falkow, S. (1991). Tyrosine phosphate hydrolysis of host proteins by an essential *Yersinia* virulence determinant. *Proc. Natl. Acad. Sci. U. S. A.* **88**, 1187–1191.

Bourdet-Sicard, R., Rudiger, M., Jockusch, B. M., *et al.* (1999). Binding of the *Shigella* protein IpaA to vinculin induces F-actin depolymerization. *EMBO J.* **18**, 5853–5862.

Bourdet-Sicard, R., Egile, C., Sansonetti, P. J., and Tran Van Nhieu, G. (2000). Diversion of cytoskeletal processes by *Shigella* during invasion of epithelial cells. *Microbes Infect.* **2**, 813–819.

Bowen, D., Rocheleau, T. A., Blackburn, M., *et al.* (1998). Insecticidal toxins from the bacterium *Photorhabdus luminescens*. *Science* **280**, 2129–2132.

Boyd, E. F. and Hartl, D. L. (1998). *Salmonella* virulence plasmid: modular acquisition of the spv virulence region by an F-plasmid in *Salmonella enterica* subspecies I and insertion into the chromosome of subspecies II, IIIa, IV and VII isolates. *Genetics* **149**, 1183–1190.

Browne, S. H., Lesnick, M. L., and Guiney, D. G. (2002). Genetic requirements for salmonella-induced cytopathology in human monocyte-derived macrophages. *Infect. Immun.* **70**, 7126–7135.

Brumell, J. H. and Grinstein, S. (2004). *Salmonella* redirects phagosomal maturation. *Curr. Opin Microbiol.* **7**, 78–84.

Brussow, H., Canchaya, C., and Hardt, W. D. (2004). Phages and the evolution of bacterial pathogens: from genomic rearrangements to lysogenic conversion. *Microbiol. Mol. Biol. Rev.* **68**, 560–602.

Buchwald, G., Friebel, A., Galan, J. E., *et al.* (2002). Structural basis for the reversible activation of a Rho protein by the bacterial toxin SopE. *EMBO J.* **21**, 3286–3295.

Chang, J., Chen, J., and Zhou, D. (2005). Delineation and characterization of the actin nucleation and effector translocation activities of *Salmonella* SipC. *Mol. Microbiol.* **55**, 1379–1389.

Chen, C. Y., Eckmann, L., Libby, S. J., *et al.* (1996). Expression of *Salmonella* typhimurium rpoS and rpoS-dependent genes in the intracellular environment of eukaryotic cells. *Infect. Immun.* **64**, 4739–4743.

Chen, H., Bernstein, B. W., and Bamburg, J. R. (2000). Regulating actin-filament dynamics in vivo. *Trends Biochem. Sci.* **25**, 19–23.

Chen, L. M., Hobbie, S., and Galan, J. E. (1996). Requirement of CDC42 for *Salmonella*-induced cytoskeletal and nuclear responses. *Science* **274**, 2115–2118.

Clark, M. A., Hirst, B. H., and Jepson, M. A. (1998). M-cell surface beta1 integrin expression and invasin-mediated targeting of *Yersinia pseudotuberculosis* to mouse Peyer's patch M cells. *Infect. Immun.* **66**, 1237–1243.

Clerc, P. L., Ryter, A., Mounier, J., and Sansonetti, P. J. (1987). Plasmid-mediated early killing of eucaryotic cells by *Shigella flexneri* as studied by infection of J774 macrophages. *Infect. Immun.* **55**, 521–527.

Collazo, C. M., and Galan, J. E. (1997). The invasion-associated type III system of *Salmonella typhimurium* directs the translocation of Sip proteins into the host cell. *Mol. Microbiol.* **24**, 747–756.

Comerci, D. J., Martinez-Lorenzo, M. J., Sieira, R., Gorvel, J. P., and Ugalde, R. A. (2001). Essential role of the VirB machinery in the maturation of the *Brucella abortus*-containing vacuole. *Cell Microbiol.* **3**, 159–168.

Cossart, P. (2000). Actin-based motility of pathogens: the Arp2/3 complex is a central player. *Cell Microbiol.* **2**, 195–205.

Cossart, P. and Sansonetti, P. J. (2004). Bacterial invasion: the paradigms of enteroinvasive pathogens. *Science* **304**, 242–248.

Criss, A. K. and Casanova, J. E. (2003). Coordinate regulation of *Salmonella* enterica serovar Typhimurium invasion of epithelial cells by the Arp2/3 complex and Rho GTPases. *Infect. Immun.* **71**, 2885–2891.

Dehio, C., Prevost, M. C., and Sansonetti, P. J. (1995). Invasion of epithelial cells by *Shigella flexneri* induces tyrosine phosphorylation of cortactin by a pp60c-src-mediated signalling pathway. *EMBO J.* **14**, 2471–2482.

Delrue, R. M., Martinez-Lorenzo, M., Lestrate, P., *et al.* (2001). Identification of *Brucella* spp. genes involved in intracellular trafficking. *Cell Microbiol.* **3**, 487–497.

Dukuzumuremyi, J. M., Rosqvist, R., Hallberg, B., *et al.* (2000). The *Yersinia* protein kinase A is a host factor inducible RhoA/Rac-binding virulence factor. *J. Biol. Chem.* **275**, 35 281–35 290.

Dumenil, G., Olivo, J. C., Pellegrini, S., *et al.* (1998). Interferon alpha inhibits a Src-mediated pathway necessary for *Shigella*-induced cytoskeletal rearrangements in epithelial cells. *J. Cell Biol.* **143**, 1003–1012.

Dumenil, G., Sansonetti, P., and Tran Van Nhieu, G. (2000). Src tyrosine kinase activity down-regulates Rho-dependent responses during *Shigella* entry into epithelial cells and stress fibre formation. *J. Cell Sci.* **113 (Pt 1)**, 71–80.

Etienne-Manneville, S. and Hall, A. (2002). Rho GTPases in cell biology. *Nature* **420**, 629–635.

Fierer, J., Hatlen, L., Lin, J. P., *et al.* (1990). Successful treatment using gentamicin liposomes of *Salmonella dublin* infections in mice. *Antimicrob. Agents Chemother.* **34**, 343–348.

Fierer, J., Krause, M., Tauxe, R., and Guiney, D. (1992). *Salmonella typhimurium* bacteremia: association with the virulence plasmid. *J. Infect. Dis.* **166**, 639–642.

Finlay, B. B., Ruschkowski, S., and Dedhar, S. (1991). Cytoskeletal rearrangements accompanying salmonella entry into epithelial cells. *J. Cell Sci.* **99 (Pt 2)**, 283–296.

Francis, C. L., Ryan, T. A., Jones, B. D., Smith, S. J., and Falkow, S. (1993). Ruffles induced by *Salmonella* and other stimuli direct macropinocytosis of bacteria. *Nature* **364**, 639–642.

Friebel, A., Ilchmann, H., Aelpfelbacher, M., *et al.* (2001). SopE and SopE2 from *Salmonella* typhimurium activate different sets of RhoGTPases of the host cell. *J. Biol. Chem.* **276**, 34 035–34 040.

Fu, Y. and Galan, J. E. (1999). A *Salmonella* protein antagonizes Rac-1 and Cdc42 to mediate host-cell recovery after bacterial invasion. *Nature* **401**, 293–297.

Galan, J. E. (2001). *Salmonella* interactions with host cells: type III secretion at work. *Annu. Rev. Cell Dev. Biol.* **17**, 53–86.

Galyov, E. E., Hakansson, S., Forsberg, A., and Wolf-Watz, H. (1993). A secreted protein kinase of *Yersinia pseudotuberculosis* is an indispensable virulence determinant. *Nature* **361**, 730–732.

Garcia-del Portillo, F. and Finlay, B. B. (1994). Salmonella invasion of nonphagocytic cells induces formation of macropinosomes in the host cell. *Infect. Immun.* **62**, 4641–4645.

Garcia-del Portillo, F., Zwick, M. B., Leung, K. Y., and Finlay, B. B. (1993a). *Salmonella* induces the formation of filamentous structures containing lysosomal membrane glycoproteins in epithelial cells. *Proc. Natl. Acad. Sci. U.S.A.* **90**, 10 544–10 548.

Garcia-del Portillo, F., Zwick, M. B., Leung, K. Y., and Finlay, B. B. (1993b). Intracellular replication of *Salmonella* within epithelial cells is associated with filamentous structures containing lysosomal membrane glycoproteins. *Infect. Agents Dis.* **2**, 227–231.

Garner, M. J., Hayward, R. D., and Koronakis, V. (2002). The *Salmonella* pathogenicity island 1 secretion system directs cellular cholesterol redistribution during mammalian cell entry and intracellular trafficking. *Cell Microbiol.* **4**, 153–165.

Ginocchio, C., Pace, J., and Galan, J. E. (1992). Identification and molecular characterization of a *Salmonella typhimurium* gene involved in triggering the internalization of salmonellae into cultured epithelial cells. *Proc. Natl. Acad. Sci. U. S. A.* **89**, 5976–5980.

Goldberg, M. B. (2001). Actin-based motility of intracellular microbial pathogens. *Microbiol. Mol. Biol. Rev.* **65**, 595–626.

Gouin, E., Welch, M. D., and Cossart, P. (2005). Actin-based motility of intracellular pathogens. *Curr. Opin. Microbiol.* **8**, 35–45.

Guan, K. L. and Dixon, J. E. (1990). Protein tyrosine phosphatase activity of an essential virulence determinant in *Yersinia. Science* **249**, 553–556.

Guiney, D. G. and Lesnick, M. (2005). Targeting of the actin cytoskeleton during infection by *Salmonella* strains. *Clin. Immunol.* **114**, 248–255.

Guiney, D. G., Libby, S., Fang, F. C., Krause, M., and Fierer, J. (1995). Growth-phase regulation of plasmid virulence genes in *Salmonella. Trends Microbiol.* **3**, 275–279.

Gulig, P. A. and Doyle, T. J. (1993). The *Salmonella typhimurium* virulence plasmid increases the growth rate of salmonellae in mice. *Infect. Immun.* **61**, 504–511.

Gulig, P. A., Doyle, T. J., Hughes, J. A., and Matsui, H. (1998). Analysis of host cells associated with the Spv-mediated increased intracellular growth rate of *Salmonella typhimurium* in mice. *Infect. Immun.* **66**, 2471–2485.

Hall, A. (1998). Rho GTPases and the actin cytoskeleton. *Science* **279**, 509–514.

Hamid, N., Gustavsson, A., Andersson, K., *et al.* (1999). YopH dephosphorylates Cas and Fyn-binding protein in macrophages. *Microb. Pathog.* **27**, 231–242.

Hapfelmeier, S., Ehrbar, K., Stecher, B., Barthel, M., Kremer, M., and Hardt, W. D. (2004). Role of the *Salmonella* pathogenicity island 1 effector proteins SipA, SopB, SopE, and SopE2 in *Salmonella enterica subspecies* 1 Serovar Typhimurium colitis in streptomycin-pretreated mice. *Infect. Immun.* **72**, 795–809.

Hardt, W. D., Chen, L. M., Schuebel, K. E., Bustelo, X. R., and Galan, J. E. (1998). *S. typhimurium* encodes an activator of Rho GTPases that induces membrane ruffling and nuclear responses in host cells. *Cell* **93**, 815–826.

Hayward, R. D. and Koronakis, V. (1999). Direct nucleation and bundling of actin by the SipC protein of invasive *Salmonella. EMBO J.* **18**, 4926–4934.

Hayward, R. D. and Koronakis, V. (2002). Direct modulation of the host cell cytoskeleton by *Salmonella* actin-binding proteins. *Trends Cell Biol.* **12**, 15–20.

Hayward, R. D., Cain, R. J., McGhie, E. J., *et al.* (2005). Cholesterol binding by the bacterial type III translocon is essential for virulence effector delivery into mammalian cells. *Mol. Microbiol.* **56**, 590–603.

Heesemann, J., Gaede, K., and Autenrieth, I. B. (1993). Experimental *Yersinia enterocolitica* infection in rodents: a model for human yersiniosis. *APMIS* **101**, 417–429.

Hernandez, L. D., Hueffer, K., Wenk, M. R., and Galan, J. E. (2004). Salmonella modulates vesicular traffic by altering phosphoinositide metabolism. *Science* **304**, 1805–1807.

Higgs, H. N. and Pollard, T. D. (2001). Regulation of actin filament network formation through ARP2/3 complex: activation by a diverse array of proteins. *Annu. Rev. Biochem.* **70**, 649–676.

Hilbi, H., Moss, J. E., Hersh, D., *et al.* (1998). Shigella-induced apoptosis is dependent on caspase-1 which binds to IpaB. *J. Biol. Chem.* **273**, 32 895–32 900.

Hobbie, S., Chen, L. M., Davis, R. J., and Galan, J. E. (1997). Involvement of mitogen-activated protein kinase pathways in the nuclear responses and cytokine production induced by *Salmonella typhimurium* in cultured intestinal epithelial cells. *J. Immunol.* **159**, 5550–5559.

Holden, D. W. (2002). Trafficking of the *Salmonella* vacuole in macrophages. *Traffic* **3**, 161–169.

Horwitz, M. A. (1983). The Legionnaires' disease bacterium (*Legionella pneumophila*) inhibits phagosome-lysosome fusion in human monocytes. *J. Exp. Med.* **158**, 2108–2126.

Hueck, C. J. (1998). Type III protein secretion systems in bacterial pathogens of animals and plants. *Microbiol. Mol. Biol. Rev.* **62**, 379–433.

Hurley, B. P. and McCormick, B. A. (2003). Translating tissue culture results into animal models: the case of *Salmonella* typhimurium. *Trends Microbiol.* **11**, 562–569.

Iriarte, M. and Cornelis, G. R. (1998). YopT, a new *Yersinia* Yop effector protein, affects the cytoskeleton of host cells. *Mol. Microbiol.* **29**, 915–929.

Jepson, M. A., Collares-Buzato, C. B., Clark, M. A., Hirst, B. H., and Simmons, N. L. (1995). Rapid disruption of epithelial barrier function by *Salmonella* typhimurium is associated with structural modification of intercellular junctions. *Infect. Immun.* **63**, 356–359.

Jepson, M. A., Kenny, B., and Leard, A. D. (2001). Role of sipA in the early stages of *Salmonella* typhimurium entry into epithelial cells. *Cell Microbiol.* **3**, 417–426.

Jockusch, B. M., Bubeck, P., Giehl, K., *et al.* (1995). The molecular architecture of focal adhesions. *Annu. Rev. Cell Dev. Biol.* **11**, 379–416.

Joshi, A. D., Sturgill-Koszycki, S., and Swanson, M. S. (2001). Evidence that Dot-dependent and -independent factors isolate the *Legionella pneumophila* phagosome from the endocytic network in mouse macrophages. *Cell Microbiol.* **3**, 99–114.

Juris, S. J., Rudolph, A. E., Huddler, D., Orth, K., and Dixon, J. E. (2000). A distinctive role for the *Yersinia* protein kinase: actin binding, kinase activation, and cytoskeleton disruption. *Proc. Natl. Acad. Sci. U. S. A.* **97**, 9431–9436.

Kaniga, K., Tucker, S., Trollinger, D., and Galan, J. E. (1995). Homologs of the Shigella IpaB and IpaC invasins are required for *Salmonella* typhimurium entry into cultured epithelial cells. *J. Bacteriol.* **177**, 3965–3971.

Kaniga, K., Uralil, J., Bliska, J. B., and Galan, J. E. (1996). A secreted protein tyrosine phosphatase with modular effector domains in the bacterial pathogen *Salmonella typhimurium*. *Mol. Microbiol.* **21**, 633–641.

Kespichayawattana, W., Rattanachetkul, S., Wanun, T., Utaisincharoen, P., and Sirisinha, S. (2000). *Burkholderia pseudomallei* induces cell fusion and actin-associated membrane protrusion: a possible mechanism for cell-to-cell spreading. *Infect. Immun.* **68**, 5377–5384.

Knodler, L. A., Finlay, B. B., and Steele-Mortimer, O. (2005). The *Salmonella* effector protein SopB protects epithelial cells from apoptosis by sustained activation of Akt. *J. Biol. Chem.* **280**, 9058–9064.

Kubori, T. and Galan, J. E. (2003). Temporal regulation of salmonella virulence effector function by proteasome-dependent protein degradation. *Cell* **115**, 333–342.

Kueltzo, L. A., Osiecki, J., Barker, J., *et al.* (2003). Structure–function analysis of invasion plasmid antigen C (IpaC) from *Shigella flexneri*. *J. Biol. Chem.* **278**, 2792–2798.

Kurita, A., Gotoh, H., Eguchi, M., *et al.* (2003). Intracellular expression of the *Salmonella* plasmid virulence protein, SpvB, causes apoptotic cell death in eukaryotic cells. *Microb. Pathog.* **35**, 43–48.

Lafont, F., Tran Van Nhieu, G., Hanada, K., Sansonetti, P., and van der Goot, F. G. (2002). Initial steps of *Shigella* infection depend on the cholesterol/sphingolipid raft-mediated CD44-IpaB interaction. *EMBO J.* **21**, 4449–4457.

Lawrence, J. G. (2005). Horizontal and vertical gene transfer: the life history of pathogens. *Contrib. Microbiol.* **12**, 255–271.

Lesnick, M. L., Reiner, N. E., Fierer, J., and Guiney, D. G. (2001). The *Salmonella* spvB virulence gene encodes an enzyme that ADP-ribosylates actin and destabilizes the cytoskeleton of eukaryotic cells. *Mol. Microbiol.* **39**, 1464–1470.

Libby, S. J., Lesnick, M., Hasegawa, P., Weidenhammer, E., and Guiney, D. G. (2000). The *Salmonella* virulence plasmid spv genes are required for cytopathology in human monocyte-derived macrophages. *Cell Microbiol.* **2**, 49–58.

Lilic, M., Galkin, V. E., Orlova, A., *et al.* (2003). Salmonella SipA polymerizes actin by stapling filaments with nonglobular protein arms. *Science* **301**, 1918–1921.

Machesky, L. M., Mullins, R. D., Higgs, H. N., *et al.* (1999). Scar, a WASp-related protein, activates nucleation of actin filaments by the Arp2/3 complex. *Proc. Natl. Acad. Sci. U. S. A.* **96**, 3739–3744.

Marcus, S. L., Wenk, M. R., Steele-Mortimer, O., and Finlay, B. B. (2001). A synaptojanin-homologous region of *Salmonella* typhimurium SigD is essential for inositol phosphatase activity and Akt activation. *FEBS Lett.* **494**, 201–207.

Marra, A. and Isberg, R. R. (1997). Invasin-dependent and invasin-independent pathways for translocation of *Yersinia pseudotuberculosis* across the Peyer's patch intestinal epithelium. *Infect. Immun.* **65**, 3412–3421.

McCormick, B. A., Colgan, S. P., Delp-Archer, C., Miller, S. I., and Madara, J. L. (1993). *Salmonella* typhimurium attachment to human intestinal epithelial monolayers: transcellular signalling to subepithelial neutrophils. *J. Cell Biol.* **123**, 895–907.

McGhie, E. J., Hayward, R. D., and Koronakis, V. (2001). Cooperation between actin-binding proteins of invasive *Salmonella*: SipA potentiates SipC nucleation and bundling of actin. *EMBO J.* **20**, 2131–2139.

McGhie, E. J., Hayward, R. D., and Koronakis, V. (2004). Control of actin turnover by a salmonella invasion protein. *Mol. Cell.* **13**, 497–510.

Menard, R., Sansonetti, P. J., and Parsot, C. (1993). Nonpolar mutagenesis of the ipa genes defines IpaB, IpaC, and IpaD as effectors of *Shigella flexneri* entry into epithelial cells. *J. Bacteriol.* **175**, 5899–5906.

Mengaud, J., Ohayon, H., Gounon, P., Mege, R. M., and Cossart, P. (1996). E-cadherin is the receptor for internalin, a surface protein required for entry of *L. monocytogenes* into epithelial cells. *Cell* **84**, 923–932.

Mirold, S., Ehrbar, K., Weissmüller, A., *et al.* (2001). *Salmonella* host cell invasion emerged by acquisition of a mosaic of separate genetic elements, including *Salmonella* pathogenicity island 1 (SPI1), SPI5, and *sopE2*. *J. Bacteriol.* **183**, 2348–2358.

Molendijk, A. J., Ruperti, B., and Palme, K. (2004). Small GTPases in vesicle trafficking. *Curr. Opin. Plant Biol.* **7**, 694–700.

Mounier, J., Vasselon, T., Hellio, R., Lesourd, M., and Sansonetti, P. J. (1992). *Shigella flexneri* enters human colonic Caco-2 epithelial cells through the basolateral pole. *Infect. Immun.* **60**, 237–248.

Mounier, J., Laurent, V., Hall, A., *et al.* (1999). Rho family GTPases control entry of *Shigella flexneri* into epithelial cells but not intracellular motility. *J. Cell Sci.* **112 (Pt 13)**, 2069–2080.

Murli, S., Watson, R. O., and Galan, J. E. (2001). Role of tyrosine kinases and the tyrosine phosphatase SptP in the interaction of *Salmonella* with host cells. *Cell Microbiol.* **3**, 795–810.

Nagai, H., Kagan, J. C., Zhu, X., Kahn, R. A., and Roy, C. R. (2002). A bacterial guanine nucleotide exchange factor activates ARF on *Legionella* phagosomes. *Science* **295**, 679–682.

Navarro, L., Alto, N. M., and Dixon, J. E. (2005). Functions of the *Yersinia* effector proteins in inhibiting host immune responses. *Curr. Opin. Microbiol.* **8**, 21–27.

Neutra, M. R., Frey, A., and Kraehenbuhl, J. P. (1996). Epithelial M cells: gateways for mucosal infection and immunization. *Cell* **86**, 345–348.

Neutra, M. R., Mantis, N. J., Frey, A., and Giannasca, P. J. (1999). The composition and function of M cell apical membranes: implications for microbial pathogenesis. *Semin. Immunol.* **11**, 171–181.

Nhieu, G. T., and Sansonetti, P. J. (1999). Mechanism of *Shigella* entry into epithelial cells. *Curr. Opin. Microbiol.* **2**, 51–55.

Niebuhr, K., Giuriato, S., Pedron, T., *et al.* (2002). Conversion of PtdIns(4,5)P(2) into PtdIns(5)P by the S. flexneri effector IpgD reorganizes host cell morphology. *EMBO J.* **21**, 5069–5078.

Nobes, C. D., and Hall, A. (1995). Rho, rac, and cdc42 GTPases regulate the assembly of multimolecular focal complexes associated with actin stress fibers, lamellipodia, and filopodia. *Cell* **81**, 53–62.

Norris, F. A., Wilson, M. P., Wallis, T. S., Galyov, E. E., and Majerus, P. W. (1998). SopB, a protein required for virulence of *Salmonella dublin*, is an inositol phosphate phosphatase. *Proc. Natl. Acad. Sci. U. S. A.* **95**, 14 057–14 059.

Osiecki, J. C., Barker, J., Picking, W. L., *et al.* (2001). IpaC from *Shigella* and SipC from *Salmonella* possess similar biochemical properties but are functionally distinct. *Mol. Microbiol.* **42**, 469–481.

Otto, H., Tezcan-Merdol, D., Girisch, R., *et al.* (2000). The spvB gene-product of the *Salmonella enterica* virulence plasmid is a mono(ADP-ribosyl)transferase. *Mol. Microbiol.* **37**, 1106–1115.

Persson, C., Carballeira, N., Wolf-Watz, H., and Fallman, M. (1997). The PTPase YopH inhibits uptake of *Yersinia*, tyrosine phosphorylation of p130Cas and FAK, and the associated accumulation of these proteins in peripheral focal adhesions. *EMBO J.* **16**, 2307–2318.

Picking, W. L., Coye, L., Osiecki, J. C., *et al.* (2001). Identification of functional regions within invasion plasmid antigen C (IpaC) of *Shigella flexneri*. *Mol. Microbiol.* **39**, 100–111.

Prehoda, K. E., Scott, J. A., Mullins, R. D., and Lim, W. A. (2000). Integration of multiple signals through cooperative regulation of the N-WASP-Arp2/3 complex. *Science* **290**, 801–806.

Rescigno, M., Urbano, M., Valzasina, B., *et al.* (2001). Dendritic cells express tight junction proteins and penetrate gut epithelial monolayers to sample bacteria. *Nat. Immunol.* **2**, 361–367.

Rosqvist, R., Forsberg, A., Rimpilainen, M., Bergman, T., and Wolf-Watz, H. (1990). The cytotoxic protein YopE of *Yersinia* obstructs the primary host defence. *Mol. Microbiol.* **4**, 657–667.

Sakaguchi, T., Kohler, H., Gu, X., McCormick, B. A., and Reinecker, H. C. (2002). *Shigella flexneri* regulates tight junction-associated proteins in human intestinal epithelial cells. *Cell Microbiol.* **4**, 367–381.

Sasakawa, C., Adler, B., Tobe, T., *et al.* (1989). Functional organization and nucleotide sequence of virulence region-2 on the large virulence plasmid in *Shigella flexneri* 2a. *Mol. Microbiol.* **3**, 1191–1201.

Scherer, C. A., Cooper, E., and Miller, S. I. (2000). The *Salmonella* type III secretion translocon protein SspC is inserted into the epithelial cell plasma membrane upon infection. *Mol. Microbiol.* **37**, 1133–1145.

Schlumberger, M. C., Friebel, A., Buchwald, G., *et al.* (2003). Amino acids of the bacterial toxin SopE involved in G-nucleotide exchange on Cdc42. *J. Biol. Chem.* **278**, 27 149–27 159.

Schwarz-Linek, U., Hook, M., and Potts, J. R. (2004). The molecular basis of fibronectin-mediated bacterial adherence to host cells. *Mol. Microbiol.* **52**, 631–641.

Sechi, A. S. and Wehland, J. (2004). ENA/VASP proteins: multifunctional regulators of actin cytoskeleton dynamics. *Front. Biosci.* **9**, 1294–1310.

Shao, F., Merritt, P. M., Bao, Z., Innes, R. W., and Dixon, J. E. (2002). A *Yersinia* effector and a *Pseudomonas* avirulence protein define a family of cysteine proteases functioning in bacterial pathogenesis. *Cell* **109**, 575–588.

Shao, F., Vacratsis, P. O., Bao, Z., *et al.* (2003). Biochemical characterization of the *Yersinia* YopT protease: cleavage site and recognition elements in Rho GTPases. *Proc. Natl. Acad. Sci. U. S. A.* **100**, 904–909.

Skoudy, A., Mounier, J., Aruffo, A., *et al.* (2000). CD44 binds to the *Shigella* IpaB protein and participates in bacterial invasion of epithelial cells. *Cell Microbiol.* **2**, 19–33.

Stamm, L. M., Morisaki, J. H., Gao, L. Y., *et al.* (2003). *Mycobacterium marinum* escapes from phagosomes and is propelled by actin-based motility. *J. Exp. Med.* **198**, 1361–1368.

Stebbins, C. E. and Galan, J. E. (2000). Modulation of host signaling by a bacterial mimic: structure of the *Salmonella* effector SptP bound to Rac1. *Mol. Cell.* **6**, 1449–1460.

Stebbins, C. E. and Galan, J. E. (2001). Structural mimicry in bacterial virulence. *Nature* **412**, 701–705.

Steele-Mortimer, O., Meresse, S., Gorvel, J. P., Toh, B. H., and Finlay, B. B. (1999). Biogenesis of *Salmonella typhimurium*-containing vacuoles in epithelial cells involves interactions with the early endocytic pathway. *Cell Microbiol.* **1**, 33–49.

Steele-Mortimer, O., Brumell, J. H., Knodler, L. A., Meresse, S., Lopez, A., and Finlay, B. B. (2002). The invasion-associated type III secretion system of *Salmonella enterica* serovar Typhimurium is necessary for intracellular proliferation and vacuole biogenesis in epithelial cells. *Cell Microbiol.* **4**, 43–54.

Stender, S., Friebel, A., Linder, S., *et al.* (2000). Identification of SopE2 from *Salmonella typhimurium*, a conserved guanine nucleotide exchange factor for Cdc42 of the host cell. *Mol. Microbiol.* **36**, 1206–1221.

Suetsugu, S., Miki, H., and Takenawa, T. (2002). Spatial and temporal regulation of actin polymerization for cytoskeleton formation through Arp2/3 complex and WASP/WAVE proteins. *Cell. Motil. Cytoskeleton* **51**, 113–122.

Takenawa, T. and Miki, H. (2001). WASP and WAVE family proteins: key molecules for rapid rearrangement of cortical actin filaments and cell movement. *J. Cell Sci.* **114**, 1801–1809.

Terebiznik, M. R., Vieira, O. V., Marcus, S. L., *et al.* (2002). Elimination of host cell PtdIns(4,5)P(2) by bacterial SigD promotes membrane fission during invasion by Salmonella. *Nat. Cell Biol.* **4**, 766–773.

Tezcan-Merdol, D., Nyman, T., Lindberg, U., *et al.* (2001). Actin is ADP-ribosylated by the *Salmonella enterica* virulence-associated protein SpvB. *Mol. Microbiol.* **39**, 606–619.

Tran Van Nhieu, G., Ben-Ze'ev, A., and Sansonetti, P. J. (1997). Modulation of bacterial entry into epithelial cells by association between vinculin and the *Shigella* IpaA invasin. *EMBO J.* **16**, 2717–2729.

Tran Van Nhieu, G., Caron, E., Hall, A., and Sansonetti, P. J. (1999). IpaC induces actin polymerization and filopodia formation during *Shigella* entry into epithelial cells. *EMBO J.* **18**, 3249–3262.

Uchiya, K., Tobe, T., Komatsu, K., *et al.* (1995). Identification of a novel virulence gene, virA, on the large plasmid of *Shigella*, involved in invasion and intercellular spreading. *Mol. Microbiol.* **17**, 241–250.

Unsworth, K. E., Way, M., McNiven, M., Machesky, L., and Holden, D. W. (2004). Analysis of the mechanisms of *Salmonella*-induced actin assembly during invasion of host cells and intracellular replication. *Cell Microbiol.* **6**, 1041–1055.

Van der Goot, F. G., Tran van Nhieu, G., Allaoui, A., Sansonetti, P., and Lafont, F. (2004). Rafts can trigger contact-mediated secretion of bacterial effectors via a lipid-based mechanism. *J. Biol. Chem.* **279**, 47 792–47 798.

Von Pawel-Rammingen, U., Telepnev, M. V., Schmidt, G., *et al.* (2000). GAP activity of the *Yersinia* YopE cytotoxin specifically targets the Rho pathway: a mechanism for disruption of actin microfilament structure. *Mol. Microbiol.* **36**, 737–748.

Wallis, T. S. and Galyov, E. E. (2000). Molecular basis of *Salmonella*-induced enteritis. *Mol. Microbiol.* **36**, 997–1005.

Welch, M. D., DePace, A. H., Verma, S., Iwamatsu, A., and Mitchison, T. J. (1997). The human Arp2/3 complex is composed of evolutionarily conserved subunits and is localized to cellular regions of dynamic actin filament assembly. *J. Cell Biol.* **138**, 375–384.

Winder, S. J. and Ayscough, K. R. (2005). Actin-binding proteins. *J. Cell Sci.* **118**, 651–654.

Wong, K. W. and Isberg, R. R. (2005). Emerging views on integrin signaling via Rac1 during invasin-promoted bacterial uptake. *Curr. Opin. Microbiol.* **8**, 4–9.

Wood, M. W., Rosqvist, R., Mullan, P. B., Edwards, M. H., and Galyov, E. E. (1996). SopE, a secreted protein of *Salmonella dublin*, is translocated into the target eukaryotic cell via a sip-dependent mechanism and promotes bacterial entry. *Mol. Microbiol.* **22**, 327–338.

Yoshida, S., Katayama, E., Kuwae, A., *et al.* (2002). Shigella deliver an effector protein to trigger host microtubule destabilization, which promotes Rac1 activity and efficient bacterial internalization. *EMBO J.* **21**, 2923–2935.

Zhang, S., Santos, R. L., Tsolis, R. M., *et al.* (2002). The Salmonella *enterica* Serotype Typhimurium effector proteins SipA, SopA, SopB, SopD, and SopE2 act in concert to induce diarrhea in calves. *Infect. Immun.* **70**, 3843–3855.

Zhou, D., Mooseker, M. S., and Galan, J. E. (1999a). Role of the *S. typhimurium* actin-binding protein SipA in bacterial internalization. *Science* **283**, 2092–2095.

Zhou, D., Mooseker, M. S., and Galan, J. E. (1999b). An invasion-associated *Salmonella* protein modulates the actin-bundling activity of plastin. *Proc. Natl. Acad. Sci. U. S. A.* **96**, 10 176–10 181.

Zhou, D., Chen, L. M., Hernandez, L., Shears, S. B., and Galan, J. E. (2001). A *Salmonella* inositol polyphosphatase acts in conjunction with other bacterial effectors to promote host cell actin cytoskeleton rearrangements and bacterial internalization. *Mol. Microbiol.* **39**, 248–260.

Zigmond, S. H. (2004). Formin-induced nucleation of actin filaments. *Curr. Opin. Cell Biol.* **16**, 99–105.

CHAPTER 9

NF-κB-dependent responses activated by bacterial–epithelial interactions

Bobby J. Cherayil

INTRODUCTION

The epithelial surfaces of the skin and the intestinal, respiratory, and reproductive tracts constitute the outer frontiers of the body and are exposed continuously to the myriad microorganisms present in the external environment. Like any good frontier guard, the cells of these epithelia must carry out two important functions – establish barriers against microbial intruders and raise the alarm if the barriers are breached. The nuclear factor kappa B (NF-κB) family of transcription factors plays a vital role in these functions by controlling the expression of a number of genes involved in antimicrobial defense and in the inflammatory response. Central to this role is the ability of NF-κB to be regulated by cellular signaling pathways that are activated by a wide variety of microorganisms. This sensitivity to microbial signals allows NF-κB function to be modulated appropriately in response to any changes in the flora that is in contact with the epithelium.

This chapter reviews what is currently known about the major NF-κB-dependent inflammatory responses elicited in mammalian epithelial cells as a result of interactions with bacteria. We start with an overview of the NF-κB family and its basic regulation. In subsequent sections, we discuss the mechanisms of modulation of NF-κB function by bacteria-derived signals, the consequent alterations in cell function, and the clinical abnormalities that can result from genetic defects in NF-κB activation.

Bacterial–Epithelial Cell Cross-Talk: Molecular Mechanisms in Pathogenesis, ed. Beth A. McCormick. Published by Cambridge University Press.

THE NF-κB FAMILY AND ITS REGULATION

The NF-κB family consists of five ubiquitously expressed, structurally related transcription factors – Rel A (p65), Rel B, c-Rel, p50 (NF-κB1, derived from constitutive cotranslational proteasome-dependent processing of the p105 precursor), and p52 (NF-κB2, derived from the inducible proteasome-mediated processing of the p100 precursor) (Hayden and Ghosh, 2004). These proteins share a conserved 300-amino-acid Rel homology domain located near their amino-termini and exist as homo- or heterodimers of each other. Under resting conditions, the NF-κB dimers are localized predominantly to the cytosol by virtue of their binding to one of five inhibitor kappa B (IκB) proteins (IκBα, IκBβ, IκBε, IκBγ, BCL-3) and therefore are not able to drive transcription of genes located in the nucleus.

When an appropriate stimulus is delivered, the resulting intracellular signals lead to the activation of a high-molecular-weight enzymatic complex known as the IκB kinase (IKK). IKK consists of two catalytic subunits, IKKα and IKKβ, and a regulatory subunit, IKKγ, also known as NF-κB essential modifier (NEMO) (Ghosh and Karin, 2002). Once activated, IKK phosphorylates IκB at two specific serine residues located near the amino-terminus. The phosphorylated IκB is recognized and polyubiquitinated by the SCF (Skp1-culin-F-box) family of ubiquitin ligases, the beta-transducin-repeat-containing protein (β-TrCP) subunit of which binds directly to the phosphorylated recognition sequence (Ben-Neriah, 2002). Polyubiquitinated IκB is targeted to the proteasome, where it undergoes proteolytic degradation, thus liberating the NF-κB dimer. The nuclear localization signal on NF-κB that is unmasked by the degradation of IκB targets the transcription factor to the nucleus, where it binds to specific DNA sequences (consensus: GGGRNNYYCC, N = any base, R = purine, Y = pyrimidine, G = guanine, C = cytosine) present in the promoters and enhancers of multiple genes involved in innate and adaptive immunity. DNA binding occurs via the amino-terminal Rel homology domains. Rel A, Rel B, and c-Rel, but not p50 or p52, contain transcriptional activation domains (TAD) near their carboxy-termini that are involved in the recruitment of transcriptional coactivators and the displacement of repressors. Thus, the binding of NF-κB dimers containing these subunits to the appropriate DNA sites leads to an increase in the rate of transcription of the target genes, an effect that can be augmented by phosphorylation of the TADs (Chen and Greene, 2004). Since homodimers of p50 and p52 lack TADs, they cannot induce transcription, and in fact they can act as repressors when bound to NF-κB sites. One of the targets of NF-κB is the gene encoding IκBα. As a result, any signal that activates NF-κB also

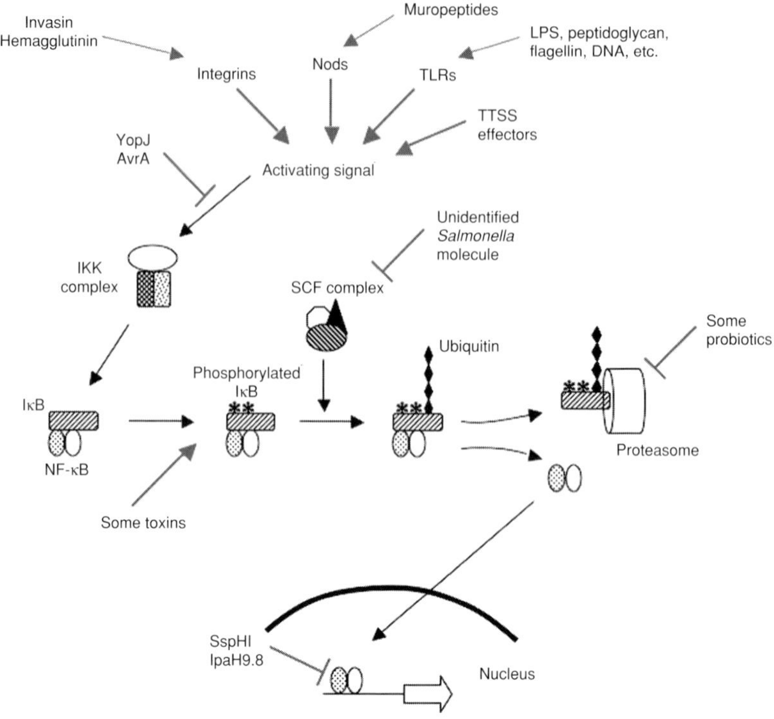

Figure 9.1 The basic steps in the activation of nuclear factor kappa B (NF-κB) are shown in black. The activating and inhibitory effects of bacterial molecules on this process are indicated in blue. See also Color plate 5.

IκB, inhibitor kappa B; IKK, inhibitor kappa B kinase; LPS, lipopolysaccharide; Nod, nucleotide-binding oligomerization domain; SCF, Sκp1-culin-F-box; TLR, Toll-like receptor; TTSS, type III secretion system.

leads to the increased expression of this inhibitor. Since IκBα contains a nuclear export signal, it can bind to NF-κB in the nucleus and take it out of this organelle and away from target DNA sites, thus terminating the NF-κB-dependent response (Huang *et al.*, 2000; Johnson *et al.*, 1999). The existence of this IκBα-mediated negative feedback loop provides a built-in regulatory mechanism that ensures an appropriate intensity and duration of response.

The events described above and shown in Figure 9.1 constitute the classical pathway of NF-κB activation. The alternative pathway, in which p100 is inducibly processed to p52, is activated only by a restricted number of stimuli, most of which are related to lymphoid organ development (Patke *et al.*, 2004; Senftleben *et al.*, 2001), although it should be mentioned that prolonged treatment with lipopolysaccharide (LPS) has been shown to lead to induction of p100 processing (Mordmuller *et al.*, 2003). Since this alternative pathway is not presently known to be involved directly in the epithelial response to bacteria, it will not be considered further here.

ACTIVATION OF NF-κB BY BACTERIA-DERIVED SIGNALS

Interactions between bacteria and epithelial cells can lead to activation or, less commonly, inhibition of NF-κB. Such modulation of NF-κB function can occur via effects on one or more components of the pathway shown in Figure 9.1. Generally, the first step in this process is the sensing of bacteria by cellular proteins that recognize microbial molecules, sometimes referred to as pathogen-associated molecular patterns (PAMPs), a topic that is dealt with in detail in Chapter 4. The discussion below focuses on the cascade of molecular events that is triggered by the initial bacterial-cellular interaction and that ultimately leads to an increase or decrease in NF-κB-dependent transcription. It should be kept in mind, however, that our understanding of the intermediate steps in the cascade is incomplete and is still evolving in many instances.

The Toll-like receptor pathway

The Toll-like receptors (TLRs) constitute a family of transmembrane proteins that recognize and respond to a variety of microbial components such as LPS, bacterial lipopeptides, flagellin, and bacterial and viral nucleic acids (Kopp and Medzhitov, 2003). Since the ligand-sensing domains of these receptors are located on the outside of cellular membranes, the TLRs respond to microorganisms that are located either in the extracellular environment or within a membrane-bound intracellular compartment. Upon activation by the corresponding ligands, the cytoplasmic domains of all the TLRs studied to date bind to an adaptor protein known as MyD88, leading to activation of signals that are responsible for early robust stimulation of the IKK complex and NF-κB (Kawai *et al.*, 1999). MyD88 also functions in signals activated by the interleukin 1 (IL-1) receptor. Some TLRs interact with additional cytoplasmic adaptors: Mal (also known as Toll interleukin receptor domain-containing adaptor protein, TIRAP) functions with MyD88 downstream of TLR-2 and TLR-4 (Yamamoto *et al.*, 2002); Toll interleukin receptor domain-containing adaptor-inducing interferon (TRIF) associates with the cytoplasmic domains of TLR-3 and TLR-4 to mediate delayed NF-κB activation independently of MyD88 (Yamamoto *et al.*, 2003a); and TRIF-related adaptor molecule (TRAM) acts with TRIF downstream of TLR-4 (Yamamoto *et al.*, 2003b).

Subsequent steps in the MyD88-dependent pathway of NF-κB activation involve the recruitment of additional signaling proteins, including members of the interleukin 1 receptor-associated kinase (IRAK) family of serine-threonine kinases (Janssens and Beyaert, 2003) and tumor necrosis factor receptor-associated factor 6 (TRAF6), a ubiquitin ligase (Chung *et al.*, 2002). Interestingly, although IRAK4 has been shown to be essential for both TLR and IL-1 receptor signaling, its kinase activity is dispensable (Suzuki *et al.*, 2002). Presumably, IRAK functions simply as an adaptor to bring TRAF6 into the signaling complex. TRAF6 has been shown to be essential for both IL-1 receptor and TLR signaling, but the exact role that it plays is still unclear (Lomaga *et al.*, 1999). It has been suggested that the serine-threonine kinase transforming growth factor-beta-activated kinase 1 (TAK1) and adaptor proteins TAK-binding protein (TAB) 1 and TAB2 link TRAF6 to activation of IKK, perhaps by ubiquitination-mediated events (Kanayama *et al.*, 2004; Sun *et al.*, 2004). A TRAF-binding protein, TRAF-interacting protein with a forkhead-associated domain (TIFA), has also been implicated in this process (Ea *et al.*, 2004). There is considerable evidence implicating ubiquitination in the regulation of IKK activity by TRAF proteins (Brummelkamp *et al.*, 2003; Kovalenko *et al.*, 2003; Zhou *et al.*, 2004). However, the roles of TRAF6-dependent ubiquitination and the proposed intermediaries between TRAF6 and IKK in this process are yet to be substantiated. In fact, knockout of the TAB1 and TAB2 genes does not indicate essential roles in TLR signaling (Komatsu *et al.*, 2002; Sanjo *et al.*, 2003). Another TRAF6-interacting protein, evolutionarily conserved signaling intermediate in Toll pathway (ECSIT), has also been implicated in TLR signaling to NF-κB (Kopp *et al.*, 1999), possibly by modulating the function of the kinase mitogen-activated protein kinase extracellular growth factor-regulated kinase kinase kinase 3 (MEKK3) (Huang, Yang *et al.*, 2004), but again the physiological importance of these molecules is yet to be established. Thus, the events downstream of TRAF6 that lead to activation of IKK and NF-κB remain an enigma.

The details of the MyD88-independent pathway activated by TLR-3 and TLR-4 that leads to delayed activation of NF-κB are also unclear. Receptor-interacting protein 1 (RIP1), a kinase that binds to TRIF, has been implicated in this pathway (Meylan *et al.*, 2004), but its precise role remains undetermined. RIP1 may function as an adaptor or scaffolding protein, since its kinase activity does not appear to be required for activation of IKK (Lee *et al.*, 2004).

The Nod pathway

The Nod family of proteins serves a function complementary to the TLRs by sensing bacterial molecules present in the cytosol of cells (Chamaillard *et al.*, 2003). Nod1 is expressed in the intestinal epithelium under basal conditions. Nod2, which is often considered to be restricted to myeloid cells, can also be induced in epithelial cells that are exposed to inflammatory mediators (Gutierrez *et al.*, 2002; Rosenstiel *et al.*, 2003). Both Nod1 and Nod2 contain leucine-rich repeat domains that are involved in sensing components of bacterial peptidoglycan – the Gram-negative specific γ-D-glutamyl-mesodiaminopimelic acid in the case of Nod1 and muramyl dipeptide (MDP), a structure present in both Gram-negative and Gram-positive peptidoglycan, in the case of Nod2. The interaction with the corresponding ligand results in the recruitment of the serine-threonine kinase RIP2, which appears to function, much like RIP1 in TLR signaling, in a kinase-independent fashion to link Nod1 and Nod2 to IKK (Inohara *et al.*, 2000). Studies indicate that the mechanism by which RIP2 activates IKK involves polyubiquitination of the regulatory IKKγ subunit without inducing its degradation (Abbott *et al.*, 2004).

Because of their cytosolic location, the Nod proteins are likely to be particularly important in responding to organisms such as *Listeria* and *Shigella*, which escape into the cytosol from the membrane-bound compartment in which they initially reside. However, extracellular bacteria, or bacteria located within phagosomes, may also activate the Nods, since exogenously added muropeptides can find their way into the cytosol by an unknown mechanism. In addition, a study of *Helicobacter pylori* infection of gastric epithelial cells suggests that Nod1 agonists may be delivered into the cytosol via the bacterial type IV secretion system (Viala *et al.*, 2004). There is a great deal of interest in trying to understand precisely how the Nod proteins sense and respond to bacterial infection because of the demonstration that mutations in Nod2 in humans are associated with an increased risk of Crohn's disease.

Bacterial toxins

Several enteric bacterial pathogens secrete toxins that influence epithelial cell function by mechanisms that, in general, do not directly involve NF-κB activation (Fasano, 2002). However, there are some instances where NF-κB activation in epithelial cells has been implicated in toxin action. For example, the toxin A of *Clostridium difficile* has been shown to activate NF-κB by a process involving mitochondrial damage and the generation of reactive oxygen intermediates (ROIs) (He *et al.*, 2002). ROIs have been shown to activate casein kinase 2, which then directly phosphorylates IκBα, inducing its degradation (Hayden and Ghosh, 2004). Similarly, the enterotoxin of

Bacteroides fragilis has been shown to activate NF-κB in epithelial cells by a mechanism that is yet to be worked out (Kim *et al.*, 2002). In addition, many of the intestinal abnormalities caused by bacterial toxins result from effects on cell types other than enterocytes, including macrophages and enteric nerves (Pothoulakis and Lamont, 2001), and some of these effects involve NF-κB (Jefferson *et al.*, 1999; Sakiri *et al.*, 1998).

Some bacteria, such as *Salmonella, Shigella*, and *Yersinia*, produce proteins that are not classified as toxins but are secreted and introduced into the cytosol of epithelial cells via a specialized secretory apparatus, the type III secretion system (TTSS). Many of these proteins, sometimes referred to as effector proteins, induce cytoskeletal changes that are involved in entry of the bacteria into the host cell (Brumell *et al.*, 1999), but some have also been shown to activate pro-inflammatory signals, including the NF-κB pathway (Hobbie *et al.*, 1997; Huang, Werne *et al.*, 2004; Viboud *et al.*, 2003). In the case of the *Salmonella* effectors SopE and SopE2, activation of mammalian Rho GTPases appears to be the first step in a signaling cascade that may ultimately influence NF-κB function downstream of IκBα degradation (Hobbie *et al.*, 1997; Huang, Werne *et al.*, 2004).

Other modes of NF-κB activation by bacteria

The outer surfaces of bacteria contain a number of molecules that are involved in interactions with host cells. Some of these are PAMPs, such as LPS, that are sensed directly by the TLRs, as discussed in detail earlier in this chapter. In addition, proteins such as pilin, a component of adhesive pili, and porin, a major constituent of the outer membrane of Gram-negative bacteria, have been shown to activate intracellular signaling cascades. The porin of *Neisseria gonorrhoeae* activates NF-κB in uroepithelial cells in a TLR-2-dependent fashion, but it is not clear whether this process involves sensing directly by the TLR or via an intermediary receptor (Massari *et al.*, 2003). Porins of *Salmonella* and *Shigella* have also been shown to activate NF-κB, although not in epithelial cells (Galdiero *et al.*, 2002; Ray and Biswas, 2005). The pilins of *Pseudomonas aeruginosa* and *Staphylococcus aureus* activate NF-κB in respiratory epithelial cells via binding to their receptor, asialoGM1; interestingly, TLRs have been implicated in this process (Adamo *et al.*, 2004; DiMango *et al.*, 1998). Other bacterial surface molecules that have been shown to activate NF-κB include the filamentous hemagglutinin of *Bordetella pertussis* and invasin of *Yersinia enterocolitica*, both of which bind to mammalian integrins, heterodimeric cell-surface proteins that are involved in cell–cell and cell–matrix interactions (Grassl *et al.*, 2003; Ishibashi and Nishikawa, 2003).

The *Yersinia* adhesin YadA may use a similar mode of cell activation (Schmid *et al.*, 2004). The exact mechanism linking bacteria–integrin interactions to activation of NF-κB remains to be elucidated, but it is likely to involve the Rac1 GTPase and the p21-activated kinase (Grassl *et al.*, 2003; Juliano *et al.*, 2004).

FUNCTIONS ACTIVATED BY NF-κB IN EPITHELIAL CELLS

The activation of NF-κB in epithelial cells results in multiple functional responses. The deployment of mechanisms with direct bacteriostatic or bacteriocidal effects and the production of inflammatory mediators that recruit cells of the immune system are two important means of eliminating microbial invaders. Clearly included in the first category is the induction of antimicrobial peptides, such as the defensins and cathelicidins (Lehrer, 2004; Yang *et al.*, 2001), which inhibit the growth of several kinds of bacteria. The production of toxic free radicals secondary to increased expression of inducible nitric oxide (NO) synthase (iNOS) has antimicrobial effects in phagocytic cells (Fang, 1997), and it may play a similar role in epithelial cells. In addition to these mechanisms that act directly on bacteria, epithelial expression of a variety of secreted mediators such as interleukin 8 (IL-8) and other chemoattractants results in the recruitment of neutrophils, macrophages, dendritic cells, and lymphocytes, all of which contribute to bacterial killing or the generation of an adaptive immune response. Paradoxically, there is evidence to indicate that NF-κB activation may lead to inhibition of inflammation in some situations. Finally, an inevitable side effect of many of the pro-inflammatory NF-κB-dependent responses is the generation, either by the epithelial cells themselves or by the recruited leukocytes, of molecules that promote tissue damage. NF-κB, activated by contact with bacteria or by the inflammatory mediators induced by bacteria, has a role in limiting this damage by turning on expression of genes that prevent epithelial cell apoptosis (Kucharczak *et al.*, 2003). We will discuss these different types of NF-κB-dependent response in greater detail below.

Epithelial production of antibacterial molecules

One of the most evolutionarily ancient functions of NF-κB in innate immune defense is the induction of antimicrobial peptides (Hoffmann, 2003). This function has been shown to be activated by bacteria in both respiratory and intestinal epithelial cells. The function manifests as the increased production of β-defensin 2 (hBD-2), a secreted cationic peptide with activity against several pulmonary and enteric pathogens (Harder et al., 2000; O'Neil *et al.*, 1999). NF-κB-dependent induction of hBD-2 in epithelial cells has been shown to occur in response to various bacterial components, including flagellin, lipoproteins, and DNA, and is mediated by the corresponding TLRs (Birchler *et al.*, 2001; Ogushi *et al.*, 2001; Platz *et al.*, 2004). hBD-2 has also been shown to be upregulated by a secreted protease of *Porphyromonas gingivalis* via a protease-activated G-protein-coupled receptor on gingival epithelial cells (Chung *et al.*, 2004). Cathelicidins constitute another class of broad-spectrum antimicrobial peptide that can be induced in epithelial cells by bacteria (Bals *et al.*, 1998), but whether NF-κB is involved in this process remains to be determined.

Paneth cells, one of the major epithelial cell lineages of the intestine, are found in close apposition to the crypts of the small bowel and are an important source of molecules that have direct antibacterial action (Ayabe *et al.*, 2004). They express a number of α-defensins (cryptdins) that are secreted rapidly into the lumen of the crypt in response to contact with whole bacteria or bacterial products such as LPS, lipoteichoic acid, and muramyl dipeptide (MDP) (Ayabe *et al.*, 2000). This response appears to be regulated largely at the level of secretion, since preformed cryptdins are present as abundant constituents of Paneth cell secretory granules in the basal state. However, NF-κB may play an indirect role in production of cryptdins, since the proteolytic processing of cryptdin precursors to the mature functional forms requires the enzyme matrilysin, which can be induced in epithelial cells by bacteria via an NF-κB-dependent mechanism (Wilson *et al.*, 1999; Wroblewski *et al.*, 2003). In addition to their direct antibacterial effects, some cryptdins may also contribute to the induction of the local inflammatory response by acting on enterocytes in a paracrine manner to increase IL-8 expression (Lin *et al.*, 2004). This response involves NF-κB activation and is dependent on the ability of the cryptdin to form pores in the cellular membrane. A similar phenomenon has been observed in bronchial epithelial cells (Sakamoto *et al.*, 2005). Besides the cryptdins, Paneth cells express the antibacterial enzyme lysozyme, which degrades cell-wall peptidoglycan (Ouellette, 1997). Although there is very little information on the regulation of lysozyme expression in Paneth cells, studies on the gene in chickens indicate that it is upregulated by bacterial signals in an NF-κB-dependent fashion (Regenhard *et al.*, 2001; van Phi, 1996).

iNOS has been shown to be upregulated via NF-κB activation in intestinal epithelial cells in response to non-pathogenic bacteria and several enteropathogens, including *Salmonella*, *Shigella*, and *H. pylori* (Islam *et al.*, 1997; Obonyo *et al.*, 2002; Resta-Lenert and Barrett, 2002; Witthoft *et al.*, 1998). Although iNOS-dependent generation of reactive nitrogen intermediates has several direct toxic effects on microorganisms in phagocytic cells (Fang, 1997), it is not clear whether these effects are also relevant to epithelial function. iNOS was found to be appreciably upregulated in the colonic epithelium following infection with the murine enteropathogen *Citrobacter rodentium*, but lack of the enzyme had only a minor effect on the growth of the bacteria (Vallance *et al.*, 2002). Increased iNOS expression does have some clear effects on epithelial cell function, including enhanced chloride secretion (Resta-Lenert and Barrett, 2002) and disruption of cytoskeletal integrity and mucosal barrier properties (Banan *et al.*, 2004). iNOS has also been implicated in the regulation of programmed cell death, with both pro- and anti-apoptotic effects being shown in different studies (Kim *et al.*, 2003; Miyazawa *et al.*, 2003). Despite the apparently detrimental consequences, there is evidence to indicate that production of NO has protective value in some types of intestinal inflammation and that the beneficial effects of some types of probiotics may be related to their ability to increase levels of this molecule (Lamine *et al.*, 2004; Vallance *et al.*, 2004).

Epithelial production of pro-inflammatory mediators

A number of cytokines and chemokines have been shown to be produced by epithelial cells in response to bacterial infection, both in tissue culture studies and in vivo (Strober, 1998). A partial list is given in Table 9.1, but this should be viewed as an illustrative spectrum, rather than a comprehensive compilation, of the responses that are elicited. Most of these studies utilized representative epithelial cell lines, although confirmatory evidence from primary epithelial cells has been provided in some cases. The involvement of NF-κB in the response has been shown in many instances by use of pharmacological or genetic inhibitors. Epithelial production of these cytokines and chemokines results in the recruitment of cells of the immune system from the circulation to the site of infection. Cells such as neutrophils and monocytes that are recruited by IL-8 and monocyte chemotactic protein 1 (MCP-1), respectively, aid in clearing the infection by phagocytosing and killing the bacteria. At the same time, the recruitment of dendritic cells and lymphocytes by the chemokines CCL20 and RANTES facilitates the activation of bacteria-specific adaptive immune responses.

Anti-inflammatory effects of NF-κB

Although NF-κB is generally considered to be a major activator of pro-inflammatory genes, it may also have some anti-inflammatory effects. Homozygous knockout of the gene encoding NF-κB p50 results in enhancement of the colitis induced by *Helicobacter hepaticus* infection in mice,

Table 9.1 *Sampling of cytokine/chemokine responses elicited by bacteria in various epithelial cell types**

Epithelial cell type	Bacteria	Cytokine/chemokine
Intestinal		
Primary human gingival	*Porphyromonas*	IL-8, MCP-1
T84, Caco-2, HT-29, Int407, primary human colonic	*Salmonella, Shigella, Listeria, Yersinia, Vibrio, Clostridium difficile* toxin	IL-8, MCP-1, TNFα, GROα, GROγ, ENA78, RANTES, CCL20
KATO-3, ST42, AGS	*Helicobacter pylori*	IL-8, IL-18
Respiratory		
A-549	*Pseudomonas, Legionella,* group B *Streptococcus*	IL-6, IL-8, IL-10, TNFα
Primary human bronchial	*Haemophilus influenzae*	IL-6, IL-8, TNFα
Urinary		
A-498, J82	*Escherichia coli*	IL-1α, IL-6, IL-8
Reproductive		
HeLa	*Salmonella, Listeria*	IL-8

* Information compiled from Agace *et al.* (1993); Chang *et al.* (2004); Crabtree *et al.* (1994); Day *et al.* (2004); Eckmann *et al.* (1993); Jung *et al.* (1995); Khair *et al.* (1994); Kusumoto *et al.* (2004); Mahida *et al.* (1996); McCormick *et al.* (1993); Mikamo *et al.* (2004); Palfreyman *et al.* (1997); Sarkar and Chaudhuri (2004); Sierro *et al.* (2001); Yang *et al.* (1997).

particularly when the mutation is combined with loss of one of the NF-κB p65 genes (Erdman *et al.*, 2001). This observation suggests that at least some of the NF-κB genes are involved in turning off inflammation. The mechanism of this anti-inflammatory effect is not clear, but it may be related to the fact that p50 homodimers, which lack the ability to activate transcription, can block expression of some genes and have been implicated in the development of a state of non-responsiveness to bacterial products (Cherayil, 2003).

NF-κB and epithelial cell apoptosis

One of the side effects of the inflammatory response is cell death and consequent tissue damage. Since NF-κB regulates the expression of several genes involved in controlling apoptosis, its activation during inflammation may help to limit cell injury and death (Kucharczak *et al.*, 2003). In states of chronic inflammation, this anti-apoptotic effect of NF-κB may promote the development of tumors, an idea that may help to explain the increased risk of intestinal malignancy in people with inflammatory bowel disease (IBD) (Itzkowitz and Yio, 2004). The key role played by NF-κB in inflammation-associated tumorigenesis has been highlighted by the observation that knockout of the IKKβ gene selectively in intestinal epithelial cells significantly reduced the incidence of tumors in a mouse model of colitis-related neoplasia (Greten *et al.*, 2004). The same study showed that knockout of this gene selectively in myeloid cells decreased the size, but not the incidence, of tumors, suggesting that inflammatory mediators released by these cells may promote tumor growth.

INHIBITION OF NF-κB BY BACTERIA

Since NF-κB turns on the expression of a number of genes that are ultimately detrimental to microbial survival and multiplication, it should come as no surprise that bacteria have evolved mechanisms to inhibit NF-κB function. These mechanisms constitute an important strategy by which pathogens establish and propagate themselves in host tissues. They may also be relevant to the state of detente that exists between the host and the commensal microbial flora. Inhibition of pro-inflammatory responses has been linked to the beneficial effects of probiotic organisms, sparking a lot of interest in detailed elucidation of the process in order to harness it for therapeutic purposes.

One of the first instances of NF-κB repression mediated by a bacterial product was provided by YopJ, an effector protein of the enteropathogen *Yersinia pseudotuberculosis* that is introduced into host cells via a TTSS (Schesser *et al.*, 1998). The mechanism of this effect involves a cysteine protease activity of YopJ that cleaves ubiquitin-like molecules and interrupts signals leading to mitogen-activated protein kinase and NF-κB activation (Orth *et al.*, 1999, 2000). AvrA, a *Salmonella* homolog of YopJ, also inhibits NF-κB activation, but the mechanism appears to be quite distinct since AvrA does not have protease activity (Collier-Hyams *et al.*, 2002). *Salmonella* may also use other means to inhibit NF-κB. Inhibition of the signaling pathway between the steps of IκBα phosphorylation and ubiquitination by an uncharacterized *Salmonella* molecule has been described (Neish *et al.*, 2000). In addition, a leucine-rich repeat effector protein, SspH1, has been shown to inhibit NF-κB activation in the nucleus by an unknown mechanism (Haraga and Miller, 2003). A homolog in *Shigella flexneri*, IpaH9.8, may act similarly.

It is not clear whether the commensal microflora actively inhibits inflammatory responses. The lack of apparent reactivity to these organisms may simply reflect the fact that they can be restricted to their normal habitat by protective mechanisms that do not depend on inciting inflammation. Nevertheless, the commensal flora does have numerous effects on epithelial function (Hooper and Gordon, 2001), and it is possible that dampening of inflammatory responses may be one of the outcomes. Probiotic organisms, many of which are components of the normal intestinal microflora, have been shown to ameliorate some types of inflammatory conditions, but whether this is related to direct inhibition of pro-inflammatory signals is open to debate. Studies have suggested that the beneficial effects of probiotics may derive from interactions between their DNA and TLR-9 (Jijon *et al.*, 2004; Rachmilewitz *et al.*, 2004). TLR-9, like other TLRs, usually activates NF-κB and can induce an inflammatory response. If it has the opposite effect in relation to probiotics, there may be something unique about either probiotic DNA or the function of TLR-9 in the intestine. Indeed, there is evidence to indicate that TLR signals activated by the commensal flora promote reparative and healing functions in the intestinal epithelium (Rakoff-Nahoum *et al.*, 2004), and work has demonstrated that TLR-9-induced interferon β (IFNβ) protects against intestinal inflammation (Katakura *et al*, 2005). Finally, one additional anti-inflammatory effect of probiotics may be related to inhibition of the proteasome, with consequent stabilization of IκB (Petrof *et al.*, 2004).

CLINICAL IMPLICATIONS OF ABNORMAL NF-κB ACTIVATION DURING BACTERIAL–EPITHELIAL INTERACTIONS

Knockout studies in mice have demonstrated quite clearly that disruption of the pathways that lead to NF-κB activation results in abnormalities in the response to infection. Rare immunodeficiency disorders in humans involving genetic defects in NEMO (IKKγ), IκBα, and IRAK4 also lend support to this idea (Ku *et al.*, 2005). In addition, one relatively common clinical problem that is particularly pertinent to bacterial–epithelial interactions is the chronic intestinal inflammation associated with Crohn's disease and ulcerative colitis. Although the precise pathogenesis of these conditions is unclear, there is considerable evidence to indicate that they are caused by abnormal inflammatory and immune responses to the commensal gut flora (Bouma and Strober, 2003). The idea that an abnormality in responding to enteric bacteria may be a basic predisposing factor in the development of chronic intestinal inflammation has received additional support from the discovery that Nod2 mutations are associated with susceptibility to Crohn's disease (Bouma and Strober, 2003). Interestingly, at least some of the Nod2 mutations identified in this disease have been shown to interfere with the ability of the protein to activate NF-κB in transfection-based tissue-culture studies (Ogura *et al.*, 2001). Although this result may seem difficult to reconcile with the evidence of activated NF-κB-dependent inflammatory responses in the diseased tissues, it has been explained on the basis of defective NF-κB-dependent antimicrobial defense mechanisms. According to this idea, the inability to adequately control the growth of normally innocuous intestinal bacteria leads to the state of chronic inflammation associated with the disease. The demonstration of abnormally low expression of Paneth cell defensins in both humans and mice with mutations in Nod2 is consistent with this model of pathogenesis (Kobayashi *et al.*, 2005; Wehkamp *et al.*, 2004). Indeed, the detection of decreased levels of enterocyte-expressed β-defensins in forms of Crohn's disease that are not linked to Nod2 mutations, as well as in ulcerative colitis, suggests that impaired defensin production may be a common underlying mechanism in IBD (Wehkamp *et al.*, 2005). On the other hand, there is some evidence to suggest that the pathogenesis of Crohn's disease may be related to hyperactivation of NF-κB-dependent inflammatory responses by at least certain types of Nod2 mutations (Maeda *et al.*, 2005). Clearly, much additional work will be required to clarify the exact role of Nod2 in the pathogenesis of chronic intestinal inflammation, but its association with Crohn's disease emphasizes the importance of maintaining an appropriate level of NF-κB activation in response to contact with bacteria at epithelial surfaces.

Recognition of the key role played by NF-κB in initiating and perpetuating the chronic inflammatory state of conditions such as Crohn's disease and ulcerative colitis has prompted evaluation of NF-κB inhibitors as approaches to treating these problems. Indeed, salicylates, including well-established medications for IBD such as sulfasalazine and mesalamine, have been shown to inhibit NF-κB (Egan *et al.*, 1999; Kopp and Ghosh, 1994; Wahl *et al.*, 1998). A more specific inhibitor, anti-sense oligonucleotides against p65, has been shown to decrease inflammation in an animal model of colitis, but its use in humans is yet to be studied (Neurath *et al.*, 1996). The beneficial effects of probiotics in some types of IBD may also involve direct or indirect modulation of NF-κB-dependent responses (Sartor, 2004). With any new therapeutic approach to NF-κB inhibition, it would be prudent to proceed cautiously given the importance of this transcription factor in protective antibacterial mechanisms.

CONCLUSION

The well-known sentiment that "good fences make good neighbors" (Frost, 1979) could be considered to be as true of bacterial–epithelial interactions as it is of human relationships. Mammals live surrounded by an immense number of bacteria. In general, these microorganisms do not pose a threat, and indeed they can provide some benefits as long as they remain on the "right" side of the epithelial surfaces of the body. In the absence of a completely impervious surface, peaceful coexistence with the bacteria depends on the deployment of defense mechanisms that are finely tuned to the microbial milieu. Under normal circumstances, these mechanisms are designed to deal with organisms with very low potential for pathogenicity. On occasion, however, when the epithelium is threatened by more dangerous invaders, additional defensive measures must be called into play quickly. The operation of such a dynamic functional barrier requires constant sensing of microorganisms at epithelial surfaces and the generation of the most appropriate response. By its ability to be regulated in different ways by a diverse set of microbial stimuli, and by its central role in controlling a variety of protective mechanisms, NF-κB functions as the perfect mediator of the host–bacterial equilibrium. Further elucidation of the function of this transcription factor, particularly with respect to interactions with commensal and pathogenic bacteria, will undoubtedly provide new insights into a variety of diseases that are caused by either too weak or too vigorous a response to these organisms and may suggest new approaches to treating the abnormalities.

ACKNOWLEDGMENTS

I am grateful to Dr Amitabha Chaudhuri for his constructive comments on the manuscript. I regret that limitations of space did not allow me to cite all of the numerous publications pertinent to this wide-ranging topic. Work in my laboratory is supported by the National Institutes of Health via R01 AI48815 and R21 AI065619.

REFERENCES

Abbott, D. W., Wilkins, A., Asara, J. M., and Cantley, L. C. (2004). The Crohn's disease protein, Nod2, requires RIP2 in order to induce ubiquitinylation of a novel site on NEMO. *Curr. Biol.* **14**, 2217–2227.

Adamo, R., Sokol, S., Soong, G., Gomez, M. I., and Prince, A. (2004). *Pseudomonas aeruginosa* flagella activate airway epithelial cells through asialoGM1 and

Toll-like receptor 2 as well as Toll-like receptor 5. *Am. J. Respir. Cell Mol. Biol.* **30**, 627–634.

Agace, W., Hedges, S., Andersson, U., *et al.* (1993). Selective cytokine production by epithelial cells following exposure to *Escherichia coli. Infect. Immun.* **61**, 602–609.

Ayabe, T., Satchell, D. P., Wilson, C. L., *et al.* (2000). Secretion of microbicidal α-defensins by intestinal Paneth cells in response to bacteria. *Nat. Immunol.* **1**, 113–118.

Ayabe, T., Ashida, T., Kohgo, Y., and Kono, T. (2004). The role of Paneth cells and their antimicrobial peptides in innate host defense. *Trends Microbiol.* **12**, 394–398.

Bals, R., Wang, X., Zasloff, M., and Wilson, J. M. (1998). The peptide antibiotic LL-37/hCAP-18 is expressed in epithelia of the human lung where it has broad antimicrobial activity at the airway surface. *Proc. Natl. Acad. Sci. U. S. A.* **95**, 9541–9546.

Banan, A., Zhang, L. J., Shaikh, M., *et al.* (2004). Novel effect of NF-κB activation: carbonylation and nitration injury to cytoskeleton and disruption of monolayer barrier in intestinal epithelium. *Am. J. Physiol. Cell Physiol.* **287**, C1139–C1151.

Ben-Neriah, Y. (2002). Regulatory functions of ubiquitination in the immune system. *Nat. Immunol.* **3**, 20–26.

Birchler, T., Siebl, R., Buchner, K., *et al.* (2001). Human Toll-like receptor 2 mediates induction of the antimicrobial peptide human β-defensin 2 in response to bacterial lipoprotein. *Eur. J. Immunol.* **31**, 3131–3137.

Bouma, G. and Strober, W. (2003). The immunological and genetic basis of inflammatory bowel disease. *Nat. Rev. Immunol.* **3**, 521–533.

Brumell, J. H., Steele-Mortimer, O., and Finlay, B. B. (1999). Bacterial invasion: force feeding by *Salmonella. Curr. Biol.* **9**, R277–R280.

Brummelkamp, T. R., Nijman, S. M., Dirac, A. M., and Bernards, R. (2003). Loss of the cylindromatosis tumor suppressor inhibits apoptosis by activating NF-κB. *Nature* **424**, 797–801.

Chamaillard, M., Girardin, S. E., Viala, J., and Philpott, D. J. (2003). Nods, Nalps and Naip: intracellular regulators of bacterial-induced inflammation. *Cell. Microbiol.* **5**, 581–592.

Chang, B., Amemura-Maekawa, J., Kura, F., Kawamura, I., and Watanabe, H. (2004). Expression of IL-6 and TNFα in human alveolar epithelial cells is induced by invading, but not by adhering, *Legionella pneumophila. Microb. Pathog.* **37**, 295–302.

Chen, L. F. and Greene, W. C. (2004). Shaping the nuclear action of NF-κB. *Nat. Rev. Mol. Cell. Biol.* **5**, 392–401.

Cherayil, B. J. (2003). How not to get bugged by bugs: mechanisms of cellular tolerance to microorganisms. *Curr. Opin. Gastroenterol.* **19**, 572–577.

Chung, J. Y., Park, Y. C., Ye, H., and Wu, H. (2002). All TRAFs are not created equal: common and distinct molecular mechanisms of TRAF-mediated signal transduction. *J. Cell Sci.* **115**, 679–688.

Chung, W. O., Hansen, S. R., Rao, D., and Dale, B. A. (2004). Protease-activated receptor signaling increases epithelial antimicrobial peptide expression. *J. Immunol.* **174**, 5165–5170.

Collier-Hyams, L. S., Zeng, H., Sun, J., *et al.* (2002). Cutting edge: *Salmonella* AvrA effector inhibits the key proinflammatory, anti-apoptotic NF-κB pathway. *J. Immunol.* **169**, 2846–2850.

Crabtree, J. E., Farmery, S. M., Lindley, I. J., *et al.* (1994). CagA/cytotoxic strains of *Helicobacter pylori* and interleukin-8 in gastric epithelial cell lines. *J. Clin. Pathol.* **47**, 945–950.

Day, A. S., Su, B., Ceponis, P. J., *et al.* (2004). *Helicobacter pylori* infection induces interleukin-18 production in gastric epithelial (AGS) cells. *Dig. Dis. Sci.* **49**, 1830–1835.

DiMango, E., Ratner, A. J., Bryan, R., Tabibi, S., and Prince, A. (1998). Activation of NF-κB by adherent *Pseudomonas aeruginosa* in normal and cystic fibrosis respiratory epithelial cells. *J. Clin. Invest.* **101**, 2598–2605.

Ea, C. K., Sun, L., Inoue, J., and Chen, Z. J. (2004). TIFA activates IκB kinase (IKK) by promoting oligomerization and ubiquitination of TRAF6. *Proc. Natl. Acad. Sci. U. S. A.* **101**, 15 318–15 323.

Eckmann, L., Kagnoff, M. F., and Fierer, J. (1993). Epithelial cells secrete the chemokine interleukin-8 in response to bacterial entry. *Infect. Immun.* **61**, 4569–4574.

Egan, L. J., Mays, D. C., Huntoon, C. J., *et al.* (1999). Inhibition of interleukin-1 stimulated NF-κB RelA/p65 phosphorylation by mesalamine is accompanied by decreased transcriptional activity. *J. Biol. Chem.* **274**, 26 448–26 453.

Erdman, S., Fox, J. G., Dangler, C. A., Feldman, D., and Horwitz, B. H. (2001). Typhlocolitis in NF-κB-deficient mice. *J. Immunol.* **166**, 1443–1447.

Fang, F. (1997). Mechanisms of nitric oxide-related antimicrobial activity. *J. Clin. Invest.* **99**, 2818–2825.

Fasano, A. (2002). Toxins and the gut: role in human disease. *Gut* **50**, III9–III14.

Frost, R. (1979). Mending wall. In *The Poetry of Robert Frost: The Collected Poems, Complete and Unabridged*, ed. E. C. Lathem. New York: Henry Holt and Co., pp. 147–148.

Galdiero, M., Vitiello, M., Sanzari, E., *et al.* (2002). Porins from *Salmonella enterica* serovar Typhimurium activate the transcription factors AP-1 and NF-κB through the Raf-1-MAP kinase cascade. *Infect. Immun.* **70**, 558–568.

Ghosh, S. and Karin, M. (2002). Missing pieces in the NF-κB puzzle. *Cell* **109**, S81–S96.

Grassl, G. A., Kracht, M., Wiedemann, A., *et al.* (2003). Activation of NF-κB and IL-8 by *Yersinia enterocolitica* invasin protein is conferred by Rac1 and MAP kinase cascades. *Cell. Microbiol.* **5**, 957–971.

Greten, F. R., Eckmann, L., Greten, T. F., *et al.* (2004). IKKβ links inflammation and tumorigenesis in a mouse model of colitis-associated cancer. *Cell* **118**, 285–296.

Gutierrez, O., Pipaon, C., Inohara, N., *et al.* (2002). Induction of Nod2 in myelomonocytic and intestinal epithelial cells via NF-κB activation. *J. Biol. Chem.* **277**, 41 701–41 705.

Haraga, A. and Miller, S. I. (2003). A *Salmonella enterica* serovar Typhimurium translocated leucine-rich repeat effector protein inhibits NF-κB-dependent gene expression. *Infect. Immun.* **71**, 4052–4058.

Harder, J., Meyer-Hoffert, U., Teran, L. M., *et al.* (2000). Mucoid *Pseudomonas aeruginosa*, TNFα, and IL-1β, but not IL-6, induce human β-defensin-2 in respiratory epithelia. *Am. J. Respir. Cell. Mol. Biol.* **22**, 714–721.

Hayden, M. S. and Ghosh, S. (2004). Signaling to NF-κB. *Genes Dev.* **18**, 2195–2224.

He, D., Sougioultzis, S., Hagen, S., *et al.* (2002). *Clostridium difficile* toxin A triggers human colonocyte IL-8 release via mitochondrial oxygen radical generation. *Gastroenterology* **122**, 1048–1057.

Hobbie, S., Chen, L. M., Davis, R. J., and Galan, J. E. (1997). Involvement of mitogen-activated protein kinase pathways in the nuclear responses and cytokine production induced by *Salmonella typhimurium* in cultured intestinal epithelial cells. *J. Immunol.* **159**, 5550–5559.

Hoffmann, J. A. (2003). The immune response of *Drosophila*. *Nature* **426**, 33–38.

Hooper, L. V. and Gordon, J. I. (2001). Commensal host-bacterial relationships in the gut. *Science* **292**, 1115–1118.

Huang, F. C., Werne, A., Li, Q., *et al.* (2004). Cooperative interactions between flagellin and SopE2 in the epithelial interleukin-8 response to *Salmonella enterica* serovar Typhimurium infection. *Infect. Immun.* **72**, 5052–5062.

Huang, Q., Yang, J., Lin, Y., *et al.* (2004). Differential regulation of interleukin 1 receptor and Toll-like receptor signaling by MEKK3. *Nat. Immunol.* **5**, 98–103.

Huang, T. T., Kudo, N., Yoshida, M., and Miyamoto, S. (2000). A nuclear export signal in the N-terminal regulatory domain of IκBα controls cytoplasmic localization of inactive NF-κB/IκBα complexes. *Proc. Natl. Acad. Sci. U. S. A.* **97**, 1014–1019.

Inohara, N., Koseki, T., Lin, J., *et al.* (2000). An induced proximity model for NF-κB activation in the Nod1/RICK and RIP signaling pathways. *J. Biol. Chem.* **275**, 27 823–27 831.

Ishibashi, Y. and Nishikawa, A. (2003). Role of NF-κB in the regulation of intercellular adhesion molecule 1 after infection of human bronchial epithelial cells by *Bordetella pertussis*. *Microb. Pathog.* **35**, 169–177.

Islam, D., Veress, B., Bardhan, P. K., Lindberg, A. A., and Christenson, B. (1997). In situ characterization of inflammatory responses in the rectal mucosae of patients with shigellosis. *Infect. Immun.* **65**, 739–749.

Itzkowitz, S. H. and Yio, X. (2004). Inflammation and cancer IV. Colorectal cancer in inflammatory bowel disease: the role of inflammation. *Am. J. Physiol. Gastrointest. Liver Physiol.* **287**, G7–G17.

Janssens, S. and Beyaert, R. (2003). Functional diversity and regulation of different interleukin-1 receptor-associated kinase (IRAK) family members. *Mol. Cell* **11**, 293–302.

Jefferson, K. K., Smith, M. F., and Bobak, D. A. (1999). Roles of intracellular calcium and NF-κB in the *Clostridium difficile* toxin A-induced up-regulation and secretion of IL-8 from human monocytes. *J. Immunol.* **163**, 5183–5191.

Jijon, H., Backer, J., Diaz, H., *et al.* (2004). DNA from probiotic bacteria modulates murine and human epithelial and immune function. *Gastroenterology* **126**, 1358–1373.

Johnson, C., Van Antwerp, D., and Hope, T. J. (1999). An N-terminal nuclear export signal is required for the nucleo-cytoplasmic shuttling of IκBα. *EMBO J.* **18**, 6682–6693.

Juliano, R. L., Reddig, P., Alahari, S., *et al.* (2004). Integrin regulation of cell signalling and motility. *Biochem. Soc. Trans.* **32**, 443–446.

Jung, H. C., Eckmann, L., Yang, S. K., *et al.* (1995). A distinct array of proinflammatory cytokines is expressed in human colon epithelial cells in response to bacterial invasion. *J. Clin. Invest.* **95**, 55–65.

Kanayama, A., Seth, R. B., Sun, L., *et al.* (2004). TAB2 and TAB3 activate the NF-κB pathway through binding to polyubiquitin chains. *Mol. Cell* **15**, 535–548.

Katakura, K., Lee, J., Rachmilewitz, D., *et al.* (2005). Toll-like receptor 9-induced type I IFN protects mice from experimental colitis. *J. Clin. Invest.* **115**, 695–702.

Kawai, T., Adachi, O., Ogawa, T., Takeda, K., and Akira, S. (1999). Unresponsiveness of MyD88-deficient mice to endotoxin. *Immunity* **11**, 115–122.

Khair, O. A., Devalia, J. L., Abdelaziz, M. M., *et al.* (1994). Effect of *Haemophilus influenzae* endotoxin on the synthesis of IL-6, IL-8, TNFα and expression of ICAM-1 in cultured human bronchial epithelial cells. *Eur. Respir. J.* **7**, 2109–2116.

Kim, J. M., Cho, S. J., Oh, Y. K., *et al.* (2002). NF-κB activation pathway in intestinal epithelial cells is a major regulator of chemokine gene expression and neutrophil migration induced by *Bacteroides fragilis* enterotoxin. *Clin. Exp. Immunol.* **130**, 59–66.

Kim, J. M., Kim, J. S., Jung, H. C., *et al.* (2003). *Helicobacter pylori* infection activates NF-κB signaling pathway to induce iNOS and protect human gastric epithelial cells from apoptosis. *Am. J. Physiol. Gastrointest. Liver Physiol.* **285**, G1171–G1180.

Kobayashi, K. S., Chamaillard, M., Ogura, Y., *et al.* (2005). Nod2-dependent regulation of innate and adaptive immunity in the intestinal tract. *Science* **307**, 731–734.

Komatsu, Y., Shibuya, H., Takeda, N., *et al.* (2002). Targeted disruption of the TAB1 gene causes embryonic lethality and defects in cardiovascular and lung morphogenesis. *Mech. Dev.* **119**, 239–249.

Kopp, E. and Ghosh, S. (1994). Inhibition of NF-κB by sodium salicylate and aspirin. *Science* **265**, 956–959.

Kopp, E., and Medzhitov, R. (2003). Recognition of microbial infection by Toll-like receptors. *Curr. Opin. Immunol.* **15**, 396–401.

Kopp, E., Medzhitov, R., Carothers, J., *et al.* (1999). ECSIT is an evolutionarily conserved intermediate in the Toll/IL-1 signal transduction pathway. *Genes Dev.* **13**, 2059–2071.

Kovalenko, A., Chable-Bessia, C., Cantarella, G., *et al.* (2003). The tumor suppressor CYLD negatively regulates NF-κB signaling by deubiquitination. *Nature* **424**, 801–805.

Ku, C. L., Yang, K., Bustamente, J., *et al.* (2005). Inherited disorders of human Toll-like receptor signaling: immunological implications. *Immunol. Rev.* **203**, 10–20.

Kucharczak, J., Simmons, M. J., Fan, Y., and Gelinas, C. (2003). To be or not to be: NF-κB is the answer – the role of Rel/NF-κB in the regulation of apoptosis. *Oncogene* **22**, 8961–8982.

Kusumoto, Y., Hirano, H., Saitoh, K., *et al.* (2004). Human gingival epithelial cells produce chemotactic factors interleukin-8 and monocyte chemoattractant protein-1 after stimulation with *Porphyromonas gingivalis* via Toll-like receptor 2. *J. Periodontol.* **75**, 370–379.

Lamine, F., Fioramonti, J., Bueno, L., *et al.* (2004). Nitric oxide released by *Lactobacillus farciminis* improves TNBS-induced colitis in rats. *Scand. J. Gastroenterol.* **39**, 37–45.

Lee, T. H., Shank, J., Cusson, N., and Kelliher, M. A. (2004). The kinase activity of Rip1 is not required for tumor necrosis factor α induced IκB kinase or p38 MAP kinase activation or for ubiquitination of Rip1 by Traf2. *J. Biol. Chem.* **279**: 33 185–33 191.

Lehrer, R. I. (2004). Primate defensins. *Nat. Rev. Microbiol.* **2**, 727–738.

Lin, P. W., Simon, P. O., Gewirtz, A. T., *et al.* (2004). Paneth cell cryptdins act in vitro as apical paracrine regulators of the innate inflammatory response. *J. Biol. Chem.* **279**, 19 902–19 907.

Lomaga, M., Yeh, W. C., Sarosi, I., *et al.* (1999). TRAF6 deficiency results in osteopetrosis and defective interleukin-1, CD40 and LPS signaling. *Genes Dev.* **13**, 1015–1024.

Maeda, S., Hsu, L.-C., Liu, H., *et al.* (2005). Nod2 mutation in Crohn's disease potentiates NF-κB activity and IL-1β processing. *Science* **307**, 734–738.

Mahida, Y. R., Makh, S., Hyde, S., Gray, T., and Borriello, S. P. (1996). Effect of *Clostridium difficile* toxin A on human intestinal epithelial cells: induction of interleukin-8 production and apoptosis after cell detachment. *Gut* **38**, 337–347.

Massari, P., Ram, S., Macleod, H., and Wetzler, L. M. (2003). The role of porins in neisserial pathogenesis and immunity. *Trends Microbiol.* **11**, 87–93.

McCormick, B. A., Colgan, S. P., Delp-Archer, C., Miller, S. I., and Madara, J. L. (1993). *Salmonella typhimurium* attachment to human intestinal epithelial monolayers: transcellular signalling to sub-epithelial neutrophils. *J. Cell Biol.* **123**, 895–907.

Meylan, E., Burns, K., Hofmann, K., *et al.* (2004). RIP1 is an essential mediator of Toll-like receptor 3-induced NF-κB activation. *Nat. Immunol.* **5**, 503–507.

Mikamo, H., Johri, A. K., Paoletti, L. C., Madoff, L. C., and Onderdonk, A. B. (2004). Adherence to, invasion by, and cytokine production in response to serotype VIII group B streptococci. *Infect. Immun.* **72**, 4716–4722.

Miyazawa, M., Suzuki, H., Masaoka, T., *et al.* (2003). Suppressed apoptosis in the inflamed gastric mucosa of *Helicobacter pylori*-colonized iNOS-knockout mice. *Free Radic. Biol. Med.* **34**, 1621–1630.

Mordmuller, B., Krappmann, D., Esen, M., Wegener, E., and Scheidereit, C. (2003). Lymphotoxin and lipopolysaccharide induce NF-κB p52 generation by a co-translational mechanism. *EMBO Rep.* **4**, 82–87.

Neish, A. S., Gewirtz, A. T., Zeng, H., *et al.* (2000). Prokaryotic regulation of epithelial responses by inhibition of IκBα ubiquitination. *Science* **289**, 1560–1563.

Neurath, M. F., Pettersson, S., Meyer zum Buschenfelde, K. H., and Strober, W. (1996). Local administration of antisense phosphorothioate oligonucleotides to the p65 subunit of NF-κB abrogates established experimental colitis in mice. *Nat. Med.* **2**, 998–1004.

Obonyo, M., Guiney, D. G., Harwood, J., Fierer, J., and Cole, S. P. (2002). Role of gamma interferon in *Helicobacter pylori* induction of inflammatory mediators during murine infection. *Infect. Immun.* **70**, 3295–3299.

Ogura, Y., Bonen, D. K., Inohara, N., *et al.* (2001). A frameshift mutation in Nod2 associated with susceptibility to Crohn's disease. *Nature* **411**, 603–606.

Ogushi, K., Wada, A., Niidome, T., *et al.* (2001). *Salmonella enteritidis* FliC (flagella filament protein) induces human β-defensin 2 mRNA production by Caco-2 cells. *J. Biol. Chem.* **276**, 30 521–30 526.

O'Neil, D. A., Porter, E. M., Elewaut, D., *et al.* (1999). Expression and regulation of the human beta-defensins hBD-1 and hBD-2 in intestinal epithelium. *J. Immunol.* **163**, 6718–6724.

Orth, K., Palmer, L. E., Bao, Z. Q., *et al.* (1999). Inhibition of the mitogen-activated protein kinase kinase superfamily by a *Yersinia* effector. *Science* **285**, 1920–1923.

Orth, K., Xu, Z., Mudgett, M. B., *et al.* (2000). Disruption of signaling by *Yersinia* effector YopJ, a ubiquitin-like protein protease. *Science* **290**, 1594–1597.

Ouellette, A. J. (1997). Paneth cells and innate immunity in the crypt microenvironment. *Gastroenterology* **113**, 1779–1784.

Palfreyman, R. W., Watson, M. L., Eden, C., and Smith, A. W. (1997). Induction of biologically active interleukin-8 from lung epithelial cells by *Burkholderia (Pseudomonas) cepacia* products. *Infect. Immun.* **65**, 617–622.

Patke, A., Mecklenbrauker, I., and Tarakhovsky, A. (2004). Survival signaling in resting B cells. *Curr. Opin. Immunol.* **16**, 251–255.

Petrof, E. O., Kojima, K., Ropeleski, M. J., *et al.* (2004). Probiotics inhibit NF-κB and induce heat shock proteins in colonic epithelial cells through proteasome inhibition. *Gastroenterology* 2004, 1474–1487.

Platz, J., Beisswenger, C., Dalpke, A., *et al.* (2004). Microbial DNA induces a host defense reaction of human respiratory epithelial cells. *J. Immunol.* **173**, 1219–1223.

Pothoulakis, C. and Lamont, J. T. (2001). Microbes and microbial toxins: paradigms for microbial-mucosal interactions II: the integrated response of the intestine to *Clostridium difficile* toxins. *Am. J. Physiol. Gastrointest. Liver Physiol.* **280**, G178–G183.

Rachmilewitz, D., Katakura, K., Karmeli, F., *et al.* (2004). Toll-like receptor 9 signaling mediates the anti-inflammatory effects of probiotics in murine experimental colitis. *Gastroenterology* **126**, 520–528.

Rakoff-Nahoum, S., Paglino, J., Eslami-Varzaneh, F., Edberg, S., and Medzhitov, R. (2004). Recognition of commensal microflora by Toll-like receptors is required for intestinal homeostasis. *Cell* **118**, 229–241.

Ray, A. and Biswas, T. (2005). Porin of *Shigella dysenteriae* enhances Toll-like receptors 2 and 6 of mouse peritoneal B-2 cells and induces the expression of IgM, IgG2a and IgA. *Immunology* **114**, 94–100.

Regenhard, P., Goethe, R., and van Phi, L. (2001). Involvement of PKA, PKC, and Ca2+ in LPS-activated expression of the chicken lysozyme gene. *J. Leukoc. Biol.* **69**, 651–658.

Resta-Lenert, S. and Barrett, K. E. (2002). Enteroinvasive bacteria alter barrier and transport properties of human intestinal epithelium: role of iNOS and COX-2. *Gastroenterology* **122**, 1070–1087.

Rosenstiel, P., Fantini, M., Brautigam, K., *et al.* (2003). TNFα and IFNγ regulate the expression of the Nod2 (CARD15) gene in human intestinal epithelial cells. *Gastroenterology* **124**, 1001–1009.

Sakamoto, N., Mukae, H., Fujii, T., *et al.* (2005). Differential effects of alpha- and beta-defensin on cytokine production by cultured human bronchial epithelial cells. *Am. J. Physiol. Lung Cell. Mol. Physiol.* **288**, L508–L513.

Sakiri, R., Ramagowda, B., and Tesh, V. L. (1998). Shiga toxin type 1 activates TNFα gene transcription and nuclear translocation of the transcriptional activators NF-κB and AP-1. *Blood* **92**, 558–566.

Sanjo, H., Takeda, K., Tsujimura, T., *et al.* (2003). TAB2 is essential for prevention of apoptosis in fetal liver but not for interleukin-1 signaling. *Mol. Cell. Biol.* **23**, 1231–1238.

Sarkar, M. and Chaudhuri, K. (2004). Association of adherence and motility in interleukin-8 induction in human intestinal epithelial cells by *Vibrio cholerae*. *Microbes Infect.* **6**, 676–685.

Sartor, R. B. (2004). Therapeutic manipulation of the enteric microflora in inflammatory bowel diseases: antibiotics, probiotics and prebiotics. *Gastroenterology* **126**, 1620–1633.

Schesser, K., Spiik, A. K., Dukuzumuremyi, J. M., *et al.* (1998). The yopJ locus is required for *Yersinia*-mediated inhibition of NF-κB activation and cytokine expression: yopJ contains a eukaryotic SH2-like domain that is essential for its repressive activity. *Mol. Microbiol.* **28**, 1067–1079.

Schmid, Y., Grassl, G. A., Buhler, O. T., *et al.* (2004). *Yersinia enterocolitica* adhesin A induces production of IL-8 in epithelial cells. *Infect. Immun.* **72**, 6780–6789.

Senftleben, U., Cao, Y., Xiao, G., *et al.* (2001). Activation by IKKα of a second, evolutionarily conserved, NF-κB signaling pathway. *Science* **293**, 1495–1499.

Sierro, F., Dubois, B., Coste, A., *et al.* (2001). Flagellin stimulation of intestinal epithelial cells triggers CCL20-mediated migration of dendritic cells. *Proc. Natl. Acad. Sci. U. S. A.* **98**, 13 722–13 727.

Strober, W. (1998). Interactions between epithelial cells and immune cells in the intestine. *Ann. N. Y. Acad. Sci.* **859**, 37–45.

Sun, L., Deng, L., Ea, C. K., Xia, Z. P., and Chen, Z. J. (2004). The TRAF6 ubiquitin ligase and TAK1 kinase mediate IKK activation by BCL10 and MALT1 in T lymphocytes. *Mol. Cell* **14**, 289–301.

Suzuki, N., Suzuki, S., Duncan, G. S., *et al.* (2002). Severe impairment of interleukin-1 and Toll-like receptor signaling in mice lacking IRAK-4. *Nature* **416**, 750–756.

Vallance, B. A., Deng, W., De Grado, M., *et al.* (2002). Modulation of inducible nitric oxide synthase expression by the attaching and effacing bacterial pathogen *Citrobacter rodentium* in infected mice. *Infect. Immun.* **70**, 6424–6435.

Vallance, B. A., Dijkstra, G., Qiu, B., *et al.* (2004). Relative contributions of NOS isoforms during experimental colitis: endothelial-derived NOS maintains mucosal integrity. *Am. J. Physiol. Gastrointest. Liver Physiol.* **287**, G865–G874.

Van Phi, L. (1996). Transcriptional activation of the chicken lysozyme gene by NF-κB p65 (RelA) and c-Rel, but not by NF-κB p50. *Biochem. J.* **313**, 39–44.

Viala, J., Chaput, C., Boneca, I. G., *et al.* (2004). Nod1 responds to peptidoglycan delivered by the *Helicobacter pylori* cag pathogenicity island. *Nat. Immunol.* **5**, 1166–1174.

Viboud, G. I., So, S. S., Ryndak, M. B., and Bliska, J. B. (2003). Proinflammatory signaling stimulated by the type III translocation factor YopB is counteracted by multiple effectors in epithelial cells infected with *Yersinia pseudotuberculosis. Mol. Microbiol.* **47**, 1305–1315.

Wahl, C., Liptay, S., Adler, G., and Schmid, R. M. (1998). Sulfasalazine: a potent and specific inhibitor of NF-κB. *J. Clin. Invest.* **101**, 1163–1174.

Wehkamp, J., Harder, J., Weichenthal, M., *et al.* (2004). Nod2 (CARD15) mutations in Crohn's disease are associated with diminished mucosal α-defensin expression. *Gut* **53**, 1658–1664.

Wehkamp, J., Schmid, M., Fellermann, K., and Stange, E. F. (2005). Defensin deficiency, intestinal microbes, and the clinical phenotypes of Crohn's disease. *J. Leukoc. Biol.* **77**, 460–465.

Wilson, C. L., Ouellette, A. J., Satchell, D. P., *et al.* (1999). Regulation of intestinal alpha-defensin activation by the metalloproteinase matrilysin in innate host defense. *Science* **286**, 113–117.

Witthoft, T., Eckmann, L., Kim, J. M., and Kagnoff, M. F. (1998). Enteroinvasive bacteria directly activate expression of iNOS and NO production in human colon epithelial cells. *Am. J. Physiol.* **275**, G564–G571.

Wroblewski, L. E., Noble, P. J., Pagliocca, A., *et al.* (2003). Stimulation of MMP-7 (matrilysin) by *Helicobacter pylori* in human gastric epithelial cells: role in epithelial cell migration. *J. Cell Sci.* **116**, 3017–3026.

Yamamoto, M., Sato, S., Hemmi, H., *et al.* (2002). Essential role for TIRAP in activation of the signaling cascade shared by TLR2 and TLR4. *Nature* **420**, 324–329.

Yamamoto, M., Sato, S., Hemmi, H., *et al.* (2003a). Role of adaptor TRIF in the MyD88-independent Toll-like receptor signaling pathway. *Science* **301**, 640–643.

Yamamoto, M., Sato, S., Hemmi, H., *et al.* (2003b). TRAM is specifically involved in the Toll-like receptor 4-mediated MyD88-independent signaling pathway. *Nat. Immunol.* **4**, 1144–1150.

Yang, D., Chertov, O., and Oppenheim, J. J. (2001). Participation of mammalian defensins and cathelicidins in anti-microbial immunity: receptors and activities of human defensins and cathelicidin (LL-37). *J. Leukoc. Biol.* **69**, 691–697.

Yang, S. K., Eckmann, L., Panja, A., and Kagnoff, M. F. (1997). Differential and regulated expression of C-X-C, C-C, and C-chemokines by human colon epithelial cells. *Gastroenterology* **113**, 1214–1223.

Zhou, H., Wertz, I., O'Rourke, K., *et al.* (2004). Bcl10 activates the NF-κB pathway through ubiquitination of NEMO. *Nature* **427**, 167–171.

CHAPTER 10

NF-κB-independent responses activated by bacterial–epithelial interactions: the role of arachidonic acid metabolites

Beth A. McCormick and Randall J. Mrsny

INTRODUCTION

Lipid membranes and the individual lipids that comprise them were initially considered to solely provide eukaryotic cells with organized hydrophobic barriers used to separate cytoplasmic and extracellular environments. Additional studies demonstrated that these lipid bilayer structures also acted as boundaries for discrete intracellular structures, e.g. mitochondria, endosomes, and endoplasmic reticulum. Although this capacity to separate aqueous compartments clearly is an essential feature of normal cell structure and function, more recent studies have demonstrated that lipid components in these bilayer membranes also provide cells with substrates to produce a spectrum of intra- and extracellular messengers. Metabolism of membrane lipid components has been shown to produce bioactive lipids that participate in numerous signaling mechanisms. Many of these bioactive lipids, such as prostaglandins, leukotrienes, hydroperoxy acids, hepoxilins, lipoxins, and thromboxanes, are derived from the metabolic processing of arachidonic acid.

Arachadonic acid, a 20-carbon fatty acid that contains four carbon–carbon double bonds, is the precursor substrate used for the production of a large family of bioactive lipids known as eicosanoids (Fitzpatrick and Soberman, 2001; Lieb, 2001) (Figure 10.1). By itself, arachidonic acid can act as a second messenger by its ability to interact with GTP-binding proteins (Abramson *et al.*., 1991), inhibit GTPase-activating protein regulated by RAS (Ras-GAP) function (Han *et al.*, 1991), cause the release of Ca^{2+} ions stored in the sarcoplasmic reticulum (Dettbarn and Palade, 1993), and modulate protein kinase C (PKC) activity (Khan *et al.*, 1995). The various arachadonic acid

Bacterial–Epithelial Cell Cross-Talk: Molecular Mechanisms in Pathogenesis, ed. Beth A. McCormick. Published by Cambridge University Press.

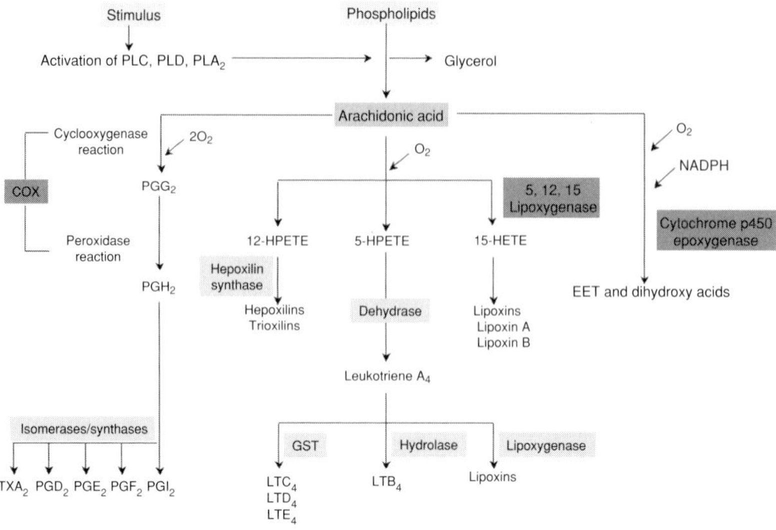

Figure 10.1 Major metabolic pathways involved in arachidonic acid metabolism. Arachidonic acid is stored in the cell membrane of virtually all cells and is released in response to stimuli such as phospholipase C (PLC), phospholipase D (PLD), and phospholipase A_2 (PLA_2). The cyclooxygenase (COX) pathway results in the formation of prostaglandin G_2 (PGG_2) from arachidonic acid through a COX reaction. During a subsequent peroxidase reaction, PGG_2 is converted to prostaglandin H_2 (PGH_2) by undergoing a two-electron reduction. PGH_2 can then serve as a substrate for cell-specific isomerases and synthases producing eicosanoids such as prostacyclin (PGI_2), thromboxane A_2 (TXA_2), and a host of prostaglandins (PGE_2, PGF_2, PGD_2). The lipoxygenase pathway forms hydroperoxyeicosatetraenoic acids (HPETEs) and dihydroxyeicosatetraenoic acid (DEA) by lipoxygenase and subsequently converts these to leukotrienes (LTC_4, LTD_4, LTE_4, LTB_4) by hydrolase glutathione S-transferase (GST), and by lipoxygenases to form lipoxins. Lipoxins can also be formed directly from 15-lipoxygenase (15-LO) via DEA. The epoxygenase pathway forms epoxyeicosatrienoic acid (EET) and dihydroxy acids by cytochrome P450 epoxygenase.

NADPH, nicotinamide adenine dinucleotide phosphate (reduced).

metabolites that are eicosanoids, however, have been shown to be extremely potent, often initiating cellular responses at nanomolar concentrations. Subtle differences in the nature of the eicosanoid produced from arachidonic acid can lead to strikingly different cellular outcomes and stability to subsequent metabolic breakdown. For example, thromboxane A_2 (TXA_2) is a labile platelet-aggregating factor (Moncada and Vane, 1978), while prostaglandin E_1 (PGE_1) inhibits platelet aggregation (Ney *et al.*, 1991), and yet both are produced from arachidonic acid and by a metabolic pathway that shares several intermediates (Kirtland, 1988; Sinzinger *et al.*, 1990). Similarly, leukotriene A_4 (LTA_4) can stimulate the mobilization of Ca^{2+} (Luscinskas *et al.*, 1990), while lipoxin A_4 can block Ca^{2+} mobilization (Grandordy *et al.*, 1990), and these molecules differ by the conversion of an epoxide moiety to a pair of hydroxyl groups. Thus, regulation of enzymatic pathways that use arachidonic acid and its metabolites can produce bioactive lipids that allow for highly selective cell functions.

Arachidonic acid does not typically reside at high concentrations in cells but rather appears to be generated de novo in response to some stimulus via phospholipase A_2 (PLA_2) activity (Leslie, 2004). PLA_2 is a (typically) cytosolic hydrolase that adds water across the ester linkage at the sn-2 position of membrane phospholipids, resulting in the release of a free fatty acid, e.g. arachidonic acid, and the production of a lysophospholipid. In general, PLA_2 can hydrolyze phospholipids at the sn-2 position that have a range of head-group components, including choline, serine, and ethanolamine (Akiba and Sato 2004; Bahnson, 2005; Hirabayashi *et al.*, 2004). Alternative regulatory pathways, such as mitogen-activated protein kinase actions and Ca^{2+} stimulation, can regulate PLA_2 activity and arachidonic acid release (Gijon and Leslie, 1999), establishing several methods by which to regulate this crucial mechanism for eicosanoid production (Fitzpatrick and Soberman, 2001). Thus, a complex arrangement of functions including PLA_2 activity, arachidonic acid, and enzymes involved in arachidonic acid metabolism producing various bioactive lipids regulates a variety of cell functions.

Arachadonic acid metabolites have been shown to drive pro-inflammatory events through their capacity to activate and attract cells associated with inflammation such as neutrophils and eosinophils and to vasodilate and increase vascular permeability. Although some bioactive lipids have been shown to produce opposing actions such as vasoconstriction and thus function as an anti-inflammatory molecule, in general these pro-inflammatory bioactive lipids are extremely labile and can act to drive inflammatory outcomes for only the duration of time that they are being actively produced. Such a system of potentially counteracting anti-inflammatory bioactive lipids and the rapid degradation of highly potent pro-inflammatory bioactive lipids provides one scenario whereby cells can establish a set point that is balanced between pro-inflammatory and anti-inflammatory outcomes. This dynamic arrangement should protect tissues from undesirable aspects of inflammatory events but yet be poised to respond rapidly to stimuli that warrant cellular responses involving inflammation (reviewed by Sansonetti (2004)). Since epithelial cells are constantly bombarded by a wide range of pathogens, stabilization of this set point can be particularly problematic. Disruption of this normal balance, in particular with respect to arachidonic acid metabolites, has been correlated with diseases associated with chronic inflammation, such as Crohn's disease (Bouchelouche *et al.*, 1995; Casellas *et al.*, 1994; Engel *et al.*, 1988; Ikehata *et al.*, 1992). Thus, interactions between bacteria and epithelial cells define a critical dynamic that modulate this set point of pro-/anti-inflammatory agents, and this set point could easily involve regulatory events involving eicosanoid synthesis achieved through arachidonic metabolism along with other factors known to affect inflammation, such as interleukins and cytokines.

PRO-INFLAMMATORY RESPONSES: BACTERIAL–EPITHELIAL INTERACTIONS

The intestinal epithelium and *Salmonella* Typhimurium

Salmonella enterica serotype Typhimurium, a Gram-negative bacterium of the family Enterobacteriaceae, causes a variety of diseases in humans and other animal hosts. Non-typhoidal serovars of *Salmonella* Typhimurium produce an acute gastroenteritis, characterized by intestinal pain and usually non-bloody diarrhea, and are a serious public-health problem in developing countries (Hook, 1990). The interaction of *S.* Typhimurium with the intestinal mucosa results in the classic induction of an early inflammatory response characterized by infiltration of neutrophils through the intestinal mucosa and into the intestinal lumen. This event culminates in the formation of an intestinal crypt abscess (Kumar *et al.*, 1982; McGovern and Slavutin, 1979; Rout *et al.*, 1974). The action of these neutrophils on the epithelium and the subsequent loss of barrier function are thought to play key roles in mediating the clinical manifestations of *S.* Typhimurium-induced enteritis. Neutrophil infiltration, for example, is associated with necrosis of the upper mucosa and large areas of the terminal ileum and colon. This damage likely occurs as a result of neutrophils releasing substances that provoke tissue injury, such as proteases, myeloperoxidase, and nicotinamide adenine dinucleotide phosphate (reduced) (NADPH) oxidase. Impairment of the intestinal epithelium and resulting loss in barrier function coupled with the ability of neutrophils to stimulate chloride secretion in the intestinal epithelium contributes to heightened levels of fluid secretion and, hence, the development of diarrhea. The role of fluid secretion in host defense remains to be determined definitively. However, the principle involves rapid and increased flow of fluid over the epithelial surface, which actively flushes organisms from the intestinal lumen. In the case of *S.* Typhimurium, such a flushing response by the host is presumably required to clear the infection and prevent systemic spread of the pathogen. Circumstantial evidence on which such a concept was built is based on the observation that infection by enteric bacteria may be prolonged by antimotility agents (DuPont and Hornick, 1973). The advantage of the flushing response to the microbe is the increased potential for transmission to other hosts.

Identification of the mechanisms that govern the induction of this early inflammatory response have been the subject of intense investigation. To elucidate some of the key host–pathogen communication networks involved in recruiting neutrophils across the intestinal epithelium, our prior work has recognized that the intestinal epithelium is not merely a barrier to neutrophil movement but rather, in response to *S*. Typhimurium, intestinal epithelial cells direct neutrophil movement via the polarized secretion of chemoattractants (McCormick, *et al.*, 1993, 1995; Mrsny *et al.*, 2004). Specifically, *S*. Typhimurium activates the transcription factor nuclear factor κB (NF-κB), resulting in the epithelial synthesis and basolateral release of a potent neutrophil chemokine, interleukin-8 (IL-8) (McCormick *et al.*, 1995). Such basolateral secretion of IL-8 recruits neutrophils through the lamina propria to the subepithelial space but is not involved in neutrophil migration across epithelial tight junctions (TJ), the final step of crypt abscess formation. In fact, the absolute insufficiency of basolateral chemokines, such as IL-8, alone to direct such neutrophil transmigration is illustrated by the many stimuli, such as purified flagellin, tumor necrosis factor alpha (TNFα), and carbachol, that potently induce these chemokines but do not result in neutrophil movement across the epithelium (Gewirtz *et al.*, 2001). Thus, although proficient at initially luring neutrophils to the subepithelial region, the basolateral gradient of IL-8 is not necessary or sufficient to guide neutrophils across the intestinal epithelium into the intestinal lumen.

Based on the above observations, it appears that additional factors might be necessary for such events to occur. As well as producing chemokines, intestinal epithelial cells have been found to produce eicosanoids that play key roles in host defense. Eicosanoids are produced from the release of arachidonic acid from the cell membrane and its enzymatic conversion by various cyclooxygenases and lipoxygenases. Noteworthy in this regard is our work concerning the neutrophil chemoattractant previously termed pathogen-elicited epithelial chemoattractant (PEEC) (McCormick, *et al.*, 1998), which we identified as the eicosinoid hepoxilin A_3 (HXA_3) (Mrsny *et al.*, 2004). HXA_3 is generated in response to pathogenic infection when arachidonic acid is converted into 12-hydroperoxyeicosatetraenoic acid (12-HpETE) by 12-lipoxygenase (12-LO), followed by its transformation into HXA_3 by the putative enzyme hepoxilin synthase. 12-LO inhibitors impede the ability of intestinal epithelial cells to generate HXA_3, thereby suppressing neutrophil migration across intestinal epithelia (Mrsny *et al.*, 2004). Release of HXA_3 has been demonstrated in intestinal epithelial cells infected with *Salmonella*. The type III secretion system (TTSS) effector protein SipA is the bacterial molecule that promotes the response (Lee *et al.*, 2000). HXA_3 is released from the apical surface of *Salmonella*-infected intestinal epithelial cells and successfully generates a gradient progressing from the mucosal side of the epithelium, strongly attracting neutrophils into the lumen (Mrsny *et al.*, 2004). Therefore, IL-8 and other neutrophil chemottractants released basolaterally most likely work in concert with the apically released HXA_3 and guide neutrophils across the intestinal epithelium.

Although HXA_3 has been shown to induce chemotaxis of neutrophils at concentrations as low as 30–40 nM (Sutherland *et al.*, 2000), our work was the first to demonstrate that HXA_3 can be secreted from epithelial cells, specifically from the apical surface, and that this secretion is regulated by conditions that contribute to inflammation. In the intestinal epithelium, HXA_3 is generated independent of NF-κB, since inhibitors of this pathway have no effect on *Salmonella*-induced neutrophil migration across model intestinal epithelia. Instead, the regulation of *S.* Typhimurium neutrophil transmigration is governed by a novel mechanism involving the GTPase ADP-ribosylation factor 6 (ARF6) (Criss *et al.*, 2001). In this model, *S.* Typhimurium contacting the apical surface of polarized epithelial cells elicits a signal through the *S.* Typhimurium bacterial effector of the TTSS SipA, which recruits an ARF6 guanine-exchange factor, such as ARF nucleotide binding site opener (ARNO), to the apical plasma membrane. ARNO facilitates ARF6 activation at the apical membrane, which in turn stimulates phospholipase D (PLD) recruitment to and activity at this site. The PLD product, phosphatidic acid, is metabolized by a phosphohydrolase into diacylglycerol, which recruits cytosolic protein kinase C alpha (PKC-α) to the apical membrane. Activated PKC-α phosphorylates downstream targets that are responsible for the production and apical release of HXA_3, which drives transepithelial movement of neutrophils. At present, ARF6 is the first example of a molecular switch specifically modulating the neutrophil transmigration aspect of *S.* Typhimurium pathogenesis independently of bacterial internalization or the release of other proinflammatory mediators such as IL-8.

ARF6 is a member of the Arf subgroup of GTPases of the Ras superfamily, which were first defined as cofactors necessary for the cholera toxin-catalyzed ADP ribosylation of the α_s subunit of the heterotrimeric G-proteins. ARF6, the only class III ARF, is novel in its localization to the plasma membrane and endosomal structures in non-polarized cells (D'Souza-Shorey *et al.*, 1995) and is highly expressed in polarized cells, where it localizes primarily to the apical brush border and apical early endosomes (Altschuler *et al.*, 1999; Londono *et al.*, 1999). Although ARF6 was initially identified as a modulator of vesicular traffic and cortical actin cytoarchitecture, more recent evidence suggests that ARF6 is also a regulator of signal-transduction cascades initiated by a variety of external stimuli, and it appears to be a key signal transducer downstream of G-protein-coupled receptors (Venkateswarlu and Cullen, 2000). Even though the identity of the cellular receptor for *Salmonella* and/or its effector SipA is not known, these observations suggest that bacterial adherence to the apical membrane of polarized epithelial cells promotes ARF6 in an analogous manner.

Increased intestinal fluid secretion is an important host innate defense mechanism and, coupled with intestinal motility, plays an important role in flushing enteric bacteria from the intestinal tract. Infection of intestinal epithelial cells with *Salmonella* spp. results in increased epithelial cell chloride secretion. One pathway in which this occurs involves the upregulated expression of cyclooxygenase 2 (COX-2), with the subsequent increase in production of the prostanoid prostaglandin E_2 (PGE_2). Prostaglandins are formed from free arachidonic acid through the conversion of arachidonic acid to prostaglandin H (PGH), which is catalyzed by COX. PGH is thereafter converted by specific synthases to PGE, PGF, thromboxanes, or prostacyclins (Eberhart and DuBois, 1995; Smith and DeWitt, 1996) (Figure 10.1). Several studies have implicated prostaglandins in *Salmonella*-induced enteropathogenic responses that may influence chloride secretory pathways. One study determined that pretreatment of rabbits with indomethacin, an inhibitor of prostaglandin biosynthesis, abolishes fluid secretion in ligated loops (Wallis *et al.*, 1990). Furthermore, PGE_1 stimulates chloride secretion in polarized cells, and PGE analog were found to induce diarrhea in vivo. Another study evaluated the role of intestinal epithelium in the secretory response after infection with *Salmonella* (Eckmann *et al.*, 1997). In this investigation, infection of intestinal epithelial cells with *Salmonella* resulted in a rapid upregulation in prostaglandin H synthase (PGHS) expression, which subsequently stimulated the production of PGE_2 and $PGF_{2\alpha}$ and consequently led to an increase in chloride secretion (Eckmann *et al.*, 1997). Therefore, PGE_2 released by intestinal epithelial cells in response to *Salmonella* infection acts in an autocrine/paracrine manner and stimulates chloride secretion by exploiting the cyclic adenosine monophosphate (cAMP)-mediated pathway. Other invasive bacteria, such as *Shigella dysenteriae*, *Salmonella dublin*, *Salmonella typhi*, *Yersinia enterocolitica*, and enteroinvasive

Escherichia coli, but not non-pathogenic bacteria, have also been found to induce the expression of PGHS-2 and its products PGE_2 and $PGF_{2\alpha}$, resulting in chloride secretion.

The gastric mucosa and *Helicobacter*

Helicobacter pylori is a Gram-negative organism that colonizes the gastric mucosa where, without appropriate therapy, it may persist in life with associated widespread gastric inflammation (Nabwera and Logan, 1999). Although the majority of infected individuals have asymptomatic chronic gastritis, a small percentage of individuals will go on to develop peptic ulcer disease within months or years and gastric cancer over decades (Goldstone *et al.*, 1996). Current evidence substantiates that *Helicobacter* species are the indigenous biota of mammalian stomachs and *H. pylori* is the human-specific inhabitant, having been present for at least tens of thousands of years. Consequently, the dual hallmarks of the interaction between *H. pylori* and humans are (i) its persistence during the life of the host and (ii) the host's responses to its continuing presence. Even though this conflict appears paradoxical, both the host and the microbe have adapted to each other by engaging a long-standing dynamic equilibrium. As a demonstration of this unique relationship, the role of prostaglandins in *H. pylori* infection is highlighted here.

Prostaglandins are found in the stomach of many species, including humans, and are synthesized by COX from arachidonic acid. Prostaglandins are not stored within tissues, but with an appropriate stimulus they can be actively synthesized and released from the mucosal cells into the gastric lumen. COX exists in at least two isoforms: COX-1 is expressed in the gastrointestinal tract, as well as in most other organs, and is constitutively expressed to maintain the gastric mucosal integrity, primarily under physiological conditions as a housekeeping enzyme. By contrast, COX-2 is undetectable or expressed at very low levels in most tissues but is significantly induced at sites of inflammation, such as *H. pylori*-induced gastric mucosal inflammation (Fu *et al.*, 1999; Gilroy *et al.*, 1995). Thus, an important parameter for increased prostaglandin production in response to *H. pylori* infection appears to be the induction of COX-2 expression.

In the stomach, PGE_2 plays an important role in the maintenance of gastric mucosal integrity and cytoprotection by modulating diverse cellular functions, which prevent the gastric mucosa from becoming inflamed and necrotic (i.e. stimulating mucus secretion and bicarbonate production, stimulating the synthesis of surface-active phospholipids, enhancing the mucous gel thickness, and inhibiting acid secretion) (Wallace, 1992). It has been well documented that gastric mucosal PGE_2 production is higher in *H. pylori*-positive patients than in *H. pylori*-negative patients (Avunduk *et al.*, 1991; Jackson, *et al.*, 2000). This presents an interesting challenge to the organism, since although *H. pylori* infection is associated with gastric ulcers and cancers at the same time it can also induce the production of cytoprotective PGE_2. Such a counterintuitive strategy for *H. pylori* infection might actually be a clever mechanism in which the organism enhances the gastric mucous-layer protective functions for its own fortification, thereby permitting an established colonization to occur that induces severe gastritis, leading to acute gastric disease and increased risk of gastric cancer.

In 1989, a strain-specific *H. pylori* gene, *cagA*, was identified (Cover *et al.*, 1990). This has been recognized as a marker for strains that confer increased risk of peptic ulcer disease and gastric cancer (Blaser *et al.*, 1995; Crabtree *et al.*, 1991). No homologs are known for *cagA* in other *Helicobacter* species or in other bacteria, suggesting that it reflects a human gastric-specific gene. The *cagA* gene is harbored within a 40-kilobase region of DNA, which represents a pathogenicity island (PAI) and encodes 31 genes. Upon contact with the gastric epithelium, PAI-encoded components contribute in a specialized type IV machinery that translocates the CagA protein into the eukaryotic target cell, where it is phosphorylated on tyrosine residues (Odenbreit *et al.*, 2000). This breakthrough led to the question of whether the increased levels of PGE_2 could be an initial step in *H. pylori* $CagA^+$ infection to enhance the gastric mucous-layer protective functions, favoring colonization of the gastric mucosa. In the only clinical study so far to address this issue, it was found that in age- and sex-matched patients without atrophy of the gastric mucosa, those infected with $CagA^+$ *H. pylori* had significantly higher levels of PGE_2 as compared with either *H. pylori*-negative patients or $CagA^-$ *H. pylori*-infected patients (Al-Marhoon *et al.*, 2004). Moreover, increased levels of PGE_2 correlated with increased *H. pylori* density. Therefore, it is possible that in the presence of $CagA^+$ *H. pylori* infection, higher levels of PGE_2 may play a role in the establishment of *H. pylori* colonization, leading to severe gastric disease.

There are at least two potential mechanisms by which *H. pylori* could increase PGE_2 production. As discussed above, one mechanism for increased prostaglandin production in response to *H. pylori* infection is an induction of COX-2 expression. Such a mechanism is substantiated by the finding that *H. pylori* infection induces the expression of COX-2 in the gastric mucosa, whereas eradication of *H. pylori* results in reduced gastric antral mucosal COX-2 expression (Cryer, 2001). The second model is based on a mechanism involving interleukin 1 (IL-1). The foundation of this model is centered on the observation that *H. pylori* infection stimulates the release of the chemokine IL-8 (Sharma *et al.*, 1995). Release of IL-8 promotes the infiltration of neutrophils and macrophages within the gastric mucosa, which in turn produces IL-1, leading to acute gastritis (Crabtree, 1998). Interestingly, IL-1 has been found to stimulate PGE_2, leading to mucin secretion and gastric cytoprotection against infection (Dinarello, 1986).

A rate-limiting step in the control of prostaglandin production is the release of arachidonic acid from membrane phospholipids known to occur by distinct pathways. The predominant avenue involves the activation of phospholipase A_2 (PLA_2), while others involve the action of phospholipase C or D (Exton, 1994; Leslie, 1997; Nishizuka, 1992). Evidence suggests that colonization of epithelial cells by *H. pylori* induces the release of PGE_2 and arachidonic acid by activation of the cytosolic PLA_2 ($cPLA_2$) through a mechanism involving pertussis toxin-sensitive heterotrimeric $G_{\alpha}/G_{\alpha o}$ proteins and the p38 stress-activated kinase cascade (Pomorski *et al.*, 2001).

Among the various types of mammalian PLA_2, the $cPLA_2$ isoform plays a key role in the release of arachidonic acid from cell membranes. Arachidonic acid can then go on to serve as a substrate for the production of prostaglandins. $cPLA_2$ has high substrate specificity for arachidonic acid at the sn-2 position of phospholipids and requires both elevated Ca^{2+} intracellular concentration and a phosphorylation step for activation (Hong and Deykin, 1981; Leslie, 1997). Therefore, release of arachidonic acid for prostaglandin production by activation of $cPLA_2$ could play an important role in mucosal defense to bacterial infection, since gastric epithelial cells are the first site of contact with *H. pylori*. However, prolonged activation of $cPLA_2$ is likely to be damaging to the gastric epithelia by excessive degradation of membrane phospholipids releasing arachidonic acid and lysophospholipids. Therefore, the identification of *H. pylori*-specific signaling pathways leading to the induction of putative anti-inflammatory immune response mediators is of substantial interest for therapeutic intervention to overcome *H. pylori*-induced diseases.

H. pylori infection is the most important environmental factor implicated in the etiology of gastric carcinoma. Among the target genes upregulated by *H. pylori*, COX-2 also appears to play an important role in the progression to gastric cancer. It has been shown that expression and activity of P-glycoprotein (P-gp) are modulated by COX-2. This relationship was initially recognized in breast cancer (Ratnasinghe *et al.*, 2001) and has been subsequently confirmed in vivo (Patel *et al.*, 2002). P-gp, the main product of the multi-drug resistance 1 gene (MDR1) belongs to the superfamily of ATP-binding cassette transporters and actively effluxes a wide range of toxic agents and structurally diverse amphipathic drugs from cells. Therefore, P-gp expression represents one of the most important mechanisms for the failure of chemotherapeutic treatment of cancer.

In the gastric mucosa, the expression of COX-2 is upregulated by *H. pylori* infection, and several reports have documented overexpression of COX-2 in gastric precancerous lesions and in gastric cancer (Tatsuguchi, *et al.*, 2000; Wambura *et al.*, 2002). The COX-2 tumor-promoting effect is mediated partly through the production of PGE_2, the product of microsomal prostaglandin E synthase 1 (mPGES1). mPGES1 is a member of the membrane-associated proteins in eicosanoid and glutathione metabolism superfamily; it catalyzes the isomeration of endoperoxide PGH_2 into the 9-keto and 11-hydroxy groups of PGE_2 (Fitzpatrick and Soberman 2001; Thoren and Jakobsson, 2000). A possible link of mPGES1 with tumorgenesis has been suggested by the finding that it is constitutively expressed in several tumors that also express COX-2 (Kamei *et al.*, 2003).

The inflammatory response to *H. pylori* infection is presumably the trigger for the elevated levels of COX-2, mPGES1, and P-gp. Indeed, in one study none of the patients with *H. pylori*-negative chronic dyspepsia showed expression of these proteins. Furthermore, high levels of expression of COX-2 and mPGES1 were detected in up to 78% of the intestinal type gastric cancer samples, whereas P-gp, which is generally undetectable in normal gastric mucosa, is overexpressed in up to 50% of gastric cancers (Nardone *et al.*, 2004). Such overexpression of P-gp could account for the failure of eradication therapy. Failure of eradication therapy, which occurs in more than 30% of cases, is an emerging problem that is not explained completely by bacterial sensitivity to antibiotic therapy and patient compliance. Thus, intrinsic or acquired cellular resistance to antimicrobial agents may be involved, presumably through the expression of the MDR1 gene. In sum, these observations may help to explain one of the possible molecular pathways by which *H. pylori* could be implicated in gastric cancer development. Furthermore, understanding the mechanism of COX-2 and mPGES1 expression may also provide attractive targets for developing strategies for gastric cancer chemoprevention and chemotherapy.

Bacterial infections involving the lung epithelium

Cystic fibrosis

Cystic fibrosis (CF) is the most prevalent lethal autosomal recessive disorder in the caucasian population, affecting 1 in 2500 newborns. Patients with CF express a typical phenotype characterized by pancreatic insufficiency, ileal hypertrophy, and recurrent pulmonary infections that ultimately lead to pulmonary failure and death (Lyczak *et al.*, 2002). However, the chronic pulmonary infection with *Pseudomonas aeruginosa* allows the basic pathologic process in CF to be designated an infectious disease. Ultimately, 80–95% of patients with CF succumb to respiratory failure brought on by chronic bacterial infection and concomitant airway inflammation (Lyczak *et al.*, 2002).

In 1989, the gene whose mutation results in CF was identified and cloned (Kerem *et al.*, 1989; Riordan *et al.*, 1989). The product of the gene, the cystic fibrosis transmembrane conductance regulator (CFTR), was characterized as an ATP-gated chloride channel regulated by cAMP-dependent protein kinase phosphorylation (Li *et al.*, 1989). This discovery fostered an explosion of research efforts that have led to a greater understanding of the molecular mechanisms underlying the various phenotypic manifestations of the disease. Yet, despite the significant advances made in CF research in recent years, the mechanism by which a mutation in the *CFTR* gene leads to the presentation of this disease remains unclear. Although a decrease in the apical membrane CFTR-dependent chloride conductance might explain some of the pathological manifestations observed in CF, such as viscous secretions, it explains neither the increased inflammation in the lung nor the membrane-recycling defects associated with cystic fibrosis.

A number of reports have supported the contention that essential-fatty-acid abnormalities observed in patients with CF may play a fundamental role in the symptoms and progression of the disease (Strandvik, 1992). Indeed, abnormal turnover of essential fatty acids has been reported by several groups, and increased release of arachidonic acid has been described in different in vitro and in vivo systems. For instance, among newborn infants with CF and in prospective studies of infants with CF identified by neonatal screening, low linoleic acid concentrations are present at birth and are more pronounced in infants who present with meconium ileus (Li *et al.*, 1989). Furthermore, alteration in levels of certain fatty acids has been shown to affect the lung response to infection. As an example, animals placed on diets deficient in essential fatty acids acquire a bronchitis in which macrophage function in general, but the ability to mount a specific lymphocytic response to *P. aeruginosa* is impaired (Craig-Schmidt *et al.*, 1986; Harper *et al.*, 1982).

Arachidonic acid, an agonist of inflammatory pathways and a stimulant of mucosal secretion, is also elevated in the phospholipid fraction from bronchial alveolar lavage fluid in people with CF (Gilljam *et al.*, 1986). Although the increased inflammation and elevated arachidonic acid levels observed in CF have long been thought as secondary to infection, this conclusion has been challenged (Heeckeren *et al.*, 1997). Heeckeren *et al.* (1997) demonstrated that instillation of agarose beads coated with *P. aeruginosa* into the lungs of *cftr*$^{-/-}$ mice resulted in increased inflammation and mortality compared with that observed in wild-type mice. These findings suggest that the lungs of *cftr*$^{-/-}$ mice are primed for inflammation and that the increase in arachidonic acid and inflammation observed in *cftr*$^{-/-}$ mice may be a primary event and not secondary to infection. This finding and the observation that mice with a targeted deletion of the *cftr* gene have an increased ratio of phospholipid-bound arachidonic acid relative to phospholipid-bound docosahexaenoic acid (DHA) in the lung, pancreas, and ileum (Freedman *et al.*, 2002), has led to speculation that abnormalities in the metabolism of arachidonic acid and DHA may be primary in CF. Furthermore, DHA may be reversing a specific abnormality related to CFTR dysfunction, since oral administration of DHA results in a selective decrease in eicosanoids in *cftr*$^{-/-}$ mice that is not observed in wild-type mice (Freedman *et al.*, 2002). To this end, the elevated levels of phospholipid-bound arachidonic acid may be responsible for the increased production of eicosanoids. Thus, DHA, by competing for incorporation at the sn-2 position in the membrane phospholipids, decreases arachidonic acid levels, thereby lowering the production of eicosanoids (Freedman *et al.*, 2002).

Although the mechanism by which CFTR regulates fatty-acid metabolism is unknown, it is thought that the low DHA levels may also be essential for the excessive inflammatory response in CF. DHA is converted to docosatrienes and 17S series resolvins, which are potent anti-inflammatory mediators normally generated during the resolution of inflammation (Hong *et al.*, 2003; Serhan *et al.*, 2002). Therefore, the low DHA levels in people with CF may explain, at least in part, the inflammatory state associated with this disease and may have important therapeutic implications. Interestingly, oral administration of high doses of DHA to *cftr*$^{-/-}$ mice was found to correct the fatty-acid abnormality, reverse the histologic changes in the pancreas and ileum, and decrease polymorphonuclear cell (PMN) levels in mice with *Pseudomonas* lipopolysaccharide (LPS)-induced pneumonia (Freedman *et al.*, 2002). Other studies have shown that defective secretion of lipoxin A_4 into CF airways may represent another important defect in regulating pulmonary *P. aeruginosa* infections in people with CF (Karp *et al.*, 2004).

In CF, the chloride-transport defect in airway epithelium and submucosal glands is somehow translated into chronic bacterial infection and excessive inflammation, which are the proximate causes of lung destruction and, ultimately, death of the patient. One of the pathogenic hallmarks of CF is the accumulation of large numbers of neutrophils (i.e. PMNs) in the lumen of the lower airway (McIntosh, 2002; Pizurki *et al.*, 2000; Weiss, 1989). Neutrophils present in the lumen can aid in eradicating offending bacteria via potent killing mechanisms, including release of proteases and reactive oxygen species. However, these mechanisms are non-specific and can lead to lung tissue damage, which, if excessive, contributes to the pathology of the disease (Burns *et al.*, 2003).

In order to reach the airway lumen, neutrophils are required to travel through several distinct tissue compartments within the alveolar wall (Smart and Casale, 1994; Wagner and Roth, 2000). This process involves the integrated actions of cytokines and adhesion molecules with specificity for particular ligands, and highly timed and compartmentalized secretion of various neutrophil-specific chemokines, such as IL-8 (DiMango *et al.*, 1995; Smart and Casale, 1994; Wagner and Roth, 2000). Despite significant progress in this area of study, the specific molecular mechanism governing neutrophil migration through each compartment to the lumen remains to be defined. A review by Burns *et al.* (2003) provides an instructive breakdown of issues governing the overall neutrophil-recruitment process from the capillaries to the airway lumen, which involves transendothelial migration, transmatrix migration, migration through the interstitium adjacent to fibroblasts, and transepithelial transmigration. Although it is clear that several different species relevant to lung inflammation demonstrate the ability to induce neutrophil transmigration across lung epithelial cells, how this process is orchestrated mechanistically remains an important unanswered question.

New information is beginning to shed light on this complex process. One successful approach to study this process has been to construct reductionist in vitro models for a distinct phase in the neutrophil recruitment process, i.e. transepithelial migration. Several reports using this tactic have determined that *P. aeruginosa* and/or products of *P. aeruginosa* are capable of initiating the secretion of the neutrophil chemokine IL-8 (Hurley *et al.*, 2004; Hybiske *et al.*, 2004; Pizurki *et al.*, 2000; Weiss, 1989). This phenomenon is believed to be important in the neutrophil-recruitment process, but a report by Hurley *et al.* (2004) suggests that although polarized monolayers of lung epithelia produce IL-8 in response to *P. aeruginosa* infection, IL-8 does not appear to be responsible for mediating the transepithelial migration of the neutrophils. This conclusion is based on two key observations: (i) the majority of IL-8 produced in response to *P. aeruginosa* is secreted into the basolateral chamber of the polarized lung monolayers, which would be counter to the direction of neutrophil migration; and (ii) neutralizing antibodies to IL-8 have no affect on *P. aeruginosa*-induced neutrophil transmigration. This result is consistent with a report of the inability of IL-8-neutralizing antibodies to prevent *Chlamydia pneumoniae*-induced neutrophil transmigration across primary human airway epithelial cells (Jahn *et al.*, 2000). Based on this evidence, it appears that IL-8 does not contribute to neutrophil migration across infected lung epithelial monolayers. However, since IL-8 does contribute to overall neutrophil recruitment during lung infection and inflammation (DiMango *et al.*, 1995; Smart and Casale, 1994; Wagner and Roth, 2000) it is likely that the primary role played by IL-8 during neutrophil infiltration of lung tissue is to guide neutrophils through the interstitium up to the epithelium rather than to guide neutrophils across the epithelium into the airspace.

How, then, do neutrophils migrate across the lung epithelium during *P. aeruginosa* infection? One concept is derived from observations describing a novel inflammatory pathway first reported for the enteric pathogen *S. typhimurium*. In this earlier study, it was demonstrated that the chemoattractant responsible for *S. typhimurium*-induced neutrophil migration across intestinal epithelial monolayers is the eicosinoid HXA_3. HXA_3 is an arachidonic acid metabolite produced by the 12-LO pathway, and it has been shown to stimulate Ca^{2+} release within neutrophils as well to possess neutrophil chemotactic activity (McCormick *et al.*, 1998; Mrsny *et al.*, 2004). Similar to the intestinal mucosa, interaction of *P. aeruginosa* with airway epithelial cells was shown to activate PKC, resulting in the production and secretion of HXA_3 through the action of the 12-LO enzymatic pathway. Consequently, lung epithelial cells produce HXA_3 in response to bacterial infection, and this neutrophil chemoattractant is responsible for guiding neutrophils across model lung epithelia.

Bacterial pneumonia

Pneumonia is the most common infectious disease leading to death in industrialized countries (Garibaldi, 1985). Over 40% of cases are due to infection with *Streptococcus pneumoniae*. High mortality has been reported (Finch, 2001). At the same time, antibiotic-resistant strains have emerged (Heffelfinger *et al.*, 2000). Although some pathogenic factors have been identified (Tuomanen *et al.*, 1995), little is known about the activation of signaling cascades in target cells.

Leukotrienes (LT), like the well-known prostaglandins and thromboxanes, constitute a family of lipid mediators with potent biological activities. In the biosynthesis of LTs, the enzyme 5-lipoxygenase (5-LO) converts arachidonic acid into the unstable epoxide LTA_4, a central intermediate in LT biosynthesis (Figure 10.1) (Ford-Hutchinson *et al.*, 1994) . LTA_4 may be hydrolyzed into LTB_4 by the enzyme LTA_4 hydrolase (Ford-Hutchinson *et al.*, 1980) or conjugated with glutathione to form LTC_4, a reaction catalyzed by a specific LTC_4 synthase (Nicholson *et al.*, 1993). Therefore, two distinct biochemical pathways leading to LTB_4 and LTC_4 synthesis compete for LTA_4 as a substrate. LTC_4 is actively transported across the plasma membrane by MRP1; extracellularly, LTD_4 and LTE_4 are formed by the action of gamma-glutamyl-transpeptidase and dipeptidase, respectfully (Leier *et al.*, 1994). Of note, the cysteinyl-containing LTs (LTC_4, LTD_4, LTE_4) have been shown to contract smooth muscles, particularly in the peripheral airways, and are regarded as pivotal mediators of bronchial asthma. In the microcirculation, these LTs have also been reported to increase the permeability in post-capillary venules, which leads to extravasation of plasma (Byrum *et al.*, 1999). Most recently, the three cysteinyl LTs have been implicated as mediators of host defense against bacterial infection, especially in the lung (Schultz *et al.*, 2001).

MRP1 is a member of the ABC transporter superfamily and is best known for the cellular excretion of anti-cancer drugs. However, glutathione-S-conjugates of lipophilic xenobiotics and the cysteinyl LTs feature prominently among the physiological substrates of this molecular pump. Therefore, MRP1 is an important regulator of cysteinyl LT bioavailability (Ishikawa, 1982). Although the importance of MRP1-dependent LT extrusion for host defense against pulmonary infection remains ill-defined, the generation of *mrp1*$^{-/-}$ mice has helped to characterize the function of this cellular pump in vivo. For instance, *mrp1*$^{-/-}$ mice appear to be hypersensitive to the anti-cancer drug etoposide and also have an impaired response to arachidonic-acid-induced inflammatory stimulus (Wijnholds *et al.*, 1997). In addition, bone-marrow-derived mast cells from *mrp1*$^{-/-}$ mice have been reported to exhibit a marked reduction in the capacity to excrete LTC_4, a defect that has been associated with intracellular accumulation of LTC_4, confirming the essential role of MRP1 in the secretion of cysteinyl LTs. Schultz *et al.* (2001) evaluated the pathophysiologic relevance of this defect in a well-established model of pneumonia induced by *S. pneumoniae*, the most frequently isolated pathogen in patients with community-acquired pneumonia. In this study, it was determined that mice lacking functional MRP1 are less sensitive to pneumonia caused by *S. pneumoniae* through a mechanism related to the function of MRP1 as a cellular pump for LTC_4 and that involved an increase in LTB_4 secretion (Schultz *et al.*, 2001).

The local production of LTs, in particular LTB_4, has been considered important for an effective host defense against invading pathogens in the pulmonary compartment. The alveolar macrophage is a major source of LTB_4 in the lung (Bigby and Holtzman, 1987), where this lipid mediator can stimulate microbicidal activities of phagocytic cells (Bailie *et al.*, 1996). 5-LO$^{-/-}$ mice, which have a general deficiency of all LTs, displayed an enhanced lethality from *Klebsiella* pneumonia in association with an increased outgrowth of bacteria in the lungs (Bailie *et al.*, 1996). However, *mrp1*$^{-/-}$ mice behaved in an opposite manner to 5-LO$^{-/-}$ mice during pneumonia induced by *S. pneumoniae*. In particular, Schultz *et al.* (2001) confirmed that by treating *mrp1*$^{-/-}$ mice with a 5-LO-activating protein (FLAP) inhibitor, the increase in resistance to *S. pneumoniae* infection was related to a selective disruption of the cysteinyl LT pathway. Indeed, administration of the FLAP inhibitor abolished the survival advantage of *mrp1*$^{-/-}$ mice, indicating that the relative protection of these mice was mediated by a function of MRP1 related to LT metabolism. Furthermore, *mrp1*$^{-/-}$ mice were unable to release significant quantities of LTC_4 in bronchoalveolar lavage fluids (BALF) during pneumonia (Schultz *et al.*, 2001).

High levels of LTB_4 have been measured in lung tissue and BALF derived from animals infected with *P. aeruginosa* or *Klebsiella pneumonia*, as well as BALF from patients with pneumonia. Therefore, somewhat unexpectedly, LTB_4 concentrations in BALF were higher in *mrp1*$^{-/-}$ mice as compared with wild-type mice following early induction of pneumonia. Such increased levels of LTB_4 are apparently essential for the protection of *mrp1*$^{-/-}$ mice, since treatment with the LTB_4 receptor antagonist LTB_4-dimethyl amide reversed the survival advantage of *mrp1*$^{-/-}$ mice (Schultz *et al.*, 2001). Along the same lines, an investigation found higher levels of LTB_4 in colon homogenates of *mrp1*$^{-/-}$ mice than in wild-type mice in a model of 2,4,6-trinitrobenzene sulfonic acid-induced colitis. Thus, an emerging concept is that the increased intracellular LTC_4 levels due to the absence of MRP1 enhance LTB_4 generation, which in turn leads to enhanced resistance of pneumoncoccal pneumoniae infection. Such an enhancement in LTB_4 synthesis might be explained by the fact that intracellular LTC_4 accumulation may give rise to product inhibition of LTC_4 synthase, thus removing substrate competition between LTC_4 synthase and LTA_4 hydrolase (the LTB_4-producing enzyme) for LTA_4 (the precursor for both LTB_4 and LTC_4), yielding enhanced LTB_4 synthesis (Schultz *et al.*, 2001).

ARACHIDONIC ACID METABOLITES

Several classes of bioactive lipids appear to dominate events involving bacterial-mediated pro-inflammatory events through their effects on epithelial cell and neutrophil cell function; including prostaglandins, leukotrienes, and hepoxilins (Figure 10.1). Arachidonic acid is an omega-6-fatty acid, since it contains four double bonds, with the one closest to the CH_3 terminus being six carbons from the end. As has been discussed, actions of such bioactive lipids can occur through controlling the release of the arachidonic acid by

PLA_2, regulation of metabolic pathway enzymes, and activation of cellular secretion systems. Since most bioactive lipids are degraded rapidly, control of their function by regulation of catabolism mechanisms does not appear to be a dominant aspect of controlling their actions. For example, many of these bioactive lipids are formed through the enzymatic introduction of an –OOH function positioned at carbon 5, 12, or 15 by specific lipoxygenase activities (Figure 10.1). Once positioned at a specific carbon, this –OOH moiety can be transformed into an epoxide structure that positions a single oxygen atom to share bonding interactions with two adjacent carbon atoms simultaneously. Such an epoxide structure is extremely labile not only to non-specific addition of water (particularly at acidic pH) but also to the actions of epoxide hydrolases present within cells. The resulting dialcohol frequently demonstrates a dramatic decline in biological activity relative to the peroxide-containing molecule.

Arachidonic acid and the bioactive lipids produced from arachidonic acid metabolism have chemical characteristics that allow them to partition between hydrophobic lipid bilayers and aqueous environments. With such potential for dynamic partitioning, how can cells modulate the production of these extremely potent and highly unstable molecules? For instance, a bioactive lipid may be inactivated before its ability to generate a biological outcome if it is synthesized in the wrong part of the cell. To reduce this possibility and to improve efficiency, cells appear to organize events such as arachidonic acid liberation through PLA_2 activation, arachidonic acid metabolism by lipoxygenases, etc., and stimulation of cellular efflux pathways. Specifically, recruitment of PLA_2 activity from the cytosol to the membrane bilayers can be regulated through phosphorylation events, interleukins, and intracellular Ca^{2+} signaling events (Balsinde *et al.*, 1999; Leslie, 2004; Pascual *et al.*, 2003). Localization of substrate-specific enzymes, either in discrete cells or in tissues, can modulate the actions of arachadonic acid metabolites by regulating sites of their synthesis (Seeds and Bass, 1999). Some enzymes involved in arachidonic acid metabolism can be localized to discrete intracellular sites, such as to the nuclear envelope in the case of 5-lipoxygenase (Chen and Funk, 2001). Release of nascent bioactive lipids generated from arachidonic acid metabolites can be regulated through efflux pathways, as in the case of prostaglandin secretion through MRP4 (Reid *et al.*, 2003). Similarly, LTC_4 secretion appears to occur through MRP1 (Robbiani *et al.*, 2000) or MRP2 (Kawabe *et al.*, 1999).

Bioactive lipids derived from arachidonic acid establish a network of cellular responses that can act in both pro-inflammatory and anti-inflammatory ways. The production and actions of these bioactive lipids can be complex. For example, although LTA_4 can mobilize Ca^{2+} in neutrophils (Luscinskas *et al.*, 1990), it is an unstable molecule in cells and acts primarily as an intermediate in the biosysnthesis of LTB_4 and LTC_4. Subsequent to its formation, LTC_4 can be modified to form LTD_4 and then to LTE_4 in order to produce potent actions on a variety of cell types (Arm and Lee, 1994). While the LTA_4–LTE_4 synthetic pathway produces a series of molecules that can mobilize cellular Ca^{2+} levels, LTA_4 can also be metabolized to form lipoxin A_4, which can block Ca^{2+} mobilization (Grandordy *et al.*, 1990). Thus, it is critical that specific pathways of arachidonic acid metabolism are regulated tightly in order to ensure that the proper response is generated from a particular stimulus. Some of this regulation can be achieved through cell-specific mechanisms that involve the discrete expression of receptor response elements, or a particular complement of synthetic enzymes in specific cells or tissues, or specific cellular response components, or even some combination of these mechanisms (Turk *et al.*, 1982). Thus, specific responses to any event that results in the release of arachidonic acid can be regulated by myriad factors (Kurahashi *et al.*, 2003; Monjazeb *et al.*, 2002), and it is not surprising that a number of bacterial pathogens have now been shown to produce factors that specifically disrupt or shift the complex balance of functions established by bioactive lipids produced by epithelial cells.

BACTERIA THAT POSSESS ARACHIDONIC-ACID-METABOLIZING ENZYMES

Although polyunsaturated fatty acids such as arachidonic acid are widespread in mammalian cell membranes, most bacteria are believed to lack lipoxygenases and their polyunsaturated fatty acid substrates (Moss *et al.*, 1972; Wilkinson, 1988). However, it has been discovered that the *P. aeruginosa* strain PA1169 encodes a bacterial lipoxygenase (LoxA) that converts arachidonic acid into 15-hydroxy-eicosatetraenoic acid (15-HETE) (Vance *et al.*, 2004). Until this report, the secretable LoxA enzyme activity was generally believed to be absent from bacteria.

As a rule, LOs catalyze the stereospecific abstraction of hydrogen and insertion of molecular oxygen at a specific fatty-acid carbon–carbon double bond position to form lipid hydroperoxidases that are reduced rapidly to alcohols or transformed further to potent mediators. LoxA appears to differ from mammalian LOs by virtue of an N-terminal signal sequence that targets LoxA for secretion (Vance *et al.*, 2004). The only other secreted LO described to date is the manganese-containing LO of the plant pathogen referred to as "take-all fungus." Nearly all LoxA activity seems to be secreted, at least to the periplasm, with some activity further secreted to the extracellular environment in a manner dependent on the Xcp type II secretion apparatus. By contrast, the majority of 15-HETE produced is cell-free (Vance *et al.*, 2004).

Although *P. aeruginosa* is perhaps best known for the chronic lung infections that are the most significant cause of morbidity and mortality in people with CF, *P. aeruginosa* is also a significant cause of serious infections in immunocompromised cancer patients, burn patients, catheterized patients, and other hospitalized individuals. Thus, *P. aeruginosa* 15-LO seems well-positioned to act on exogenous human-derived substrates, thereby potentially modulating the local inflammatory responses during *P. aeruginosa* infection. Given that the cytotoxin Exo U (exo enzyme U) secreted by certain *P. aeruginosa* strains has phospholipase activity (Sato *et al.*, 2003), it is tempting to speculate that the action of ExoU may result in increased availability of arachidonic acid at local sites of *P. aeruginosa* infection. In addition, several other bacterial PLAs exist (Songer, 1997).

These observations raise the possibility that production of anti-inflammatory lipid mediators may be a general strategy by which pathogens regulate the host–pathogen relationship. At present, there are limited examples to substantiate this notion. Apart from *P. aeruginosa*, only four other microorganisms seem to contain LO-like sequences – *Nitrosomonas europaea*, an obligate chemolithoautotroph; *Anabaena* sp. strain PCC 7120, a cyanobacterium; *Sorangium cellulosum*, a soil bacterium; and *Nostoc punctiforme*, a cyanobacterium related closely to *Anabaena* sp). However, none of these putative bacterial LOs has been characterized. In addition, the parasite *Toxoplasma gondii* has been found to possess a 15-LO; such exogenous 15-LO was found to be anti-inflammatory in vivo (Bannenberg *et al.*, 2004). It is also interesting that *P. aeruginosa* can secrete a 15-LO activity that could act to modulate host defense and inflammatory events by altering the biosynthesis of bioactive lipids generated by lung epithelial cells in people with CF (Vance *et al.*, 2004).

The absence of LO from most bacteria raises the interesting possibility that LoxA might have been acquired horizontally from eukaryotes, although the mechanism of such a putative horizontal transfer remains in question. Nevertheless, the acquisition of eukaryotic-like enzymes by *P. aeruginosa* is not unprecedented, because a eukaryotic-like PLD has also been described in *P. aeruginosa* (Wilderman *et al.*, 2001). Why some microorganisms are equipped to synthesize eukaryotic anti-inflammatory chemical mediators is an important question that awaits further investigation.

CONCLUSION

A variety of factors contribute to the complex course of inflammation. In the early phase of inflammation, excessive amounts of cytokines and inflammatory mediators are released. These factors activate, in addition to other pathways, the lipid-synthesis pathway, which plays a crucial role in the pathogenesis of many chronic and acute inflammatory illnesses. As exemplified in this chapter, interactions between bacteria and epithelial cells underscore a critical dynamic that modulates the balance between pro- and anti-inflammatory events. Understanding how host–pathogen interactions tip the balance of regulatory events that involve the synthesis of eicosanoids achieved through arachidonic acid metabolism will shed new light on the topic of disease pathogenesis, which will be essential for the development of novel drug strategies aimed at thwarting many inflammatory-based diseases.

ACKNOWLEDGMENTS

We thank Drs Bryan Hurley and Karen Mumy for their constructive comments on the manuscript. Dr McCormick is supported by National Institutes of Health Grants DK56754 and DK33506.

REFERENCES

Abramson, S. B., Leszczynska-Piziak, J., and Weissmann, G. (1991). Arachidonic acid as a second messenger: interaction with a TP-binding protein of human neutrophils. *J. Immunol.* **147**, 231–236.

Akiba, S. and Sato, T. (2004). Cellular function of calcium-independent Phospholipase A2. *Biol. Pharm. Bull.* **27**, 1174–1178.

Al-Marhoon, M. S., Nunn, S., and Soames, R. W. (2004). CagA+ *Helicobacter pylori* induces greater levels of prostaglandin E2 than CagA(−) strains. *Prostaglandins Other Lipid Mediat.* **7**, 181–189.

Altschuler, Y., Liu, S., Katz, L., *et al.* (1999). ADP-ribosylation factor 6 and endocytosis at the apical surface of Madin-Darby canine kidney cells. *J. Cell Biol.* **147**, 7–12.

Arm, J. P. and Lee, T. H. (1994). Evidence for a specific role of leukotriene E_4 in asthma and airway hyperresponsiveness. *Adv. Prostaglandin Thromboxane Leukot. Res.* **22**, 227–240.

Avunduk, C., Suliman, M., Gang, D., Polakowski, N., and Eastwood, G. L. (1991). Gastroduodenal mucosal prostaglandin generation in patients with

Helicobacter pylori before and after treatment with bismuth subsalicylate. *Dig. Dis. Sci.* **36**, 431–434.

Bahnson, B. J. (2005). Structure, function and interfacial allosterism in phospholipase A2: insight from the anion-assisted dimmer. *Arch. Biochem. Biophys.* **433**, 96–106.

Bailie, M. B., Staniford, T. J., Laichalk, L. L., *et al.* (1996). Leukotriene deficient mice manifest enhanced lethality from *Kleibsiella* pneumonia in association with decreased alveolar macrophage phagocytic and bacterial activities. *J. Immunol.* **157**, 5221–5224.

Balsinde, J., Balboa, M. A., Insel, P. A., and Dennis, E. A. (1999). Regulation and inhibition of phospholipase A2. *Annu. Rev. Pharmacol. Toxicol.* **39**, 175–189.

Bannenberg, G. L., Aliberti, J., Hong, S., Sher, A., and Serhan, C. N. (2004). Exogenous pathogen and plant 15-lipoxygenase initiate endogenous lipoxin A4 biosynthesis. *J. Exp. Med.* **199**, 515–523.

Bigby, T. D. and Holtzman, M. J. (1987). Enhanced 5-lipoxygenase activity in lung macrophages compared to monocytes from animal subjects. *J. Immunol.* **138**, 1546–1550.

Blaser, M. J., Perez-Perez, G. I., Kleanthous, H., *et al.* (1995). Infection with *Helicobacter pylori* strains possessing *cagA* associated with an increased risk of developing adenocarcinoma of the stomach. *Cancer Res.* **55**, 2111–2115.

Bouchelouche, P. N., Berild, D., Nielson, O. H., Elmgreen, J., and Poulsen, H. S. (1995). Leukotriene B_4 receptor levels and intracellular calcium signaling in polymorphonuclear leukocytes from patients with Crohn's disease. *Eur. J. Gastroenterol. Hepatol.* **7**, 349–356.

Burns A. R., Smith, C. W., and Walker, D. C. (2003). Unique structural features that influence neutrophil emigration into the lung. *Physiol. Rev.* **83**, 309–336.

Byrum, R. S., Goulet, J. L., Snouwaert, J. N., Griffiths, J. R., and Koller, B. H. (1999). Determination of the contribution of cysteinyl leukotrienes and leukotriene B_4 in acute inflammatory responses using 5-lipoxygenase- and leukotriene A_4 hydrolase-deficient mice. *J. Immunol.* **163**, 6810–6819.

Casellas, F., Guarner, F., Antolin, M., *et al.* (1994). Abnormal leukotriene C4 released by unaffected jejunal mucosa in patients with inactive Crohn's disease. *Gut* **35**, 517–522.

Chen, X. S. and Funk, C. D. (2001). The N-terminal beta barrel domain of 5-lipoxygenase is essential for nuclear membrane translocation. *J. Biol. Chem.* **276**, 811–818.

Cover, T. L., Dooley, C. P., and Blaser, M. J. (1990). Characterization and human serologic response to proteins in *Helicobacter pylori* broth culture supernatants with vacuolizing cytotoxin activity. *Infect. Immun.* **58**, 603–610.

Crabtree, J. E. (1998). Role of cytokines in the pathogenesis of *Helicobacter pylori*-induced mucosal damage. *Dig. Dis. Sci.* **43**, 46S–55S.

Crabtree, J. E., Taylor, J. D., Wyatt, J. I., *et al.* (1991). Mucosal IgA regulation of *Helicobacter pylori* 120 K Da protein, peptic ulceration, and gastric pathology. *Lancet* **338**, 332–335.

Craig-Schmidt, M. C., Faircloth, S. A., Teer, P. A., Weete, J. D., and Wu, C.-Y. (1986). The essential fatty acid deficient chicken as a model for cystic fibrosis. *Am. J. Clin. Nutr.* **44**, 816–824.

Criss, A. K., Silva, M., Casanova, J. E., and McCormick, B. A. (2001). Regulation of *Salmonella*-induced neutrophil transmigration by epithelial ADP-ribosylation factor 6. *J. Biol. Chem.* **276**, 48 431–48 439.

Cryer, B. (2001). Mucosal defense and repair: role of prostaglandins in the stomach and duodenum. *Gastronterol. Clin. North Am.* **30**, 877–894.

Dettbarn, C. and Palade, P. (1993). Arachidonic acid-induced Ca^{2+} release from isolated sarcaoplasmic reticulum. *Biochem. Pharmacol.* **45**, 1301–1309.

DiMango E, Zar, H. J., Bryan, R., and Prince, A. (1995). Diverse *Pseudomonas aeruginosa* gene products stimulate respiratory epithelial cells to produce interleukin-8. *J. Clin. Invest.* **96**, 2204–2210.

Dinarello, C. A. (1986). Multiple biological activities of human recombinant interleukin-1. *Immunobiology* **172**, 301–315.

D'Souza-Shorey, C, Li, G., Colombo, M. I., and Stahl, P. D. (1995). A regulatory role for ARF6 in receptor-mediated endocytosis. *Science* **267**, 1175–1178.

DuPont, H. L. and Hornick, R. B. (1973). Adverse effect of lomotil therapy in shigellosis. *J. Am. Med. Assoc.* **226**, 1525–1528.

Eberhart, C. E. and DuBois, R. N. (1995). Eicosanoids and the gastrointestinal tract. *Gastroenterology* **109**, 285–301.

Eckmann, L., Stenson, W. F., Savidge, T. C., *et al.* (1997). Role of intestinal epithelial cells in the host secretory response to infection by invasive bacteria. *J. Clin. Invest.* **100**, 296–309.

Engel, L. D., Pasquinelli, K. L., Leone, S. A., *et al.* (1988). Abnormal lymphocyte profiles and leukotriene B_4 in a patient with Crohn's disease and severe periodontitis. *J. Periodontol.* **59**, 841–847.

Exton, J. H. (1994). Phosphoinositide phospholipases and G proteins in hormone action. *Annu. Rev. Physiol.* **56**, 349–369.

Finch, R. (2001). Community-acquired pneumonia: the evolving challenge. *Clin. Microbiol. Infect.* **7**, 30–38.

Fitzpatrick, F. A. and Soberman, R. (2001). Regulated formation of eicosinoids. *J. Clin. Invest.* **107**, 1347–1351.

Ford-Hutchinson, A. W., Bray, M. A., Doig, M. V., Shipley, M. E., and Smith, M. J. (1980). Leukotriene B_4, a potent chemokinetic and aggregating substance released from polymorphonuclear leukocytes. *Nature* **286**, 264–265.

Ford-Hutchinson, A. W., Gresser, M., and Young, R. N. (1994). 5-Lipoxygenase. *Annu. Rev. Biochem.* **63**, 383–417.

Freedman, S. D., Katz, M. H., Parker, E. M., *et al.* (1999). A membrane lipid imbalance plays a role in the phenotypic expression of cystic fibrosis in cftr−/− mice. *Proc. Natl. Acad. Sci. U. S. A.* **96**, 13 995–14 000.

Freedman, S. D., Weinstein, D., Blanco, P. G., *et al.* (2002). Characterization of LPS-induced lung inflammation in cftr−/− mice and the effect of docosahexaenoic acid. *J. Appl. Physiol.* **92**, 2169–2176.

Fu, S., Ramanujam, K. S., Wong, A., *et al.* (1999). Increased expression and cellular localization of inducible nitric oxide synthase and cyclooxygenase 2 in *Helicobacter pylori* gastritis. *Gastroenterology* **116**, 1319–1329.

Garibaldi, R. A. (1985). Epidemiology of community-acquired respiratory tract infections in adults: incidence, etiology, and impact. *Am. J. Med.* **78**, 32–37.

Gewirtz, A. T., Simon, P. O., Scmitt, C. K., *et al.* (2001). *Salmonella typhimurium* translocates flagellin across intestinal epithelia, inducing a proinflammatory response. *J. Clin. Invest.* **107**, 99–109.

Gijon, M. A. and Leslie, C. C. (1999). Regulation of arachidonic acid release and cytosolic phospholipase A_2 activation. *J. Leukoc. Biol.* **65**, 330–336.

Gilljam, H., Strandvik, B., Ellin, A., and Wiman, L. G. (1986). Increased mole fraction of arachidonic acid in bronchial phospholipids in patients with cystic fibrosis. *Scand. J. Clin. Lab. Invest.* **46**, 511–518.

Gilroy, D. W., Tomlinson, A., and Willoughby, D. A. (1995). Differential effects of inhibition of isoforms of cyclooxygenase (COX-1, COX-2) in chronic inflammation. *Inflamm. Res.* **47**, 79–85.

Goldstone, A. R., Quirke, P., and Dixon, M. F. (1996). *Helicobacter pylori* infection and gastric cancer. *J. Pathol.* **179**, 129–137.

Grandordy, B. M., Lacroix, H., Mavoungou, E., *et al.* (1990). Lipoxin A_4 inhibits phosphoinositide hydrolysis in human neutrophils. *Biochem. Biophys. Res. Commun.* **167**, 1022–1029.

Han, J., McCormick, F., and Macara, I. G. (1991). Regulation of Ras-GAP and the neurofibromatosis-gene product by eicosanoids. *Science* **252**, 576–579.

Harper, T. B., Chase, H. R., Henson, J., and Henson, P. M. (1982). Essential fatty acid deficiency in the rabbit as a model of nutritional impairment in cystic fibrosis. *In vitro* and *in vivo* effects on lung defense mechanisms. *Am. Rev. Respir. Dis.* **126**, 540–547.

Heeckeren, A., Walenga, R., Konstan, M. W., *et al.* (1997). Excessive inflammatory response of cystic fibrosis mice to bronchopulmonary infection with *Pseudomonas aeruginosa. J. Clin. Invest.* **100**, 2810–2815.

Heffelfinger, J. D., Dowell, S. F., Jorgenssen, J. H., *et al.* (2002). Management of community-acquired pneumonia in the era of pneumococcal resistance: a report from the Drug-Resistant *Streptococcus pneumoniae* Therapeutic Working Group. *Arch. Intern. Med.* **160**, 1399–1408.

Hirabayashi, T., Maurayama, T., and Shimizu, T. (2004). Regulatory mechanism and physiological role of cytosolic phospholipase A2. *Biol. Pharm. Bull.* **27**, 1168–1173.

Hong, S. L. and Deykin, D. (1981). The activation of phosphoinositol-hydrolyzing phospholipase A2 during prostaglandin synthesis in transformed mouse BALB/3T3 cells. *J. Biol. Chem.* **256**, 5215–5219.

Hong, S., Gronert, K., Devchand, P. R., Moussignac, R.-L., and Serhan, C. N. (2003). Novel docosatrienes and 17S-resolvins generated from docosahexaenoic acid in murine brain, human blood, and glial cells: autocoids and anti-inflammation. *J. Biol. Chem.* **278**, 14 677–14 687.

Hook, E. W. (1990). *Salmonella* species (including typhoid fever). In *Principles and Practice of Infectious Diseases*, ed. G. L. Mandell, R. G. Douglas, and J. E. Bennet. New York: Churchill Livingston, pp. 1700–1716.

Hurley, B. P., Siccardi, D., Mrsny, R. J., and McCormick, B. A. (2004). PMN transepithelial migration induced by *Pseudomonas aeruginosa* requires the eicosinoid hepoxilin A_3. *J. Immunol.* **173**, 5712–5720.

Hybiske K, Ichikawa, J. K., Huang, V., Lory, S. J., and Machen, T. E. (2004). Cystic fibrosis airway epithelial cell polarity and bacterial flagellin determine host response to *Pseudomonas aeruginosa. Cell Microbiol.* **6**, 49–63.

Ikehata, A., Hiwatashi, N., Kinouchi, Y., *et al.* (1992). Effect of intravenously infused eicosapentaenoic acid on the leukotriene generation in patients with active Crohn's disease. *Am. J. Clin. Nutr.* **56**, 938–942.

Ishikawa, T. (1992). The ATP-dependent glutathione S-conjugate export pump. *Trends Biochem. Sci.* **17**, 463–468.

Jackson, L. M., Wu, K. C., Mahida, Y. R., Jenkins, D., and Hawkey, C. J. (2000). Cyclooxygenase (COX) 1 and 2 in normal, inflamed, and ulcerated human gastric mucosa. *Gut* **47**, 762–770.

Jahn, H. U., Krull, M., Wuppermann, F. N., *et al.* (2000). Infection and activation of airway epithelial cells by Chlamydia. *J. Infect. Dis.* **182**, 1678–1684.

Kamei, D., Murakami, M., Nakatani, Y., *et al.* (2003). Potential role of microsomal prostaglandin E synthase-1 in tumorigenesis. *J. Biol. Chem.* **267**, 6428–6432.

Karp, C. L., Flick, L. M., Park, K. W., *et al.* (2004). Defective lipoxin-mediated anti-inflammatory activity in the cystic fibrosis pathway. *Nat. Immunol.* **5**, 388–392.

Kawabe, T., Chen, Z. S., Wada, M., *et al.* (1999). Enhanced transport of anticancer agents and leukotriene C_4 by the human canalicular multispecific organic anion transporter cMOAT/MRP2. *FEBS Lett.* **456**, 327–331.

Kerem, B., Rommens, J. M., Buchanan, J. A., *et al.* (1989). Identification of the cystic fibrosis gene: genetic analysis. *Science* **245**, 1073–1080.

Khan, W. A., Blobe, G. C., and Hannun, Y. A. (1995). Arachidonic acid and free fatty acids as second messengers and the role of protein kinase C. *Cell Signal.* **7**, 171–184.

Kirtland, S. J. (1988). Prostaglandin E1: a review. *Prostaglandins Leukot. Essent. Fatty Acids* **32**, 165–174.

Kumar, N. B., Nostrant, T. T., and Appelman, H. D. (1982). The histopathologic spectrum of acute self-limited colitis (acute infectious type colitis). *Am. J. Surg. Path.* **6**, 523–529.

Kurahashi, K., Nishihashi, T., Trandafir, C. C., *et al.* (2003). Diversity of endothelium-derived vasocontracting factors: arachidonic acid metabolites. *Acta. Pharmacol. Sin.* **24**, 1065–1069.

Lai H.-C., Kosorok, M. R., Laxova, A., *et al.* (2000). Nutritional status of patients with cystic fibrosis with meconium ileus: a comparison with patients without meconium ileus and diagnosed early through neonatal screening. *Pediatrics* **105**, 53–61.

Lee, C. A., Silva, M., Siber, A. M., *et al.* (2000). A secreted *Salmonella* protein induces a proinflammatory response in epithelial cells, which promotes neutrophil migration. *Proc. Natl. Acad. Sci. U. S. A.* **97**, 12 283–12 288.

Leier, I., Jedlitschky, G., Buchholz, U., *et al.* (1994). The MRP gene encodes an ATP-dependent export pump for leukotriene C4 and structurally related conjugates. *J. Biol. Chem.* **269**, 27 807–27 810.

Leslie, C. C. (1997). Properties and regulation of cytosolic phospholipase A_2. *J. Biol. Chem.* **272**, 16 709–16 712.

Leslie, C. C. (2004). Regulation of the specific release of arachidonic acid by cytosolic phospholipase A_2. *Prostaglandins Leukot. Essent. Fatty Acids* **70**, 373–376.

Li, M., McCann, J. D., Anderson, M. P., *et al.* (1989). Regulation of chloride channels by protein kinase C in normal and cystic fibrosis airway epithelia. *Science* **244**, 1353–1356.

Lieb, J. (2001). Eicosanoids: the molecules of evolution. *Med. Hypotheses* **56**, 686–693.

Londono, I., Marshansky, V., Bourgoin, S., Vinay, P., and Bendayan, M. (1999). Expression and distribution of adenosine diphosphate ribosylation factors in the rat kidney. *Kidney Int.* **55**, 1407–1416.

Luscinskas, F. W., Nicolaou, K. C., Webber, S. E., *et al.* (1990). Ca^{2+} mobilization with leukotriene A4 and epoxytetraenes in human neutrophils. *Biochem. Pharmacol.* **39**, 355–365.

Lyczak, J. B., Cannon, C. L., and Pier, G. B. (2002). Lung infections associated with cystic fibrosis. *Clin. Microbiol. Rev.* **15**, 194–222.

McCormick, B. A., Colgan, S. P., Archer, C. D., Miller, S. I., and Madara, J. L. (1993). *Salmonella typhimurium* attachment to human intestinal epithelial monolayers: transcellular signalling to subepithelial neutrophils. *J. Cell Biol.* **123**, 895–907.

McCormick, B., Hofman, P., Kim, J., *et al.* (1995). Surface attachment of *Salmonella typhimurium* to intestinal epithelia imprints the subepithelial matrix with gradients chemotactic for neutrophils. *J. Cell Biol.* **131**, 1599–1608.

McCormick, B. A., Parkos, C. A., Colgan, S. P., Carnes, D. K., and Madara, J. L. (1998). Apical secretion of a pathogen-elicited epithelial chemoattractant (PEEC) activity in response to surface colonization of intestinal epithelia by *Salmonella typhimurium*. *J. Immunol.* **160**, 455–466.

McGovern, V. J. and Slavutin, L. J. (1979). Pathology of *Salmonella* colitis. *Am. J. Surg. Pathol.* **3**, 483–490.

McIntosh, K. (2002). Community-acquired pneumonia in children. *N. Engl. J. Med.* **346**, 429–437.

Moncada, S. and Vane, J. R. (1978). Pharmacology and endogenous roles of prostaglandin endoperoxides, thromboxane A2, and prostacyclin. *Pharmacol. Rev.* **30**, 293–331.

Monjazeb, A. M., Clay, C. E., High, K. P., and Chilton, F. H. (2002). Antineoplastic properties of arachidonic acid and its metabolites. *Prostaglandins Leukot. Essent. Fatty Acids* **66**, 5–12.

Moss, C. W., Samuels, S. B., and Weaver, R. E. (1972). Cellular fatty acid composition of selected *Pseudomonas* species. *Appl. Microbiol.* **24**, 596–598.

Mrsny, R. J., Gewirtz, A. T., Siccardi, D., *et al.* (2004). Identification of hepoxilin A_3 in inflammatory events: a required role in neutrophil migration across the intestinal epithelia. *Proc. Natl. Acad. Sci. U. S. A.* **101**, 7421–7426.

Nabwera, H. M. and Logan, R. P. (1999). Epidemiology of *Helicobacter pylori* infection: transmission, translocation, and extragastric reservoirs. *J. Physiol. Pharmacol.* **50**, 711–722.

Nardone, G., Rocco, A., Vaira, D., *et al.* (2004). Expression of COX-2, mPGE-synthase, MDR-1 (P-GP), and Bcl-xL: a molecular pathway of *H. pylori*-related gastric carcinogenesis. *J. Pathol.* **202**, 305–312.

Ney, P., Braun, M., Szymanski, C., Bruch, L., and Schror, K. (1991). Antiplatelet, antineutrophil and vasodilating properties of 13, 14-dihydro-PGE1 (PGE0): an *in vivo* metabolite of PGE1 in man. *Eicosanoids* **4**, 117–184.

Nguyen, T. and Gupta, S. (1997). Leukotriene C4 secretion from normal murine mast cells by a probenecid-sensitive and multidrug resistance-associated protein independent mechanism. *J. Immunol.* **158**, 4916–4920.

Nicholson, D. W., Ali, A., Vailancourt, J. P., *et al.* (1993). Purification to homogeneity and the N-terminal sequences of human leukotriene C_4 synthase: a homodimeric glutathione S-transferase composed of 18-kDa subunits. *Proc. Natl. Acad. Sci. U. S. A.* **90**, 2015–2019.

Nishizuka, Y. (1992). Intracellular signaling by hydrolysis of phospholipids and activation of protein kinase C. *Science* **258**, 607–614.

Odenbreit, S., Puls, J., Sedlmaier, B., *et al.* (2000). Translocation of *Helicobacter pylori* CagA into gastric epithelial cells by type 1V secretion. *Science* **287**, 1497–1500.

Pascual, R. M., Awsare, B. K., Farber, S. A., *et al.* (2003). Regulation of phospholipase A_2 by interleukin-1 in human airway smooth muscle. *Chest* **123**, 433S–434S.

Patel, V. A., Dunn, M. J., and Sorokin, A. (2002). Regulation of MDR-1 (P-glycoprotein) by cyclooxygenase-2. *J. Biol. Chem.* **277**, 38 915–38 920.

Pizurki L, Morris, M. A., Chanson, M., *et al.* (2000). CFTR does not affect PMN migration across cystic fibrosis airway epithelial monolayers. *Am. J. Pathol.* **156**, 1407–1416.

Pomorski, T., Meyer, T. F., and Naumann, M. (2001). *Helicobacter pylori*-induced prostaglandin E_2 synthesis involves activation of cytosolic phospholipase A_2 in epithelial cells. *J. Biol. Chem.* **276**, 804–810.

Ratnasinghe, D., Dashner, P. J., Anver, M. R., *et al.* (2001). Cyclooxygenase-2, P-glycoprotein-170 and drug resistance: is chemoprevention against multidrug resistance possible? *Anticancer Res.* **21**, 2141–2148.

Reid, G., Wielinga, P., Zelcer, N., *et al.* (2003). The human multidrug resistance protein MRP4 functions as a prostaglandin efflux transporter and is inhibited by nonsteroidal anti-inflammatory drugs. *Proc. Natl. Acad. Sci. U. S. A.* **100**, 9244–9249.

Riordan, J. R., Rommens, J. M., Karen, B. S., *et al.* (1989). Identification of the cystic fibrosis gene: cloning and characterization of complementary DNA. *Science* **245**, 1066–1073.

Robbiani, D. F., Finch, R. A., Jager, D., *et al.* (2000). The leukotriene C(4) transporter MRP1 regulates CCL19 (MIP-beta, ELC)-dependent mobilization of dendritic cells to lymph nodes. *Cell* **103**, 757–768.

Rout, W. R., Formal, S. B., Dammin, G. J., and Giannella, R. A. (1974). Pathophysiology of *Salmonella* diarrhea in the Rhesus monkey: intestinal transport, morphological and bacteriological studies. *Gastroenterology* **67**, 59–70.

Sansonetti, P. J. (2004). War and peace at mucosal surfaces. *Nat. Rev. Immunol.* **4**, 953–964.

Sato, H. Frank, D. W., Hillard, C. J., *et al.* (2003). The mechanism of action of the *Pseudomonas aeruginosa*-encoded type III cytotoxin, ExoU. *EMBO J.* **22**, 2959–2969.

Schultz, M. J., Wijnholds, J., Peppelenbosch, M. P., *et al.* (2001). Mice lacking the multi-drug resistance protein-1 are resistant to *Streptococcus pneumoniae*-induced pneumonia. *J. Immunol.* **166**, 4059–4064.

Seeds M. C. and Bass, D. A. (1999). Regulation and metabolism of arachidonic acid. *Clin. Rev. Allergy Immunol.* **17**, 5–26.

Serhan, C. N., Hong, S., Gronert, K., *et al.* (2002). Resolvins: a family of bioactive products of omega-3 fatty acid transformation circuits initiated by aspirin treatment that counter proinflammation signals. *J. Exp. Med.* **196**, 1025–1037.

Sharma, S. A., Tummuru, M. K., Miller, G. C., and Blaser, M. J. (1995). Interleukin-8 response of gastric epithelial cell lines to *Helicobacter pylori* stimulation *in vitro*. *Infect. Immun.* **63**, 1681–1687.

Sinzinger, H., O'Grady, J., Demers, L. M., *et al.* (1990). Thromboxane in cardiovascular disease. *Eicosanoids* **3**, 59–64.

Smart, S. J. and Casale, T. B. (1994). Pulmonary epithelial cells facilitate TNF-alpha-induced neutrophil chemotaxis: a role for cytokine networking. *J. Immunol.* **152**, 4087–4094.

Smith, W. L. and DeWitt, D. L. (1996). Prostaglandin endoperoxidae H synthase-1 and -2. *Adv. Immunol.* **62**, 167–215.

Songer, J. G. (1997). Bacterial phospholipases and their role in virulence. *Trends Microbiol.* **5**, 156–161.

Strandvik, B. (1992). Long chain fatty acid metabolism and essential fatty acid deficiency with special emphasis on cystic fibrosis. In *Polyunsaturated Fatty Acids in Human Nutrition*, ed. U. Bracco and R. J. Decklelbaum. New York: Raven Press, pp. 159–167.

Sutherland, M., Schewe, T., and Nigam, S. (2000). Biological actions of the free acid of hepoxilin A_3 on human neutrophils. *Biochem. Pharmacol.* **59**, 435–440.

Tatsuguchi, A., Sakamoto, C., Wada, K., *et al.* (2000). Localization of cyclooxygenase 1 and cyclooxygenase 2 in *Helicobacter pylori* related gastritis and gastric ulcer tissues in humans. *Gut* **46**, 782–789.

Thoren, S. and Jakobsson, P. J. (2000). Coordinate up- and down regulation of glutathione-dependent prostaglandin E synthase and cyclooxygenase-2 in A549 cells: inhibition by NS-398 and leukotriene C_4. *Eur. J. Biochem.* **267**, 6428–6432.

Tuomanen, E. I., Austrian, R., and Masure, H. R. (1995). Pathogenesis of pneumococcal infection. *N. Engl. J. Med.* **322**, 1280–1284.

Turk, J., Maas, R. L., Brash, A. R., Roberts, L. J., and Oates J. A. (1982). Arachidonic acid 15-lipoxygenase products from human eosinophils. *J. Biol. Chem.* **257**, 7068–7076.

Vance, R. E., Hong, S., Gronert, K., Serhan, C. N., and Mekalanos, J. J. (2004). The opportunistic pathogen *Pseudomonas aeruginosa* carries a secretable arachidonate 15-lipoxygenase. *Proc. Natl. Acad. Sci. U. S. A.* **101**, 2135–2139.

Venkateswarlu, K. and Cullen, P. J. (2000). Signaling via ADP-ribosylation factor 6 lies downstream of phosphatidylinositide 3-kinase. *Biochem. J.* **345**, 719–724.

Wagner, J. G. and Roth, R. A. (2000). Neutrophil migration mechanisms, with an emphasis on the pulmonary vasculature. *Pharmacol. Rev.* **52**, 349–374.

Wallace, J. L. (1992). Prostaglandins, NSAIDs, and cytoprotection. *Gastroenterol. Clin. North Am.* **21**, 631–641.

Wallis, T. S., Vaughan, A. T. M., Clarke, G. J., *et al.* (1990). The role of leucocytes in the induction of fluid secretion by *Salmonella typhimurium*. *J. Med. Microbiol.* **31**, 27–35.

Wambura, C., Aoyama, N., Shirasaka, D., *et al.* (2002). Effect of *Helicobacter pylori*-induced cyclooxygenase-2 on gastric epithelial cell kinetics: implication for gastric carcinogenesis. *Helicobacter* **7**, 129–138.

Weiss, S. (1989). Tissue destruction by neutrophils. *N. Engl. J. Med.* **320**, 365–376.

Wijnholds, J., Evers, R., van Leusden, M. R., *et al.* (1997). Increased sensitivity to anti-cancer drugs and decreased inflammatory response in mice lacking the multi-drug resistance associated protein. *Nat. Med.* **3**, 1275–1279.

Wilderman, P. J., Vasil, A. I., Johnson, Z., and Vasil, M. L. (2001). Genetic and biochemical analyses of a eukaryotic-like phospholipase D of *Pseudomonas aeruginosa* suggest horizontal acquisition and a role for persistence in a chronic pulmonary infection model. *Mol. Microbiol.* **39**, 291–303.

Wilkinson, S. G. (1988). In *Microbial Lipids*, ed. C. Ratledge and S. G. Wilkinson. London: Academic Press, pp. 299–488.

Part IV Exploitation of host niches by pathogenic bacteria: mechanisms and consequences

CHAPTER 11

Lung infections

Marisa I. Gómez and Alice S. Prince

INTRODUCTION

The airway epithelium represents a primary site for the introduction and deposition of potentially pathogenic microorganisms into the body through inspired air. The ciliated epithelium lining the airways possesses several mechanisms to prevent colonization by inhaled bacteria and, despite repeated exposures to a wide variety of organisms, the lower respiratory tract usually remains sterile. The airway is defined anatomically as the upper respiratory tract, which includes the nasal sinuses and the nasopharynx, and the lower respiratory tract, which begins at the larynx and continues to the trachea, before dividing into the smaller airways until they reach the alveoli. The luminal surface of the airways is lined by a layer of epithelial cells. In the conducting airways, these cells are pseudostratified columnar epithelial cells, which become simple cuboidal epithelium as the branches extend to the alveoli (Diamond *et al.*, 2000). The respiratory epithelium is an essential barrier that features tight intercellular apical junctions between the cells, a superficial liquid layer or film that contains mucous-gland and goblet-cell secretions, immunoglobulins, and lysozyme, components that are propelled and cleared by cilia.

INNATE HOST DEFENSES AGAINST BACTERIAL LUNG PATHOGENS

In the upper airways, the nose functions as a filter by trapping large particulate matter (>10 μm) in nasal hair or on the surface of the turbinates

Bacterial–Epithelial Cell Cross-Talk: Molecular Mechanisms in Pathogenesis, ed. Beth A. McCormick. Published by Cambridge University Press.

and septum. Smaller particles, including bacteria ranging in size from 2 μm to 10 μm, are inhaled and deposited in the lower conducting airways. The dichotomous branches of the respiratory tree gradually impose resistance in order to slow down the velocity of air molecules and permit further cleansing. The airways between the larynx and the respiratory bronchioles are lined with ciliated columnar epithelium covered by a mucous film 5–100 μm thick. This mucous film produced by mucous glands and goblet cells entraps microorganisms on the epithelial surface, which are propelled up the airway by ciliary motion or coughing.

The alveolar space has several mechanisms to engage particles and microbes that have eluded the complex cleansing mechanisms in the upper airways and have reached the alveolar surface. Airway cells secrete a large array of molecules with antimicrobial activity. These molecules include small cationic antimicrobial peptides such as β-defensins (BD) and LL-37 and larger antimicrobial proteins such as lysozyme, lactoferrin and leukocyte proteinase inhibitor (SLPI) (Hiemstra, 2001). The hydrophilic surfactant proteins A (SP-A) and D (SP-D) play important roles in host defense mechanisms of the lung (Sano and Kuroki, 2005). These proteins belong to a collectin subgroup in which lectin domains are associated with collagenous structures. Collectins include mannose-binding lectin and are considered to function as part of the innate immune system. SP-A and SP-D interact with various microorganisms and pathogen-derived components. They act as opsonins by binding and agglutinating pathogens; they also possess direct inhibitory effects on bacterial growth (Lawson and Reid, 2000). SP-A and SP-D also associate with immune cells, and their direct interaction with macrophages results in modulation of phagocytosis or the production of reactive oxygen species. Moreover, by associating with cell-surface pattern-recognition receptors, SP-A and SP-D regulate inflammatory cellular responses such as the release of lipopolysaccharides (LPS)-induced pro-inflammatory cytokines (Lawson and Reid, 2000). Studies with transgenic mice lacking surfactant proteins have showed that SP-A-null mice have delayed microbial clearance after intratracheal administration of group B *Streptococcus* (LeVine *et al.*, 1997), *Haemophilus influenzae* (LeVine *et al.*, 2000), and *Pseudomonas aeruginosa* (LeVine *et al.*, 1998).

ETIOLOGY AND EPIDEMIOLOGY OF BACTERIAL PNEUMONIA

Bacterial pneumonia occurs when there is a breakdown in the normal host defenses, particularly disruption of the mucosal barrier, when organisms are extremely virulent or a large inoculum is introduced, or when key components of the immune system are dysfunctional. There are several routes of pathogen acquisition involved in the pathophysiology of pneumonia.

Streptococcus pneumoniae is the most common pathogen involved in community-acquired pneumonia, followed by *H. influenzae*, *Staphylococcus aureus*, and *Legionella* sp. (Andrews *et al.*, 2003). Ninety percent of community-acquired pneumonias involve organisms that descend from the oropharynx into the lower respiratory tract. Other less frequent routes of pathogen acquisition include hematogenous spread (*Staphylococcus*) and contiguous spread. Most people aspirate to some degree while sleeping, and oropharyngeal secretions may enter the lower respiratory tract; however, due to the numerous defense mechanisms that exist in the airways, especially mucociliary clearance, most aspirated material is of no clinical significance. However, alterations in mucosal barriers, such as impaired ciliary action, mechanical trauma, and inflammatory changes induced by viral infection, predispose the lower airways to pneumonia. In addition, impairment of the immune system (in either humoral or cell-mediated immunity) or phagocytic function facilitates colonization at the lower respiratory tract.

The most frequent cause of hospital-acquired pneumonia is mechanical ventilation in hospitalized patients. Ventilator-associated pneumonia is commonly caused by Gram-negative bacteria, such as *P. aeruginosa*, *Klebsiella pneumoniae*, and *Acinetobacter baumannii*, and methicillin-resistant *S. aureus* (MRSA) (Shaw, 2005). These species have the ability to form biofilms and adhere to plastic, which facilitates the colonization of the endotracheal tube and subsequent lung infection. Once the patient is intubated, the natural barrier between the oropharynx and the trachea is bypassed and the epithelium is damaged as a result of the mechanical injury associated with endotracheal intubation. These conditions favor attachment and growth of bacteria and allow for greatly increased bacterial density. Intubation and sedation of the patient impairs normal cough-mediated clearance and facilitates the entry of colonizing pathogens through micro- and macro-aspiration of infected oral and gastric contents (Craven, 2000). Among Gram-positive bacteria, *S. aureus* is a major cause of pneumonia in hospitalized patients and is becoming increasingly resistant to antibiotics. Some 40–60% of all hospital *S. aureus* isolates are resistant to methicillin, and intermediate to high levels of resistance to vancomycin have also been described (Craven, 2000; Lindsay and Holden, 2004).

Defective mucociliary clearance also contributes to the development of pneumonia in cystic fibrosis (CF). In CF, dysfunction of the cystic fibrosis transmembrane conductance regulator (CFTR), a chloride channel, in airway epithelium and submucosal glands leads to dehydrated secretions and predisposes the respiratory tract to infection and chronic inflammation. This is manifested early in life by airway obstruction and recurrent infections of the lung and paranasal sinuses. The CF lung is particularly susceptible to *P. aeruginosa*, and this organism plays a critical role in the development and progression of pulmonary disease in people with CF. *S. aureus* and *H. influenzae* also contribute to CF-associated infection in people with CF (Goss and Rosenfeld, 2004).

BACTERIAL VIRULENCE FACTORS INVOLVED IN THE PATHOGENESIS OF PNEUMONIA

Bacteria cause pneumonia by eliciting the mobilization and activation of phagocytes – usually polymorphonuclear cells (PMNs) – into the airways. Although the PMNs function to eradicate infection, they also impede air exchange. Thus, the balance between efficient phagocytosis of inspired bacteria and efficient air exchange is physiologically critical. Intact bacteria and shed bacterial components can activate epithelial pro-inflammatory signaling. Components of the professional immune system (T-cells, macrophages, dendritic cells) express cytokines and chemokines in order to deal with the inhaled organisms. However, the airway epithelial cell, which is the most abundant in the lung and the most likely to come into contact with inhaled organisms, makes a substantial contribution to the recruitment and activation of phagocytic cells into the airways.

Lung pathogens possess several virulence factors involved in adhesion and colonization of the airway epithelium. The virulence factors of three major respiratory pathogens are summarized in Table 11.1. Bacterial adhesins play a key role in colonization because they allow the bacteria to attach to airway cells. *P. aeruginosa* pili mediate epithelial adherence and are important in the pathogenesis of airway infection, particularly during invasive infection. Piliated *P. aeruginosa*, but not *pil* mutants, can colonize neonatal mice and cause pulmonary inflammatory responses (Tang *et al.*, 1995). The secretion of many *P. aeruginosa* toxins, which act within eukaryotic cells, also requires pilin-mediated attachment (Feldman *et al.*, 1998; Hauser *et al.*, 1998). Streptococcal and staphylococcal surface adhesins can bind to host cell-matrix components and to cellular receptors upregulated during lung inflammation (Bogaert *et al.*, 2004; Foster and McDevitt, 1994).

For successful colonization, bacteria also need to acquire iron from host tissues, where it is bound tightly to transferrin or lactoferrin, which is present in the airways (Xiao and Kisaalita, 1997). Pathogens have developed a complex regulatory system to compete with lactoferrin for iron. *P. aeruginosa* siderophores pyochelin and pyoverdin (Vasil and Ochsner, 1999; Xiao and

Table 11.1 *Virulence factors in respiratory pathogens*

Function	Virulence factor	Bacteria
Adhesion and colonization	Cell-wall-associated adhesins	*Streptococcus pneumoniae*
		Staphylococcus aureus
	Pili	*Pseudomonas aeruginosa*
	Flagella	*P. aeruginosa*
	Biofilm	*P. aeruginosa*
		S. aureus
	Siderophores	*P. aeruginosa*
		S. aureus
	Neuraminidase	*P. aeruginosa*
		S. pneumoniae
	Hyaluronidase	*S. pneumoniae*
Evasion of immune system	Capsule	*S. pneumoniae*
		S. aureus
	Complement inhibition	*S. aureus*
		P. aeruginosa
		S. pneumoniae
	IgA protease	*S. aureus*
		S. pneumoniae
Induction of inflammation	Pneumolysin	*S. pneumoniae*
	Pili	*P. aeruginosa*
	Flagella	*P. aeruginosa*
	LPS	Gram-negative bacteria
	Lipoproteins	*S. aureus*
		S. pneumoniae
	Protein A	*S. aureus*
Lung damage and invasion	Type III secretion system (ExoS, ExoT, ExoU, ExoY)	*P. aeruginosa*
	Exotoxin A	*P. aeruginosa*
	Coagulase	*S. aureus*
	Protease	*S. aureus*
	Hemolysin	*S. aureus*
	Cell-wall choline	*S. pneumoniae*
	Pneumolysin	*S. pneumoniae*

IgA, immunoglobulin A; LPS, lipopolysaccharide.

Kisaalita, 1997) and the SirABC transporter in *S. aureus* (Dale *et al.*, 2004) are examples of these iron-uptake systems.

The pulmonary pathogens *P. aeruginosa* and *S. aureus* can live either in free planktonic form or in biofilms (Parsek and Singh, 2003; Yarwood and Schlievert, 2003). Biofilms are highly structured communities that coat surfaces such as plastic catheters and the mucosal surfaces of the airways. The coordinated expression of diverse groups of genes within this community of bacteria is directed by small diffusible molecules called quorum sensors. At low density, bacteria live in the planktonic form, but as the number of organisms increases, the secreted quorum sensors accumulate. Once a critical density of these molecules is achieved, they diffuse back into the organisms, where, along with transcriptional activators, they direct the expression of virulence genes that allow the bacteria to evade the host response and survive as a community. Biofilm production has an important role in bacterial persistence in the lung, especially in chronic diseases such as CF. In CF and other chronic infections, bacterial adaptation to the environment results in the selection of organisms that are more persistent and less invasive, and the biofilm mode of growth plays a key role in this adaptation (Costerton *et al.*, 1999; Hoiby *et al.*, 2001; Singh *et al.*, 2000, 2002). Within the lungs, bacteria proliferate in both biofilms and the planktonic form of growth.

During colonization of the lung, the airway epithelium senses the presence of bacteria and initiates the inflammatory response. Lung injury associated with bacterial infection is usually the result of both the direct destructive effects of the organism on the lung parenchyma and damage due to exuberant host inflammatory responses.

AIRWAY EPITHELIAL CELL RESPONSE TO BACTERIA

Airway epithelial cells orchestrate the activation of many signaling cascades in response to bacterial ligands through surface and intracellular receptors (Figure 11.1, Table 11.2). In addition to generating signals for the recruitment of immune cells, airway epithelial cells also synthesize and secrete many effector proteins themselves.

Bacterial induction of antimicrobial peptides and chemical defenses

Antimicrobial peptides in the lung are mainly produced and secreted by epithelial and phagocytic cells. The expression of antimicrobial peptide genes is regulated tightly. Some peptides are produced constitutively, such as

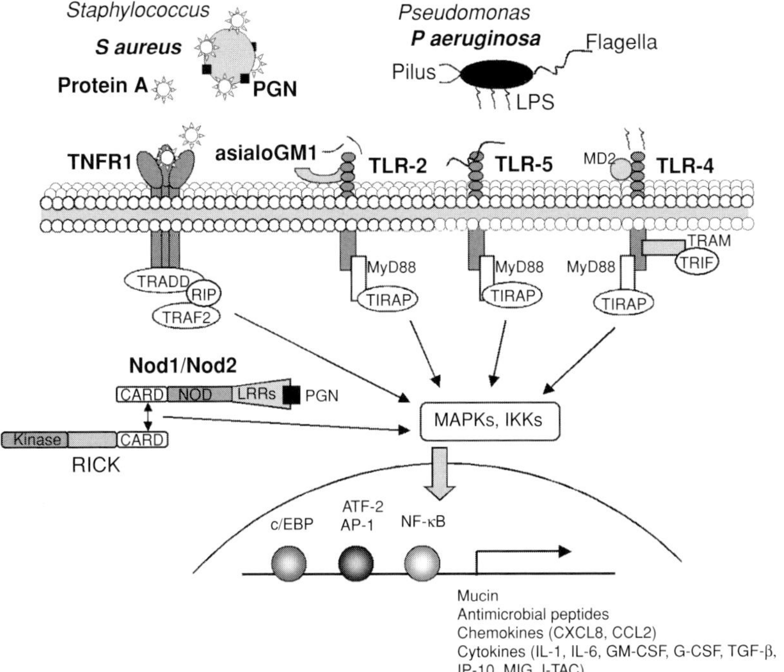

Figure 11.1 Airway epithelial responses to bacterial ligands. Bacterial ligands (flagella, pilus, lipopolysaccharide (LPS), protein A, peptidoglycan (PGN)) are recognized by surface receptors (asialoGM1, Toll-like receptors (TLRs), tumor necrosis factor receptor 1 (TNFR1)) or intracellular receptors (nucleotide-binding oligomerization domain protein, Nod). Signaling cascades are initiated through adaptor proteins (MyD88/TIRAP, TRAM/TRIF, TRADD/RIP, RICK). MAPK- and IKK-dependent translocation of transcription factors leads to transcription of inflammatory mediators.

G-CSF, granulocyte colony-stimulating factor; GM-CSF, granulocyte/macrophage colony-stimulating factor; IL-1, interleukin 1; IL-6, interleukin 6; RICK, RIP-like interacting caspase-like apoptosis-regulatory protein kinase; RIP, receptor-interacting protein; TGF-β, transforming growth factor, beta; TIRAP, TIR domain-containing adaptor protein; TRADD, tumor necrosis factor receptor 1-associated death domain protein; TRAM, TRIF-related adaptor molecule; TRIF, TIR domain-containing adaptor inducing interferon beta.

human β-defensin 1 (hBD-1), but the expression of others is increased upon contact of cells with bacterial products or pro-inflammatory mediators. It has been shown that expression of hBD-2, hBD-3, hBD-4, the cathelicidin-derived peptide LL-37, and several other antimicrobial peptides are induced in vivo during inflammatory or infectious diseases, such as pneumonia (Hiratsuka

Table 11.2 *Airway epithelial cell receptors and bacterial ligands*

Receptor	Bacterial ligand
AsialoGM1	Flagella, pili, Gram-positive and Gram-negative bacteria
TLR-2	Lipoteicoic acid, lipoproteins, flagella, lipoarabinomannan, phenol soluble modulins
TLR-4	LPS, pneumolysin
TLR-5	Flagella
TNFR1	Protein A
NOD1/2	Peptidoglycan

LPS, lipopolysaccharide; NOD, nucleotide-binding oligomerization domain protein; TLR, Toll-like receptor; TNFR1, tumor necrosis factor receptor 1.

et al., 1998) and CF (Bals *et al.*, 2001). In vitro studies have shown upregulation of hBD-2 by primary airway epithelial cells in response to *P. aeruginosa* LPS and inflammatory cytokines such as interleukin (IL)-1β and tumor necrosis factor alpha (TNF-α) (Becker *et al.*, 2000; Singh *et al.*, 1998). Antimicrobial peptides have a broad spectrum of activity against Gram-positive and Gram-negative bacteria and show synergistic activity with other host defense molecules, such as lysosyme and lactoferrin (Bals and Hiemstra, 2004).

Other inducible host defense molecules include mucins and reactive nitrogen species such as nitric oxide (NO) (Rochelle *et al.*, 1998). NO is produced by inducible nitric oxide synthase (iNOS), which is upregulated in response to LPS. Mucin concentration in broncheoalveolar lavage (BAL) is also increased in response to LPS and flagella, and *MUC2* and *MUC5A* gene expression is upregulated by LPS and Gram-positive and Gram-negative bacteria (Dohrman *et al.*, 1998).

Inflammation and bacterial clearance

Epithelial cells signal the presence of bacterial components and secrete pro-inflammatory cytokines and chemokines that recruit immune cells to the site of infection and activate them. Several bacterial components are highly immunostimulatory, such as flagella, lipoproteins, and staphylococcal protein A (Table 11.1). In response to bacteria, airway epithelial cells secrete the neutrophil chemokine CXCL-8 and cytokines such as IL-6, IL-1β, granulocyte–macrophage colony-stimulating factor (GM-CSF), granulocyte colony-stimulating factor (G-CSF), transforming growth factor beta (TGF-β), interferon (IFN)-induced protein of 10 kDa (IP-10), monokine induced by interferon gamma (INF-γ) (MIG), and IFN-inducible T-cell alpha-chemoattractant (I-TAC). IL-1 is the first cytokine produced in response to bacteria along with TNF-α, which is produced mainly by alveolar macrophages in the lung. Although TNF-α has a protective effect in animal models of *P. aeruginosa* infection, IL-1 appears to have a deleterious role during *P. aeruginosa* pneumonia (Strieter *et al.*, 2003). IL-1-receptor-null mice inoculated intranasally with *P. aeruginosa* were found to have greater bacterial clearance in their lungs and reduced bacteremia, as compared with wild-type mice (Schultz *et al.*, 2002), suggesting that reduction in IL-1 activity improves host defense against *P. aeruginosa* pneumonia. Both IL-1 and TNF-α signaling induce activation of nuclear factor κB (NF-κB), which promotes pro-inflammatory chemokine expression. Neutrophil chemokines (CXC), in particular CXCL8, play a critical role in recruitment and maintenance of leukocytes during infection. Animal models of pneumonia have demonstrated an increase in CXCs in BAL of infected mice; blockade of CXC receptors results in reduced neutrophil infiltration and clearance of bacteria in the lung and increased mortality (Mehrad and Standiford, 1999; Strieter *et al.*, 2002). Transgenic mice engineered for enhanced expression of KC (the mouse equivalent of CXCL8) and CXCL1 (Groα/MIP-2α) have improved survival during bacterial pneumonia (Tsai *et al.*, 1998). Epithelial cells also secrete the CC chemokine monocyte chemottractant protein 1 (CCL2), which plays an important role in orchestrating the polarization of the inflammatory response during the initiation of the adaptive immune response (Strieter *et al.*, 2003). The cytokines G-CSF and GM-CSF are also expressed by airway epithelial cells and are important in activating PMNs at the site of infection and enhancing their survival by inhibition of apoptosis (Saba *et al.*, 2002). GM-CSF-deficient mice display significantly increased susceptibility to streptococcal infection (LeVine *et al.*, 1999). In addition to their role in cell recruitment, epithelial cytokines can have an autocrine and paracrine inductive effect on the airway by maintenance of high levels of antimicrobial peptides.

As part of the inflammatory response, airway epithelial cells also express adhesion molecules, including intercellular adhesion molecule 1 (ICAM-1), to allow the adhesion of recruited neutrophils. Neutrophils are the main cells involved in the recognition, phagocytosis, and clearance of bacteria. This is accomplished by opsonization through Fc-mediated binding or antigen recognition using complement receptors. The pathogen is ingested and killed in the PMN phagosome through the expression of peptides and reactive oxygen intermediates. Thus, neutrophils are critically important in the phagocytosis and killing of bacteria. However, their own lysis and release of elastase is a potent stimulus of epithelial CXCL8, which promotes a cycle of continued inflammation (Nakamura *et al.*, 1992).

HOW DO EPITHELIAL CELLS SENSE THE PRESENCE OF BACTERIA IN THE AIRWAYS?

Airway epithelial cells are polarized and form tight junctions. Unlike other mucosal surfaces, the lower airways are sterile and exposure to bacterial compounds triggers an immediate response. Intact bacteria are rarely in direct contact with airway epithelial cells, which are well protected by mucins. Following epithelial damage, bacteria may gain access to the epithelial surface, or shed bacterial components may stimulate epithelial responses. The distribution of surface receptors is different than in other cells. Receptors must be exposed apically in order to recognize pathogens, or bacterial products present in the lumen of the airways. These receptors are less abundant than immune cells, which prevents high inflammatory responses that impede lung function. However, upon repeated bacterial stimulation, more receptors are recruited to the apical surface, where they initiate the inflammatory response required to clear the infection.

Toll-like receptors

Cells of the innate immune system, including epithelial cells, use pattern-recognition molecules to bind conserved products that are present on microorganisms. Pattern-recognition molecules are present in secretions and circulate in soluble form, such as mannan-binding lectin (MBL), or they are transmembrane molecules that mediate direct cellular responses to microbial exposure. The Toll-like receptors (TLR) constitute an intensely studied family of pattern-recognition receptors. Human TLRs are type I transmembrane proteins with an extracellular domain, a transmembrane domain, and an intracellular domain. The extracellular domains include arrays of leucine-rich repeats (LRR), and the cytoplasmic portions show high similarity to that of the IL-1 receptor family, termed Toll/IL-1 receptor (TIR) domain. In the lungs, TLRs are expressed by professional immune cells, such as resident alveolar macrophages and newly recruited dendritic and T-cells, as well as by epithelial and endothelial cells (Takeda *et al.*, 2003).

TLRs sense components of both Gram-positive and Gram-negative bacteria. Stimulation of TLRs by microbial components triggers activation of signaling pathways originating from a cytoplasmic TIR domain. The TIR domain-containing adaptor MyD88 is essential for induction of inflammatory cytokines and chemokines. Upon stimulation, MyD88 recruits interleukin 1 receptor-activated kinase 4 (IRAK-4) to TLRs and facilitates IRAK-4-mediated phosphorylation of interleukin 1 receptor-activated kinase 1 (IRAK-1). Activated IRAK-1 then associates with tumor necrosis factor receptor-associated factor 6 (TRAF6), leading to activation of activating protein 1 (AP-1) transcription factors through mitogen-activated protein (MAP) kinases and nuclear translocation of transcription factor NF-κB (Yamamoto and Akira, 2004).

Although regulation of cytokine and chemokine production is one of the best-characterized roles of TLR signaling, it is clear that TLRs also regulate expression of antimicrobial peptides. CD14, a part of the TLR-4 receptor complex, is essential in the LPS-induced expression of hBD-2 on tracheobronchial epithelial cells (Becker *et al.*, 2000). Similarly, TLR-2 regulates the expression of hBD-2 in response to bacterial lipoproteins in A549 lung epithelial cells (Birchler *et al.*, 2001).

Toll-like receptor expression in lung epithelium

Eleven TLRs have been recognized to date (Takeda and Akira, 2004). TLRs 1–10 are expressed in airway epithelial cells (Greene *et al.*, 2005; Muir *et al.*, 2004). However, not all of these TLRs participate in the response to inhaled organisms. TLRs play important roles in recognizing specific microbial components derived from pathogens, including bacteria, fungi, protozoa, and viruses. Among the different TLRs, TLR-2, TLR-4, and TLR-5 play a major role in signaling bacterial components in the lung.

TLR-2 in airway infection

TLR-2 recognizes a variety of microbial components. These include lipoproteins/lipopeptides from various pathogens, lipoteicoic acid from Gram-positive bacteria, lipoarabinomannan from mycobacteria, and a phenol-soluble modulin from *Staphylococcus epidermidis*. In airway epithelial cells, TLR-2 is also involved in early responses to *P. aeruginosa* flagella (Adamo *et al.*, 2004).

In the airways, TLR-2 forms a receptor complex with the asialoganglioside gangliotetraosylceramide (Galb1,2GalNacb1,4Galb1, 4Gal1Cer) (asialoGM1) on the apical surface of epithelial cells within the context of lipid rafts (Soong *et al.*, 2004). This glycolipid has an exposed GalNacb1–4Gal moiety that serves as a receptor for bacterial pili (DiMango *et al.*, 1998), flagella (Feldman *et al.*, 1998), and a large number of pulmonary pathogens, including those commonly associated with airway infection: *S. pneumoniae, S. aureus,* and *P. aeruginosa* (Krivan *et al.,* 1988). TLR-2 is present on the apical surface of polarized cells with tight junctions and mobilized into specialized lipid raft microdomains containing caveolin-1 after bacterial stimulation. The role of TLR-2 in initiating pro-inflammatory signaling in professional immune cells and as airway epithelial cells in vitro has been established (Greene *et al.,* 2005; Muir *et al.,* 2004; Soong *et al.,* 2004; Takeda and Akira, 2004). TLR-2 and asialoGM1 initiate signaling in response to *S. aureus* and *P. aeruginosa,* leading to the activation of NF-κB and CXCL8 production in a MyD88-dependent manner. The lipid raft microdomain seems to be essential for signaling, as suggested by the effects of filipin in inhibiting activation of CXCL8 expression in response to bacteria.

TLR-2 mRNA expression is upregulated in the lungs during both Gram-positive and Gram-negative infections (Kajikawa *et al.,* 2005; Knapp *et al.,* 2004; Power *et al.,* 2004), but its contribution to bacterial clearance during pneumonia is not completely understood. Initial studies involving systemic infection in TLR-2-null mice indicated a role for TLR-2 in eradicating Gram-positive bacteria (Takeuchi *et al.,* 2000). However, the situation in the lung seems to be different. The response of TLR-2-null mice to intranasally inoculated *S. pneumoniae* did not differ significantly from that in wild-type mice (Knapp *et al.,* 2004). Cytokine and chemokine production and inflammation were modestly reduced in TLR-2-null mice, but there was no difference in bacterial clearance. TLR-2 expression in macrophages contributes to inhibition of *Legionella pneumophilia* growth (Akamine *et al.,* 2005). This is consistent with previous reports of *L. pneumophilia* LPS signaling through TLR-2 and not TLR-4 (Girard *et al.,* 2003). More studies are required in order to establish whether similar mechanisms are triggered in the lung by *L. pneumophila.* A study using aerosolized *S. aureus* again demonstrated the involvement of TLR-2 in the production of inflammatory cytokines (TNF-α, IL-1β) and chemokines (KC, MIP-2) and PMN recruitment. However, *S. aureus* clearance was not affected in MyD88-null mice (Skerrett *et al.,* 2004b). Thus, other signaling pathways in addition to TLR2/ MyD88 signaling clearly are involved in responses to pulmonary pathogens.

TLR-4 in airway infection

TLR-4 is an essential receptor for LPS recognition by professional immune cells (Hoshino *et al.,* 1999; Poltorak *et al.,* 1998). Although TLR-4 is abundant in airway epithelial cells, it does not appear to function in signaling responses in vitro to *P. aeruginosa* (Muir *et al.,* 2004; Soong *et al.,* 2004). Airway epithelial cells, like other mucosal epithelia, are not particularly responsive to LPS as compared with myeloid cells, even when all of the required coreceptors and LPS-binding proteins are provided (DiMango *et al.*, 1995; Guillot *et al.*, 2004). This low responsiveness is likely due to the intracellular localization of TLR-4 in airway epithelial cells, which prevents excessive inflammation during the course of pulmonary infection (Guillot *et al.*, 2004). The lack of TLR-4 involvement in epithelial responses to LPS in vitro does not imply that the lung itself is unresponsive. In one study, nuclear translocation of NF-κB in response to inhaled LPS was observed in the bronchiolar epithelium of wild-type mice. This response was not observed in transgenic mice expressing a dominant negative inhibitory protein of nuclear factor kappa B (IKBα) in airway epithelium (Skerrett *et al.*, 2004a). The pulmonary response to LPS provided systemically (rather than by inhalation) can be mediated by TLR-4 expressed by pulmonary endothelial cells (Andonegui *et al.*, 2003).

TLR-4 expression is increased during Gram-negative infection (Kajikawa *et al.*, 2005; Knapp *et al.*, 2004; Power *et al.*, 2004), and this TLR plays an important role in the overall defenses of the lung against *P. aeruginosa*. TLR-4/MyD88 signaling has been shown to be critical for the induction of inflammatory cytokines (TNF-α, IL-1β) and chemokines (KC, MIP-2), PMN recruitment to the lungs, and bacterial clearance in mouse models of pneumonia (Power *et al.*, 2004; Skerrett *et al.*, 2004b). Although early responses to *P. aeruginosa* are TLR-4/MyD88-dependent, a later response mediated by TLR-2 has been proposed (Power *et al.*, 2004). TLR-4 contributes to a protective innate immune response to *H. influenzae* (Wang *et al.*, 2002) and *K. pneumoniae* (Branger *et al.*, 2004; Schurr *et al.*, 2005). In addition to recognition of Gram-negative pathogens, TLR-4 can play a modest role in the protective immune responses to pneumococcal pneumonia (Branger *et al.*, 2004), and it has been demonstrated that this receptor is involved in pneumolysin signaling (Malley *et al.*, 2003). Whether the protective role of TLR-4 during bacterial pneumonia is due to signaling in professional immune cells or epithelial cells in the lungs remains to be elucidated.

TLR-5 in airway infection

TLR-5 recognizes flagellin, the principal component of flagella, from both Gram-positive and Gram-negative bacteria (Hayashi *et al.*, 2001; Smith *et al.*, 2003). The expression of flagella is particularly important for bacterial colonization of the lung, and Fla mutants are less virulent in a mouse model of pneumonia (Feldman *et al.*, 1998). Flagellin is highly immunogenic and activates pro-inflammatory responses in immune cells of myeloid origin and epithelial cells (Wyant *et al.*, 1999). Epithelial responses to flagella have been well characterized in the gastrointestinal tract, where flagella interact with TLR-5 to activate pro-inflammatory gene expression (Eaves-Pyles *et al.*, 2001). TLR-5 is basolaterally distributed in the gut epithelium and thus is activated only by invasive pathogens (Gewirtz *et al.*, 2001). In airway epithelial cell lines, flagella induce signaling through the induction of Ca^{2+} fluxes (McNamara *et al.*, 2001), a response not previously associated with TLR-5 but typical of ligands that activate asialoGM1 TLR-2 signaling (Ratner *et al.*, 2001). TLR-5, although expressed in airway epithelial cells, is not abundant on the apical surface but can be recruited following exposure to flagella (Adamo *et al.*, 2004). Direct binding studies and confocal microscopy data indicate that flagella bind to apically displayed asialoGM1 on the airway cell surface. This initial interaction activates an epithelial response that results in TLR-5 transcription and mobilization to the apical surface, where it can mediate further signaling (Adamo *et al.*, 2004). TLR-2, which is available on the apical surface of airway cells, can contribute to the initial signaling of flagella through asialoGM1.

Multiple signaling pathways involving different TLRs are triggered in response to bacterial stimulation. Individual TLRs can activate distinct signaling cascades, depending on the adaptor proteins involved. Although most TLR pro-inflammatory signaling in the lungs is MyD88-dependent, MyD88-independent signaling through TRAM and TRIF in response to TLR-4 ligands is also involved in late NF-κB activation (Kawai *et al.*, 1999). This MyD88-independent/TRIF-dependent cascade regulates production of IFN-γ and IFN-inducible genes (Toshchakov *et al.*, 2002). Studies with knockout mice have revealed that MyD88-null mice have a more severe phenotype than null mice for any of the individual TLRs (Feng *et al.*, 2003). This suggests that multiple TLRs contribute to the host response to certain organisms or that other receptors not yet described can signal through MyD88 and participate in bacterial responses.

Signaling through tumor necrosis factor receptor 1

The TNF-α receptor, tunor necrosis receptor 1 (TNFR1), is distributed widely on the airway epithelia. This is the major signaling pathway for the induction of inflammation by *S. aureus*. The central role of staphylococcal protein A signaling through TNFR1 in the pathogenesis of *S. aureus* pneumonia has been recently demonstrated (Gómez *et al.*, 2004). The lack of either protein A (*spa* null mutants) or TNFR1 (TNFR1-null mice) expression resulted in decreased inflammatory responses and reduced bacterial virulence in a mouse model of pneumonia. The absence of TNFR1-dependent PMN recruitment prevents morbidity due to the pathological consequences of excessive PMN accumulation into the airways. *S. aureus* pneumonia is induced only in mice that express TNFR1, which suggests that, for this pathogen, protein A–TNFR1 signaling is more important than TLR-2 recognition of other cell-wall components and can explain why studies with TLR-2 knockout mice or MyD88 knockout mice did not reveal a prominent role for the TLR-2/MyD88 pathway in models of *S. aureus* pneumonia (Knapp *et al.*, 2004; Skerrett *et al.*, 2004b).

Other receptors: nucleotide-binding oligomerization domain proteins

In addition to TLRs that recognize microbial components in extracellular compartments or on the luminal side of intracellular vesicles, mammalian cells have other surveillance mechanisms to recognize bacteria and other infectious microorganisms in the cytosol of infected cells. These cytosolic receptors are the nucleotide-binding oligomerization domain (Nod) proteins (Inohara *et al.*, 2004), which contain amino-terminal alpha-helix-rich or TIR domains, a central NOD domain, and carboxyl-terminal leucine-rich repeats (LRRs). It is well established that Nod1 and Nod2 function as pattern-recognition molecules and induce signaling pathways upon recognition of bacterial pathogen-associated molecular patterns (PAMPs). Nod1 and Nod2 contain amino-terminal caspase-recruitment domains (CARDs) linked to a centrally placed NOD domain. Nod2 is expressed primarily by immune cells, while Nod1 expression is ubiquitous. Nod1 recognizes peptidoglycan containing mesodiaminopimelate acid found mainly in Gram-negative bacteria (Chamaillard *et al.*, 2003; Girardin *et al.*, 2003a). Nod2 mediates responsiveness to the muramyldipeptide MurNac-L-Ala-D-iso-Gln (MDP) conserved in peptidoglycans of all bacteria (Girardin *et al.*, 2003b; Inohara *et al.*, 2003). The role of Nod2 in *S. pneumoniae* signaling has been demonstrated (Opitz *et al.*, 2004; Schmeck *et al.*, 2004). *S. pneumoniae* can transiently invade epithelial and endothelial cells and signal through the Nod receptors. Lung expression of Nod2 was upregulated during *S. pneumoniae* infection in mice. In addition, in vitro experiments showed that NF-κB activation induced by *S. pneumoniae* depends on Nod2, and this signaling is mediated by IRAK and IRAK2. Although more studies are required, these results suggest that signaling through Nod receptors is the major pathway through which *S. pneumoniae* induces inflammation in the lungs.

SIGNALING PATHWAYS INVOLVED IN CHEMOKINE AND CYTOKINE PRODUCTION BY EPITHELIAL CELLS

The signaling pathways activated through TLRs and TNFR1 in airway epithelial cells by bacterial pathogens resemble the cascades activated via these receptors in immune cells (Hehlgans and Pfeffer, 2005; Takeda and Akira, 2004). TLR signaling is mediated by MyD88, IRAK, and TRAF6, and all these molecules are recruited to the receptor complex with asialoGM1 in lipid raft domains (Soong *et al.*, 2004). Activation of this pathway leads to the nuclear translocation of NF-κB and transcription of pro-inflammatory genes. Protein A signaling through TNFR1 resembles TNF signaling, with recruitment of tumor necrosis factor receptor 1-associated death domain protein (TRADD), receptor-interacting protein (RIP), and TRAF2 to the receptor and activation of p38 and JNK MAPK and activating transcription factor 2 (ATF-2) phosphorylation and translocation to the nucleus (Gómez *et al.*, 2004).

The pulmonary pathogens *S. aureus* and *P. aeruginosa* activate Ca^{2+} fluxes in epithelial cells upon contact with specific receptors. Recognition of asialoGM1/TLR-2 on airway cells initiates 100 nM Ca^{2+} fluxes, sufficient to activate NF-κB and generate CXCL8 and GM-CSF expression (Ratner *et al.*, 2001; Saba *et al.*, 2002). Several other Ca^{2+}-dependent transcription factors are also activated by bacterial ligands, leading to local cytokine expression and mucin production (McNamara *et al.*, 2001). Peptidoglycan activates the leucine zipper containing transcription factors cyclic adenosine monophosphate (cAMP)-responsive element-binding protein (CREB)/ATF and AP-1 (Gupta *et al.*, 1999). CREB is expected to sense changes in cyclic nucleotides released at the surface of the airway in response to Ca^{2+} fluxes. In addition, CREB functions as a coactivator of CCAAT/enhancer binding protein (C/EBP), which regulates the expression of IL-6 (Kovacs *et al.*, 2003).

REGULATION OF INFLAMMATION BY EPITHELIAL CELLS: RECEPTOR SHEDDING

Airway epithelial cells regulate pro-inflammatory signaling. The pulmonary pathogens *S. aureus* and *P. aeruginosa* induce transcription, mobilization to the cell surface, and activation of TNF-α converting enzyme (TACE) in airway epithelial cells (Gómez *et al.*, 2004, 2005). TACE or ADAM 17 is a member of the ADAM (a disintegrin and metalloprotease) family of proteases and is involved in the release of a number of superficial proteins, including the TNF, epidermal growth factor (EGF), and IL-6 receptors (Mezyk *et al.*, 2003). TACE has an important role in the regulation of inflammation. Following bacterial exposure, TACE cleaves TNFR1 from the surface of airway epithelial cells (Gómez *et al.*, 2004); the shed soluble TNFR1 serves to neutralize free TNF-α (produced mainly by immune cells) and protein A in the airway lumen and to prevent further epithelial activation through loss of TNFR1 from the cell. Soluble TNFR1 exerts immunoregulatory functions by induction of apoptosis in monocytes through reverse signaling via membrane-bound TNF-α (Waetzig et al., 2005).

Bacterial activation of TACE induces shedding of the IL-6 receptor alpha from epithelial cells (Gómez *et al.*, 2005). Epithelial responsiveness to IL-6 is dependent upon the presence of two receptors, gp130 and IL-6R-α (gp80) (Bauer *et al.*, 1989; Heinrich *et al.*, 2003). Shed soluble IL-6R-α binds to IL-6, forming a ligand–receptor complex that interacts with membrane-bound gp130 in a high-affinity interaction termed "trans-signaling". This interaction initiates CCL2 expression by epithelial cells, which heralds the shift from acute inflammation (PMN recruitment) to a resolution phase with macrophage/monocyte signaling and clearance of apoptotic PMNs (Amano *et al.*, 2004; Hurst *et al.*, 2001). Shed IL-6R-α induces a decrease in CXCL8 production (Hurst *et al.*, 2001; Marin *et al.*, 2001), probably due to signal transducer and activator of transcription 5 (STAT5)-dependent inhibition of NF-κB (Luo and Yu-Lee, 2000).

Bacterial pathogens can induce both pro-inflammatory and anti-inflammatory signaling in airway epithelial cells, which suggests that mucosal epithelial cells have a primary role in regulating their own signaling capabilities as well as responding to cytokines from exogenoos sources. Shed epithelial receptors decrease the pro-inflammatory signaling induced by immune cells in the lung.

LUNG DAMAGE AND BACTERIAL INVASION OF THE AIRWAY EPITHELIUM

Under specific circumstances, airway infection can lead to invasion, bacteremia, and mortality. *P. aeruginosa* express a type III secretion system that is a major determinant of virulence, allowing the bacteria to inject toxins into host cells. The type III secretion system is associated with acute invasive infection and requires pilin-mediated bacterial–epithelial cell contact (Feldman *et al.*, 1998; Garrity-Ryan *et al.*, 2004; Hauser *et al.*, 1998). This system consists of three components: the secretion apparatus, the translocation or targeting apparatus, and the secreted toxins (effector proteins) and cognate chaperons (Gauthier *et al.*, 2003). *P. aeruginosa* secretes four effector proteins, ExoS, ExoT, ExoU, and ExoY. ExoS can induce inflammatory responses through TLR-2 and TLR-4 (Epelman *et al.*, 2004). ExoU is a potent cytotoxin, and its injection in mammalian cells causes irreversible damage to cellular membranes and rapid necrotic death (McMorran *et al.*, 2003; Sato and Frank, 2004). These toxins also interfere with the cytoskeleton and the normal formation of tight junctions and enable organisms to invade paracellularly (Rajan *et al.*, 2000).

Several animal models have demonstrated the importance of type III secretion proteins in acute *P. aeruginosa* infections. In a mouse model of pneumonia, intravenous administration of antibodies against PcrV, a protein involved in translocation of type III secreted toxins, resulted in survival of the animals (Shime *et al.*, 2001). Anti-PcrV antibodies also significantly reduced lung injury, bacteremia, and plasma TNF-α levels in a rabbit model of pneumonia. Type III secretion protein phenotype analysis may help to distinguish respiratory tract colonization from potentially lethal infection. Antibodies against type III secretory proteins may be useful as adjuvant therapy in people with *P. aeruginosa* infections that demonstrate the type III secretory phenotype (Sadikot *et al.*, 2005).

SUMMARY

Bacterial pathogens in the lung interact with airway epithelial cells by expressing numerous ligands that elicit inflammatory responses through epithelial surface exposed receptors. Failure of the normal innate clearance mechanisms, mucociliary clearance, and the antimicrobial activities of the airway secretions enables organisms to persist in the airway lumen. Both adherent bacteria and shed products are potent stimuli for epithelial proinflammatory chemokine and cytokine production. This serves to recruit PMNs from the circulation into the airways. Recruitment of PMNs to the lung is critical to eradicate respiratory pathogens but is not innocuous to the host. Inflammation is detrimental to the major function of the airway in maintaining an open conduit for gas exchange, and much more so than at other mucosal surfaces. The balance between efficient phagocytosis of inspired bacteria and airway compromise is physiologically critical and determines the outcome of lung infections.

REFERENCES

Adamo, R., Sokol, S., Soong, G., Gómez, M. I., and Prince, A. (2004). *Pseudomonas aeruginosa* flagella activate airway epithelial cells through asialoGM1 and toll-like receptor 2 as well as toll-like receptor 5. *Am. J. Respir. Cell. Mol. Biol.* **30**, 627–634.

Akamine, M., Higa, F., Arakaki, N., *et al.* (2005). Differential roles of Toll-like receptors 2 and 4 in in vitro responses of macrophages to *Legionella pneumophila*. *Infect. Immun.* **73**, 352–361.

Amano, H., Morimoto, K., Senba, M., *et al.* (2004). Essential contribution of monocyte chemoattractant protein-1/C-C chemokine ligand-2 to resolution and repair processes in acute bacterial pneumonia. *J. Immunol.* **172**, 398–409.

Andonegui, G., Bonder, C. S., Green, F., *et al.* (2003). Endothelium-derived Toll-like receptor-4 is the key molecule in LPS-induced neutrophil sequestration into lungs. *J. Clin. Invest.* **111**, 1011–1020.

Andrews, J., Nadjm, B., Gant, V., and Shetty, N. (2003). Community-acquired pneumonia. *Curr. Opin. Pulm. Med.* **9**, 175–180.

Bals, R. and Hiemstra, P. S. (2004). Innate immunity in the lung: how epithelial cells fight against respiratory pathogens. *Eur. Respir. J.* **23**, 327–333.

Bals, R., Weiner, D. J., Meegalla, R. L., Accurso, F., and Wilson, J. M. (2001). Salt-independent abnormality of antimicrobial activity in cystic fibrosis airway surface fluid. *Am. J. Respir. Cell. Mol. Biol.* **25**, 21–25.

Bauer, J., Bauer, T. M., Kalb, T., *et al.* (1989). Regulation of interleukin 6 receptor expression in human monocytes and monocyte-derived macrophages: comparison with the expression in human hepatocytes. *J. Exp. Med.* **170**, 1537–1549.

Becker, M. N., Diamond, G., Verghese, M. W., and Randell, S. H. (2000). CD14-dependent lipopolysaccharide-induced beta-defensin-2 expression in human tracheobronchial epithelium. *J. Biol. Chem.* **275**, 29 731–29 736.

Birchler, T., Seibl, R., Buchner, K., *et al.* (2001). Human Toll-like receptor 2 mediates induction of the antimicrobial peptide human beta-defensin 2 in response to bacterial lipoprotein. *Eur. J. Immunol.* **31**, 3131–3137.

Bogaert, D., De Groot, R., and Hermans, P. W. (2004). *Streptococcus pneumoniae* colonisation: the key to pneumococcal disease. *Lancet Infect. Dis.* **4**, 144–154.

Branger, J., Knapp, S., Weijer, S., *et al.* (2004). Role of Toll-like receptor 4 in Gram-positive and Gram-negative pneumonia in mice. *Infect. Immun.* **72**, 788–794.

Chamaillard, M., Hashimoto, M., Horie, Y., *et al.* (2003). An essential role for NOD1 in host recognition of bacterial peptidoglycan containing diaminopimelic acid. *Nat. Immunol.* **4**, 702–707.

Costerton, J. W., Stewart, P. S., and Greenberg, E. P. (1999). Bacterial biofilms: a common cause of persistent infections. *Science* **284**, 1318–1322.

Craven, D. E. (2000). Epidemiology of ventilator-associated pneumonia. *Chest* **117**, 186S–187S.

Dale, S. E., Sebulsky, M. T., and Heinrichs, D. E. (2004). Involvement of SirABC in iron-siderophore import in *Staphylococcus aureus*. *J. Bacteriol.* **186**, 8356–8362.

Diamond, G., Legarda, D., and Ryan, L. K. (2000). The innate immune response of the respiratory epithelium. *Immunol. Rev.* **173**, 27–38.

DiMango, E., Zar, H. J., Bryan, R., and Prince, A. (1995). Diverse *Pseudomonas aeruginosa* gene products stimulate respiratory epithelial cells to produce interleukin-8. *J. Clin. Invest.* **96**, 2204–2210.

DiMango, E., Ratner, A. J., Bryan, R., Tabibi, S., and Prince, A. (1998). Activation of NF-kappaB by adherent *Pseudomonas aeruginosa* in normal and cystic fibrosis respiratory epithelial cells. *J. Clin. Invest.* **101**, 2598–2605.

Dohrman, A., Miyata, S., Gallup, M., *et al.* (1998). Mucin gene (MUC 2 and MUC 5AC) upregulation by Gram-positive and Gram-negative bacteria. *Biochim. Biophys. Acta* **1406**, 251–259.

Eaves-Pyles, T., Murthy, K., Liaudet, L., *et al.* (2001). Flagellin, a novel mediator of *Salmonella*-induced epithelial activation and systemic inflammation: I kappa B alpha degradation, induction of nitric oxide synthase, induction of proinflammatory mediators, and cardiovascular dysfunction. *J. Immunol.* **166**, 1248–1260.

Epelman, S., Stack, D., Bell, C., *et al.* (2004). Different domains of *Pseudomonas aeruginosa* exoenzyme S activate distinct TLRs. *J. Immunol.* **173**, 2031–2040.

Feldman, M., Bryan, R., Rajan, S., *et al.* (1998). Role of flagella in pathogenesis of *Pseudomonas aeruginosa* pulmonary infection. *Infect. Immun.* **66**, 43–51.

Feng, C. G., Scanga, C. A., Collazo-Custodio, C. M., *et al.* (2003). Mice lacking myeloid differentiation factor 88 display profound defects in host resistance and immune responses to *Mycobacterium avium* infection not exhibited by Toll-like receptor 2 (TLR2)- and TLR4-deficient animals. *J. Immunol.* **171**, 4758–4764.

Foster, T. J. and McDevitt, D. (1994). Surface-associated proteins of *Staphylococcus aureus*: their possible roles in virulence. *FEMS Microbiol. Lett.* **118**, 199–205.

Garrity-Ryan, L., Shafikhani, S., Balachandran, P., *et al.* (2004). The ADP ribosyltransferase domain of *Pseudomonas aeruginosa* ExoT contributes to its biological activities. *Infect. Immun.* **72**, 546–558.

Gauthier, A., Thomas, N. A., and Finlay, B. B. (2003). Bacterial injection machines. *J. Biol. Chem.* **278**, 25 273–25 276.

Gewirtz, A. T., Navas, T. A., Lyons, S., Godowski, P. J., and Madara, J. L. (2001). Cutting edge: bacterial flagellin activates basolaterally expressed TLR5 to induce epithelial proinflammatory gene expression. *J. Immunol.* **167**, 1882–1885.

Girard, R., Pedron, T., Uematsu, S., *et al.* (2003). Lipopolysaccharides from *Legionella* and *Rhizobium* stimulate mouse bone marrow granulocytes via Toll-like receptor 2. *J. Cell Sci.* **116**, 293–302.

Girardin, S. E., Boneca, I. G., Carneiro, L. A., *et al.* (2003a). Nod1 detects a unique muropeptide from Gram-negative bacterial peptidoglycan. *Science* **300**, 1584–1587.

Girardin, S. E., Boneca, I. G., Viala, J., *et al.* (2003b). Nod2 is a general sensor of peptidoglycan through muramyl dipeptide (MDP) detection. *J. Biol. Chem.* **278**, 8869–8872.

Gómez, M. I., Lee, A., Reddy, B., *et al.* (2004). *Staphylococcus aureus* protein A induces airway epithelial inflammatory responses by activating TNFR1. *Nat. Med.* **10**, 842–848.

Gómez, M. I., Sokol, S., Muir, A. B., *et al.* (2005). Bacterial induction of TNF-alpha converting enzyme expression and IL-6 receptor alpha shedding regulates airway inflammatory signaling. *J. Immunol.* **175**, 1930–1936.

Goss, C. H. and Rosenfeld, M. (2004). Update on cystic fibrosis epidemiology. *Curr. Opin. Pulm. Med.* **10**, 510–514.

Greene, C. M., Carroll, T. P., Smith, S. G., *et al.* (2005). TLR-induced inflammation in cystic fibrosis and non-cystic fibrosis airway epithelial cells. *J. Immunol.* **174**, 1638–1646.

Guillot, L., Medjane, S., Le-Barillec, K., *et al.* (2004). Response of human pulmonary epithelial cells to lipopolysaccharide involves Toll-like receptor 4 (TLR4)-dependent signaling pathways: evidence for an intracellular compartmentalization of TLR4. *J. Biol. Chem.* **279**, 2712–2718.

Gupta, D., Wang, Q., Vinson, C., and Dziarski, R. (1999). Bacterial peptidoglycan induces CD14-dependent activation of transcription factors CREB/ATF and AP-1. *J. Biol. Chem.* **274**, 14 012–14 020.

Hauser, A. R., Fleiszig, S., Kang, P. J., Mostov, K., and Engel, J. N. (1998). Defects in type III secretion correlate with internalization of *Pseudomonas aeruginosa* by epithelial cells. *Infect. Immun.* **66**, 1413–1420.

Hayashi, F., Smith, K. D., Ozinsky, A., *et al.* (2001). The innate immune response to bacterial flagellin is mediated by Toll-like receptor 5. *Nature* **410**, 1099–1103.

Hehlgans, T. and Pfeffer, K. (2005). The intriguing biology of the tumour necrosis factor/tumour necrosis factor receptor superfamily: players, rules and the games. *Immunology* **115**, 1–20.

Heinrich, P. C., Behrmann, I., Haan, S., *et al.* (2003). Principles of interleukin (IL)-6-type cytokine signalling and its regulation. *Biochem. J.* **374**, 1–20.

Hiemstra, P. S. (2001). Epithelial antimicrobial peptides and proteins: their role in host defence and inflammation. *Paediatr. Respir. Rev.* **2**, 306–310.

Hiratsuka, T., Nakazato, M., Date, Y., *et al.* (1998). Identification of human beta-defensin-2 in respiratory tract and plasma and its increase in bacterial pneumonia. *Biochem. Biophys. Res. Commun.* **249**, 943–947.

Hoiby, N., Krogh Johansen, H., Moser, C., *et al.* (2001). *Pseudomonas aeruginosa* and the in vitro and in vivo biofilm mode of growth. *Microbes Infect.* **3**, 23–35.

Hoshino, K., Takeuchi, O., Kawai, T., *et al.* (1999). Cutting edge: Toll-like receptor 4 (TLR4)-deficient mice are hyporesponsive to lipopolysaccharide: evidence for TLR4 as the Lps gene product. *J. Immunol.* **162**, 3749–3752.

Hurst, S. M., Wilkinson, T. S., McLoughlin, R. M., *et al.* (2001). Il-6 and its soluble receptor orchestrate a temporal switch in the pattern of leukocyte recruitment seen during acute inflammation. *Immunity* **14**, 705–714.

Inohara, N., Ogura, Y., Fontalba, A., *et al.* (2003). Host recognition of bacterial muramyl dipeptide mediated through NOD2. Implications for Crohn's disease. *J. Biol. Chem.* **278**, 5509–5512.

Inohara, N., Chamaillard, M., McDonald, C., and Nunez, G. (2004). NOD-LRR proteins: role in host–microbial interactions and inflammatory disease. *Annu. Rev. Biochem.* **74**, 355–383.

Kajikawa, O., Frevert, C. W., Lin, S. M., *et al.* (2005). Gene expression of Toll-like receptor-2, Toll-like receptor-4, and MD2 is differentially regulated in rabbits with *Escherichia coli* pneumonia. *Gene* **344**, 193–202.

Kawai, T., Adachi, O., Ogawa, T., Takeda, K., and Akira, S. (1999). Unresponsiveness of MyD88-deficient mice to endotoxin. *Immunity* **11**, 115–122.

Knapp, S., Wieland, C. W., van't Veer, C., *et al.* (2004). Toll-like receptor 2 plays a role in the early inflammatory response to murine pneumococcal pneumonia but does not contribute to antibacterial defense. *J. Immunol.* **172**, 3132–3138.

Kovacs, K. A., Steinmann, M., Magistretti, P. J., Halfon, O., and Cardinaux, J. R. (2003). CCAAT/enhancer-binding protein family members recruit the coactivator CREB-binding protein and trigger its phosphorylation. *J. Biol. Chem.* **278**, 36 959–36 965.

Krivan, H. C., Roberts, D. D., and Ginsburg, V. (1988). Many pulmonary pathogenic bacteria bind specifically to the carbohydrate sequence GalNAc beta 1–4Gal found in some glycolipids. *Proc. Natl. Acad. Sci. U. S. A.* **85**, 6157–6161.

Lawson, P. R. and Reid, K. B. (2000). The roles of surfactant proteins A and D in innate immunity. *Immunol. Rev.* **173**, 66–78.

LeVine, A. M., Bruno, M. D., Huelsman, K. M., *et al.* (1997). Surfactant protein A-deficient mice are susceptible to group B streptococcal infection. *J. Immunol.* **158**, 4336–4340.

LeVine, A. M., Kurak, K. E., Bruno, M. D., *et al.* (1998). Surfactant protein-A-deficient mice are susceptible to *Pseudomonas aeruginosa* infection. *Am. J. Respir. Cell. Mol. Biol.* **19**, 700–708.

LeVine, A. M., Reed, J. A., Kurak, K. E., Cianciolo, E., and Whitsett, J. A. (1999). GM-CSF-deficient mice are susceptible to pulmonary group B streptococcal infection. *J. Clin. Invest.* **103**, 563–569.

LeVine, A. M., Whitsett, J. A., Gwozdz, J. A., *et al.* (2000). Distinct effects of surfactant protein A or D deficiency during bacterial infection on the lung. *J. Immunol.* **165**, 3934–3940.

Lindsay, J. A. and Holden, M. T. (2004). *Staphylococcus aureus*: superbug, super genome? *Trends Microbiol.* **12**, 378–385.

Luo, G. and Yu-Lee, L. (2000). Stat5b inhibits NFkappaB-mediated signaling. *Mol. Endocrinol.* **14**, 114–123.

Malley, R., Henneke, P., Morse, S. C., *et al.* (2003). Recognition of pneumolysin by Toll-like receptor 4 confers resistance to pneumococcal infection. *Proc. Natl. Acad. Sci. U. S. A.* **100**, 1966–1971.

Marin, V., Montero-Julian, F. A., Gres, S., *et al.* (2001). The IL-6-soluble IL-6Ralpha autocrine loop of endothelial activation as an intermediate between acute and chronic inflammation: an experimental model involving thrombin. *J. Immunol.* **167**, 3435–3442.

McMorran, B., Town, L., Costelloe, E., *et al.* (2003). Effector ExoU from the type III secretion system is an important modulator of gene expression in lung epithelial cells in response to *Pseudomonas aeruginosa* infection. *Infect. Immun.* **71**, 6035–6044.

McNamara, N., Khong, A., McKemy, D., *et al.* (2001). ATP transduces signals from ASGM1, a glycolipid that functions as a bacterial receptor. *Proc. Natl. Acad. Sci. U. S. A.* **98**, 9086–9091.

Mehrad, B. and Standiford, T. J. (1999). Role of cytokines in pulmonary antimicrobial host defense. *Immunol. Res.* **20**, 15–27.

Mezyk, R., Bzowska, M., and Bereta, J. (2003). Structure and functions of tumor necrosis factor-alpha converting enzyme. *Acta Biochim. Pol.* **50**, 625–645.

Muir, A., Soong, G., Sokol, S., *et al.* (2004). Toll-like receptors in normal and cystic fibrosis airway epithelial cells. *Am. J. Respir. Cell. Mol. Biol.* **30**, 777–783.

Nakamura, H., Yoshimura, K., McElvaney, N. G., and Crystal, R. G. (1992). Neutrophil elastase in respiratory epithelial lining fluid of individuals with cystic fibrosis induces interleukin-8 gene expression in a human bronchial epithelial cell line. *J. Clin. Invest.* **89**, 1478–1484.

Opitz, B., Puschel, A., Schmeck, B., *et al.* (2004). Nucleotide-binding oligomerization domain proteins are innate immune receptors for internalized *Streptococcus pneumoniae*. *J. Biol. Chem.* **279**, 36 426–36 432.

Parsek, M. R. and Singh, P. K. (2003). Bacterial biofilms: an emerging link to disease pathogenesis. *Annu. Rev. Microbiol.* **57**, 677–701.

Poltorak, A., He, X., Smirnova, I., *et al.* (1998). Defective LPS signaling in C3H/HeJ and C57BL/10ScCr mice: mutations in Tlr4 gene. *Science* **282**, 2085–2088.

Power, M. R., Peng, Y., Maydanski, E., Marshall, J. S., and Lin, T. J. (2004). The development of early host response to *Pseudomonas aeruginosa* lung infection is critically dependent on myeloid differentiation factor 88 in mice. *J. Biol. Chem.* **279**, 49 315–49 322.

Rajan, S., Cacalano, G., Bryan, R., *et al.* (2000). *Pseudomonas aeruginosa* induction of apoptosis in respiratory epithelial cells: analysis of the effects of cystic fibrosis transmembrane conductance regulator dysfunction and bacterial virulence factors. *Am. J. Respir. Cell. Mol. Biol.* **23**, 304–312.

Ratner, A. J., Bryan, R., Weber, A., *et al.* (2001). Cystic fibrosis pathogens activate Ca^{2+}-dependent mitogen-activated protein kinase signaling pathways in airway epithelial cells. *J. Biol. Chem.* **276**, 19 267–19 275.

Rochelle, L. G., Fischer, B. M., and Adler, K. B. (1998). Concurrent production of reactive oxygen and nitrogen species by airway epithelial cells in vitro. *Free Radic. Biol. Med.* **24**, 863–868.

Saba, S., Soong, G., Greenberg, S., and Prince, A. (2002). Bacterial stimulation of epithelial G-CSF and GM-CSF expression promotes PMN survival in CF airways. *Am. J. Respir. Cell. Mol. Biol.* **27**, 561–567.

Sadikot, R. T., Blackwell, T. S., Christman, J. W., and Prince, A. S. (2005). Pathogen–host interactions in *Pseudomonas aeruginosa* pneumonia: the state of the art. *Am. J. Respir. Crit. Care Med.* **171**, 1209–1223.

Sano, H. and Kuroki, Y. (2005). The lung collectins, SP-A and SP-D, modulate pulmonary innate immunity. *Mol. Immunol.* **42**, 279–287.

Sato, H. and Frank, D. W. (2004). ExoU is a potent intracellular phospholipase. *Mol. Microbiol.* **53**, 1279–1290.

Schmeck, B., Zahlten, J., Moog, *et al.* (2004). *Streptococcus pneumoniae*-induced p38 MAPK-dependent phosphorylation of RelA at the interleukin-8 promotor. *J. Biol. Chem.* **279**, 53 241–53 247.

Schultz, M. J., Rijneveld, A. W., Florquin, S., *et al.* (2002). Role of interleukin-1 in the pulmonary immune response during *Pseudomonas aeruginosa* pneumonia. *Am. J. Physiol. Lung Cell. Mol. Physiol.* **282**, L285–290.

Schurr, J. R., Young, E., Byrne, P., *et al.* (2005). Central role of toll-like receptor 4 signaling and host defense in experimental pneumonia caused by Gram-negative bacteria. *Infect. Immun.* **73**, 532–545.

Shaw, M. J. (2005). Ventilator-associated pneumonia. *Curr. Opin. Pulm. Med.* **11**, 236–241.

Shime, N., Sawa, T., Fujimoto, J., *et al.* (2001). Therapeutic administration of anti-PcrV F(ab′)(2) in sepsis associated with *Pseudomonas aeruginosa. J. Immunol.* **167**, 5880–5886.

Singh, P. K., Jia, H. P., Wiles, K., *et al.* (1998). Production of beta-defensins by human airway epithelia. *Proc. Natl. Acad. Sci. U. S. A.* **95**, 14 961–14 966.

Singh, P. K., Schaefer, A. L., Parsek, M. R., *et al.* (2000). Quorum-sensing signals indicate that cystic fibrosis lungs are infected with bacterial biofilms. *Nature* **407**, 762–764.

Singh, P. K., Parsek, M. R., Greenberg, E. P., and Welsh, M. J. (2002). A component of innate immunity prevents bacterial biofilm development. *Nature* **417**, 552–555.

Skerrett, S. J., Liggitt, H. D., Hajjar, A. M., *et al.* (2004a). Respiratory epithelial cells regulate lung inflammation in response to inhaled endotoxin. *Am. J. Physiol. Lung Cell. Mol. Physiol.* **287**, L143–152.

Skerrett, S. J., Liggitt, H. D., Hajjar, A. M., and Wilson, C. B. (2004b). Cutting edge: myeloid differentiation factor 88 is essential for pulmonary host defense against *Pseudomonas aeruginosa* but not *Staphylococcus aureus. J. Immunol.* **172**, 3377–3381.

Smith, K. D., Andersen-Nissen, E., Hayashi, F., *et al.* (2003). Toll-like receptor 5 recognizes a conserved site on flagellin required for protofilament formation and bacterial motility. *Nat. Immunol.* **4**, 1247–1253.

Soong, G., Reddy, B., Sokol, S., Adamo, R., and Prince, A. (2004). TLR2 is mobilized into an apical lipid raft receptor complex to signal infection in airway epithelial cells. *J. Clin. Invest.* **113**, 1482–1489.

Strieter, R. M., Belperio, J. A., and Keane, M. P. (2002). Cytokines in innate host defense in the lung. *J. Clin. Invest.* **109**, 699–705.

Strieter, R. M., Belperio, J. A., and Keane, M. P. (2003). Host innate defenses in the lung: the role of cytokines. *Curr. Opin. Infect. Dis.* **16**, 193–198.

Takeda, K. and Akira, S. (2004). TLR signaling pathways. *Semin. Immunol.* **16**, 3–9.

Takeda, K., Kaisho, T., and Akira, S. (2003). Toll-like receptors. *Annu. Rev. Immunol.* **21**, 335–376.

Takeuchi, O., Hoshino, K., and Akira, S. (2000). Cutting edge: TLR2-deficient and MyD88-deficient mice are highly susceptible to *Staphylococcus aureus* infection. *J. Immunol.* **165**, 5392–5396.

Tang, H., Kays, M., and Prince, A. (1995). Role of *Pseudomonas aeruginosa* pili in acute pulmonary infection. *Infect. Immun.* **63**, 1278–1285.

Toshchakov, V., Jones, B. W., Perera, P. Y., *et al.* (2002). TLR4, but not TLR2, mediates IFN-beta-induced STAT1alpha/beta-dependent gene expression in macrophages. *Nat. Immunol.* **3**, 392–398.

Tsai, W. C., Strieter, R. M., Wilkowski, J. M., *et al.* (1998). Lung-specific transgenic expression of KC enhances resistance to *Klebsiella pneumoniae* in mice. *J. Immunol.* **161**, 2435–2440.

Vasil, M. L. and Ochsner, U. A. (1999). The response of *Pseudomonas aeruginosa* to iron: genetics, biochemistry and virulence. *Mol. Microbiol.* **34**, 399–413.

Waetzig, G. H., Rosenstiel, P., Arlt, A., *et al.* (2005). Soluble tumor necrosis factor (TNF) receptor-1 induces apoptosis via reverse TNF signaling and autocrine transforming growth factor-beta1. *FASEB J.* **19**, 91–93.

Wang, X., Moser, C., Louboutin, J. P., *et al.* (2002). Toll-like receptor 4 mediates innate immune responses to *Haemophilus influenzae* infection in mouse lung. *J. Immunol.* **168**, 810–815.

Wyant, T. L., Tanner, M. K., and Sztein, M. B. (1999). *Salmonella typhi* flagella are potent inducers of proinflammatory cytokine secretion by human monocytes. *Infect. Immun.* **67**, 3619–3624.

Xiao, R. and Kisaalita, W. S. (1997). Iron acquisition from transferrin and lactoferrin by *Pseudomonas aeruginosa* pyoverdin. *Microbiology* **143 (Pt 7)**, 2509–2515.

Yamamoto, M. and Akira, S. (2004). [TIR domain-containing adaptors regulate TLR-mediated signaling pathways.] *Nippon Rinsho* **62**, 2197–2203.

Yarwood, J. M. and Schlievert, P. M. (2003). Quorum sensing in *Staphylococcus* infections. *J. Clin. Invest.* **112**, 1620–1625.

CHAPTER 12

Interaction of *Helicobacter pylori* with the gastric mucosa

D. Scott Merrell

INTRODUCTION

Microbes populate virtually every square inch of the earth's surface. This success is due in large part to the numerous adaptation strategies they have developed to facilitate survival/persistence. The practical requirement for microbial resilience is particularly true of pathogenic microorganisms, as they must cope with host environments that are often actively adversarial. This is the case with *Helicobacter pylori*, which colonizes in the unlikely niche of the human stomach. Once in the stomach, *H. pylori* establishes a chronic infection in a manner that shows many similarities to what we typically think of as behavior of the host-adapted flora – "with an eye towards persistence rather than towards causing disease" (Merrell and Falkow, 2004). This is evidenced by the fact that severe disease typically takes decades to develop. This delayed development of overt pathology suggests that there is a balance shift that causes colonization to go awry and leads to disease. This shift is likely due to a combination of physiological and genetic factors for both participants of the host–pathogen interaction. On the bacterial front, *H. pylori* interacts with gastric mucosal cells and expresses a repertoire of factors that result in alterations in host cell signaling. These changes in host cell signaling are likely ultimately responsible for *H. pylori*-induced disease. Thus, *H. pylori* represents a model organism in terms of its ability to chronically exploit the gastric niche as well as to manipulate gastric epithelial cells (Figure 12.1).

Bacterial–Epithelial Cell Cross-Talk: Molecular Mechanisms in Pathogenesis, ed. Beth A. McCormick. Published by Cambridge University Press.

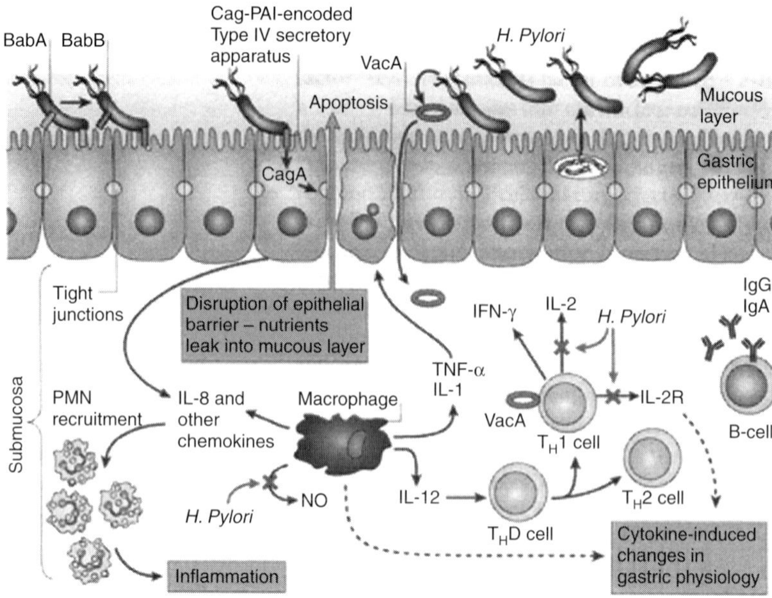

Figure 12.1 Persistent *Helicobacter pylori* infection involves interplay between bacterial and host factors. *H. pylori* binds to a subpopulation of gastric epithelial cells using BabA and other adhesions. In strains that carry the Cag pathogenicity island (PAI), a type IV secretion apparatus is used to inject CagA into the host cell. This results in increased expression of interleukin (IL) 8 (IL-8) and other chemokines that recruit polymorphonuclear cells (PMN) to the site of infection. Additionally, CagA associates with tight-junction proteins and disrupts the epithelial barrier and alters various downstream signaling pathways. VacA induces apoptosis in epithelial cells by targeting mitochondria. In the chronic phase of *H. pylori* infection, T-cells and B-cells infiltrate and are targeted by VacA to stop proper antigen presentation and T cell proliferation. Reproduced with permission from D. M. Monack, A. Mueller, and S. Falkow. *Nature Reviews Microbiology* **2**, 747–65. Copyright © 2004 Macmillan Magazines Ltd.

IFN-γ, interferon gamma; IgA, immunoglobulin A; IgG, immunoglobulin G; NO, nitric oxide; TFN-α, tumor necrosis factor alpha.

HELICOBACTER: THE NEW KID ON THE BLOCK

From the perspective of the scientific community, *H. pylori* is a relatively new player in the arena of bacterial pathogenesis. Of course, this has less to do with the emergence of the pathogen than with the fact that we failed to detect the bacterium until relatively recently. It was not until 1982 that Marshall and Warren discovered *H. pylori* in biopsies from people suffering from antral gastritis. Marshall and Warren (1984) subsequently went on to propose a causal relationship between *H. pylori* infection and gastritis. This suggestion was considered controversial for a number a years. However, subsequent studies showed that the majority of children suffering from gastritis also tested positive for *H. pylori* carriage. Remarkably, this gastritis was resolved by antibiotic treatment and eradication of *H. pylori* from the gastric mucosa (Drumm *et al.*, 1987; Valle *et al.*, 1991; Yeung *et al.*, 1990). These findings pushed *H. pylori* to the forefront of the consciousness of the scientific community and stimulated a flurry of research on the etiology of gastric disease that continues today.

Since its discovery, *H. pylori* has been shown to persistently infect more than 50% of the world's population (Matysiak-Budnik and Megraud, 1997). Moreover, it is now recognized that *H. pylori* is a significant cause of worldwide morbidity and mortality and colonization is associated with numerous forms of gastric disease. In its mildest form, chronic infection by *H. pylori* causes sustained inflammation of the gastric mucosa. This is likely due to enhanced secretion of interleukin 8 (IL-8) by gastric epithelial cells, which leads to the influx of polymorphonuclear leukocytes and lymphocytes (reviewed in Dunn *et al.*, 1997; Ernst and Gold, 2000). In addition, atrophy and increased gastric epithelial cell turnover often occur. These stages of increased cellular proliferation are often associated with increased permeability across the epithelial membrane barrier, and it is perhaps this factor that leads to increased access of luminal acid and pepsin to the underlying tissue and the subsequent development of ulcer disease (Borch *et al.*, 1998; Curtis and Gall, 1992; Ernst and Gold, 2000; Rabassa *et al.*, 1996; Vera *et al.*, 1997). *H. pylori* infection is associated with duodenal and gastric ulcers, with 90% and 75% of these disease cases, respectively, attributed to gastric colonization by the organism (Covacci *et al.*, 1999).

In addition to its role in the development of peptic ulcers, it is now well established that *H. pylori* infection is causally associated with two forms of gastric malignancy, mucosa-associated lymphoid tissue (MALT) lymphoma and adenocarcinoma (Blaser, 1998; Dunn *et al.*, 1997; Parsonnet *et al.*, 1991). With the high incidence of *H. pylori* infection, it is perhaps not surprising that gastric cancer is the world's second most common cause of cancer-related morbidity and mortality (Neugut *et al.*, 1996). The relationship between *H. pylori* infection and the development of gastric carcinoma was proposed following epidemiological, cross-sectional, and prospective studies of *H. pylori* infections that showed that infection is associated with a 2.7- to 12-fold increased risk of developing gastric cancer (Blaser and Parsonnet, 1994; Blaser *et al.*, 1995). Because of this, *H. pylori* was classified as a class I carcinogen in 1994 by the World Health Organization, and to date it remains the only bacterium to receive such status (IARC Working Group on the Evaluation of Carcinogenic Risks to Humans, 1994).

Although we lack a thorough understanding of the process of infection and pathogenesis of *H. pylori*, progress has been made in elucidating some key components of these processes. For instance, although we have not identified an environmental reservoir for the bacterium and do not know the specifics of bacterial transmission, we do know that the pathogen uses a repertoire of genes that are necessary for survival and colonization within the gastric niche. Subsets of these genes play specific roles in survival in the face of the harsh environmental conditions found within the stomach, while other subsets are important for proper localization of the bacterium to the mucous layer overlaying the gastric epithelium.

HELICOBACTER PYLORI AND THE INNATE IMMUNE SYSTEM

Because it requires survival during extreme changes in pH, *H. pylori* acid resistance (AR) has been the focus of considerable study and has been shown to be linked intricately to the bacterium's ability to produce copious amounts of the enzyme urease (Mobley *et al.*, 1995). Urease functions by hydrolyzing urea to carbon dioxide and ammonia. This activity helps maintain a proton-motive force across the inner membrane of the bacterium (Meyer-Rosberg *et al.*, 1996; Scott *et al.*, 1998) and is essential for colonization and persistence in the stomach (Andrutis *et al.*, 1995; Eaton and Krakowka, 1994; Ferrero *et al.*, 1992; Tsuda *et al.*, 1994). It was previously believed that cell-surface localized urease created a neutral microenvironment that was conducive to bacterial survival (Phadnis *et al.*, 1996), but more recent work has indicated that the intracellular enzyme appears to play a more important role in AR (Scott *et al.*, 1998). The urea substrate gains access to the cytoplasmically localized enzyme via the activity of an inner-membrane proton-gated urea-specific channel formed by UreI (Weeks *et al.*, 2000). The UreI pore opens as the pH drops below 6.5 and urea moves into the bacterial cytoplasm where the urease enzyme is active.

Another front-line strategy used by the body to fight bacterial infection is the production of nitric oxide (NO). Reactive nitrogen intermediates are effective antimicrobial agents, and *H. pylori* has been shown to be sensitive to chemical sources of NO (Dykhuizen *et al.*, 1998; Kuwahara *et al.*, 2000). Additionally, it is known that *H. pylori* infection induces significant increases in inducible nitric oxide synthase (iNOS) in macrophages and in gastric tissues (Fu *et al.*, 1999; Wilson *et al.*, 1996). Despite this, *H. pylori* thrives within the gastric mucosa. This suggests strongly that the bacterium has developed mechanisms to avoid NO-dependent killing. Indeed, this is the case, and a mechanistic explanation has been provided by the elucidation that a bacterial-produced enzyme competes with the eukaryotic iNOS for an essential substrate required for NO production (Gobert *et al.*, 2001). Encoded by *rocF*, arginase is associated with the bacterial cell envelope and competes with the host iNOS for L-arginine. L-Arginine is converted to urea and L-ornithine rather than NO. The importance of this competition for substrate is evidenced by the fact that a *rocF* mutant of *H. pylori* is efficiently killed in a NO-dependent manner, whereas the wild-type bacteria survive (Gobert *et al.*, 2001). Moreover, the *rocF* mutant is mildly attenuated for colonization in a murine model of infection (McGee *et al.*, 1999), suggesting that the ability to inhibit NO production via competition for a limiting substrate is important for survival and persistence in vivo.

An additional mechanism for counteracting the host innate immune system involves the apparent lack of immunostimulatory properties of *H. pylori* via the Toll-like receptors (TLRs). Bacterial lipopolysaccharide (LPS) is typically recognized by TLR-4 and causes a robust pro-inflammatory response. Work has shown that primary stomach epithelial cells and gastric epithelial cell lines do not react to the LPS from *H. pylori* (Smith *et al.*, 2003). Additionally, it has been shown that recognition of *H. pylori* flagellin via TLR-5 is markedly different from that observed with other Gram-negative pathogens. *H. pylori* flagellins do not signal via TLR-5 to stimulate an innate immune response (Lee *et al.*, 2003). The lack of stimulation via these traditional pathways is interesting in light of the fact that in vivo, *H. pylori* is reported to cause a strong inflammatory response via nuclear factor κB (NF-κB) activation, and this has been suggested to be required for establishment of a chronic infection (Rhen *et al.*, 2003). Thus, the bacterium has chosen to bypass the innate TLR system and engages inflammatory mediators of its own choosing.

LOCATION, LOCATION, LOCATION

After entering the stomach lumen, *H. pylori* penetrates and colonizes the mucous layer overlaying the gastric epithelium. Studies investigating the spatial distribution of *H. pylori* within the gastric mucous layer of Mongolian gerbils showed that localization of *H. pylori* was markedly different from that of the related *Helicobacter* species, *Helicobacter felis* (Schreiber *et al.*, 2004). Whereas *H. felis* avoids close proximity to the murine tissue surface, the majority of *H. pylori* were found swimming in the layer immediately adjacent to the epithelial cells (less than 5 μm from the cell surface) or adhering to the cells. This study went on to define the gradients used by *Helicobacter* within the gastric mucus to determine orientation. Analysis of the urea/ammonium gradient, the bicarbonate/CO_2 gradient, and the luminal and arterial pH gradient showed that orientation of bacteria in the mucus layer could be altered by the simultaneous reduction of arterial pH and bicarbonate concentration (Schreiber *et al.*, 2004). Thus, *H. pylori* uses the gastric mucus pH gradient in order to achieve proper orientation in the stomach. This may help to partially explain the prior observation that a large number of genes involved in motility are regulated by low pH and that overall motility and swimming velocity are increased by acidic pH (Merrell *et al.*, 2003).

Not surprisingly, localization to the proper gastric niche requires flagella and motility. Thus, both factors are required for colonization of *H. pylori* in numerous models of infection (Eaton *et al.*, 1992, 1996; McGee *et al.*, 2002; Ottemann and Lowenthal, 2002). In addition, chemotaxis – the ability of microorganisms to move in response to chemical cues – seems to play a role in vivo, as non-chemotactic mutants exhibit various degrees of attenuation (Andermann *et al.*, 2002; Terry *et al.*, 2005). Interesting new studies suggest that components of the chemotaxis system also play a novel role in the host inflammatory response in response to chronic *H. pylori* colonization (McGee *et al.*, 2005). Chemoreceptors, also known as methyl-accepting chemotaxis proteins, affect flagellar rotation via monitoring environmental cues. Upon receiving a signal, chemoreceptors transmit the ligand-binding information to a signal transduction cascade that directly controls the direction of flagellar rotation. *H. pylori* is predicted to encode four chemoreceptors, *tlpA*, *tlpB*, *tlpC*, and *hlyB*, and mutational analysis has shown that, of these, *tlpA* and *tlpC* are required for efficient colonization within the murine stomach. In contrast, a *tlpB* mutant colonizes at levels that are similar to the wild type (McGee *et al.*, 2005). However, the *tlpB* mutant showed a substantial decrease in gastric inflammation in comparison with wild-type *H. pylori*. In addition, the distribution of infiltrating immune cells was skewed in the *tlpB* mutant colonization assays. Whereas wild-type bacteria induce inflammation that is significantly enriched for lymphocytes but contains fewer neutrophils, the *tlpB* mutant induced infiltration of approximately equal numbers of both lymphocytes and neutrophils (McGee *et al.*, 2005). This strongly suggests that TlpC is involved in the normal immune response, although the mechanistic nature of this involvement is not immediately evident.

Although the majority of *H. pylori* remain within the gastric mucus, a subpopulation of the bacteria bind the surface of the mucus cells. Interaction at the cell surface takes place at several levels. Carbohydrate structures, such as the fucosylated Lewis b (Le^b) histo-blood group antigen and the sialyl Lewis x (s-Le^X) glycosphingolipid serve as receptors for *H. pylori*. Binding of these carbohydrate substrates is mediated by the *H. pylori* adhesions BabA and SabA, respectively (Ilver *et al.*, 1998; Mahdavi *et al.*, 2002). The BabA and SabA adhesions belong to a large family of *H. pylori* outer-membrane proteins (Hop) that show extensive homology (Alm *et al.*, 2000). BabA has been shown to be deleted or mutated during chronic colonization of a Rhesus macaque model of *H. pylori* infection (Solnick *et al.*, 2004). In some instances, the *babA* gene was replaced by *babB*, an uncharacterized Hop that shows strong homology to *babA*. Similarly, a subset of clinical isolates of *H. pylori* were shown to have deleted *babA* and duplicated *babB* (Solnick *et al.*, 2004), indicating that *H. pylori* regulates Hop gene expression in humans through similar mechanisms and suggesting that this regulation may facilitate adherence to the gastric epithelium and promote chronic infection. The importance of the Hops in chronic infection has been further supported by two additional discoveries. First, some Le^b non-binding strains actually carry a silent copy of the *babA* gene, and the non-binding phenotype is meta-stable; a portion of the non-binding population gains the virulence-associated Le^b binding phenotype (Backstrom *et al.*, 2004). This meta-stability is due to recombination of the silent *babA* gene into the *babB* locus to create an expressed and functional BabB/A chimeric adhesin. Expression of this chimeric protein is regulated further via phase variation through slipped-strand mispairing (Backstrom *et al.*, 2004). Second, binding of the fucosylated blood group antigens via BabA has been shown to undergo diversification based on population structure (Aspholm-Hurtig *et al.*, 2004). This was elucidated due to the fact that more than 95% of the *H. pylori* strains that are Le^b-binders are "generalists," i.e. they bind A, B, and O blood-group antigens equally well. On the other hand, in a South American Amerindian population where there is a unique predominance of the O blood group, 60% of the adherent strains bind specifically to the O blood group antigen; they have become "specialists." This specialization is mediated at the level of diversifying selection in the *babA* sequence (Aspholm-Hurtig *et al.*, 2004). Such intricate levels of regulation of the Le^b binding phenotype highlight the importance of this adhesin–receptor interaction within the context of a natural infection.

Although we typically think of *H. pylori* as an extracellular pathogen, work by several groups indicates that a subset of *H. pylori* cells have the ability to survive intracellularly (Amieva *et al.*, 2002; Oh *et al.*, 2005; Semino-Mora *et al.*, 2003). Previous analysis of human biopsy samples revealed *H. pylori* cells that appeared to be intracellular (Bode *et al.*, 1988; Foliguet *et al.*, 1989), but these observations were hindered by an inability to determine whether these bacteria were viable. However, utilizing a tissue culture model, Amieva *et al.* (2002) showed that *H. pylori* induces the formation of numerous large vacuoles within the cell monolayer and that many of these vacuoles contained *H. pylori*. Time-lapse video microscopy and gentamicin protection assays proved viability of the intracellular organisms and that they had the ability to exit the vacuoles and reinfect the tissue culture monolayer (Amieva *et al.*, 2002). Concurrent studies on human biopsies collected from *H. pylori* patients with mild to severe disease symptoms showed that *H. pylori* could be visualized within gastric epithelial cells in the areolar connective tissue and within the cytoplasm of goblet cells (Semino-Mora *et al.*, 2003). These cells were viable, as in situ hybridization showed coexpression of the virulence genes *cagA* (discussed below) and *babA2*. Patients with the most severe manifestations of disease showed significantly higher expression of *cagA*, presenting the intriguing possibility that intracellular survival and expression of *cagA* and other virulence factors might exacerbate disease progression. Finally, scanning confocal and transmission electron microscopic studies of *H. pylori*-infected gnotobiotic transgenic mice revealed intracellular bacterial collections in a subset of multi- and oligopotential epithelial progenitor cells (Oh *et al.*, 2005). Taken together, these studies suggest that *H. pylori* has the ability to enter a bacterial-induced compartment within the host cell, to reside intracellularly for a period of time, and then to exit the host cell. Depending on the duration of this period, this intracellular component of the bacterial lifecycle has important implications for the disease process. As disease severity appears to correlate with the level of expression of virulence genes, intracellular persistence may directly facilitate *H. pylori*-induced carcinogenesis. Additionally, intracellular persistence is significant when one considers that antibiotic therapies that are designed for clearance of bacterial infection might not be thorough enough to eliminate intracellular organisms, thus leaving a microbial reservoir that would be capable of reseeding the bacterial infection to the gastric epithelium.

VIRULENCE FACTORS AND THEIR TARGETS: VACA

The immediate steps following localization to the gastric epithelium that lead to long-term colonization and disease are poorly understood. With respect to the latter, presumably the bacteria produce virulence factors that are crucial to the onset of disease. Although the identity and function of many of these virulence factors remain a mystery, highly pathogenic strains are known to produce a powerful vacuolating cytotoxin (VacA) (reviewed in Cover and Blanke, 2005; Salama *et al.*, 2002). Encoded by the *vacA* gene, VacA is a secreted protein that has been purified as an oligomeric toxin that dissociates into monomers upon exposure to low pH (Cover and Blaser, 1992; Cover,

1996; Lupetti *et al.*, 1996; Yahiro *et al.*, 1997). The secreted toxin is internalized by eukaryotic cells and exerts its action in the cytoplasm (McClain *et al.*, 2000). Toxic activity of VacA has been shown in vitro to produce vacuolar degeneration of gastric epithelial cells (Cover and Blaser, 1992; Leunk *et al.*, 1988). This degeneration results in the accumulation of large non-functional endosomal-lysosomal hybrids that contain Rab7 and lgp110 (Molinari *et al.*, 1997; Papini *et al.*, 1994). Among other things, this disruption of vesicular membranes interferes with antigen processing in immune cells and prevents lysosomal degradation of surface receptors in epithelial cells (Molinari *et al.*, 1998). The toxic activity of VacA has been shown to be linked to the formation of chloride-conducting channels in lipid bilayers (Iwamoto *et al.*, 1999; Tombola *et al.*, 1999), and more recent work has suggested that VacA produces transmembrane pores that allow for selective diffusion of urea across the membrane (Tombola *et al.*, 2001).

The true biological implication of VacA intoxication on the disease process remains elusive, but all strains of *H. pylori* that have been isolated from humans have been shown to contain *vacA* (Cover and Blanke, 2005). This suggests strongly that VacA plays a role in establishment or maintenance of chronic infection within the human host. Additionally, vacuolization of cells in human biopsy samples has been observed (Caselli *et al.*, 1989; el-Shoura, 1995), and intragastric administration of purified toxin results in epithelial damage in a murine model of infection (Telford *et al.*, 1994). Finally, *vacA* mutants have an infectious dose (ID_{50}) that is 100-fold higher than an isogenic wild-type strain and show a significant colonization defect when coinfected with a wild-type strain (Salama *et al.*, 2001). Taken together, these data suggest that VacA plays a role in establishing colonization. Although the nature of this role is not immediately evident, exciting work has begun to characterize the molecular targets of VacA, and several lines of evidence now indicate that the toxin plays a role in persistence due to its ability to affect basic components of the eukaryotic cell as well as the host immune system.

In addition to overt vacuolar degeneration, for which VacA was first identified, VacA has been shown to localize to the mitochondria of intoxicated cells (Galmiche *et al.*, 2000; Willhite and Blanke, 2004). This is significant due to the fact that in addition to their role in central metabolism, mitochondria function as central sensors and executioners of apoptotic and necrotic cell death within eukaryotic cells (Blanke, 2005). The consequences of VacA localization to mitochondria have been shown to include a decrease in mitochondrial transmembrane potential, release of cytochrome c, reduced cellular ATP concentrations, and impaired cell-cycle progression (Kimura *et al.*, 1999; Willhite *et al.*, 2003). These findings are consistent with a net alteration of mitochondrial membrane permeability and are likely to be due to the ability of VacA to form channels in the mitochondrial membrane (Cover and Blanke, 2005). Two lines of evidence support this hypothesis. First, mutant forms of VacA that are defective in their ability to form membrane channels are also defective for their ability to induce cytochrome c release. Second, chemicals that block the formation of VacA-induced membrane channels also block mitochondrial cytochrome c release (Willhite *et al.*, 2003; Willhite and Blanke, 2004). Directly targeting the mitochondrial membrane could bypass the mitochondrial apoptotic checkpoints and, therefore, could help facilitate *H. pylori* persistence (Blanke, 2005).

In keeping with a role in colonization and a potential role in persistence, VacA has been shown to modulate immune cell function at several levels. Specifically, VacA has been shown to contribute to the formation of large vesicular compartments in *H. pylori*-infected macrophages (Allen *et al.*, 2000; Zheng and Jones, 2003). In these phagosomes, VacA promotes recruitment and retention of the tryptophan-aspartate-containing coat protein (TACO) and disrupts phagosome maturation (Allen *et al.*, 2000; Zheng and Jones, 2003). Alteration of phagosome maturation likely has profound implications for *H. pylori* persistence by impairing phagocytic killing of the bacterium (Cover and Blanke, 2005). Additionally, VacA has been reported to interfere with antigen presentation by B-lymphocytes (Molinari *et al.*, 1998). The molecular mechanism of interference was shown to be due to incomplete proteolytic processing and antigen presentation by newly synthesized major histocompatibility complex (MHC) class II molecules (Molinari *et al.*, 1998). Improper presentation of *H. pylori* antigens and inhibition of phagosome function may partially explain the ability of the bacterium to chronically infect in the face of a robust inflammatory response.

Another line of evidence that suggests that VacA may directly modulate immune cell function comes from findings that show that VacA effects T-lymphocyte function (Boncristiano *et al.*, 2003; Gebert *et al.*, 2003; Sundrud *et al.*, 2004). In cultured Jurkat T-cells, VacA inhibits production of interleukin 2 (IL-2), which is required for T-cell viability and proliferation. This suppression of IL-2 expression is linked to VacA-mediated inhibition of activation of nuclear factor of activated T-cells (NFAT). Typically, activated NFAT translocates to the nuclease, where it acts as a global transcription factor for modulation of immune response genes. In keeping with this, intoxication of Jurkat T-cells by VacA alters the expression of more than 100 genes (Gebert *et al.*, 2003). Similar studies conducted on primary T-cells indicate that although VacA also inhibits T-cell proliferation in these cells, it does so independent of the affect of VacA on NFAT translocation and IL-2 expression (Sundrud *et al.*, 2004). Overall, the ability of VacA to both inhibit antigen presentation by B-cells and T-cell proliferation in response to any *H. pylori* antigens that may be presented would likely have profound affects on *H. pylori* persistence in vivo.

THE CAG PATHOGENICITY ISLAND AND CAGA

Perhaps the most significant advance in understanding *H. pylori* pathogenesis came as a result of the discovery of a pathogenicity island (PAI) that encodes a type IV secretion apparatus in some strains of *H. pylori* (Akopyants *et al.*, 1998; Censini *et al.*, 1996). This PAI is composed of a 40-kilobase locus that contains at least 27 genes. *H. pylori* strains carrying the PAI are far more likely to be associated with serious manifestation of *H. pylori* infection (Xiang *et al.*, 1995). However, it should be noted that strains lacking the PAI are still associated with chronic gastritis and, occasionally, with more severe disease (Covacci *et al.*, 1999). A number of open reading frames contained on the PAI show homology to genes that encode secretion machinery in other pathogenic bacteria (Covacci *et al.*, 1999). Mutations in *cagE, cagF, cagG, cagH, cagL,* and *cagI* were shown early on to abrogate the strong IL-8 induction caused by *H. pylori* infection of cultured gastric cancer cells (AGS) (Segal, 1997; Segal *et al.*, 1997). A more recent systematic study of each of the genes encoded within the PAI showed that mutations in 14 of the 27 genes resulted in a loss of IL-8 induction (gene designations: HP0522, HP0523, HP0525, HP0527–532, HP0537, HP0539, HP0541, HP0544, HP0546) (Fischer *et al.*, 2001).

The origin of the PAI is unknown, but studies have shown that components of the island function in the delivery of CagA, which is encoded by HP0547 and is also part of the island, to the host cell (Fischer *et al.*, 2001). It has been known for a number of years that most people infected with PAI strains develop antibodies to CagA. More recently it became evident that CagA was delivered to the cytoplasm or plasma membrane of the eukaryotic cell. This was elucidated after the observation that PAI and CAG+ strains induced tyrosine phosphorylation of a 145-kDA protein within the infected host cell; strains lacking CAG did not (Asahi *et al.*, 2000; Odenbreit *et al.*, 2000; Segal *et al.*, 1999; Stein *et al.*, 2000). Mass-spectrometric analysis of this 145-kDa protein revealed that it was CagA, and confocal microscopy showed that the bacterial CagA protein was inserted into the plasma membrane of host cells and, in some cases, is phosphorylated (Segal *et al.*, 1999). It has since been demonstrated that Src-like protein tyrosine kinases are responsible for this phosphorylation event (Selbach *et al.*, 2002; Stein *et al.*, 2002). Other studies have shown that of the 27 genes encoded on the PAI, 18 are absolutely required for the efficient translocation and/or phosphorylation of CagA into the host cell (Fischer *et al.*, 2001).

Bacterial attachment and subsequent injection of CagA into the host cell have been shown to induce a striking morphological change in the infected cell. This so-called "hummingbird phenotype" is marked by elongation and spreading of cells (Segal *et al.*, 1999). CagA was shown to be necessary and sufficient for these morphological changes, as human cultured gastric adenocarcinoma (AGS) cells transfected with *cagA* expression constructs displayed the marked elongation and spreading (Higashi *et al.*, 2002b). Tyrosine phosphorylation of CagA is required for this event, as expression of phosphorylation-deficient CagA results in no apparent changes to cell morphology, although it localizes properly to the plasma membrane (Higashi *et al.*, 2002b).

In addition to showing the necessary and sufficient nature of CagA in mediating the morphological changes associated with *H. pylori* attachment to host cells, this study also identified one of the first targets of CagA within the eukaryotic cell. Segal *et al.* (1999) and Higashi *et al.* (2002b) both noted that CagA-mediated morphological changes showed striking similarity to changes induced by exposure to hepatocyte growth factor (HGF). It had been shown previously that a cytoplasmic tyrosine phosphatase, SHP-2, played a major role in the HGF-induced changes (Kodama *et al.*, 2000). Higashi *et al.* (2002b) showed conclusively that CagA binds to SH2-domain-containing tyrosine phosphatase 2 (SHP-2) within the context of gastric epithelial cells in a tyrosine-phosphorylation-dependent manner and that immunodepletion of SHP-2 from AGS cell lysates effectively cleared all phosphorylated CagA from the lysates, thus suggesting stoichiometric binding of the two components. SHP-2 activity was also shown to be required for induction of cell elongation, as phosphatase-defective SHP-2 and inhibition of phosphatase activity with the inhibitor calpeptin both negated the ability of CagA to induce morphological cell changes. Finally, Higashi *et al.* (2002b) demonstrated that even in the absence of CagA, membrane-targeted SHP-2 resulted in the indistinguishable induction of cell elongation, thus suggesting that the true mode of action of CagA was sequestration of SHP-2 at the plasma membrane and deregulation of phosphatase activity.

CagA VARIABILITY

Early observations indicated that different *H. pylori* isolates showed variability in the migration of CagA on sodium dodecyl sulfate (SDS) gels, and it is now understood that a number of distinct *cagA* alleles exist (reviewed in Hatakeyama, 2004). Evidence has begun to suggest that this genetic diversity

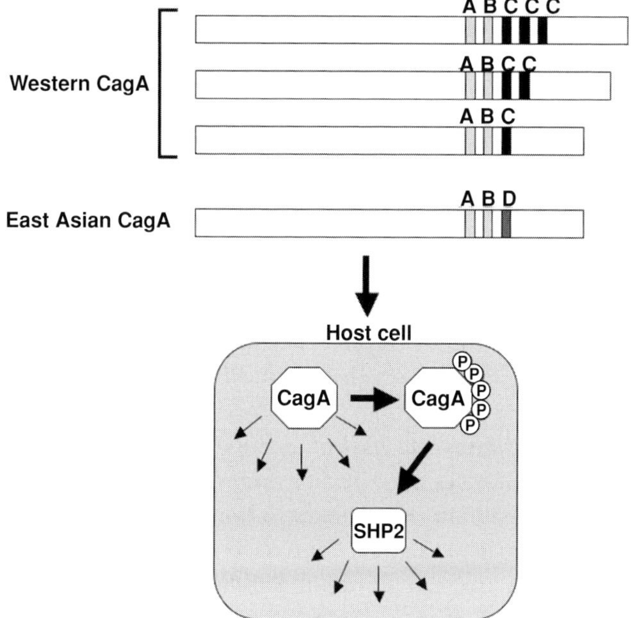

Figure 12.2 Geographical sequence variation in the CagA gene of *Helicobacter pylori*. The *cagA* coding sequence shows genetic diversity that is characterized by sequence variations in the area surrounding the EPIYA motif as well as the number of EPIYA motifs found in a strain. Four distinct EPIYA sites exist (-A, -B, -C, -D). Western strains are most often isolated in North America, Europe, and Australia and contain EPIYA-A, EPIYA-B, and from one to three copies of EPIYA-C. Conversely, East Asian strains are typically isolated in Japan, Korea, and China and contain EPIYA-A, EPIYA-B, and a single copy of EPIYA-D. Upon injection into the host cell, CagA in its non-phosphorylated form alters a subset of host cell-signaling pathways. Based on the number and type of EPIYA motifs found in the CagA protein, various levels of phosphorylation occur, which then directly affect the number and intensity of perturbation of additional host cell-signaling pathways.

may have profound affects on *H. pylori*-induced disease etiology. Variability in *cagA* sequence exists primarily in the carboxy-terminal end of the gene in the region that encodes the site of CagA phosphorylation (Figure 12.2). Phosphorylation occurs at conserved tyrosine residues contained within a repeated five-amino-acid sequence (Glu–Pro–Ile–Tyr–Alu) known as the EPIYA motif (Higashi *et al.*, 2002a, b). The sequence that surrounds the EPIYA motif shows sequence divergence depending on the *H. pylori* strain. Based on analysis of numerous CagA sequences, four distinct EPIYA sites have been identified and named EPIYA-A, EPIYA-B, EPIYA-C, and EPIYA-D. Strains isolated from various areas of the world have been shown to encode distinct combinations of these sites. All CagA-positive strains of *H. pylori* typically encode both an -A and a -B site in combination with either a -C or a -D site. Based on the most common geographic distribution of strains that contain these particular EPIYA site combinations, CagA alleles are called Western and East Asian. Western CagA is prevalent in North America, Europe, and Australia and contains EPIYA-A, -B, and -C sites (the -C site is repeated up to three times in some Western isolates). East Asian CagA is prevalent in Japan, Korea, and China and contains EPIYA-A, -B, and -D sites (Covacci *et al.*, 1993; Higashi *et al.*, 2002a, b; Stein *et al.*, 2002).

As with CagA and the EPIYA sites, incidence of gastric carcinoma shows geographic distribution. The highest rates of gastric carcinoma occur in China, where *H. pylori* encoding the East Asian CagA are most prevalent. Likewise, mortality caused by gastric cancer is highest in Japan and then Korea, two other countries where strains carrying the East Asian CagA allele are known to predominate (Hatakeyama, 2004). Recognition of the fact that the geographic distribution of the CagA allele and the incidence of gastric cancer seem to correlate, combined with the elucidation of SHP-2 as a target of activated CagA, has led to exciting findings that may partially explain this fact. The EPIYA-C and -D sites are the primary sites phosphorylated upon injection of CagA into the host cell (Backert *et al.*, 2001; Higashi *et al.*, 2002a; Stein *et al.*, 2002), although both EPIYA-A and EPIYA-B can be phosphorylated to a lesser degree (Hatakeyama, 2004). Formation of the CagA–SHP-2 complex requires phosphorylation at the EPIYA-C or EPIYA-D site. SHP-2 binding at the EPIYA sites occurs with varying affinity based on the EPIYA site sequence. EPIYA-D shows stronger binding to SHP-2 and, interestingly, induces greater morphogenetic activity than EPIYA-C (Higashi *et al.*, 2002a). Along with this, the grades of inflammation, severity of gastritis, and atrophy are significantly higher in people infected with East Asian CagA strains (Azuma *et al.*, 2004). For people infected with Western isolates, molecular epidemiological studies have shown that increased disease severity correlates with increased numbers of EPIYA-C sites contained within CagA (Yamaoka *et al.*, 1999). The number of EPIYA-C sites, in turn, correlates directly with the levels of tyrosine phosphorylation, SHP-2 binding, and morphological transformation seen in vitro (Higashi *et al.*, 2002a). Thus, the affinity of SHP-2 binding to CagA based upon the degree of phosphorylation and the specific sequence of the EPIYA sites may explain partially why some infected individuals develop gastric cancer and others do not (Argent *et al.*, 2004; Azuma *et al.*, 2004; Yamaoka *et al.*, 1999). That being said, it remains to be seen which specific downstream targets of SHP-2 phosphatase activity will be effected in *H. pylori*-targeted disruption of cellular activities and what role these will play in the clinical outcome of infection.

CagA AND HOST-CELL JUNCTIONS

A number of studies have attempted to expand our knowledge of host-cell targets of *H. pylori* using DNA microarrays (reviewed in Merrell and Falkow, 2003). These studies have relied primarily on the human gastric epithelial cell line AGS to assess host transcriptional changes upon infection with *H. pylori*. Although this cell line has been studied extensively, AGS cells do not polarize to form proper cell junctions and, hence, do not mimic accurately the gastric epithelial surface encountered by *H. pylori* in the human stomach. This caveat likely influences previously identified targets of *H. pylori*, a fact that has been accentuated by the characterization of *H. pylori* interaction with polarized Madin-Darby canine kidney (MDCK) monolayers (Amieva *et al.*, 2003). *H. pylori* was shown to preferentially attach to and disrupt cellular apical-junction complexes and to colocalize with the tight-junction scaffolding protein ZO-1 (Amieva *et al.*, 2003).

To assess host-cell targets of CagA in a polarized monolayer, a novel study characterized the interaction of *H. pylori* with the differentiated human intestinal epithelial cell line T84 (El-Etr *et al.*, 2004). Unlike AGS, T84 cells polarize on transwells, form a brush border, and assemble tight junctions and desmosomes. These cells also secrete mucin and can be induced to form a mucous layer (Madara *et al.*, 1987). Temporal transcriptional profiling using a panel of wild-type and isogenic *H. pylori* mutants revealed infection-specific profiles that show many distinctions from previous AGS cell profiles. Of note, there was a strong enrichment for components associated with cellular junctions. These included changes in expression of the syndecan-binding protein (a component of the adherens junctions that connect the cytoskeletal filaments between cells and between cells and the extracellular matrix), β-catenin, α-catenin, desmoglein (components of the intercellular junction), the cadherin family members desmoplakin and cadherin 5 and 17, gap junction binding proteins 2 and 3, and the tight junctional components Par3 and claudin 3 and 4 (El-Etr *et al.*, 2004).

Deregulation of α- and β-catenin is of special interest since these play roles in multiple forms of cancer (Jawhari *et al.*, 1997; Pierceall *et al.*, 1995; Rimm *et al.*, 1995). Catenins typically link E-cadherin to the actin cytoskeleton. β-Catenin binds the cytoplasmic domain of E-cadherin. However, excess non-structural β-catenin associates with the adenomatous polyposis coli (APC) multiprotein complex, a tumor suppressor implicated in the development of colon cancer (Taipale and Beachy, 2001). Normally, APC and excess β-catenin are targeted for degradation by phosphorylation by the WNT signaling pathway. However, when WNT is induced, excess β-catenin enters the nucleus and interacts with T-cell factors (TCFs) or lymphocyte-enhancer factors (LEFs), which can then act as transcription factors. Target genes of the β-catenin–TCF complex include matrix metalloproteinase-7 (Brabletz *et al.*, 1999), *Myc* (He *et al.*, 1998), and *Jun* and its downstream target urokinase-type plasminogen activator receptor (*uPar*) (Mann *et al.*, 1999). Of note, all of these genes were strongly upregulated in the T84 microarray study (El-Etr *et al.*, 2004). This is the first suggestion that *H. pylori* infection may affect WNT signaling and lead to the nuclear accumulation of β-catenin. Subsequent research will be needed to determine the significance of this within the context of *H. pylori*-induced disease.

HELICOBACTER PYLORI AND APOPTOSIS

An additional interesting feature of the previously mentioned T84 study was that many genes with predicted roles in apoptosis were differentially regulated (El-Etr *et al.*, 2004). Different classes of apoptosis genes showed virulence factor-dependent regulation that could be broken down into three distinct groups – PAI-independent, CagA phosphorylation-dependent, and CagA phosphorylation-independent. Analysis of these groups showed that all of the genes regulated in a PAI-independent manner are predicted to stimulate apoptosis and are strongly suppressed by *H. pylori* infection (El-Etr *et al.*, 2004). Delivery of the non-phosphorylateable form of CagA resulted in increased expression of the apoptosis inhibitors tumor necrosis factor alpha-induced protein3 (TNFAIP3) and BIRC2 and BIRC3, suggesting that non-phosphorylated CagA actively suppresses apoptosis. When phosphorylated CagA was delivered, there was increased expression of genes that promote apoptosis. Taken together, these data suggest that *H. pylori* possess multiple pathways for the regulation of apoptotic pathways and that the balance between promotion and inhibition of apoptosis is a delicate one. This is in keeping with the fact that the literature on the effect of *H. pylori* on apoptosis is extremely convoluted. Some studies report induction, while others report suppression (Jones *et al.*, 1997; Le'Negrate *et al.*, 2001; Neu *et al.*, 2005). Part of these discrepancies may be explained by the following model (Figure 12.3): *H. pylori* has the ability to downregulate expression of apoptotic factors in a PAI-independent manner and to induce the expression of inhibitors of apoptosis by delivery of non-phosphorylated CagA. The ultimate phosphorylation of CagA and subsequent disruption of normal signaling pathways lead to

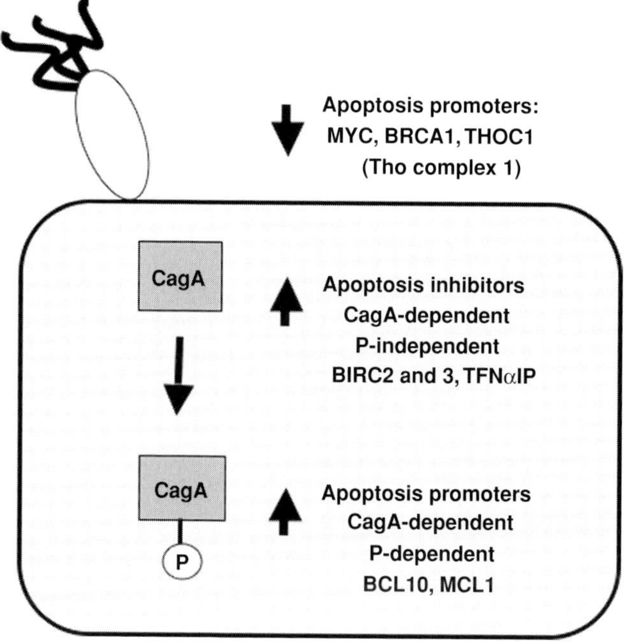

Figure 12.3 Model of *Helicobacter pylori*-dependent affects upon apoptosis. Upon interaction of *H. pylori* with epithelial cells, host genes that promote apoptosis are downregulated independent of the presence of the Cag pathogenicity island (PAI). The effector protein CagA is subsequently injected into the host cell in a PAI-dependent manner. In its non-phosphorylated form, CagA induces increased expression of numerous apoptosis inhibitors. However, subsequent to phosphorylation, CagA causes increased expression of genes that promote apoptosis. Taken together, the model suggests that *H. pylori* contains virulence-factor-dependent and -independent mechanisms by which it can modulate host-cell apoptosis. Additionally, regulation of the amount of CagA delivered to the host cell is crucial for maintaining a balance between host-cell survival and apoptosis. TNFαIP, tumor necrosis factor alpha-induced protein.

the increased expression of apoptosis factors. This suggests that the level of phosphorylated CagA found within a cell or the ability to modulate the delivery of a non-phosphorylateable form of CagA may play an important role in the ability of *H. pylori* to set up a dynamic equilibrium within the host that allows long-term persistent infection. This is particularly interesting in light of the finding that within a single human host, bacteria are found that have deleted the *cagA* gene (Israel *et al.*, 2001) or that have undergone intragenomic recombination, resulting in deletion of CagA phosphorylation sites (Aras *et al.*, 2003). Thus, bacteria may use genomic rearrangements as a means of controlling the level of apoptosis seen within the human stomach. The importance of this strategy of genetic rearrangement within the host may explain the identification of RuvC, a Holliday junction resolvase involved in recombination, as crucial for bacterial persistence (Loughlin *et al.*, 2003).

CagA AS A MULTIDOMAIN PROTEIN?

Although it is clear that phosphorylation of CagA is important in terms of effects on host–cell signaling, an increasing line of evidence is beginning to accumulate that suggests that the dogma that phosphorylation is essential to effect host cell signaling pathways is incorrect. This evidence was first suggested in work by Hirata *et al.* (2002) and Umehara *et al.* (2003), which showed that CagA regulation of serum response elements in HeLa cells and regulation of JAK/STAT signaling pathways in B-lymphocytes, respectively, are independent of the phosphorylation state of CagA. Additionally, Mimuro *et al.* (2002) showed that non-phosphorylated CagA interacts with growth factor receptor-bound protein2 (Grb2) and activates extracellular signal-regulated kinase (ERK) signaling pathways. An in-depth analysis of the role of non-phosphorylated CagA on effects in host cell signaling revealed that changes in expression of more than 30% of the genes modulated by CagA occurred irregardless of the phosphorylation state of the protein (El-Etr *et al.*, 2004). Taken together, these data suggest strongly that CagA likely has multiple distinct functional domains.

In keeping with this, more recent studies have investigated this hypothesis directly and shown that the CagA protein can be broken into distinct functional domains. In the first study, Higashi *et al.* (2005) showed that the EPIYA motif serves as a membrane-targeting signal for CagA. Various truncations in the CagA protein were constructed and transiently expressed in AGS cells. Deletion of either the N-terminal or extreme C-terminal region did not effect the ability of CagA to localize to the host cell membrane or induce cell elongation. However, deletion of the EPIYA motifs resulted in a disruption of both activities. CagA did not induce changes in cell morphology and was found to localize to the cytoplasm. The presence of a single EPIYA motif was found to be sufficient for CagA localization, as truncations that deleted various combinations of EPIYA-A, -B, and -C were found to localize properly as long as a single EPIYA site was maintained. This localization was not dependent on phosphorylation, as a single EPIYA-A, -B, or -C site, which are known to be phosphorylated to different degrees, localized properly. Additionally, substitution of the conserved tyrosine residue within all of the

EPIYA motifs did not affect CagA membrane targeting (Higashi *et al.*, 2005). These data suggest that a single EPIYA motif is necessary and sufficient for membrane localization. However, the authors were unable to demonstrate that insertion of an EPIYA motif into an unrelated cytoplasmic protein would induce membrane localization. This implies that additional regions of CagA are required for proper membrane localization.

In the second study that directly investigated functional domains of CagA, Bagnoli *et al.* (2005) provide data that present conflicting roles for the N- and C-terminal regions of CagA. Using confluent MDCK monolayers, the authors demonstrate that the N-terminus of CagA is sufficient for membrane localization. Alternatively, the C-terminal domain showed diffuse localization when expressed alone. This diffuse expression of the C-terminal domain was sufficient to induce pseudopodial activity but not sufficient to induce cell migration. When expressed alone, neither domain was sufficient to disrupt cell adhesion or cell polarity; interestingly, when coexpressed in trans, the N-terminus could direct membrane localization of itself as well as the C-terminus (Bagnoli *et al.*, 2005). The reasons for the discrepancies between this study and the Higashi *et al.* study are unclear, but they highlight the fact that additional research into the different functional domains of CagA will be required in order to elucidate the role of each within the host cell.

CONCLUSIONS

H. pylori is truly proficient in its ability to exploit the host niche. This fact is particularly well highlighted by the fact that more than 50% of the world's population is infected with this bacterium. Research has shown that within the human stomach, *H. pylori* utilizes a wide repertoire of factors in order to establish a lifelong chronic infection. These factors include those designed to specifically subvert the host innate immune system as well as those that affect the adaptive branch of immunity. Additionally, *H. pylori* directly interacts with a subset of the gastric mucosal cells and alters host cell signaling, a fact that is likely responsible for *H. pylori*-induced disease. Future research that is focused on further molecular characterization of the role of VacA and CagA in *H. pylori* pathogenesis is certain to shed insight on the biology of this important pathogen. However, one must not lose sight of the fact that strains of *H. pylori* that do not encode CagA still induce significant disease. This highlights our need to expand our understanding of the etiology of *H. pylori*-induced disease to include additional bacterial factors that are required for exploitation of the human gastric niche.

ACKNOWLEDGMENTS

We would like to acknowledge D. Monack, F. Bagnoli, M. Amieva, and S. Falkow for sharing unpublished data and for allowing us to reproduce previously published figures. Research in the laboratory of D. Scott Merrell is supported by H073LA and G173LU from the Uniformed Services University of the Health Sciences (USUHS) and US Military Cancer Institute (USMCI), respectively.

REFERENCES

Akopyants, N. S., Clifton, S. W., Kersulyte, D., *et al.* (1998). Analyses of the *cag* pathogenicity island of *Helicobacter pylori*. *Mol. Microbiol.* **28**, 37–53.

Allen, L. A., Schlesinger, L. S., and Kang, B. (2000). Virulent strains of *Helicobacter pylori* demonstrate delayed phagocytosis and stimulate homotypic phagosome fusion in macrophages. *J. Exp. Med.* **191**, 115–128.

Alm, R. A., Bina, J., Andrews, B. M., *et al.* (2000). Comparative genomics of *Helicobacter pylori*: analysis of the outer membrane protein families. *Infect. Immun.* **68**, 4155–4168.

Amieva, M. R., Salama, N. R., Tompkins, L. S., and Falkow, S. (2002). *Helicobacter pylori* enter and survive within multivesicular vacuoles of epithelial cells. *Cell. Microbiol.* **4**, 677–690.

Amieva, M. R., Vogelmann, R., Covacci, A., *et al.* (2003). Disruption of the epithelial apical-junctional complex by *Helicobacter pylori* CagA. *Science* **300**, 1430–1434.

Andermann, T. M., Chen, Y. T., and Ottemann, K. M. (2002). Two predicted chemoreceptors of *Helicobacter pylori* promote stomach infection. *Infect. Immun.* **70**, 5877–5881.

Andrutis, K. A., Fox, J. G., Schauer, D. B., *et al.* (1995). Inability of an isogenic urease-negative mutant stain of *Helicobacter mustelae* to colonize the ferret stomach. *Infect. Immun.* **63**, 3722–3725.

Aras, R. A., Lee, Y., Kim, S. K., *et al.* (2003). Natural variation in populations of persistently colonizing bacteria affect human host cell phenotype. *J. Infect. Dis.* **188**, 486–496.

Argent, R. H., Kidd, M., Owen, R. J., *et al.* (2004). Determinants and consequences of different levels of CagA phosphorylation for clinical isolates of *Helicobacter pylori*. *Gastroenterology* **127**, 514–523.

Asahi, M., Azuma, T., *et al.* (2000). *Helicobacter pylori* CagA protein can be tyrosine phosphorylated in gastric epithelial cells. *J. Exp. Med.* **191**, 593–602.

Aspholm-Hurtig, M., Dailide, G., Lahmann, M., *et al.* (2004). Functional adaptation of BabA, the H. pylori ABO blood group antigen binding adhesin. *Science* **305**, 519–522.

Azuma, T., Yamazaki, S., Yamakawa, A., *et al.* (2004). Association between diversity in the Src homology 2 domain-containing tyrosine phosphatase binding site of *Helicobacter pylori* CagA protein and gastric atrophy and cancer. *J. Infect. Dis.* **189**, 820–827.

Backert, S., Moese, S., Selbach, M., Brinkmann, V., and Meyer, T. F. (2001). Phosphorylation of tyrosine 972 of the *Helicobacter pylori* CagA protein is essential for induction of a scattering phenotype in gastric epithelial cells. *Mol. Microbiol.* **42**, 631–644.

Backstrom, A., Lundberg, C., Kersulyte, D., *et al.* (2004). Metastability of *Helicobacter pylori bab* adhesin genes and dynamics in Lewis b antigen binding. *Proc. Natl. Acad. Sci. U. S. A.* **101**, 16 923–16 928.

Bagnoli, F., Buti, L., Tompkins, L., Covacci, A., and Amieva, M. R. (2005). *Helicobacter pylori* CagA induces transition from polarized to invasive phenotypes in MDCK cells. *Proc. Natl. Acad. Sci. U. S. A.* **102**, 16 339–16 344.

Blanke, S. R. (2005). Micro-managing the executioner: pathogen targeting of mitochondria. *Trends Microbiol.* **13**, 64–71.

Blaser, M. J. (1998). *Helicobacter pylori* and gastric diseases. *Br. Med. J.* **316**, 1507–1510.

Blaser, M. J. and Parsonnet, J. (1994). Parasitism by the "slow" bacterium *Helicobacter pylori* leads to altered gastric homeostasis and neoplasia. *J. Clin. Invest.* **94**, 4–8.

Blaser, M. J., Chyou, P. H., and Nomura, A. (1995). Age at establishment of *Helicobacter pylori* infection and gastric carcinoma, gastric ulcer, and duodenal ulcer risk. *Cancer Res.* **55**, 562–565.

Bode, G., Malfertheiner, P., and Ditschuneit, H. (1988). Pathogenetic implications of ultrastructural findings in *Campylobacter pylori* related gastroduodenal disease. *Scand. J. Gastroenterol. Suppl.* **142**, 25–39.

Boncristiano, M., Paccani, S. R., Barone, S., *et al.* (2003). The *Helicobacter pylori* vacuolating toxin inhibits T cell activation by two independent mechanisms. *J. Exp. Med.* **198**, 1887–1897.

Borch, K., Sjostedt, C., Hannestad, U., *et al.* (1998). Asymptomatic *Helicobacter pylori* gastritis is associated with increased sucrose permeability. *Dig. Dis. Sci.* **43**, 749–753.

Brabletz, T., Jung, A., Dag, S., Hlubek, F., and Kirchner, T. (1999). Beta-Catenin regulates the expression of the matrix metalloproteinase-7 in human colorectal cancer. *Am. J. Pathol.* **155**, 1033–1038.

Caselli, M., Figura, N., Trevisani, L., *et al.* (1989). Patterns of physical modes of contact between *Campylobacter pylori* and gastric epithelium: implications about the bacterial pathogenicity. *Am. J. Gastroenterol.* **84**, 511–513.

Censini, S., Lange, C., Xiang, Z., *et al.* (1996). *cag*, a pathogenicity island of *Helicobacter pylori*, encodes type I-specific and disease-associated virulence factors. *Proc. Natl. Acad. Sci. U. S. A.* **93**, 14 648–14 653.

Covacci, A., Censini, S., Bugnoli, M., *et al.* (1993). Molecular characterization of the 128-kDa immunodominant antigen of *Helicobacter pylori* associated with cytotoxicity and duodenal ulcer. *Proc. Natl. Acad. Sci. U. S. A.* **90**, 5791–5795.

Covacci, A., Telford, J. L., Del Giudice, G., Parsonnet, J., and Rappuoli, R. (1999). *Helicobacter pylori* virulence and genetic geography. *Science* **284**, 1328–1333.

Cover, T. L. (1996). The vacuolating cytotoxin of *Helicobacter pylori*. *Mol. Microbiol.* **20**, 241–246.

Cover, T. L. and Blaser, M. J. (1992). Purification and characterization of the vacuolating toxin from *Helicobacter pylori*. *J. Biol. Chem.* **267**, 10 570–10 575.

Cover, T. L. and Blanke, S. R. (2005). *Helicobacter pylori* VacA: a paradigm for toxin multifunctionality. *Nat. Rev. Microbiol.* **3**, 320–332.

Curtis, G. H. and Gall, D. G. (1992). Macromolecular transport by rat gastric mucosa. *Am. J. Physiol.* **262**, G1033–1040.

Drumm, B., Sherman, P., Cutz, E., and Karmali, M. (1987). Association of *Campylobacter pylori* on the gastric mucosa with antral gastritis in children. *N. Engl. J. Med.* **316**, 1557–1561.

Dunn, B. E., Cohen, H., and Blaser, M. J. (1997). *Helicobacter pylori*. *Clin. Microbiol. Rev.* **10**, 720–741.

Dykhuizen, R. S., Fraser, A., McKenzie, H., *et al.* (1998). *Helicobacter pylori* is killed by nitrite under acidic conditions. *Gut* **42**, 334–337.

Eaton, K. A. and Krakowka, S. (1994). Effect of gastric pH on urease-dependent colonization of gnotobiotic piglets by *Helicobacter pylori*. *Infect. Immun.* **62**, 3604–3607.

Eaton, K. A., Morgan, D. R., and Krakowka, S. (1992). Motility as a factor in the colonisation of gnotobiotic piglets by *Helicobacter pylori*. *J. Med. Microbiol.* **37**, 123–127.

Eaton, K. A., Suerbaum, S., Josenhans, C., and Krakowka, S. (1996). Colonization of gnotobiotic piglets by *Helicobacter pylori* deficient in two flagellin genes. *Infect. Immun.* **64**, 2445–2448.

El-Etr, S. H., Mueller, A., Tompkins, L. S., Falkow, S., and Merrell, D. S. (2004). Phosphorylation-independent effects of CagA during interaction between *Helicobacter pylori* and T84 Polarized monolayers. *J. Infect. Dis.* **190**, 1516–1523.

El-Shoura, S. M. (1995). *Helicobacter pylori*: I. Ultrastructural sequences of adherence, attachment, and penetration into the gastric mucosa. *Ultrastruct. Pathol.* **19**, 323–333.

Ernst, P. B. and Gold, B. D. (2000). The disease spectrum of *Helicobacter pylori*: the immunopathogenesis of gastroduodenal ulcer and gastric cancer. *Annu. Rev. Microbiol.* **54**, 615–640.

Ferrero, R. L., Cussac, V., Courcoux, P., and Labigne, A. (1992). Construction of isogenic urease-negative mutants of *Helicobacter pylori* by allelic exchange. *J. Bacteriol.* **174**, 4212–4217.

Fischer, W., Puls, J., Buhrdorf, R., *et al.* (2001). Systematic mutagenesis of the *Helicobacter pylori cag* pathogenicity island: essential genes for CagA translocation in host cells and induction of interleukin-8. *Mol. Microbiol.* **42**, 1337–1348.

Foliguet, B., Vicari, F., Guedenet, J. C., De Korwin, J. D., and Marchal, L. (1989). [Scanning electron microscopic study of *Campylobacter pylori* and associated gastroduodenal lesions.] *Gastroenterol. Clin. Biol.* **13**, 65B–70B.

Fu, S., Ramanujam, K. S., Wong, A., *et al.* (1999). Increased expression and cellular localization of inducible nitric oxide synthase and cyclooxygenase 2 in *Helicobacter pylori* gastritis. *Gastroenterology* **116**, 1319–1329.

Galmiche, A., Rassow, J., Doye, A., *et al.* (2000). The N-terminal 34 kDa fragment of *Helicobacter pylori* vacuolating cytotoxin targets mitochondria and induces cytochrome c release. *EMBO J.* **19**, 6361–6370.

Gebert, B., Fischer, W., Weiss, E., Hoffmann, R., and Haas, R. (2003). *Helicobacter pylori* vacuolating cytotoxin inhibits T lymphocyte activation. *Science* **301**, 1099–1102.

Gobert, A. P., McGee, D. J., Akhtar, M., *et al.* (2001). *Helicobacter pylori* arginase inhibits nitric oxide production by eukaryotic cells: a strategy for bacterial survival. *Proc. Natl. Acad. Sci. U. S. A.* **98**, 13 844–13 849.

Hatakeyama, M. (2004). Oncogenic mechanisms of the *Helicobacter pylori* CagA protein. *Nat. Rev. Cancer* **4**, 688–694.

He, T. C., Sparks, A. B., Rago, C., *et al.* (1998). Identification of c-MYC as a target of the APC pathway. *Science* **281**, 1509–1512.

Higashi, H., Tsutsumi, R., Fujita, A., *et al.* (2002a). Biological activity of the *Helicobacter pylori* virulence factor CagA is determined by variation in the tyrosine phosphorylation sites. *Proc. Natl. Acad. Sci. U. S. A.* **99**, 14 428–14 433.

Higashi, H., Tsutsumi, R., Muto, S., *et al.* (2002b). SHP-2 tyrosine phosphatase as an intracellular target of *Helicobacter pylori* CagA protein. *Science* **295**, 683–686.

Higashi, H., Yokoyama, K., Fujii, Y., *et al.* (2005). EPIYA motif is a membrane targeting signal of *Helicobacter pylori* CagA in mammalian cells. *J. Biol. Chem.* **280**, 231–307.

Hirata, Y., Maeda, S., Mitsuno, Y., *et al.* (2002). *Helicobacter pylori* CagA protein activates serum response element-driven transcription independently of tyrosine phosphorylation. *Gastroenterology* **123**, 1962–1971.

IARC Working Group on the Evaluation of Carcinogenic Risks to Humans (1994). Schistosomes, liver flukes and *Helicobacter pylori*. *IARC Monogr. Eval. Carcinog. Risks Hum.* **61**, 1–241.

Ilver, D., Arnqvist, A., Ogren, J., *et al.* (1998). *Helicobacter pylori* adhesin binding fucosylated histo-blood group antigens revealed by retagging. *Science* **279**, 373–377.

Israel, D. A., Salama, N., Krishna, U., *et al.* (2001). *Helicobacter pylori* genetic diversity within the gastric niche of a single human host. *Proc. Natl. Acad. Sci. U. S. A.* **98**, 14 625–14 630.

Iwamoto, H., Czajkowsky, D. M., Cover, T. L., Szabo, G., and Shao, Z. (1999). VacA from *Helicobacter pylori*: a hexameric chloride channel. *FEBS Lett.* **450**, 101–104.

Jawhari, A., Jordan, S., Poole, S., *et al.* (1997). Abnormal immunoreactivity of the E-cadherin-catenin complex in gastric carcinoma: relationship with patient survival. *Gastroenterology* **112**, 46–54.

Jones, N. L., Shannon, P. T., Cutz, E., Yeger, H., and Sherman, P. M. (1997). Increase in proliferation and apoptosis of gastric epithelial cells early in the natural history of *Helicobacter pylori* infection. *Am. J. Pathol.* **151**, 1695–1703.

Kimura, M., Goto, S., Wada, A., *et al.* (1999). Vacuolating cytotoxin purified from *Helicobacter pylori* causes mitochondrial damage in human gastric cells. *Microb. Pathog.* **26**, 45–52.

Kodama, A., Matozaki, T., Fukuhara, A., *et al.* (2000). Involvement of an SHP-2-Rho small G protein pathway in hepatocyte growth factor/scatter factor-induced cell scattering. *Mol. Biol. Cell* **11**, 2565–2575.

Kuwahara, H., Miyamoto, Y., Akaike, T., *et al.* (2000). *Helicobacter pylori* urease suppresses bactericidal activity of peroxynitrite via carbon dioxide production. *Infect. Immun.* **68**, 4378–4383.

Le'Negrate, G., Ricci, V., Hofman, V., *et al.* (2001). Epithelial intestinal cell apoptosis induced by *Helicobacter pylori* depends on expression of the *cag* pathogenicity island phenotype. *Infect. Immun.* **69**, 5001–5009.

Lee, S. K., Stack, A., Katzowitsch, E., *et al.* (2003). *Helicobacter pylori* flagellins have very low intrinsic activity to stimulate human gastric epithelial cells via TLR5. *Microbes Infect.* **5**, 1345–1356.

Leunk, R. D., Johnson, P. T., David, B. C., Kraft, W. G., and Morgan, D. R. (1988). Cytotoxic activity in broth-culture filtrates of *Campylobacter pylori*. *J. Med. Microbiol.* **26**, 93–99.

Loughlin, M. F., Barnard, F. M., Jenkins, D., Sharples, G. J., and Jenks, P. J. (2003). *Helicobacter pylori* mutants defective in RuvC Holliday junction resolvase display reduced macrophage survival and spontaneous clearance from the murine gastric mucosa. *Infect. Immun.* **71**, 2022–2031.

Lupetti, P., Heuser, J. E., Manetti, R., *et al.* (1996). Oligomeric and subunit structure of the *Helicobacter pylori* vacuolating cytotoxin. *J. Cell. Biol.* **133**, 801–807.

Madara, J. L., Stafford, J., Dharmsathaphorn, K., and Carlson, S. (1987). Structural analysis of a human intestinal epithelial cell line. *Gastroenterology* **92**, 1133–1145.

Mahdavi, J., Sonden, B., Hurtig, M., *et al.* (2002). *Helicobacter pylori* SabA adhesin in persistent infection and chronic inflammation. *Science* **297**, 573–578.

Mann, B., Gelos, M., Siedow, A., *et al.* (1999). Target genes of beta-catenin-T cell-factor/lymphoid-enhancer-factor signaling in human colorectal carcinomas. *Proc. Natl. Acad. Sci. U. S. A.* **96**, 1603–1608.

Marshall, B. J. and Warren, J. R. (1984). Unidentified curved bacilli in the stomach of patients with gastritis and peptic ulceration. *Lancet* **1**, 1311–1315.

Matysiak-Budnik, T. and Megraud, F. (1997). Epidemiology of *Helicobacter pylori* infection with special reference to professional risk. *J. Physiol. Pharmacol.* **48 (Suppl 4)**, 3–17.

McClain, M. S., Schraw, W., Ricci, V., Boquet, P., and Cover, T. L. (2000). Acid activation of *Helicobacter pylori* vacuolating cytotoxin (VacA) results in toxin internalization by eukaryotic cells. *Mol. Microbiol.* **37**, 433–442.

McGee, D. J., Radcliff, F. J., Mendz, G. L., Ferrero, R. L., and Mobley, H. L. (1999). *Helicobacter pylori rocF* is required for arginase activity and acid protection in vitro but is not essential for colonization of mice or for urease activity. *J. Bacteriol.* **181**, 7314–7322.

McGee, D. J., Coker, C., Testerman, T. L., *et al.* (2002). The *Helicobacter pylori flbA* flagellar biosynthesis and regulatory gene is required for motility and virulence and modulates urease of *H. pylori* and *Proteus mirabilis*. *J. Med. Microbiol.* **51**, 958–970.

McGee, D. J., Langford, M. L., Watson, E. L., *et al.* (2005). Colonization and inflammation deficiencies in Mongolian gerbils infected by *Helicobacter pylori* chemotaxis mutants. *Infect. Immun.* **73**, 1820–1827.

Merrell, D. S. and Falkow, S. (2003). Expression profiling in *Helicobacter pylori* infection. In *Perspectives in Gene Expression*, ed. K. Appasani. Westborough: Eaton Publishing, pp. 273–303.

Merrell, D. S. and Falkow, S. (2004). Frontal and stealth attack strategies in microbial pathogenesis. *Nature* **430**, 250–256.

Merrell, D. S., Goodrich, M. L., Otto, G., Tompkins, L. S., and Falkow, S. (2003). pH regulated gene expression of the gastric pathogen *Helicobacter pylori*. *Infect. Immun.* **71**, 3529–3539.

Meyer-Rosberg, K., Scott, D. R., Rex, D., Melchers, K., and Sachs, G. (1996). The effect of environmental pH on the proton motive force of *Helicobacter pylori*. *Gastroenterology* **111**, 886–900.

Mimuro, H., Suzuki, T., Tanaka, J., *et al.* (2002). Grb2 is a key mediator of *Helicobacter pylori* CagA protein activities. *Mol. Cell* **10**, 745–755.

Mobley, H. L., Island, M. D., and Hausinger, R. P. (1995). Molecular biology of microbial ureases. *Microbiol. Rev* **59**, 451–480.

Molinari, M., Galli, C., Norais, N., *et al.* (1997). Vacuoles induced by *Helicobacter pylori* toxin contain both late endosomal and lysosomal markers. *J. Biol. Chem.* **272**, 25 339–25 344.

Molinari, M., Salio, M., Galli, C., *et al.* (1998). Selective inhibition of Ii-dependent antigen presentation by *Helicobacter pylori* toxin VacA. *J. Exp. Med.* **187**, 135–140.

Neu, B., Rad, R., Reindl, W., *et al.* (2005). Expression of tumor necrosis factor-alpha-related apoptosis-inducing ligand and its proapoptotic receptors is down-regulated during gastric infection with virulent *cagA+/vacAs1+ Helicobacter pylori* strains. *J. Infect. Dis.* **191**, 571–578.

Neugut, A. I., Hayek, M., and Howe, G. (1996). Epidemiology of gastric cancer. *Semin. Oncol.* **23**, 281–291.

Odenbreit, S., Puls, J., Sedlmaier, B., *et al.* (2000). Translocation of *Helicobacter pylori* CagA into gastric epithelial cells by type IV secretion. *Science* **287**, 1497–1500.

Oh, J. D., Karam, S. M., and Gordon, J. I. (2005). Intracellular *Helicobacter pylori* in gastric epithelial progenitors. *Proc. Natl. Acad. Sci. U. S. A.* **102**, 5186–5191.

Ottemann, K. M. and Lowenthal, A. C. (2002). *Helicobacter pylori* uses motility for initial colonization and to attain robust infection. *Infect. Immun.* **70**, 1984–1990.

Papini, E., de Bernard, M., Milia, E., *et al.* (1994). Cellular vacuoles induced by *Helicobacter pylori* originate from late endosomal compartments. *Proc. Natl. Acad. Sci. U. S. A.* **91**, 9720–9724.

Parsonnet, J., Friedman, G. D., Vandersteen, D. P., *et al.* (1991). *Helicobacter pylori* infection and the risk of gastric carcinoma. *N. Engl. J. Med.* **325**, 1127–1131.

Phadnis, S. H., Parlow, M. H., Levy, M., *et al.* (1996). Surface localization of *Helicobacter pylori* urease and a heat shock protein homolog requires bacterial autolysis. *Infect. Immun.* **64**, 905–912.

Pierceall, W. E., Woodard, A. S., Morrow, J. S., Rimm, D., and Fearon, E. R. (1995). Frequent alterations in E-cadherin and alpha- and beta-catenin expression in human breast cancer cell lines. *Oncogene* **11**, 1319–1326.

Rabassa, A. A., Goodgame, R., Sutton, F. M., *et al.* (1996). Effects of aspirin and *Helicobacter pylori* on the gastroduodenal mucosal permeability to sucrose. *Gut* **39**, 159–163.

Rhen, M., Eriksson, S., Clements, M., Bergstrom, S., and Normark, S. J. (2003). The basis of persistent bacterial infections. *Trends Microbiol.* **11**, 80–86.

Rimm, D. L., Sinard, J. H., and Morrow, J. S. (1995). Reduced alpha-catenin and E-cadherin expression in breast cancer. *Lab. Invest.* **72**, 506–512.

Salama, N. R., Otto, G., Tompkins, L., and Falkow, S. (2001). Vacuolating cytotoxin of *Helicobacter pylori* plays a role during colonization in a mouse model of infection. *Infect. Immun.* **69**, 730–736.

Salama, N. R., Ottemann, K. M., and Falkow, S. (2002). Toxins, tropisms and travels: *H. pylori* and host cells. In *Helicobacter Infections and Immunity*, ed. Y. Yamamoto, H. Friedman, and P. Hoffman. New York: Kluwer Academic/Plenum Publishers, pp. 173–201.

Schreiber, S., Konradt, M., Groll, C., *et al.* (2004). The spatial orientation of *Helicobacter pylori* in the gastric mucus. *Proc. Natl. Acad. Sci. U. S. A.* **101**, 5024–5029.

Scott, D. R., Weeks, D., Hong, C., *et al.* (1998). The role of internal urease in acid resistance of *Helicobacter pylori. Gastroenterology* **114**, 58–70.

Segal, E. D. (1997). Consequences of attachment of *Helicobacter pylori* to gastric cells. *Biomed. Pharmacother.* **51**, 5–12.

Segal, E. D., Lange, C., Covacci, A., Tompkins, L. S., and Falkow, S. (1997). Induction of host signal transduction pathways by *Helicobacter pylori. Proc. Natl. Acad. Sci. U. S. A.* **94**, 7595–7599.

Segal, E. D., Cha, J., Lo, J., Falkow, S., and Tompkins, L. S. (1999). Altered states: involvement of phosphorylated CagA in the induction of host cellular growth changes by *Helicobacter pylori. Proc. Natl. Acad. Sci. U. S. A.* **96**, 14 559–14 564.

Selbach, M., Moese, S., Hauck, C. R., Meyer, T. F., and Backert, S. (2002). Src is the kinase of the *Helicobacter pylori* CagA protein in vitro and in vivo. *J. Biol. Chem.* **277**, 6775–6778.

Semino-Mora, C., Doi, S. Q., Marty, A., *et al.* (2003). Intracellular and interstitial expression of *Helicobacter pylori* virulence genes in gastric precancerous intestinal metaplasia and adenocarcinoma. *J. Infect. Dis.* **187**, 1165–1177.

Smith, M. F., Jr, Mitchell, A., Li, G., *et al.* (2003). Toll-like receptor (TLR) 2 and TLR5, but not TLR4, are required for *Helicobacter pylori*-induced NF-kappa B activation and chemokine expression by epithelial cells. *J. Biol. Chem.* **278**, 32 552–32 560.

Solnick, J. V., Hansen, L. M., Salama, N. R., Boonjakuakul, J. K., and Syvanen, M. (2004). Modification of *Helicobacter pylori* outer membrane protein expression during experimental infection of rhesus macaques. *Proc. Natl. Acad. Sci. U. S. A.* **101**, 2106–2111.

Stein, M., Rappuoli, R., and Covacci, A. (2000). Tyrosine phosphorylation of the *Helicobacter pylori* CagA antigen after cag-driven host cell translocation. *Proc. Natl. Acad. Sci. U. S. A.* **97**, 1263–1268.

Stein, M., Bagnoli, F., Halenbeck, R., *et al.* (2002). c-Src/Lyn kinases activate *Helicobacter pylori* CagA through tyrosine phosphorylation of the EPIYA motifs. *Mol. Microbiol.* **43**, 971–980.

Sundrud, M. S., Torres, V. J., Unutmaz, D., and Cover, T. L. (2004). Inhibition of primary human T cell proliferation by *Helicobacter pylori* vacuolating toxin (VacA) is independent of VacA effects on IL-2 secretion. *Proc. Natl. Acad. Sci. U. S. A.* **101**, 7727–7732.

Taipale, J. and Beachy, P. A. (2001). The Hedgehog and Wnt signalling pathways in cancer. *Nature* **411**, 349–354.

Telford, J. L., Ghiara, P., Dell'Orco, M., *et al.* (1994). Gene structure of the *Helicobacter pylori* cytotoxin and evidence of its key role in gastric disease. *J. Exp. Med.* **179**, 1653–1658.

Terry, K., Williams, S. M., Connolly, L., and Ottemann, K. M. (2005). Chemotaxis plays multiple roles during *Helicobacter pylori* animal infection. *Infect. Immun.* **73**, 803–811.

Tombola, F., Carlesso, C., Szabo, I., *et al.* (1999). *Helicobacter pylori* vacuolating toxin forms anion-selective channels in planar lipid bilayers: possible implications for the mechanism of cellular vacuolation. *Biophys. J.* **76**, 1401–1409.

Tombola, F., Morbiato, L., Del Giudice, G., *et al.* (2001). The *Helicobacter pylori* VacA toxin is a urea permease that promotes urea diffusion across epithelia. *J. Clin. Invest.* **108**, 929–937.

Tsuda, M., Karita, M., Morshed, M. G., Okita, K., and Nakazawa, T. (1994). A urease-negative mutant of *Helicobacter pylori* constructed by allelic exchange mutagenesis lacks the ability to colonize the nude mouse stomach. *Infect. Immun.* **62**, 3586–3589.

Umehara, S., Higashi, H., Ohnishi, N., Asaka, M., and Hatakeyama, M. (2003). Effects of *Helicobacter pylori* CagA protein on the growth and survival of B lymphocytes, the origin of MALT lymphoma. *Oncogene* **22**, 8337–8342.

Valle, J., Seppala, K., Sipponen, P., and Kosunen, T. (1991). Disappearance of gastritis after eradication of *Helicobacter pylori*. A morphometric study. *Scand. J. Gastroenterol.* **26**, 1057–1065.

Vera, J. F., Gotteland, M., Chavez, E., *et al.* (1997). Sucrose permeability in children with gastric damage and *Helicobacter pylori* infection. *J. Pediatr. Gastroenterol. Nutr.* **24**, 506–511.

Weeks, D. L., Eskandari, S., Scott, D. R., and Sachs, G. (2000). A H+-gated urea channel: the link between *Helicobacter pylori* urease and gastric colonization. *Science* **287**, 482–485.

Willhite, D. C. and Blanke, S. R. (2004). *Helicobacter pylori* vacuolating cytotoxin enters cells, localizes to the mitochondria, and induces mitochondrial membrane permeability changes correlated to toxin channel activity. *Cell. Microbiol.* **6**, 143–154.

Willhite, D. C., Cover, T. L., and Blanke, S. R. (2003). Cellular vacuolation and mitochondrial cytochrome c release are independent outcomes of *Helicobacter pylori* vacuolating cytotoxin activity that are each dependent on membrane channel formation. *J. Biol. Chem.* **278**, 48 204–48 209.

Wilson, K. T., Ramanujam, K. S., Mobley, H. L., *et al.* (1996). *Helicobacter pylori* stimulates inducible nitric oxide synthase expression and activity in a murine macrophage cell line. *Gastroenterology* **111**, 1524–1533.

Xiang, Z., Censini, S., Bayeli, P. F., *et al.* (1995). Analysis of expression of CagA and VacA virulence factors in 43 strains of *Helicobacter pylori* reveals that clinical isolates can be divided into two major types and that CagA is not necessary for expression of the vacuolating cytotoxin. *Infect. Immun.* **63**, 94–98.

Yahiro, K., Niidome, T., Hatakeyama, T., *et al.* (1997). *Helicobacter pylori* vacuolating cytotoxin binds to the 140-kDa protein in human gastric cancer cell lines, AZ-521 and AGS. *Biochem. Biophys. Res. Commun.* **238**, 629–632.

Yamaoka, Y., El-Zimaity, H. M., Gutierrez, O., *et al.* (1999). Relationship between the *cagA* 3′ repeat region of *Helicobacter pylori*, gastric histology, and susceptibility to low pH. *Gastroenterology* **117**, 342–349.

Yeung, C. K., Fu, K. H., Yuen, K. Y., *et al.* (1990). *Helicobacter pylori* and associated duodenal ulcer. *Arch. Dis. Child.* **65**, 1212–1216.

Zheng, P. Y. and Jones, N. L. (2003). *Helicobacter pylori* strains expressing the vacuolating cytotoxin interrupt phagosome maturation in macrophages by recruiting and retaining TACO (coronin 1) protein. *Cell. Microbiol.* **5**, 25–40.

CHAPTER 13

Interactions of enteric bacteria with the intestinal mucosa

Samuel Tesfay, Donnie Edward Shifflett, and Gail A. Hecht

INTRODUCTION

Bacteria colonize the gastrointestinal tract as early as a few hours after birth. This relationship that develops at an early stage between humans and bacteria is shared with other mammals. Gastrointestinal epithelial cells play a crucial role in maintaining a quiescent environment while being bathed with normal flora, and yet at the same time they must possess functions that allow them to participate in immune surveillance. In addition to screening for and responding to the presence of pathogens in the intestinal lumen, gastrointestinal epithelial cells provide barrier function and transport of ions and solutes.

Enteric pathogens, as opposed to normal flora, cause disease by exploiting the host cytoskeleton or signaling pathways, which ultimately alters the physiologic functions of the intestinal epithelium. For example, pathogens can induce or suppress inflammatory responses, alter the transport of fluid, solutes, and ions, perturb the tight-junction barrier, and activate programmed cell death (apoptosis). This chapter summarizes the cross-talk between bacterial pathogens and host cells that leads to gastrointestinal symptoms.

ENTERIC PATHOGENS AND INTESTINAL EPITHELIAL CELL RECEPTORS

The interaction that occurs between pathogenic or non-pathogenic bacteria and intestinal epithelial cells begins with the adherence of bacteria to the cellular surface. This is a common mechanism by which bacteria cause

Bacterial–Epithelial Cell Cross-Talk: Molecular Mechanisms in Pathogenesis, ed. Beth A. McCormick. Published by Cambridge University Press.

disease, not only in the gastrointestinal tract but also in other systems where epithelial cells face the external environment, such as the genitourinary and respiratory systems. Adherence of bacteria to the cellular surface is essential to reduce the washout effect caused by intestinal secretion and peristalsis. Adhesion, however, may serve a purpose for both the bacteria and the epithelial cells. Direct contact enables epithelial cells to identify and localize microorganisms to the target site. Adherence also allows bacteria to gain a point of entry into the host cells or to deliver effector molecules into cells. Arrays of bacterial and cellular molecules are involved in the execution of these initial steps of molecular cross-talk.

Adherence factors expressed by the microorganisms, termed adhesins, bind to specific host-cell receptors. Numerous bacterial adhesins have been described. In addition, a variety of host-cell receptors for adhesins of several bacterial pathogens have been identified (see below). Intestinal epithelial cells (IECs) employ several receptors on their membranes to detect pathogens and signal a response. Toll-like receptors (TLR) and nucleotide-binding oligomerization domain (Nod) proteins are a family of host-cell receptors that sense the presence of bacteria and, hence, are termed pathogen-associated molecular patterns (PAMPs) or microbial-associated molecular patterns (MAMPs), since these receptors detect both pathogenic and non-pathogenic bacteria (Akira *et al.*, 2001; Girardin *et al.*, 2003). These external and internal receptors, respectively, are linked to host signaling pathways that initiate a defense response regardless of the identity of the specific organism. Therefore, enteric pathogens must evade these receptors or possess a means of shutting down subsequent signaling responses that summon inflammatory cells (Akira *et al.*, 2001; Girardin *et al.*, 2003) and employ other cellular surface receptors to cause gastrointestinal diseases. Bacteria utilize numerous protein, carbohydrate, and/or glycolipid cell-surface receptors to adhere and gain access into host cells (Hauck, 2002). Some enteric pathogens have been shown to utilize host-cell receptors, including integrins, E-cadherin, members of the immunoglobulin superfamily, and selectins in order to interact with intestinal epithelial cells and establish infection (Juliano, 2002).

One of the first-recognized examples of such exploitation of host-cell receptors regards the enteroinvasive bacteria *Yersinia*. *Yersinia* species are responsible for bubonic plague (*Y. pestis*), gastroenteritis (*Y. enterocolitica*, *Y. pseudotuberculosis*), and other extragastrointestinal diseases. Invasion of *Y. pseudotuberculosis* involves the intimate interaction of its adhesin, called invasin, encoded by the virulence gene *inv*, with β1-integrins expressed on the apical surface of M-cells in the intestinal mucosa (Clark *et al.*, 1998; Isberg and Leong, 1990; Isberg *et al.*, 1987). Invasin shares molecular homology with fibronectin, the endogenous receptor for integrins. Integrins are heterodimeric surface glycoproteins that interact with extracellular matrix proteins and are usually localized to basolateral focal adhesion sites of the IECs. *Y. pseudotuberculosis* strains harboring *inv* were able to invade human epidermoid larynx carcinoma (HEp-2) cells, and mutation of *inv* rendered the strain non-invasive (Miller and Falkow, 1988). Also, transfer of the wild-type *inv* gene, but not mutated *inv*, into non-invasive *Escherichia coli* conferred the invasive phenotype to this non-pathogenic organism (Isberg *et al.*, 1987; Miller and Falkow, 1988). Similar to the features of integrin creating a transmembrane link between the extracellular matrix and the cytoskeleton of the cells (Burridge *et al.*, 1988), *Yersinia* internalization initiated by the interaction of invasin with β1-integrin results in the recruitment of actin and actin-associated proteins (Young *et al.*, 1992). The recruitment and reorganization of actin molecules was observed using transmission electron micrographs of detergent-insoluble cytoskeleton of HEp-2 cells infected with invasin-expressing *E. coli* (Young *et al.*, 1992). In addition, invasin-stimulated uptake of bacteria was abrogated when microtubule organization was perturbed through use of microtubule-disrupting drugs, indicating the use of cytoskeletal proteins for *Yersinia* internalization (McGee *et al.*, 2003). These experiments highlight the importance of cellular receptors, such as integrins, as targets for microorganisms to gain entry into the cells. Furthermore, they demonstrate the exploitation of host cytoskeletal proteins for pathogen internalization and the utilization of molecular mimicry as a method deployed by enteric pathogens to cause several gastrointestinal diseases.

Shigella is another enteroinvasive pathogen that binds to an integrin early in the infectious process (Watarai *et al.*, 1996). *Shigella* is responsible for dysentery (shigellosis), a disease that provokes severe inflammatory diarrhea in humans and is due to colonization and destruction of the colonic mucosa. Destruction of the intestinal mucosa is the result of virulence factors produced by this enteric pathogen. Like other Gram-negative enteric pathogens, *Shigella* exerts its pathogenic effects by actively delivering these factors, effector molecules, directly into host epithelial cells through a type III secretion system (TTSS), which has been compared to a molecular syringe (Figure 13.1). This protein-secretion system spans both bacterial and host-cell membranes with an extracellular filamentous molecular conduit in order to deliver type III effector proteins into the host-cell cytoplasm (Yip *et al.*, 2005). Effector proteins IpaB, IpaC, and IpaD, encoded by the *ipa* operon present on a virulence plasmid, play a role in pathogenesis by the binding to α5β1 integrin. In vitro experiments using Chinese hamster ovary (CHO) cells indicate that the Ipa proteins interact directly with this integrin, and internalization of *Shigella*

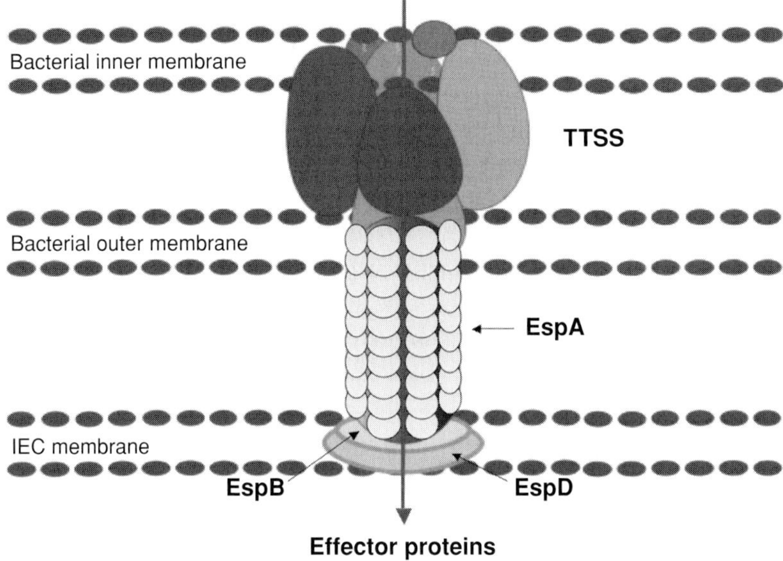

Figure 13.1 Type III secretion system (TTSS) spans the inner and outer bacterial membranes with a needle-like complex formed by EspA protein, which connects with EspB and EspD proteins that form pores in the intestinal epithelial cell (IEC) membrane.

was more efficient in CHO cell transfectants that overexpressed α5β1 integrin, suggesting that the interaction between the effector proteins and host cell-adhesion receptors promotes bacterial entry into cells (Watarai *et al.*, 1996). Furthermore, experiments utilizing polarized epithelial cells (Caco-2 cells) grown on permeable supports showed that *Shigella flexeri* invades cells more efficiently from the basolateral surface where α5β1 integrin typically resides (Mounier *et al.*, 1992). In addition to gaining access into the cells, the interaction between the effector proteins and integrin leads to the phosphorylation of tyrosine kinases and the subsequent recruitment and rearrangement of actin. Actin polymerization at the site of entry appears to be mediated by the IpaB and IpaC effector proteins and is believed to play a role in the invasive process of *Shigella* (Watarai *et al.*, 1996).

In addition to the exploitation of integrins, bacteria such as *Listeria monocytogenes* utilize other cellular adhesion molecules (CAMs) to execute their invasive properties. *Listeria monocytogenes* is a food-borne pathogen that causes severe gastroenteritis, meningoencephalitis in elderly people and immunocompromised patients, and spontaneous abortions in pregnant women (Doganay, 2003). *L. monocytogenes* enters epithelial cells via the surface protein internalin encoded by the *inlA* gene (Gaillard *et al.*, 1991). Similar to invasin of *Yersinia* conferring invasiveness to the non-invasive organisms, transformation of non-invasive *Listeria innocua* with internalin demonstrated the role of this protein in internalization into epithelial cells (Gaillard *et al.*, 1991). The IEC receptor for internalin was identified as E-cadherin, a basolateral transmembrane cell-adhesion protein that is normally involved in cell–cell interactions (Mengaud *et al.*, 1996). E-cadherins are members of a protein family of calcium-dependent cell–cell-adhesion molecules that are expressed primarily in epithelial tissues (e.g. digestive tract) and play an important role in maintenance of the intestinal epithelial structure, differentiation, and regulation of programmed cell death (Hermiston and Gordon, 1995). The extracellular domain of E-cadherin interacts with internalin, and the intracellular domain associates with actin via α- and β-catenins, which are recruited to the site of bacterial entry (Lecuit *et al.*, 2000). Like many other invasive and non-invasive enteric pathogens, *L. monocytogenes* induces the recruitment of actin, driving its intra- and intercellular movement (Theriot *et al.*, 1992).

Tight-junction proteins have also been reported to serve as microbial receptors. *Clostridium perfringens* enterotoxin (CPE) binds to the tight-junction transmembrane proteins claudin-3 and -4 and is believed to elicit diarrhea by altering the permeability of the intestinal epithelium (Sonoda *et al.*, 1999). Claudins associate with the cytoskeleton (actin) and function to maintain cell polarity, regulate paracellular water and solute transport, and serve as a barrier to luminal contents (Schneeberger and Lynch, 2004). CPE binds to claudin-3 and -4, triggering a multistep process that leads to the lysis of susceptible epithelial target cells within 10–20 min (McClane, 1996).

Interestingly, the receptors for bacterial adhesins are not limited to those expressed by host cells. Two non-invasive and closely related diarrheagenic pathogens, enteropathogenic *E. coli* (EPEC) and enterohemorrhagic *E. coli* (EHEC), have developed a sophisticated mechanism for intimate attachment to host cells. Both of these pathogens deliver their own receptors into the host cell membrane via the TTSS (DeVinney *et al.*, 2001; Kenny, DeVinney *et al.*, 1997). These receptors, which interact with outer-membrane adhesins called intimin, are called translocated intimin receptors (Tir). Not surprisingly, Tir was originally thought to be a host-cell-derived protein. Elegant studies by Kenny, DeVinney *et al.* (1997) and Diebel *et al.* (1998) showed simultaneously that Tir is actually an EPEC-derived protein. The ligand for EPEC Tir is the bacterial surface adhesin protein intimin-α, while intimin-γ serves this role for EHEC. Both ligands are encoded by the *eae* gene and mediate intimate attachment by binding to their respective receptors (Kenny, DeVinney *et al.*, 1997; Yu and Kaper, 1992). Tir and intimin are encoded on a pathogenicity island called the locus of enterocyte effacement (LEE) housed in the chromosome of EPEC and EHEC along with the genes that comprise the TTSS and effector proteins (McDaniel and Kaper, 1997). Although intimin is structurally homologous to invasin of *Yersinia*, EPEC and EHEC are not invasive organisms (Donnenberg *et al.*, 1993; Isberg *et al.*, 1987). Once stabilized in the host cell membrane, Tir induces the characteristic attaching and effacing (A/E) lesion in enterocytes (Frankel *et al.*, 1998). A/E lesions are identified as areas of cytoskeletal (primarily actin) recruitment associated with areas of intimate bacterial attachment to the cells. Phenotypically, actin-rich pedestals are formed and effacement of surrounding brush-border microvilli is seen (Frankel *et al.*, 1998; Hecht, 2001).

Tir is one effector protein that plays an early essential step in EPEC and EHEC pathogenesis by anchoring these bacteria to IECs, contributing to the loss of microvilli and formation of actin-rich pedestals. The subsequent actin rearrangement that results from EPEC Tir–intimin interactions at the IEC membrane is initiated by phosphorylation of tyrosine residue 474 (Y474) by a host tyrosine kinase (Ismaili *et al.*, 1995; Kenny, 1999). Interestingly, this phosphorylation step is absent in EHEC, and yet both pathogens induce similar cytoskeletal rearrangement (DeVinney, Stein *et al.*, 1999). Immunofluorescence studies have revealed that the cytoskeletal plaque associated with induced pedestal formation contains filamentous (F)-actin, α-actinin, ezrin, neural Wiskott–Aldrich syndrome protein (N-WASP), Arp2/3 complex, and several other cytoskeletal proteins (Freeman *et al.*, 2000; Goosney *et al.*, 2000, 2001; Kalman *et al.*, 1999; Sanger *et al.*, 1996). Even though EPEC- and EHEC-induced A/E lesions recruit similar cytoskeletal associated proteins, especially N-WASP and Arp2/3 complex, there are mechanistic differences that mediate pedestal formation by EPEC and EHEC. For instance, immunofluorescence studies of HeLa cells infected with EPEC or EHEC reveal that pedestals formed by EPEC, but not EHEC, contain the adapter proteins Grb2 and CrkII, which mediate protein–protein interactions. In addition, EPEC Tir recruits actin by directly binding the adapter protein Nck at the phosphorylated Y474 site, which then recruits and activates a member of the Wiskott–Aldrich syndrome (WAS) family of proteins, N-WASP, a regulator of the Arp2/3 actin-nucleating process (Gruenheid *et al.*, 2001).

Although EHEC also recruits WASP and Arp2/3 complexes, it does so in a Nck-independent manner, as pedestal formation occurs in the absence of Tir tyrosine phosphorylation. In fact, EHEC Tir does not possess a Y474 residue or other phosphotyrosines. Complementation experiments using plasmid-encoded EHEC and EPEC *tir* have confirmed that the Tir molecule of EHEC O157:H7 is not functionally interchangeable with EPEC Tir (Kenny, 2001) and has only 44% structural similarity to EPEC Tir (Perna *et al.*, 1998). Pedestal formation by EHEC, therefore, was determined to require additional bacterial factors. The Leong laboratory identified the non-LEE-encoded translocated effector $EspF_U$ as the factor substituting for phosphorylated Tir and Nck. $EspF_U$ was shown to interact with both Tir and N-WASP, circumventing the need for Nck adaptors that are required for pedestal formation in EPEC (Campellone *et al.*, 2004). In addition, $EspF_U$ was shown to contribute to the disruption of tight junctions by EHEC, resembling a role for the LEE-encoded EspF of EPEC, a TTSS effector protein (Viswanathan, Koutsouris *et al.*, 2004). EHEC also secretes an LEE-encoded EspF protein that is functionally equivalent to EPEC EspF when substituted in an EPEC *espF* deletion strain, as determined by decreased transepithelial electrical resistance (TER), an electrophysiological measurement of the tight-junction barrier, and the redistribution of the transmembrane tight-junction protein occludin (Viswanathan, Koutsouris *et al.*, 2004). The magnitude of these molecular differences as it relates to pedestal function and diarrheal disease is unknown but gives insight into the degree of cellular exploitation and highlights the mechanistic differences of two closely related enteric pathogens modulating different signal-transduction pathways to affect the host cytoskeleton proteins.

The similarities of Tir with CAMs as transmembrane proteins and the recruitment of the cytoskeleton to these receptors have led to the postulate that Tir may act in a manner similar to that of β-integrins. Solid-phase binding assays and T-cell adherence assays have demonstrated that the receptor-binding domain (C-terminus portion) of intimin from EPEC binds to α_4- and $\alpha_5\beta_1$-integrins (Frankel *et al.*, 1996).

Although the intimate contact of EPEC and EHEC with the IEC membrane is mediated via intimin–Tir interaction, adherence of bacteria still occurs in the absence of Tir, but distinct actin accumulation or pedestal formation is absent in Tir-deficient mutants (DeVinney, Knoechel *et al.*, 1999; DeVinney, Stein *et al.*, 1999). The interaction of intimin-α with β1-integrin has been demonstrated to potentiate a decrease in TER in the late stages of EPEC infection, a consequence of perturbation of cell polarity and redistribution of basolateral membrane proteins (integrins) initiated by Tir–intimin-mediated tight-junction disruption (Muza-Moons *et al.*, 2003). In addition to the demonstrated physiologic relevance of intimin-α interaction with a eukaryotic cell receptor (β-integrin), nucleolin has been identified as a cellular receptor for *E. coli* O157:H7 intimin-γ (Sinclair and O'Brien, 2002). *E. coli* O157:H7 is a Shiga-toxin-producing EHEC strain responsible for severe hemorrhagic colitis associated with the systemic complication of hemolytic-uremic syndrome (HUS), which is characterized by hemolytic anemia, thrombocytopenia, and renal failure (Nataro and Kaper, 1998). Nucleolin is a protein expressed at the cell surface that functions in ribosome biogenesis and cell growth. Bacterial adherence assays using polyclonal antiserum against human nucleolin have demonstrated the importance of this receptor in the early phase of EHEC O157:H7 adherence by partially blocking this process (Sinclair and O'Brien, 2002). Nucleolin binds the extracellular matrix protein laminin and is involved in the extension of filopodia, suggesting that the interaction of intimin-γ and nucleolin may trigger a similar response, thus participating in pedestal formation (Nougayrede *et al.*, 2003).

Regardless of whether enteric pathogens insert their own receptors or utilize existing IEC receptors for adherence, the exploitation of the cellular cytoskeleton is central and essential to the pathogenesis of both invasive and non-invasive organisms. The targeting of cellular adhesion molecules enables pathogens to recruit cytoskeletal proteins in order to gain access to or form intimate contact with the IECs. Adherence and subsequent recruitment of intracellular cytoskeletal proteins via CAMs is an essential first stage in disease processes for many pathogens, including *Salmonella, Shigella, Yersinia,* and *Listeria*. Other enteric pathogens, such as EPEC and EHEC, have devised methods to imitate CAMs structurally and functionally by inserting their own receptors, which ultimately leads to reorganization of the host cytoskeleton.

MOLECULAR MANIPULATION OF INTESTINAL EPITHELIAL CELLS BY SECRETED EFFECTORS OF ENTERIC PATHOGENS AND THEIR EFFECTS ON THE GUT MUCOSA

Several enteric pathogens have evolved a complex protein-secretion system termed type III to inject specific proteins (effectors) directly into the host cytosol, thereby modulating host cell functions. The important role of TTSS has been established not only for animal pathogenic bacteria but also for plant pathogens, highlighting evolutionary uniformity. Although the secretion system assembly is similar across various pathogens and is related evolutionarily to flagellar apparatus, the secreted effector proteins vary in their target and cellular effects. A TTSS has been described for *Yersinia, Shigella, Salmonella,* pathogenic *E. coli,* and other Gram-negative bacterial species, and each has the capacity to translocate a variety of effector proteins that act on the host cytoskeleton or intracellular signaling cascades.

The translocated effector proteins have been deemed virulence factors for many enteric bacterial pathogens and are responsible for the phenotypic characteristics of their pathogenesis. For instance, two effector proteins from *Shigella* species, IpaA and IpaC, promote bacterial entry into non-phagocytic cells by acting on the host-cell cytoskeleton (Nhieu and Sansonetti, 1999; Tran Van Nhieu *et al.*, 1999). *Salmonella* species effector proteins SipC and SipA also bind directly to actin and induce its nucleation and stabilize the actin filaments, respectively, allowing the pathogen to gain entry into non-phagocytic cells (Hayward and Koronakis, 1999; Zhou *et al.*, 1999). Besides the contribution of SipA to *Salmonella* entry, this effector induces a pro-inflammatory response in IECs by orchestrating the transepithelial migration of polymorphonuclear leukocytes, a process that involves a pathogen-elicited epithelial chemoattractant (hepoxilin A3) and protein kinase C (PKC)-dependent signaling (McCormick *et al.*, 1998; Mrsny *et al.*, 2004; Silva *et al.*, 2004). In addition, the *Salmonella* effector protein SopE stimulates bacterial entry by acting as an exchange factor for small GTPases belonging to the Rho subfamily of proteins (Rac-1, Cdc42), which are associated with actin polymerization (Hardt *et al.*, 1998).

Yersinia species also inject effector proteins into host cells that interfere with the cytoskeleton. YopH, YopE, and YopT effector proteins exert a negative role on cytoskeletal dynamics by reorganizing and disassembling the cytoskeleton, which contributes to the resistance of these bacteria to phagocytosis by macrophages (Black and Bliska, 1997; Grosdent *et al.*, 2002; Persson *et al.*, 1997; Zumbihl *et al.*, 1999). In addition to targeting the cytoskeleton, *Yersinia* species modulate the host inflammatory response via its effector proteins. YopP and YopJ effector proteins block the release of tumor necrosis factor alpha (TNF-α) by macrophages and interleukin 8 (IL-8) response by epithelial cells, resulting in a significant reduction in inflammation (Boland and Cornelis, 1998; Schesser *et al.*, 1998; Schulte *et al.*, 1996). Suppression of inflammation by YopP/J is a direct inhibitory effect on the mitogen-activated protein kinases c-jun-N-terminal kinase (JNK), p38, and extracellular signal-regulated kinases 1 and 2 (ERK1 and ERK2) (Boland and Cornelis, 1998; Palmer *et al.*, 1998; Ruckdeschel *et al.*, 1997). These effector proteins also induce programmed cell death in macrophages, but the suppressive effect on inflammation is independent of their ability to induce apoptosis (Mills *et al.*, 1997; Monack *et al.*, 1997; Ruckdeschel *et al.*, 1998). In contrast to the *Yersinia* effector proteins, IpaB of *Shigella* and SipB of *Salmonella* induce inflammation by stimulating an interleukin 1 (IL-1) response and cause apoptosis of macrophages, thus contributing to their invasive and destructive properties in the intestinal epithelial mucosa (Hersh *et al.*, 1999; Thirumalai *et al.*, 1997).

The enteroinvasive pathogens mentioned above utilize their effector proteins to exploit host-cell proteins and signaling pathways, resulting in cytoskeletal disruption and up- or downregulation of inflammation. The importance of the TTSS and the effector proteins is particularly evident for the non-invasive bacteria, as they cause disease by delivering the effector molecules into the cytoplasm from an extracellular position. Discoveries and advances gained by studying non-invasive EPEC and EHEC effector proteins underscore their potency in promoting diarrhea in the host. In the next section, we examine closely and compare the effects of the secreted effectors on the intestinal mucosa as they relate to pathogenesis.

ENTEROPATHOGENIC *ESCHERICHIA COLI* AND ENTEROHEMORRHAGIC *ESCHERICHIA COLI* EFFECTORS AND EFFECTS

EPEC and EHEC are phylogenetically related Gram-negative enteric pathogens that pose a significant risk to human health worldwide. EPEC is a leading cause of diarrhea among infants from developing nations worldwide and is associated with sporadic cases of diarrhea in the USA and other developed countries (Nataro and Kaper, 1998). The symptoms of EPEC infection are profuse watery diarrhea, vomiting, and low-grade fever. EHEC and its most notorious serotype O157:H7 are responsible for a distinctive gastrointestinal illness associated with ingestion of undercooked meat, characterized by severe cramping, abdominal pain, and watery diarrhea progressing to bloody diarrhea due to hemorrhagic colitis, with systemic complications of renal failure, hemolytic anemia, and thrombocytopenia (HUS) (Nataro and Kaper, 1998).

EHEC O157:H7 notoriety is attributed to the production of Shiga-toxin (Stx), a key virulence factor expressed by subtypes of EHEC. Stx is responsible for hemorrhagic colitis and HUS largely through its propensity for binding to endothelial cells. EHEC and the non-toxigenic EPEC reside in a group of pathogenic bacteria that produce the characteristic histopathological feature in the intestinal mucosa referred to as A/E lesions (Frankel *et al.*, 1998). This phenotypic feature results from the expression of virulence genes housed in the LEE pathogenicity island that encodes proteins for the TTSS, intimin, and the secreted effector proteins (McDaniel and Kaper, 1997).

Diarrhea is the most prominent symptom associated with EPEC and EHEC infection. Although Stx has been identified as the cause of hemorrhagic colitis and HUS, the mechanisms that mediate EPEC- and EHEC-induced diarrhea are still largely unknown. The effacement of microvilli in the A/E lesions may diminish the absorptive capacity of the intestinal epithelium, but it is now clear that the mechanisms underlying diarrhea associated with infection by these pathogens are complex and multifactorial and include

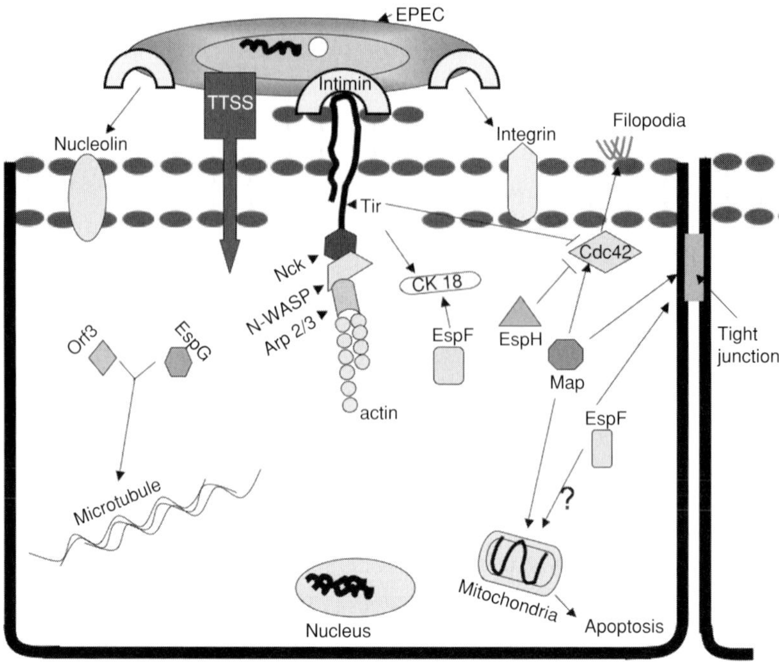

Figure 13.2 Locus of enterocyte effacement (LEE) and non-LEE effectors of enteropathogenic *Escherichia coli* (EPEC) and their intestinal epithelial cell targets. EPEC forms pedestals by recruiting actin and associated proteins by inserting the intimin receptor Tir via a type III secretion system (TTSS). Intimin also interacts with the host-membrane receptors integrin and nucleolin. The intimin–Tir interaction inhibits Cdc42-dependent filopodia formation along with EspH, while mitochondria-associated protein (Map) activates Cdc42 to form filopodia extensions. Map, along with EspF, participates in disrupting tight junctions but also targets mitochondria, playing a potential role in apoptosis. Tir and EspF also target the intermediate filament protein CK18. EspG and Orf3 participate in concert, disrupting microtubules. Modified from Dean *et al.* (2005).

perturbation of ion transport, stimulation of inflammation, and disruption of the tight-junction barrier (Figure 13.2).

EPEC- and EHEC-secreted effector proteins clearly play an active role in the disruption of intestinal epithelial structures and functions. Although most of the effector proteins appear to target the cytoskeleton of IECs, certain effectors have been demonstrated to alter inflammation and apoptosis. EPEC and EHEC TTSS-secreted effector proteins, other than Tir, are referred as *E. coli*-secreted proteins (Esps). One of the proposed mechanisms for EPEC-induced diarrhea is disruption of intestinal epithelial tight-junction barrier function. This was originally demonstrated in vitro using polarized monolayers of IECs (T84 and Caco-2 cells) infected with EPEC or EHEC. Increased permeability of these tissues was determined as a decrease in TER and an increase in the paracellular permeability to solutes (Canil *et al.*, 1993; Philpott *et al.*, 1998; Spitz *et al.*, 1995). Further studies identified EspF of EPEC (McNamara *et al.*, 2001), and later U-EspF of EHEC (Viswanathan, Koutsouris *et al.*, 2004), to be responsible for the full impact of these pathogens on intestinal epithelial tight-junction disruption and decrease in TER.

More recent work with EPEC has demonstrated that other effector molecules contribute to tight-junction disruption, including mitochondria-associated protein (Map) (Dean and Kenny, 2004) and Orf3 (Tomson, Viswanathan *et al.*, 2005). Although EspF disrupts tight-junction integrity, *espF* gene-deletion experiments showed no role for EspF in A/E lesion formation (McNamara *et al.*, 2001). Studies have also shown that EPEC disrupts tight-junction structure and function in a murine model (C57BL/6) of EPEC infection (Savkovic *et al.*, 2005; Shifflett *et al.*, 2005). As well as reducing the integrity of tight junctions, EspF has been reported to induce death of host cells. *espF* gene-transfection experiments in HeLa and African green monkey kidney fibroblast (COS) cells have demonstrated that overexpression of EspF induces epithelial cell apoptosis/necrosis (Crane *et al.*, 2001). However, infection of rabbits with the related rabbit strain EPEC (REPEC) showed no change in the level of apoptosis, or showed even less apoptosis, following infection as compared with the level seen in uninfected controls (Heczko *et al.*, 2001).

Additionally, EPEC and EHEC secrete EspA, B, D, G, and H and Map via the TTSS. The secretion of these proteins in vitro resembles conditions similar to those in the gastrointestinal tract (Kenny, Abe *et al.*, 1997). Once the TTSS has spanned the inner and outer membranes of the organism, EspA serves as a filamentous surface appendage traversing the extracellular space to form a physical link with the epithelial cells, a process that occurs early in the course of infection (Knutton *et al.*, 1989). Although the host receptor for EspA is unknown, its molecular-syringe-like structure provides a conduit for the translocation of EspB and EspD effectors, which form a pore complex in the IEC membrane at the tip of the molecular syringe (Fivaz and van der Goot, 1999; Knutton *et al.*, 1998; Shaw *et al.*, 2001).

Confocal laser scanning microscopy and cellular fractionation studies have also demonstrated EspB in the cytoplasm of infected HeLa cells (Taylor *et al.*, 1998). Hemolysis studies indicate that EPEC induces the lysis of red blood cells (RBCs) by creating a translocation pore formed by EspD alone, a phenomenon that does not require intimate bacteria–RBC membrane contact

(Shaw *et al.*, 2001). In addition to the contribution of pore formation by EspB, yeast-two hybrid assays determined that this effector protein binds and recruits the cytoskeletal associated protein α-catenin, which is essential for actin accumulation beneath the adherence site of EHEC (Kodama *et al.*, 2002). The resourcefulness of EPEC in exploiting cellular functions is demonstrated further by the induction of humoral and cell-mediated immune responses in IECs, a virulence role of the translocated EspB effector demonstrated in experimental human EPEC infections (Tacket *et al.*, 2000).

Map, another LEE-encoded effector that is delivered to the cytoplasm of epithelial cells, interacts with mitochondria, disrupting membrane potential and presumably affecting key functions such as cellular energy and regulation of apoptosis (Kenny and Jepson, 2000). In addition to its effect on mitochondria, Map independently or in conjunction with EspF decreases TER by disrupting the tight-junction barrier function (Dean and Kenny, 2004). Although Map is not involved directly in pedestal formation, overexpression of this protein appears to have an inhibitory effect on Tir–intimin-mediated pedestal configuration (Kenny *et al.*, 2002). As well as its regulatory effect on pedestal formation, Map interacts with actin and the cytoskeleton-associated host GTPase protein Cdc42 to form filopodia-like extensions, a dynamic and transient process that is downregulated by the Tir–intimin complex (Kenny *et al.*, 2002). Although the significance of filopodia formation and Cdc42 activation in EPEC pathogenesis remains unknown, EspH, has been reported to repress filopodia formation orchestrated by Map (Tu *et al.*, 2003). EspH, a cytoskeletal-modulating effector of EPEC and EHEC, localizes to the membrane and interacts with actin to promote pedestal elongation (Tu *et al.*, 2003). Therefore, deletion of espH enhances filopodia formation and fragments pedestals in HeLa cells (Tu *et al.*, 2003).

In addition to actin, microtubules, another major component of the cytoskeleton, are also targeted by EPEC and EHEC effectors. EspG and its homolog Orf3 have been shown to disrupt the microtubular network in Swiss 3T3 fibroblasts and non-polarized HeLa cells directly beneath adherent bacteria (Matsuzawa *et al.*, 2004). Both effectors interact with tubulins and stimulate microtubule destabilization, causing the release of guanine nucleotide exchange factor (GEF)-H1, RhoA activation, and an increase in actin stress fibers in these non-polarized cells (Matsuzawa *et al.*, 2004). Tomson, Viswanathan *et al.* (2005) also demonstrated that EPEC infection of HeLa cells and polarized IECs (Caco-2 and T84) disrupts microtubule structure and degrades tubulin subunits, a process that is reversible and dependent on the TTSS and the secreted protein EspG. This group established further that EspG, in conjunction with its homolog Orf3, contributes to the disruption of tight junctions, as evidenced by decreased TER in IEC monolayers infected with wild-type EPEC (Tomson, Viswanathan *et al.*, 2005a). Although the contribution of these proteins to EPEC pathogenesis remains to be determined by in vivo studies, microtubule and tight-junction disruption may lead to the mistargeting or relocalization of membrane proteins, such as those involved in active transport processes (Tomson, Viswanathan *et al.*, 2005a). Furthermore, the release of GEF-H1 from destabilized microtubules and activation of RhoA further supports their effect on cell polarity and membrane ruffling (Ren *et al.*, 1998).

The inflammatory response of IECs to infection was initially described for invasive bacterial pathogens such as *Salmonella* and *Shigella*. Using EPEC as a prototype, non-invasive pathogens were also shown to activate the major cellular inflammatory signaling pathways, including extracellular signal-regulated kinases (ERK-1/2), inhibitory kappa B (IκB), and the transcription factor nuclear factor kappa B (NF-κB), which in turn trigger IL-8 secretion (Savkovic *et al.*, 1997, 2001). In vitro and in vivo studies have also shown that EPEC infection induces the transmigration of polymorphonuclear leukocytes (neutrophils) and increases the number of lamina propria neutrophils, with occasional crypt abscesses (Savkovic *et al.*, 1996, 2005). To date, the only identified stimulator for IL-8 response is flagellin from EPEC and EHEC (Zhou *et al.*, 2003). Work with EPEC, however, suggests that a TTSS-independent non-flagellin molecule also contributes to the EPEC-mediated inflammatory response (Sharma *et al.*, 2005). In addition to triggering IL-8 release, both EPEC and EHEC exert an inhibitory effect on IL-8 response in a TTSS-dependent manner that may involve LEE and non-LEE effectors (Dean *et al.*, 2005; Hauf and Chakraborty, 2003; Tomson, Tesfay, *et al.*, 2005. Continued explorations into the host-cell effects of these secreted molecules will undoubtedly further enhance our understanding of the pathogenesis of TTSS-expressing organisms.

ENTERIC PATHOGEN EFFECTS ON INTESTINAL BARRIER FUNCTION

Tight junctions are the most apical of the intercellular junctions of epithelial cells and are the primary determinants of intestinal epithelial permeability. There are several mechanisms by which tight junctions can be influenced. One physiological regulator of tight-junction permeability is small-intestinal Na^+/nutrient cotransport. For example, initiation of Na^+/glucose transport in isolated small intestine induces increases in paracellular flux and dilation of the tight junction (Berglund *et al.*, 2001; Pappenheimer and Reiss,

1987; Turner *et al.*, 2000). Cytokines, which serve as extracellular signals in many pathophysiological conditions, also influence intestinal epithelial barrier function. Additionally, bacterial pathogens disrupt tight junctions and increase intestinal permeability; this topic is the focus of the following section.

TIGHT-JUNCTION STRUCTURE AND FUNCTION

Tight junctions of epithelial cells consist of a belt-like interconnected series of strands that associate in the intercellular space between apposing epithelial cells. Tight junctions have two physiological roles. First, they separate the apical from the basolateral plasma membrane (fence function) and are, therefore, responsible for maintaining the polarity of epithelial cells. Polarization of epithelial cells is required for the appropriate movement of ions, water, and macromolecules between compartments, as each membrane contains distinct ion channels, enzymes, and transport proteins. For example, Na^+/K^+/ATPase is localized to the basolateral surface and allows for the development of a net charge across the cell; this is the principal driving force for cellular transport processes. Second, tight junctions provide the rate-limiting barrier to paracellular transport (gate function) (Gumbiner, 1987, 1993). The mechanisms by which microbial pathogens alter intestinal epithelial tight-junction barrier function are discussed in this chapter.

Paracellular transport is a passive process that works in conjunction with transcellular active transport processes. Thus, the paracellular and transcellular routes of transport work in concert to determine the overall barrier function and net transport of ions and solutes across the intestinal epithelium. Tight junctions are dynamic structures (Anderson and Van Itallie, 1995) regulated by numerous mechanisms, including cytoskeletal tone (Madara, 1988; Madara *et al.*, 1987), signaling cascades, and pathogens (McNamara *et al.*, 2001; Muza-Moons *et al.*, 2004; Zolotarevsky *et al.*, 2002). Occludin, claudins, junction-adhesion molecules (JAM), and coxsackievirus and adenovirus receptor (CAR) are transmembrane proteins that form the tight junction sealing the paracellular space (Furuse, Fujita *et al.*, 1998; Furuse, Sasaki *et al.*, 1998; Saitou *et al.*, 1998) (Figure 13.3). Thus far, more than 20 members of the claudin family have been identified. Claudins appear to be responsible for the formation of tight-junction strands (Furuse, Fujita *et al.*, 1998; Furuse, Sasaki *et al.*, 1998; Heiskala *et al.*, 2001; Tsukita *et al.*, 2001). Although occludin is also associated with tight junctions, expression of occludin in MDCK cells does not alter the number of tight-junction strands as assessed by freeze-fracture analysis, and yet it is capable of regulating paracellular permeability (Balda *et al.*, 2000). It has been reported that the relationship

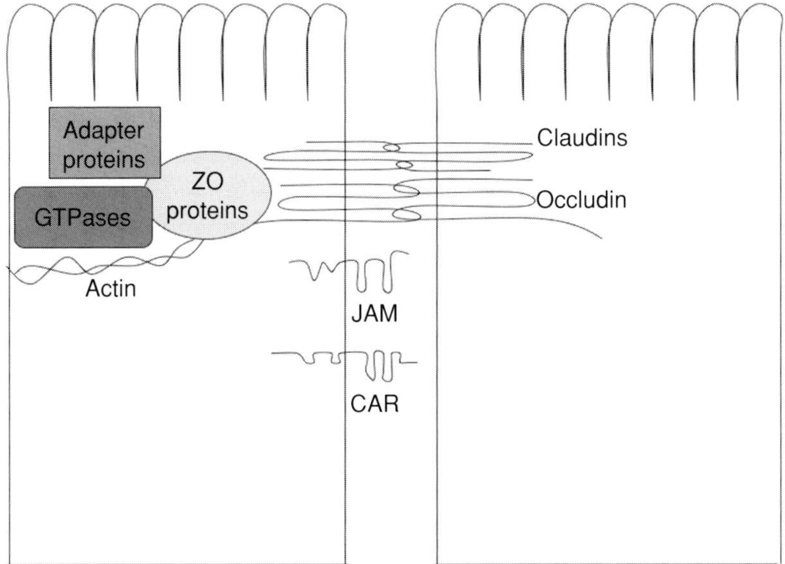

Figure 13.3 Simplified structure of the paracellular space and the tight junction. There exists a complex interaction of adapter proteins, GTPases, and Zonula occludens (ZO) proteins with the tight-junction transmembrane proteins occludin and claudins.
CAR, coxsackie virus and adenovirus receptor; JAM, junction-adhesion molecule.

between the number of strands and TER is logarithmic rather than linear (Claude and Goodenough, 1973). However, no correlation has been found between strand number and TER in MDCK cells with low TER ($\sim$100 Ω.cm^2) versus cells with high TER ($\sim$1000 Ω.cm^2) (Stevenson *et al.*, 1988). Thus, the correlation between tight-junction strand number and TER remains to be determined.

Structurally, claudins and occludin have two extracellular loops. In claudins, the first loop is larger and more hydrophobic than the second loop. Additionally, the first extracellular loop of claudin-2 and claudin-15 creates a charge-selective channel in the paracellular space (Colegio *et al.*, 2002, 2003; Furuse, Fujita *et al.*, 1998). Expression of several of the claudins is restricted to specific cell types. For example, claudin-16 is expressed in the loop of Henle of the kidney (Simon *et al.*, 1999) and claudin-11 is expressed in the myelin sheaths of the central nervous system and Sertoli cells in the testis (Morita *et al.*, 1999). Evidence for claudins regulating barrier function includes studies in MDCK cells in which overexpression of claudin-1 increased TER and decreased paracellular flux of 4- and 40-kDa FITC-labeled dextrans (Inai *et al.*, 1999). Other authors have shown that overexpression of claudin-1 enhances

TER (McCarthy *et al.*, 2000). Evidence that claudins govern charge selectivity was demonstrated by work regarding claudin-16 (paracellin-1). As mentioned previously, expression of claudin-16 is restricted to the ascending loop of Henle and is required for the paracellular reabsorption of Mg^{2+} in the kidney (Simon *et al.*, 1999). The importance of this protein in human health was revealed by identifying a defect in the gene encoding for claudin-16 in patients with Mg^{2+} deficiency. Lack of claudin-16 expression leads to decreased Mg^{2+} and Ca^{2+} reabsorption by the kidney. The ensuing increase in urinary Ca^{2+} predisposes to renal-stone formation and often these patients require transplantation. Coinciding with the idea that claudins have a role in regulating transport functions, different isoforms are expressed in different organs and may have differential expression patterns within a given organ. For example, claudin-2 is expressed almost exclusively in the crypts of the small intestine, while claudin-4 is expressed mainly in villi. These differential expression patterns are likely to be functionally important (Rahner *et al.*, 2001). As an example, overexpression of wild-type claudin-4, but not mutant claudin-4, in MDCK cells resulted in decreased permeability to Na^+ compared with Cl^- (Colegio *et al.*, 2003; Van Itallie *et al.*, 2001). Furthermore, mutation of claudin-15 created a Cl^--selective pathway compared with Na^+. Conversely, wild-type claudin-15 exhibited the opposite charge selectivity (Colegio *et al.*, 2002).

Another transmembrane protein of the tight junction is occludin. Occludin is a ~60-kDa tetraspanning membrane protein and was the first identified tight-junction transmembrane protein (Furuse *et al.*, 1993). A role for occludin in tight-junction barrier function is supported by studies showing that overexpression of occludin increased TER (Balda *et al.*, 1996; McCarthy *et al.*, 1996; Sakakibara *et al.*, 1997). Unlike claudins, however, occludin is not required for the formation of tight-junction strands (Saitou *et al.*, 1998), and occludin-deficient mice do not display any obvious structural or functional tight-junction abnormalities (Saitou *et al.*, 2000). Furthermore, occludin-deficient mice have no alteration in intestinal barrier function (Schulzke *et al.*, 2005). It is important to note that occludin is a single-protein component of the tight junction, and in occludin-deficient mice there could be a compensatory shift towards any number of the claudins regulating tight-junction permeability. Thus, the function of occludin remains unresolved.

Junction adhesion molecule 1 (JAM-1) and CAR, members of the immunoglobulin G (IgG) superfamily, are also tight-junction transmembrane proteins. JAM contains a single transmembrane domain and binds to the PDZ3 domain of zonula occludens 1 (ZO-1) (Itoh *et al.*, 2001). Interestingly, it has been speculated that JAM-1 has a role in the transepithelial or transendothelial migration of neutrophils during the inflammatory response (Martin-Padura *et al.*, 1998). CAR also recruits ZO-1 to the membrane, decreases flux rates of the paracellular marker dextran, and enhances TER (Cohen *et al.*, 2001). As the name indicates, CAR, as well as other tight-junction proteins, is a target of microbial pathogens.

Tight-junction transmembrane proteins associate with an intracellular cluster of proteins referred to as the cytoplasmic plaque, the latter being important for tight-junction assembly. The primary proteins comprising the cytosolic plaque are ZO-1 (Stevenson *et al.*, 1986), ZO-2 (Gumbiner *et al.*, 1991; Jesaitis and Goodenough, 1994), ZO-3 (Haskins *et al.*, 1998), cingulin (Citi *et al.*, 1988), and 7H6 (Zhong *et al.*, 1993). The ZO family of proteins have been shown to bind to each other (Fanning *et al.*, 1998; Haskins *et al.*, 1998), the actin cytoskeleton (Fanning *et al.*, 1998), and JAM (Ebnet *et al.*, 2000). Furthermore, the carboxyl termini of both occludin and claudins bind ZO-1, -2, and -3 (Fanning *et al.*, 1998; Haskins *et al.*, 1998; Itoh *et al.*, 1999; Wittchen *et al.*, 1999). Thus, it has been speculated that the ZO proteins are important for targeting claudins and occludin to the tight junction.

Another structural and functional component of the tight junction is the actin cytoskeleton. Evidence that actin regulates the tight junction was provided by showing that disruption of actin with cytochalasin decreased the number and size of tight-junction fibrils (Madara *et al.*, 1986; Nybom and Magnusson, 1996) and reduced epithelial barrier function. Furthermore, introduction of a constitutively activated myosin light-chain (MLC) kinase in MDCK cells caused contraction of the cytoskeleton, decreased TER, and increased paracellular flux (Hecht *et al.*, 1996).

Regulation of the tight-junction barrier is a growing topic of interest. Phosphorylation is one means of tight-junction protein regulation. Phosphorylated forms of occludin are associated with the tight junction (Andreeva *et al.*, 2001; Chen *et al.*, 2002; Farshori and Kachar, 1999; Sakakibara *et al.*, 1997). Phosphorylation of this protein occurs on serine and weakly on threonine (Tsukamoto and Nigam, 1999). However, another study has shown that occludin is phosphorylated on threonine and tyrosine residues (Yoo *et al.*, 2003). Phosphorylation of ZO-1, which occurs on tyrosine and serine residues, is also important for barrier function (Staddon *et al.*, 1995). ZO-1 phosphorylation is enhanced in MDCK monolayers exhibiting low TER as compared with MDCK monolayers with high TER (Stevenson *et al.*, 1989). Additionally, tyrosine phosphorylation of ZO-1 is associated with ZO-1 rearrangement to the apical cell borders and colocalization with perijunctional actin in A431 cells, a human epidermal carcinoma cell (Van Itallie *et al.*, 1995). Dephosphorylation of the tight-junction proteins has been shown to precede a decrease in TER (Staddon *et al.*, 1995). There is less evidence of phosphorylation of claudins, although threonine phosphorylation of claudin-5 occurs in a cyclic adenosine monophosphate (cAMP)- and protein kinase A(PKA)-dependent manner in endothelial cells (Fujibe *et al.*, 2004; Ishizaki *et al.*, 2003). Additionally, the protein kinase WNK4 phosphorylates the C-terminal domain of claudin-4 in MDCK cells (Yamauchi *et al.*, 2004). However, it is not known whether claudin phosphorylation occurs in the intestinal epithelium.

MICROBIAL PATHOGEN-INDUCED DISRUPTION OF THE TIGHT JUNCTION VIA THE ACTIN CYTOSKELETON

Many pathogens, including bacteria and viruses, disrupt tight junctions and enhance intestinal permeability. Bacterial pathogens have unique mechanisms for exploiting the dynamic nature of tight junctions to alter the movement of ions and molecules beween the lumen and the systemic circulation. One way in which EPEC disrupts tight junctions is through its effects on the actin cytoskeleton. EPEC infection decreases TER (Savkovic *et al.*, 2001; Spitz *et al.*, 1995), in part through phosphorylation of MLC (Manjarrez-Hernandez *et al.*, 1991; Yuhan *et al.*, 1997; Zolotarevsky *et al.*, 2002). The phosphorylation of MLC triggers contraction of the actomyosin ring that immediately underlies tight junctions and is believed to open the tight junction via increased tension on the lateral membrane. Further evidence that EPEC perturbation of tight junctions is due in part to effects on the actin cytoskeleton has been demonstrated pharmacologically. The EPEC-induced decrease in TER of intestinal epithelial monolayers is inhibited by addition of a specific membrane-permeant inhibitor of MLC kinase (Zolotarevsky *et al.*, 2002). EHEC-induced MLC phosphorylation has also been linked to declining barrier function, as inhibitors of this pathway attenuate the resultant decline in TER (Dahan *et al.*, 2003; Philpott *et al.*, 1998).

Another pathogen that disrupts the actin cytoskeleton and alters intestinal permeability is *Clostridium difficile. C. difficile,* the etiologic agent of antibiotic-associated colitis (Borriello, 1998), elaborates two toxins, TxA and TxB, both of which are glucosyltransferases that inactivate members of the Rho family of small GTPases, regulators of actin-dependent cellular functions. Thus, the effects of TxA and TxB include actin depolymerization and enhanced intestinal permeability (Hecht *et al.*, 1988, 1992; Liu *et al.*, 2003; Pothoulakis, 2000). Additionally, TxA- and TxB-induced cytoskeletal rearrangement is associated with restructuring of occludin, ZO-1, and ZO-2 away from the region of the tight junction (Nusrat *et al.*, 2001). The signaling mechanisms responsible for *C. difficile* TxA effects on tight junctions appear to be PKC-dependent, because TxA activates PKC, increases dissociation of ZO-1 from the tight junction, and decreases TER in T84 cells. Furthermore, these deleterious effects are inhibitable by a PKC-α antagonist (Chen *et al.*, 2002).

The *Clostridium botulinum* toxins C2 and C3 also disrupt barrier function through effects on the actin cytoskeleton. The C2 toxin ADP-ribosylates G-actin, leading to a loss of actin ATPase activity, thus preventing polymerization into F-actin and ultimately leading to disruption of the tight junction. The C3 toxin ADP-ribosylates RhoA, RhoB, and RhoC and induces disassembly of actin filaments in a manner similar to the *C. difficile* toxins. Additionally, ZO-1 is dissociated from tight junctions and correlates with increased intestinal epithelial permeability (Nusrat *et al.*, 1995).

Bacteroides fragilis toxin (BFT) is another bacterial toxin that alters intestinal epithelial barrier function. BFT is a zinc-dependent metalloprotease that exerts its effects extracellularly and alters the tight-junction structure through effects on the ZOs (Chambers *et al.*, 1997; Obiso *et al.*, 1995), E-cadherin (Wu *et al.*, 1998), and rearrangement of the actin cytoskeleton (Chambers *et al.*, 1997). Wu et al. (1998) demonstrated that BFT-induced cleavage of the cell–cell-adhesion molecule E-cadherin is ATP-independent. The ensuing intracellular degradation likely triggers the dissociation of ZO-1 and occludin from tight junctions. Furthermore, BFT also is not cell-specific, as exposure of human intestinal epithelial cells (HT-29) and canine kidney epithelium (MDCK) to BFT decreased TER, highlighting its potential importance in extraintestinal infections (Obiso *et al.*, 1995).

Another bacterial protease that affects tight-junction function is the hemagglutinin protease (HA/P) elaborated by *Vibrio cholerae*. HA/P is a metalloprotease that disrupts ZO-1 and the actin cytoskeleton, thus decreasing TER of intestinal epithelial cell monolayers (Wu *et al.*, 1996). Unable to provide evidence that HA/P was internalized into cells, Wu *et al.* (2000) examined the effects on cell-surface proteins. They discovered that HA/P cleaves occludin into two fragments in a dose- and time-dependent manner but has no effects on the structure of ZO-1, despite its reorganization and almost complete disappearance from cell–cell boundaries. Another *V. cholera* elaborated toxin, zonula occludens toxin (ZOT) (Fasano *et al.*, 1991), enhances intestinal epithelial permeability via modulation of tight junctions. ZOT significantly increases the flux of mannitol across Caco-2 cell monolayers after a 1-h exposure (Cox *et al.*, 2001) and reversibly increases rabbit intestinal permeability to insulin (Fasano and Uzzau, 1997). Furthermore, ZOT regulates the permeability of intestinal tight junctions by binding to microtubules

(Wang *et al.*, 2000). This is important, as microtubules are involved in intracellular trafficking and membrane and receptor recycling.

MICROBIAL PATHOGEN-INDUCED DISRUPTION OF THE TIGHT JUNCTION VIA EFFECTS ON TIGHT-JUNCTION PROTEINS

The effects of EPEC on tight junctions appear to be multifactorial. In addition to the previously mentioned increase in MLC phosphorylation, EPEC also has direct effects on tight-junction proteins in intestinal epithelial monolayers. Following infection with EPEC, occludin is dephosphorylated, resulting in its dissociation from the tight junctions (Simonovic *et al.*, 2000). This is important, as the phosphorylated forms of occludin associate with tight junctions and enhance the tight-junction epithelial barrier (Wong, 1997). A similar dissociation has been reported for ZO-1 after EPEC infection (Muza-Moons *et al.*, 2004; Philpott *et al.*, 1996).

Although the sequence of events of EPEC disruption of the tight junction has not been elucidated fully, it is clear that EPEC-secreted protein F (EspF) plays a major role. EspF is delivered into the host cell via the TTSS apparatus (McNamara and Donnenberg, 1998). Mutation of the *espF* gene significantly attenuates the effects of EPEC on barrier disruption (McNamara *et al.*, 2001). The mechanism by which EspF perturbs tight-junction barrier function is not understood completely, although it has been speculated that EspF must be targeted to the host mitochondria for pathogenesis (Nagai *et al.*, 2005; Nougayrede and Donnenberg, 2004). It is possible, however, that this finding is a consequence of overexpression of this protein or high multiplicity of infection (Nougayrede and Donnenberg, 2004). That this phenotype may not occur in vivo is supported by studies with rabbit EPEC showing that the degree of apoptosis in the intestine was not altered when compared with uninfected rabbits. Furthermore, EspF interactions with the intermediate filament protein CK18, and its resultant increase in solubility by interacting with 14–3–3 may also be important for EPEC pathogenesis (Viswanathan, Lukic *et al.*, 2004). It has been reported that a second effector protein, Map, also contributes to EPEC-induced impairment of barrier function in vitro (Dean and Kenny, 2004). Additionally, the effector protein Orf3 also disrupts tight-junction barrier function (Tomson, Viswanathan *et al.*, 2005). Thus, it appears that multiple EPEC effector proteins contribute to EPEC-induced impairment of intestinal barrier function.

The EPEC- and EHEC-induced perturbations of intestinal barrier function, in part through phosphorylation of MLC, was discussed above. In contrast to EPEC, EHEC may not be dependent upon EspF for disruption of barrier function. Instead, the EHEC effector protein U-EspF, which also plays a significant role in pedestal formation, serves this role (Viswanathan, Koutsouris *et al.*, 2004).

Yersinia is another enteric pathogen reported to perturb tight-junction permeability (Gogarten *et al.*, 1994; Tafazoli *et al.*, 2000). Similar to EPEC, an effort has been made to identify effector molecules responsible for this functional phenotype. Yersinia outer protein E (YopE) is, in part, responsible for this physiologic change, as a *YopE* mutant failed to disrupt actin and alter the localization of occludin and ZO-1 in MDCK monolayers (Tafazoli *et al.*, 2000). The mechanism by which YopE disrupts tight-junction barrier function remains unknown, and the participation of other *Yersinia* effector molecules has not been examined.

The ability of bacteria or their toxins to bind directly to tight-junction proteins was discussed earlier in this chapter. With regard to the physiological effects of *Clostridium perfringens* enterotoxin (CPE) on intestinal barrier function, expression of the carboxy-terminal half fragment of this protein resulted in removal of claudin-4 from the tight junction. Furthermore, tight-junction strands were disintegrated, TER was decreased, and the flux of paracellular markers was increased (Sonoda *et al.*, 1999), demonstrating a unique mechanism by which bacteria can directly alter the tight-junction structure and gain access to the basolateral surface.

Shigella is also an intriguing enteric pathogen in that it fails to bind to the apical surface of intestinal epithelium. Instead, it utilizes M-cells to gain access to the basolateral compartment, where it invades IECs. The ensuing response is the transepithelial migration of neutrophils, which disrupts tight junctions, thus allowing additional access of *Shigella* organisms to intestinal epithelial cells (Perdomo *et al.*, 1994). However, a plasmid-cured non-invasive *Shigella* strain was found to translocate across the tight junctions of T84 cells, a human IEC line (Sakaguchi *et al.*, 2002), and cause a shift of claudin-1 and ZO-1 from a Triton-X-insoluble protein fraction to a soluble fraction, indicative of dissociation of these proteins from the membrane (Sakaguchi *et al.*, 2002), suggesting that the effects on tight junctions are independent of invasion.

EFFECTS OF MICROBIAL PATHOGEN-ELICITED INFLAMMATION ON TIGHT JUNCTIONS

The effects on the intestinal epithelium of the inflammatory response to microbial infection cannot be ignored. For a more detailed review of microbial-induced inflammation, the reader is referred to Chapter 9.

However, infiltration of neutrophils and elaboration of pro-inflammatory cytokines clearly have detrimental effects on tight-junction barrier function. For example, TNF-α decreases intestinal epithelial tight-junction barrier function (Ma *et al.*, 2004; Marano *et al.*, 1998; Rodriguez *et al.*, 1995; Schmitz *et al.*, 1999). Similarly, interferon gamma (IFN-γ) perturbs intestinal epithelial tight-junction function (Fish *et al.*, 1999; Madara and Stafford, 1989; Sugi *et al.*, 2001; Youakim and Ahdieh, 1999). The mechanism by which these cytokines regulate tight-junction permeability has been discovered to be MLCK-dependent (Ma *et al.*, 2005; Wang *et al.*, 2005). This may be physiologically important for microbial pathogens such as EPEC, since EPEC-induced alteration of the tight-junction barrier is in part MLCK-dependent. Furthermore, TNF-α-induced increases in tight-junction permeability of the intestinal epithelial cell line Caco-2 are NF-κB-dependent and do not involve apoptosis (Ma *et al.*, 2004). Thus, it is evident that microbial-elicited inflammation regulates tight-junction permeability in a variety of ways.

OVERVIEW OF EPITHELIAL CELL TRANSPORT

A full understanding of pathogen–epithelial interactions would not be complete without attention to electrolyte transport processes. Although research is lacking in this area, it is well documented that pathogens can alter fluid and electrolyte transport processes of the gastrointestinal epithelium. The alteration of electrolyte transport is often through the activation or inhibition of ion channels and transporters. The two principal ions involved in electrolyte transport are Cl^- and Na^+ and these will be discussed briefly here.

Cl^- secretion is regarded as the determinant of luminal hydration. Coinciding with Cl^- secretion is the paracellular movement of Na^+. These two processes provide the osmotic gradient for movement of water into the lumen. Cl^- secretion utilizes several transporters. First, the energy source for ion transport by the cell is mediated by the Na^+/K^+/ATPase pump located on the basolateral membrane. The sodium gradient generated by this pump is utilized by the $Na^+/K^+/2Cl^-$ (NKCC1) cotransporter, also situated on the basolateral membrane, to accumulate intracellular Cl^- above its electrochemical equilibrium. Potassium recycling is mediated by basolaterally located K^+ channels such as KCNQ1 and KCNE3. Additionally, there is an apically located Ba^{++} inhibitable channel through which a small amount of K^+ is able to exit the cell. The majority of accumulated intracellular Cl^- is secreted from the cell via the cystic fibrosis transmembrane regulator (CFTR) located on the apical surface. Lesser amounts of Cl^- are transported from the cell through ClC2 channels. Additionally, there are apically located electroneutral sodium-hydrogen exchangers (NHE2, NHE3) through which Na^+ and H^+ are exchanged at a 1:1 ratio. NHE3 has been shown to be the predominant Na^+-absorbing isoform in mammalian small intestine (Shull *et al.*, 2000).

INTERFERENCE OF TRANSPORT PROCESSES BY MICROBIAL PATHOGENS

The primary second messengers responsible for activation of Cl^- secretion are cAMP/cyclic guanosine monophosphate (cGMP) and calcium. For information on calcium-mediated transport processes, the reader is referred to previously published reviews (Barrett and Keely, 2000; Keely and Barrett, 2000). Cl^- secretion mediated by cAMP has been demonstrated for various pathogens, including *E. coli*, *V. cholera*, *Salmonella*, and *Campylobacter jejuni* (Molina and Peterson, 1980; Ruiz-Palacios *et al.*, 1983). Cholera toxin, via adenylate cyclase, increases intracellular cAMP, activates cAMP-dependent protein kinase A, and subsequently phosphorylates the CFTR. The end result of CFTR activation is electrogenic Cl^- secretion. Activation of CFTR by cGMP has similar effects. Guanylin is a peptide that activates intestinal guanylate cyclase C and increases cGMP levels. Heat-stable toxins, such as those elaborated from enterotoxigenic *E. coli*, have a 50% homology with guanylin, bind to intestinal guanylate cyclase C, and result in increased cGMP, phosphorylation of CFTR, and enhanced Cl^- secretion. The importance of the intestinal guanylate cyclase C receptor for ST_a-mediated Cl^- secretion has been demonstrated in mice. Although intestinal guanylate cyclase C-null mice were resistant to stable toxin, infection with ST_a enterotoxigenic bacteria led to diarrhea and death in wild-type and heterozygous mice (Mann *et al.*, 1997). ST_a inhibits Na^+ absorption by an undefined mechanism (Lucas, 2001).

C. difficile toxins A and B also elicit Cl^- secretion, but it is unknown whether these toxins act via a cAMP-mediated process. Indirect evidence suggests that TxA induces cyclooxygenase 2 (COX-2) in lamina propria cells, with a resultant increase in prostaglandin E2 (PGE_2) levels. PGE_2, through its G-protein-coupled receptor, increases cAMP levels and PKA-dependent activation of CFTR. *C. difficile* toxins also regulate Na^+ exchange, as *C. difficile* TxB significantly inhibits NHE3 activity and is accompanied by translocation to an intramembrane compartment (Hayashi *et al.*, 2004).

EPEC infection also alters intestinal epithelial ion transport. EPEC infection of intestinal epithelial Caco-2 monolayers stimulated a rapid increase in short-circuit current (Isc), indicative of electrogenic Cl^- secretion (Collington *et al.*, 1998). In contrast, infection of T84 monolayers yielded contrasting results, as stimulation of EPEC-infected monolayers with the classic cAMP-mediated secretagog forskolin yielded an attenuated response that was not attributable to altered Cl^- secretion (Hecht and Koutsouris, 1999). In this case, it was determined that the effects were due to perturbation of bicarbonate-dependent transport processes. Furthermore, in T84 cells, cAMP-mediated Cl^- secretion was diminished by EPEC infection (Philpott *et al.*, 1996). The variability between transport properties of the different cell lines and the different models of infection potentially explains the discrepancy between these studies. However, EPEC infection also affects the transport of Na^+ by the intestinal epithelial cell monolayers Caco-2 and T84 by enhancing NHE2 activity and decreasing NHE3 activity (Hecht *et al.*, 2004).

ELECTROLYTE TRANSPORT MODIFICATIONS RESULTING FROM HOST-DERIVED PRODUCTS STIMULATED BY BACTERIAL INFECTION

Bacterial infection of the intestinal epithelium can also induce upregulation of host-derived products. These products can, in turn, regulate electrolyte transport processes. Galanin is one such host-derived product. Galanin is a 29-amino-acid neuropeptide in enteric nerves lining the gastrointestinal tract (Bauer *et al.*, 1986). It induces Cl^- secretion through activation of the galanin-1 receptor in human colonic epithelium (Benya *et al.*, 1999; Katsoulis *et al.*, 1996). Expression of the galanin-1 receptor is under the regulation of NF-κB, suggesting that it is a part of the inflammatory response. As such, NF-κB-dependent expression of the galanin-1 receptor appears to be a common response to microbial infections, including EPEC, EHEC, *Salmonella*, *Shigella*, and rotavirus. The end result is increased Cl^- secretion and fluid accumulation in the colon (Hecht *et al.*, 1999; Matkowskyj *et al.*, 2000).

Prostaglandins are host-derived factors of intestinal epithelial cells that are upregulated in response to bacterial infection. An increase in PGE_2 expression and subsequent enhancement of Cl^- secretion occurs following infection of intestinal epithelium with *Salmonella dublin*, *Salmonella typhi*, *Salmonella typhimurium*, and *Y. enterocolitica* (Eckmann *et al.*, 1997). A similar pathogen-induced Cl^- secretory response has been observed with *Cryptosporidium parvum* and *Entamoeba histolytica* (Gookin *et al.*, 2004; Laurent *et al.*, 1998; Stenson *et al.*, 2001).

A POSSIBLE LINK BETWEEN TRANSPORT AND INTESTINAL BARRIER FUNCTION

Diarrhea can be viewed as a means by which pathogens ensure their transmission to additional hosts. With regard to human disease, pathogen-induced diarrhea is a means for the host to clear infection of the pathogen. In this regard, the upregulation of prostanoids may indeed have a beneficial role. As mentioned previously, prostanoid production is often increased during infection by microbial pathogens, with a resultant elevation in intracellular cAMP and induction of PKA-dependent Cl^- secretion. Additionally, prostaglandins also inhibit Na^+ uptake across the apical surface of the epithelium. Interestingly, inhibition of prostanoid production by the non-selective cyclooxygenase inhibitor indometacin results in not only a decrease in Cl^- secretion but also a decrease in the recovery of intestinal barrier function after injury. Furthermore, PGE_2-elicited Cl^- secretion improves barrier function (Blikslager *et al.*, 1997, 1999, 2001; Little *et al.*, 2003; Moeser *et al.*, 2004; Shifflett, Bottone *et al.*, 2004; Shifflett, Jones *et al.*, 2004). In the case of *Cryptosporidium parvum*-infected ileal epithelium, nitric oxide is a proximal mediator of PGE_2 synthesis and intestinal barrier function (Gookin *et al.*, 2004). Thus, there is evidence that electrolyte transport processes mediate properties of intestinal barrier function.

CONCLUSION

The remarkable and recent advances gained by studying the molecular cross-talk between enteric pathogens and intestinal epithelial cells provide specific knowledge regarding molecular mechanisms underlying bacterial pathogenesis. The common strategies set out by multiple enteric pathogens and the unique mechanisms used to establish an infection by invasive and non-invasive pathogenic bacteria have helped us to define the host-cell regulatory functions, including ion transport, inflammation, cell polarity, and permeability. Enteric pathogens have found ways to manipulate these cellular functions by exploiting the host-cell cytoskeleton, tight-junction proteins, and signal-transduction pathways through the employment of molecular mimicry, protein-secretion machineries, and toxin deployment, which may or may not require invasion. Despite the tremendous amount of progress made over the past several years, there is still a great deal to learn about the infectious disease process and the multifaceted interactions between enteric pathogens and their hosts.

REFERENCES

Akira, S., Takeda, K., and Kaisho, T. (2001). Toll-like receptors: critical proteins linking innate and acquired immunity. *Nat. Immunol.* **2**, 675–680.

Anderson, J. M. and Van Itallie, C. M. (1995). Tight junctions and the molecular basis for regulation of paracellular permeability. *Am. J. Physiol.* **269**, G467–G475.

Andreeva, A. Y., Krause, E., Muller, E. C., Blasig, I. E., and Utepbergenov, D. I. (2001). Protein kinase C regulates the phosphorylation and cellular localization of occludin. *J. Biol. Chem.* **276**, 38 480–38 486.

Balda, M. S., Whitney, J. A., Flores, C., *et al.* (1996). Functional dissociation of paracellular permeability and transepithelial electrical resistance and disruption of the apical-basolateral intramembrane diffusion barrier by expression of a mutant tight junction membrane protein. *J. Cell. Biol.* **134**, 1031–1049.

Balda, M. S., Flores-Maldonado, C., Cereijido, M., and Matter, K. (2000). Multiple domains of occludin are involved in the regulation of paracellular permeability. *J. Cell. Biochem.* **78**, 85–96.

Barrett, K. E. and Keely, S. J. (2000). Chloride secretion by the intestinal epithelium: molecular basis and regulatory aspects. *Annu. Rev. Physiol.* **62**, 535–572.

Bauer, F. E., Adrian, T. E., Christofides, N. D., *et al.* (1986). Distribution and molecular heterogeneity of galanin in human, pig, guinea pig, and rat gastrointestinal tracts. *Gastroenterology* **91**, 877–883.

Benya, R. V., Marrero, J. A., Ostrovskiy, D. A., Koutsouris, A., and Hecht, G. (1999). Human colonic epithelial cells express galanin-1 receptors, which when activated cause Cl- secretion. *Am. J. Physiol.* **276**, G64–G72.

Berglund, J. J., Riegler, M., Zolotarevsky, Y., Wenzl, E., and Turner, J. R. (2001). Regulation of human jejunal transmucosal resistance and MLC phosphorylation by Na(+)-glucose cotransport. *Am. J. Physiol. Gastrointest. Liver Physiol.* **281**, G1487–G1493.

Black, D. S. and Bliska, J. B. (1997). Identification of p130Cas as a substrate of *Yersinia* YopH (Yop51), a bacterial protein tyrosine phosphatase that translocates into mammalian cells and targets focal adhesions. *EMBO. J.* **16**, 2730–2744.

Blikslager, A. T., Roberts, M. C., Rhoads, J. M., and Argenzio, R. A. (1997). Prostaglandins I2 and E2 have a synergistic role in rescuing epithelial barrier function in porcine ileum. *J. Clin. Invest.* **100**, 1928–1933.

Blikslager, A. T., Roberts, M. C., and Argenzio, R. A. (1999). Prostaglandin-induced recovery of barrier function in porcine ileum is triggered by chloride secretion. *Am. J. Physiol.* **276**, G28–G36.

Blikslager, A. T., Pell, S. M., and Young, K. M. (2001). PGE2 triggers recovery of transmucosal resistance via EP receptor cross talk in porcine ischemia-injured ileum. *Am. J. Physiol. Gastrointest. Liver Physiol.* **281**, G375–G381.

Boland, A. and Cornelis, G. R. (1998). Role of YopP in suppression of tumor necrosis factor alpha release by macrophages during *Yersinia* infection. *Infect. Immun.* **66**, 1878–1884.

Borriello, S. P. (1998). Pathogenesis of *Clostridium difficile* infection. *J. Antimicrob. Chemother.* **41(Suppl C)**, 13–19.

Burridge, K., Fath, K., Kelly, T., Nuckolls, G., and Turner, C. (1988). Focal adhesions: transmembrane junctions between the extracellular matrix and the cytoskeleton. *Annu. Rev. Cell. Biol.* **4**, 487–525.

Campellone, K. G., Robbins, D., and Leong, J. M. (2004). EspFUxy2 is a translocated EHEC effector that interacts with Tir and N-WASP and promotes Nck-independent actin assembly. *Dev. Cell* **7**, 217–228.

Canil, C., Rosenshine, I., Ruschkowski, S., *et al.* (1993). Enteropathogenic *Escherichia coli* decreases the transepithelial electrical resistance of polarized epithelial monolayers. *Infect. Immun.* **61**, 2755–2762.

Chambers, F. G., Koshy, S. S., Saidi, R. F., *et al.* (1997). *Bacteroides fragilis* toxin exhibits polar activity on monolayers of human intestinal epithelial cells (T84 cells) in vitro. *Infect. Immun.* **65**, 3561–3570.

Chen, M. L., Pothoulakis, C., and LaMont, J. T. (2002). Protein kinase C signaling regulates ZO-1 translocation and increased paracellular flux of T84 colonocytes exposed to *Clostridium difficile* toxin A. *J. Biol. Chem.* **277**, 4247–4254.

Citi, S., Sabanay, H., Jakes, R., Geiger, B., and Kendrick-Jones, J. (1988). Cingulin, a new peripheral component of tight junctions. *Nature* **333**, 272–276.

Clark, M. A., Hirst, B. H., and Jepson, M. A. (1998). M-cell surface beta1 integrin expression and invasin-mediated targeting of *Yersinia pseudotuberculosis* to mouse Peyer's patch M cells. *Infect. Immun.* **66**, 1237–1243.

Claude, P. and Goodenough, D. A. (1973). Fracture faces of zonulae occludentes from "tight" and "leaky" epithelia. *J. Cell. Biol.* **58**, 390–400.

Cohen, C. J., Shieh, J. T., Pickles, R. J., *et al.* (2001). The coxsackievirus and adenovirus receptor is a transmembrane component of the tight junction. *Proc. Natl. Acad. Sci. U. S. A.* **98**, 15 191–15 196.

Colegio, O. R., Van Itallie, C. M., McCrea, H. J., Rahner, C., and Anderson, J. M. (2002). Claudins create charge-selective channels in the paracellular pathway between epithelial cells. *Am. J. Physiol. Cell. Physiol.* **283**, C142–C147.

Colegio, O. R., Van Itallie, C., Rahner, C., and Anderson, J. M. (2003). Claudin extracellular domains determine paracellular charge selectivity and resistance but not tight junction fibril architecture. *Am. J. Physiol. Cell. Physiol.* **284**, C1346–C1354.

Collington, G. K., Booth, I. W., and Knutton, S. (1998). Rapid modulation of electrolyte transport in Caco-2 cell monolayers by enteropathogenic *Escherichia coli* (EPEC) infection. *Gut* **42**, 200–207.

Cox, D. S., Gao, H., Raje, S., Scott, K. R., and Eddington, N. D. (2001). Enhancing the permeation of marker compounds and enaminone anticonvulsants across Caco-2 monolayers by modulating tight junctions using zonula occludens toxin. *Eur. J. Pharm. Biopharm.* **52**, 145–150.

Crane, J. K., McNamara, B. P., and Donnenberg, M. S. (2001). Role of EspF in host cell death induced by enteropathogenic *Escherichia coli*. *Cell. Microbiol.* **3**, 197–211.

Dahan, S., Dalmasso, G., Imbert, V., *et al.* (2003). *Saccharomyces boulardii* interferes with enterohemorrhagic *Escherichia coli*-induced signaling pathways in T84 cells. *Infect. Immun.* **71**, 766–773.

Dean, P. and Kenny, B. (2004). Intestinal barrier dysfunction by enteropathogenic *Escherichia coli* is mediated by two effector molecules and a bacterial surface protein. *Mol. Microbiol.* **54**, 665–675.

Dean, P., Maresca, M., and Kenny, B. (2005). EPEC's weapons of mass subversion. *Curr. Opin. Microbiol.* **8**, 28–34.

Deibel, C., Kramer, S., Chakraborty, T., and Ebel, F. (1998). EspE, a novel secreted protein of attaching and effacing bacteria, is directly translocated into infected host cells, where it appears as a tyrosine-phosphorylated 90 kDa protein. *Mol. Microbiol.* **28**, 463–474.

DeVinney, R., Knoechel, D. G., and Finlay, B. B. (1999). Enteropathogenic *Escherichia coli*: cellular harassment. *Curr. Opin. Microbiol.* **2**, 83–88.

DeVinney, R., Stein, M., Reinscheid, D., *et al.* (1999). Enterohemorrhagic *Escherichia coli* O157:H7 produces Tir, which is translocated to the host cell membrane but is not tyrosine phosphorylated. *Infect. Immun.* **67**, 2389–2398.

DeVinney, R., Puente, J. L., Gauthier, A., Goosney, D., and Finlay, B. B. (2001). Enterohaemorrhagic and enteropathogenic *Escherichia coli* use a different Tir-based mechanism for pedestal formation. *Mol. Microbiol.* **41**, 1445–1458.

Doganay, M. (2003). Listeriosis: clinical presentation. *FEMS Immunol. Med. Microbiol.* **35**, 173–175.

Donnenberg, M. S., Yu, J., and Kaper, J. B. (1993). A second chromosomal gene necessary for intimate attachment of enteropathogenic *Escherichia coli* to epithelial cells. *J. Bacteriol.* **175**, 4670–4680.

Ebnet, K., Schulz, C. U., Meyer Zu Brickwedde, M. K., Pendl, G. G., and Vestweber, D. (2000). Junctional adhesion molecule interacts with the PDZ domain-containing proteins AF-6 and ZO-1. *J. Biol. Chem.* **275**, 27 979–27 988.

Eckmann, L., Stenson, W. F., Savidge, T. C., *et al.* (1997). Role of intestinal epithelial cells in the host secretory response to infection by invasive bacteria: bacterial entry induces epithelial prostaglandin h synthase-2 expression and prostaglandin E2 and F2alpha production. *J. Clin. Invest.* **100**, 296–309.

Fanning, A. S., Jameson, B. J., Jesaitis, L. A., and Anderson, J. M. (1998). The tight junction protein ZO-1 establishes a link between the transmembrane protein occludin and the actin cytoskeleton. *J. Biol. Chem.* **273**, 29 745–29 753.

Farshori, P. and Kachar, B. (1999). Redistribution and phosphorylation of occludin during opening and resealing of tight junctions in cultured epithelial cells. *J. Membr. Biol.* **170**, 147–156.

Fasano, A. and Uzzau, S. (1997). Modulation of intestinal tight junctions by Zonula occludens toxin permits enteral administration of insulin and other macromolecules in an animal model. *J .Clin. Invest.* **99**, 1158–1164.

Fasano, A., Baudry, B., Pumplin, D. W., *et al.* (1991). *Vibrio cholerae* produces a second enterotoxin, which affects intestinal tight junctions. *Proc. Natl. Acad. Sci. U. S. A.* **88**, 5242–5246.

Fish, S. M., Proujansky, R., and Reenstra, W. W. (1999). Synergistic effects of interferon gamma and tumour necrosis factor alpha on T84 cell function. *Gut* **45**, 191–198.

Fivaz, M. and van der Goot, F. G. (1999). The tip of a molecular syringe. *Trends Microbiol* **7**, 341–343.

Frankel, G., Lider, O., Hershkoviz, R., *et al.* (1996). The cell-binding domain of intimin from enteropathogenic *Escherichia coli* binds to beta1 integrins. *J. Biol. Chem.* **271**, 20 359–20 364.

Frankel, G., Phillips, A. D., Rosenshine, I., *et al.* (1998). Enteropathogenic and enterohaemorrhagic *Escherichia coli*: more subversive elements. *Mol. Microbiol.* **30**, 911–921.

Freeman, N. L., Zurawski, D. V., Chowrashi, P., *et al.* (2000). Interaction of the enteropathogenic *Escherichia coli* protein, translocated intimin receptor (Tir), with focal adhesion proteins. *Cell. Motil. Cytoskeleton* **47**, 307–318.

Fujibe, M., Chiba, H., Kojima, T., *et al.* (2004). Thr203 of claudin-1, a putative phosphorylation site for MAP kinase, is required to promote the barrier function of tight junctions. *Exp. Cell Res.* **295**, 36–47.

Furuse, M., Hirase, T., Itoh, M., *et al.* (1993). Occludin: a novel integral membrane protein localizing at tight junctions. *J. Cell Biol.* **123**, 1777–1788.

Furuse, M., Fujita, K., Hiiragi, T., Fujimoto, K., and Tsukita, S. (1998). Claudin-1 and -2: novel integral membrane proteins localizing at tight junctions with no sequence similarity to occludin. *J. Cell Biol.* **141**, 1539–1550.

Furuse, M., Sasaki, H., Fujimoto, K., and Tsukita, S. (1998). A single gene product, claudin-1 or -2, reconstitutes tight junction strands and recruits occludin in fibroblasts. *J. Cell. Biol.* **143**, 391–401.

Gaillard, J. L., Berche, P., Frehel, C., Gouin, E., and Cossart, P. (1991). Entry of *L. monocytogenes* into cells is mediated by internalin, a repeat protein reminiscent of surface antigens from gram-positive cocci. *Cell* **65**, 1127–1141.

Girardin, S. E., Boneca, I. G., Carneiro, L. A., *et al.* (2003). Nod1 detects a unique muropeptide from gram-negative bacterial peptidoglycan. *Science* **300**, 1584–1587.

Gogarten, W., Kockerling, A., Fromm, M., Riecken, E. O., and Schulzke, J. D. (1994). Effect of acute *Yersinia enterocolitica* infection on intestinal barrier function in the mouse. *Scand. J. Gastroenterol.* **29**, 814–819.

Gookin, J. L., Duckett, L. L., Armstrong, M. U., *et al.* (2004). Nitric oxide synthase stimulates prostaglandin synthesis and barrier function in *C. parvum*-infected porcine ileum. *Am. J. Physiol. Gastrointest. Liver Physiol.* **287**, G571–G581.

Goosney, D. L., DeVinney, R., Pfuetzner, R. A., *et al.* (2000). Enteropathogenic *E. coli* translocated intimin receptor, Tir, interacts directly with alpha-actinin. *Curr. Biol.* **10**, 735–738.

Goosney, D. L., DeVinney, R., and Finlay, B. B. (2001). Recruitment of cytoskeletal and signaling proteins to enteropathogenic and enterohemorrhagic *Escherichia coli* pedestals. *Infect. Immun.* **69**, 3315–3322.

Grosdent, N., Maridonneau-Parini, I., Sory, M. P., and Cornelis, G. R. (2002). Role of Yops and adhesins in resistance of *Yersinia enterocolitica* to phagocytosis. *Infect. Immun.* **70**, 4165–4176.

Gruenheid, S., DeVinney, R., Bladt, F., *et al.* (2001). Enteropathogenic *E. coli* Tir binds Nck to initiate actin pedestal formation in host cells *Nat. Cell. Biol.* **3**, 856–859.

Gumbiner, B. (1987). Structure, biochemistry, and assembly of epithelial tight junctions. *Am. J. Physiol.* **253**, C749–C758.

Gumbiner, B. (1993). Breaking through the tight junction barrier. *J. Cell Biol.* **123**, 1631–1633.

Gumbiner, B., Lowenkopf, T., and Apatira, D. (1991). Identification of a 160-kDa polypeptide that binds to the tight junction protein ZO-1. *Proc. Natl. Acad. Sci. U. S. A.* **88**, 3460–3464.

Hardt, W. D., Chen, L. M., Schuebel, K. E., Bustelo, X. R., and Galan, J. E. (1998). *S. typhimurium* encodes an activator of Rho GTPases that induces membrane ruffling and nuclear responses in host cells. *Cell* **93**, 815–826.

Haskins, J., Gu, L., Wittchen, E. S., Hibbard, J., and Stevenson, B. R. (1998). ZO-3, a novel member of the MAGUK protein family found at the tight junction, interacts with ZO-1 and occludin. *J. Cell. Biol.* **141**, 199–208.

Hauck, C. R. (2002). Cell adhesion receptors: signaling capacity and exploitation by bacterial pathogens. *Med. Microbiol. Immunol. (Berl.)* **191**, 55–62.

Hauf, N. and Chakraborty, T. (2003). Suppression of NF-kappa B activation and proinflammatory cytokine expression by Shiga toxin-producing *Escherichia coli*. *J. Immunol.* **170**, 2074–2082.

Hayashi, H., Szaszi, K., Coady-Osberg, N., *et al.* (2004). Inhibition and redistribution of NHE3, the apical Na+/H+ exchanger, by *Clostridium difficile* toxin B. *J. Gen. Physiol.* **123**, 491–504.

Hayward, R. D. and Koronakis, V. (1999). Direct nucleation and bundling of actin by the SipC protein of invasive *Salmonella*. *EMBO. J.* **18**, 4926–4934.

Hecht, G. (2001). Microbes and microbial toxins: paradigms for microbial-mucosal interactions. VII. Enteropathogenic *Escherichia coli*: physiological alterations from an extracellular position. *Am. J. Physiol. Gastrointest. Liver Physiol.* **281**, G1–7.

Hecht, G. and Koutsouris, A. (1999). Enteropathogenic *E. coli* attenuates secretagogue-induced net intestinal ion transport but not Cl− secretion. *Am. J. Physiol.* **276**, G781–G788.

Hecht, G., Pothoulakis, C., LaMont, J. T., and Madara, J. L. (1988). *Clostridium difficile* toxin A perturbs cytoskeletal structure and tight junction permeability of cultured human intestinal epithelial monolayers. *J. Clin. Invest.* **82**, 1516–1524.

Hecht, G., Koutsouris, A., Pothoulakis, C., LaMont, J. T., and Madara, J. L. (1992). *Clostridium difficile* toxin B disrupts the barrier function of T84 monolayers. *Gastroenterology* **102**, 416–423.

Hecht, G., Pestic, L., Nikcevic, G., *et al.* (1996). Expression of the catalytic domain of myosin light chain kinase increases paracellular permeability. *Am. J. Physiol.* **271**, C1678–C1684.

Hecht, G., Marrero, J. A., Danilkovich, A., *et al.* (1999). Pathogenic *Escherichia coli* increase Cl− secretion from intestinal epithelia by upregulating galanin-1 receptor expression. *J. Clin. Invest.* **104**, 253–262.

Hecht, G., Hodges, K., Gill, R. K., *et al.* (2004). Differential regulation of Na+/H+ exchange isoform activities by enteropathogenic *E. coli* in human intestinal epithelial cells. *Am. J. Physiol. Gastrointest. Liver Physiol.* **287**, G370–G378.

Heczko, U., Carthy, C. M., O'Brien, B. A., and Finlay, B. B. (2001). Decreased apoptosis in the ileum and ileal Peyer's patches: a feature after infection with rabbit enteropathogenic *Escherichia coli* O103. *Infect. Immun.* **69**, 4580–4589.

Heiskala, M., Peterson, P. A., and Yang, Y. (2001). The roles of claudin superfamily proteins in paracellular transport. *Traffic* **2**, 93–98.

Hermiston, M. L. and Gordon, J. I. (1995). In vivo analysis of cadherin function in the mouse intestinal epithelium: essential roles in adhesion, maintenance of differentiation, and regulation of programmed cell death. *J. Cell Biol.* **129**, 489–506.

Hersh, D., Monack, D. M., Smith, M. R., *et al.* (1999). The *Salmonella* invasin SipB induces macrophage apoptosis by binding to caspase-1. *Proc. Natl. Acad. Sci. U. S. A.* **96**, 2396–2401.

Inai, T., Kobayashi, J., and Shibata, Y. (1999). Claudin-1 contributes to the epithelial barrier function in MDCK cells. *Eur. J. Cell Biol.* **78**, 849–855.

Isberg, R. R. and Leong, J. M. (1990). Multiple beta 1 chain integrins are receptors for invasin, a protein that promotes bacterial penetration into mammalian cells. *Cell* **60**, 861–871.

Isberg, R. R., Voorhis, D. L., and Falkow, S. (1987). Identification of invasin: a protein that allows enteric bacteria to penetrate cultured mammalian cells. *Cell* **50**, 769–778.

Ishizaki, T., Chiba, H., Kojima, T., *et al.* (2003). Cyclic AMP induces phosphorylation of claudin-5 immunoprecipitates and expression of claudin-5 gene in blood–brain-barrier endothelial cells via protein kinase A-dependent and -independent pathways. *Exp. Cell Res.* **290**, 275–288.

Ismaili, A., Philpott, D. J., Dytoc, M. T., and Sherman, P. M. (1995). Signal transduction responses following adhesion of verocytotoxin-producing *Escherichia coli*. *Infect. Immun.* **63**, 3316–3326.

Itoh, M., Morita, K., and Tsukita, S. (1999). Characterization of ZO-2 as a MAGUK family member associated with tight as well as adherens junctions with a binding affinity to occludin and alpha catenin. *J. Biol. Chem.* **274**, 5981–5986.

Itoh, M., Sasaki, H., Furuse, M., *et al.* (2001). Junctional adhesion molecule (JAM) binds to PAR-3: a possible mechanism for the recruitment of PAR-3 to tight junctions. *J. Cell. Biol.* **154**, 491–497.

Jesaitis, L. A. and Goodenough, D. A. (1994). Molecular characterization and tissue distribution of ZO-2, a tight junction protein homologous to ZO-1 and the Drosophila discs-large tumor suppressor protein. *J. Cell Biol.* **124**, 949–961.

Juliano, R. L. (2002). Signal transduction by cell adhesion receptors and the cytoskeleton: functions of integrins, cadherins, selectins, and immunoglobulin-superfamily members. *Annu. Rev. Pharmacol. Toxicol.* **42**, 283–323.

Kalman, D., Weiner, O. D., Goosney, D. L., *et al.* (1999). Enteropathogenic *E. coli* acts through WASP and Arp2/3 complex to form actin pedestals. *Nat. Cell Biol.* **1**, 389–391.

Katsoulis, S., Clemens, A., Morys-Wortmann, C., *et al.* (1996). Human galanin modulates human colonic motility in vitro: characterization of structural requirements. *Scand. J. Gastroenterol.* **31**, 446–451.

Keely, S. J. and Barrett, K. E. (2000). Regulation of chloride secretion: novel pathways and messengers. *Ann. N. Y. Acad. Sci.* **915**, 67–76.

Kenny, B. (1999). Phosphorylation of tyrosine 474 of the enteropathogenic *Escherichia coli* (EPEC) Tir receptor molecule is essential for actin nucleating activity and is preceded by additional host modifications. *Mol. Microbiol.* **31**, 1229–1241.

Kenny, B. (2001). The enterohaemorrhagic *Escherichia coli* (serotype O157:H7) Tir molecule is not functionally interchangeable for its enteropathogenic *E. coli* (serotype O127:H6) homologue. *Cell. Microbiol.* **3**, 499–510.

Kenny, B. and Jepson, M. (2000). Targeting of an enteropathogenic *Escherichia coli* (EPEC) effector protein to host mitochondria. *Cell. Microbiol.* **2**, 579–590.

Kenny, B., Abe, A., Stein, M., and Finlay, B. B. (1997). Enteropathogenic *Escherichia coli* protein secretion is induced in response to conditions similar to those in the gastrointestinal tract. *Infect. Immun.* **65**, 2606–2612.

Kenny, B., DeVinney, R., Stein, M., *et al.* (1997). Enteropathogenic *E. coli* (EPEC) transfers its receptor for intimate adherence into mammalian cells. *Cell* **91**, 511–520.

Kenny, B., Ellis, S., Leard, A. D., *et al.* (2002). Co-ordinate regulation of distinct host cell signalling pathways by multifunctional enteropathogenic *Escherichia coli* effector molecules. *Mol. Microbiol.* **44**, 1095–1107.

Knutton, S., Baldwin, T., Williams, P. H., and McNeish, A. S. (1989). Actin accumulation at sites of bacterial adhesion to tissue culture cells: basis of a new diagnostic test for enteropathogenic and enterohemorrhagic *Escherichia coli*. *Infect. Immun.* **57**, 1290–1298.

Knutton, S., Rosenshine, I., Pallen, M. J., *et al.* (1998). A novel EspA-associated surface organelle of enteropathogenic *Escherichia coli* involved in protein translocation into epithelial cells. *EMBO J.* **17**, 2166–2176.

Kodama, T., Akeda, Y., Kono, G., *et al.* (2002). The EspB protein of enterohaemorrhagic *Escherichia coli* interacts directly with alpha-catenin. *Cell. Microbiol.* **4**, 213–222.

Laurent, F., Kagnoff, M. F., Savidge, T. C., Naciri, M., and Eckmann, L. (1998). Human intestinal epithelial cells respond to *Cryptosporidium parvum* infection with increased prostaglandin H synthase 2 expression and prostaglandin E2 and F2alpha production. *Infect. Immun.* **66**, 1787–1790.

Lecuit, M., Hurme, R., Pizarro-Cerda, J., *et al.* (2000). A role for alpha-and beta-catenins in bacterial uptake. *Proc. Natl. Acad. Sci. U. S. A.* **97**, 10 008–10 013.

Little, D., Dean, R. A., Young, K. M., *et al.* (2003). PI3K signaling is required for prostaglandin-induced mucosal recovery in ischemia-injured porcine ileum. *Am. J. Physiol. Gastrointest. Liver Physiol.* **284**, G46–G56.

Liu, T. S., Musch, M. W., Sugi, K., *et al.* (2003). Protective role of HSP72 against *Clostridium difficile* toxin A-induced intestinal epithelial cell dysfunction. *Am. J. Physiol. Cell Physiol.* **284**, C1073–C1082.

Lucas, M. L. (2001). A reconsideration of the evidence for *Escherichia coli* STa (heat stable) enterotoxin-driven fluid secretion: a new view of STa action and a new paradigm for fluid absorption. *J. Appl. Microbiol.* **90**, 7–26.

Ma, T. Y., Iwamoto, G. K., Hoa, N. T., *et al.* (2004). TNF-alpha-induced increase in intestinal epithelial tight junction permeability requires NF-kappa B activation. *Am. J. Physiol. Gastrointest. Liver Physiol.* **286**, G367–G376.

Ma, T. Y., Boivin, M. A., Ye, D., Pedram, A., and Said, H. M. (2005). Mechanism of TNF-{alpha} modulation of Caco-2 intestinal epithelial tight junction barrier: role of myosin light-chain kinase protein expression. *Am. J. Physiol. Gastrointest. Liver Physiol.* **288**, G422–G430.

Madara, J. L. (1988). Tight junction dynamics: is paracellular transport regulated? *Cell* **53**, 497–498.

Madara, J. L. and Stafford, J. (1989). Interferon-gamma directly affects barrier function of cultured intestinal epithelial monolayers. *J. Clin. Invest.* **83**, 724–727.

Madara, J. L., Barenberg, D., and Carlson, S. (1986). Effects of cytochalasin D on occluding junctions of intestinal absorptive cells: further evidence that the cytoskeleton may influence paracellular permeability and junctional charge selectivity. *J. Cell. Biol.* **102**, 2125–2136.

Madara, J. L., Moore, R., and Carlson, S. (1987). Alteration of intestinal tight junction structure and permeability by cytoskeletal contraction. *Am. J. Physiol.* **253**, C854–C861.

Manjarrez-Hernandez, H. A., Amess, B., Sellers, L., *et al.* (1991). Purification of a 20 kDa phosphoprotein from epithelial cells and identification as a myosin light chain: phosphorylation induced by enteropathogenic *Escherichia coli* and phorbol ester. *FEBS Lett.* **292**, 121–127.

Mann, E. A., Jump, M. L., Wu, J., Yee, E., and Giannella, R. A. (1997). Mice lacking the guanylyl cyclase C receptor are resistant to STa-induced intestinal secretion. *Biochem. Biophys. Res. Commun.* **239**, 463–466.

Marano, C. W., Lewis, S. A., Garulacan, L. A., Soler, A. P., and Mullin, J. M. (1998). Tumor necrosis factor-alpha increases sodium and chloride conductance across the tight junction of CACO-2 BBE, a human intestinal epithelial cell line. *J. Membr. Biol.* **161**, 263–274.

Martin-Padura, I., Lostaglio, S., Schneemann, M., *et al.* (1998). Junctional adhesion molecule, a novel member of the immunoglobulin superfamily that distributes at intercellular junctions and modulates monocyte transmigration. *J. Cell Biol.* **142**, 117–127.

Matkowskyj, K. A., Danilkovich, A., Marrero, J., *et al.* (2000). Galanin-1 receptor up-regulation mediates the excess colonic fluid production caused by infection with enteric pathogens. *Nat. Med.* **6**, 1048–1051.

Matsuzawa, T., Kuwae, A., Yoshida, S., Sasakawa, C., and Abe, A. (2004). Enteropathogenic *Escherichia coli* activates the RhoA signaling pathway via the stimulation of GEF-H1. *EMBO J.* **23**, 3570–3582.

McCarthy, K. M., Skare, I. B., Stankewich, M. C., *et al.* (1996). Occludin is a functional component of the tight junction. *J. Cell Sci.* **109 (Pt 9)**, 2287–2298.

McCarthy, K. M., Francis, S. A., McCormack, J. M., *et al.* (2000). Inducible expression of claudin-1-myc but not occludin-VSV-G results in aberrant tight junction strand formation in MDCK cells. *J. Cell. Sci.* **113 (Pt 19)**, 3387–3398.

McClane, B. A. (1996). An overview of *Clostridium perfringens* enterotoxin. *Toxicon* **34**, 1335–1343.

McCormick, B. A., Parkos, C. A., Colgan, S. P., Carnes, D. K., and Madara, J. L. (1998). Apical secretion of a pathogen-elicited epithelial chemoattractant activity in response to surface colonization of intestinal epithelia by *Salmonella typhimurium. J. Immunol.* **160**, 455–466.

McDaniel, T. K. and Kaper, J. B. (1997). A cloned pathogenicity island from enteropathogenic *Escherichia coli* confers the attaching and effacing phenotype on E. coli K-12. *Mol. Microbiol.* **23**, 399–407.

McGee, K., Holmfeldt, P., and Fallman, M. (2003). Microtubule-dependent regulation of Rho GTPases during internalisation of *Yersinia pseudotuberculosis. FEBS Lett.* **533**, 35–41.

McNamara, B. P. and Donnenberg, M. S. (1998). A novel proline-rich protein, EspF, is secreted from enteropathogenic *Escherichia coli* via the type III export pathway. *FEMS Microbiol. Lett.* **166**, 71–78.

McNamara, B. P., Koutsouris, A., O'Connell, C. B., *et al.* (2001). Translocated EspF protein from enteropathogenic *Escherichia coli* disrupts host intestinal barrier function. *J. Clin. Invest.* **107**, 621–629.

Mengaud, J., Ohayon, H., Gounon, P., Mege, R. M., and Cossart, P. (1996). E-cadherin is the receptor for internalin, a surface protein required for entry of *L. monocytogenes* into epithelial cells. *Cell* **84**, 923–932.

Miller, V. L. and Falkow, S. (1988). Evidence for two genetic loci in *Yersinia enterocolitica* that can promote invasion of epithelial cells. *Infect. Immun.* **56**, 1242–1248.

Mills, S. D., Boland, A., Sory, M. P., *et al.* (1997). *Yersinia enterocolitica* induces apoptosis in macrophages by a process requiring functional type III secretion and translocation mechanisms and involving YopP, presumably acting as an effector protein. *Proc. Natl. Acad. Sci. U. S. A.* **94**, 12 638–12 643.

Moeser, A. J., Haskell, M. M., Shifflett, D. E., *et al.* (2004). ClC-2 chloride secretion mediates prostaglandin-induced recovery of barrier function in ischemia-injured porcine ileum. *Gastroenterology* **127**, 802–815.

Molina, N. C. and Peterson, J. W. (1980). Cholera toxin-like toxin released by *Salmonella* species in the presence of mitomycin C. *Infect. Immun.* **30**, 224–230.

Monack, D. M., Mecsas, J., Ghori, N., and Falkow, S. (1997). *Yersinia* signals macrophages to undergo apoptosis and YopJ is necessary for this cell death. *Proc. Natl. Acad. Sci. U. S. A.* **94**, 10 385–10 390.

Morita, K., Sasaki, H., Fujimoto, K., Furuse, M., and Tsukita, S. (1999). Claudin-11/OSP-based tight junctions of myelin sheaths in brain and Sertoli cells in testis. *J. Cell Biol.* **145**, 579–588.

Mounier, J., Vasselon, T., Hellio, R., Lesourd, M., and Sansonetti, P. J. (1992). *Shigella flexneri* enters human colonic Caco-2 epithelial cells through the basolateral pole. *Infect. Immun.* **60**, 237–248.

Mrsny, R. J., Gewirtz, A. T., Siccardi, D., *et al.* (2004). Identification of hepoxilin A3 in inflammatory events: a required role in neutrophil migration across intestinal epithelia. *Proc. Natl. Acad. Sci. U. S. A.* **101**, 7421–7426.

Muza-Moons, M. M., Koutsouris, A., and Hecht, G. (2003). Disruption of cell polarity by enteropathogenic *Escherichia coli* enables basolateral membrane proteins to migrate apically and to potentiate physiological consequences. *Infect. Immun.* **71**, 7069–7078.

Muza-Moons, M. M., Schneeberger, E. E., and Hecht, G. A. (2004). Enteropathogenic *Escherichia coli* infection leads to appearance of aberrant tight junctions strands in the lateral membrane of intestinal epithelial cells. *Cell. Microbiol.* **6**, 783–793.

Nagai, T., Abe, A., and Sasakawa, C. (2005). Targeting of enteropathogenic *Escherichia coli* EspF to host mitochondria is essential for bacterial pathogenesis: critical role of the 16th leucine residue in EspF. *J. Biol. Chem.* **280**, 2998–3011.

Nataro, J. P. and Kaper, J. B. (1998). Diarrheagenic *Escherichia coli. Clin. Microbiol. Rev.* **11**, 142–201.

Nhieu, G. T. and Sansonetti, P. J. (1999). Mechanism of *Shigella* entry into epithelial cells. *Curr. Opin. Microbiol.* **2**, 51–55.

Nougayrede, J. P. and Donnenberg, M. S. (2004). Enteropathogenic *Escherichia coli* EspF is targeted to mitochondria and is required to initiate the mitochondrial death pathway. *Cell. Microbiol.* **6**, 1097–1111.

Nougayrede, J. P., Fernandes, P. J., and Donnenberg, M. S. (2003). Adhesion of enteropathogenic *Escherichia coli* to host cells. *Cell. Microbiol.* **5**, 359–372.

Nusrat, A., Giry, M., Turner, J. R., *et al.* (1995). Rho protein regulates tight junctions and perijunctional actin organization in polarized epithelia. *Proc. Natl. Acad. Sci. U. S. A.* **92**, 10 629–10 633.

Nusrat, A., von Eichel-Streiber, C., Turner, J. R., *et al.* (2001). *Clostridium difficile* toxins disrupt epithelial barrier function by altering membrane microdomain localization of tight junction proteins. *Infect. Immun.* **69**, 1329–1336.

Nybom, P. and Magnusson, K. E. (1996). Modulation of the junctional integrity by low or high concentrations of cytochalasin B and dihydrocytochalasin B is associated with distinct changes in F-actin and ZO-1. *Biosci. Rep.* **16**, 313–326.

Obiso, R. J., Jr, Lyerly, D. M., Van Tassell, R. L., and Wilkins, T. D. (1995). Proteolytic activity of the *Bacteroides fragilis* enterotoxin causes fluid secretion and intestinal damage in vivo. *Infect. Immun.* **63**, 3820–3826.

Palmer, L. E., Hobbie, S., Galan, J. E., and Bliska, J. B. (1998). YopJ of *Yersinia pseudotuberculosis* is required for the inhibition of macrophage TNF-alpha production and downregulation of the MAP kinases p38 and JNK. *Mol. Microbiol.* **27**, 953–965.

Pappenheimer, J. R. and Reiss, K. Z. (1987). Contribution of solvent drag through intercellular junctions to absorption of nutrients by the small intestine of the rat. *J. Membr. Biol.* **100**, 123–136.

Perdomo, J. J., Gounon, P., and Sansonetti, P. J. (1994). Polymorphonuclear leukocyte transmigration promotes invasion of colonic epithelial monolayer by *Shigella flexneri*. *J. Clin. Invest.* **93**, 633–643.

Perna, N. T., Mayhew, G. F., Posfai, G., *et al.* (1998). Molecular evolution of a pathogenicity island from enterohemorrhagic *Escherichia coli* O157:H7. *Infect. Immun.* **66**, 3810–3817.

Persson, C., Carballeira, N., Wolf-Watz, H., and Fallman, M. (1997). The PTPase YopH inhibits uptake of *Yersinia*, tyrosine phosphorylation of p130Cas and FAK, and the associated accumulation of these proteins in peripheral focal adhesions. *EMBO J.* **16**, 2307–2318.

Philpott, D. J., McKay, D. M., Sherman, P. M., and Perdue, M. H. (1996). Infection of T84 cells with enteropathogenic *Escherichia coli* alters barrier and transport functions. *Am. J. Physiol.* **270**, G634–G645.

Philpott, D. J., McKay, D. M., Mak, W., Perdue, M. H., and Sherman, P. M. (1998). Signal transduction pathways involved in enterohemorrhagic *Escherichia coli*-induced alterations in T84 epithelial permeability. *Infect. Immun.* **66**, 1680–1687.

Pothoulakis, C. (2000). Effects of *Clostridium difficile* toxins on epithelial cell barrier. *Ann. N. Y. Acad. Sci.* **915**, 347–356.

Rahner, C., Mitic, L. L., and Anderson, J. M. (2001). Heterogeneity in expression and subcellular localization of claudins 2, 3, 4, and 5 in the rat liver, pancreas, and gut. *Gastroenterology* **120**, 411–422.

Ren, Y., Li, R., Zheng, Y., and Busch, H. (1998). Cloning and characterization of GEF-H1, a microtubule-associated guanine nucleotide exchange factor for Rac and Rho GTPases. *J. Biol. Chem.* **273**, 34 954–34 960.

Rodriguez, P., Heyman, M., Candalh, C., Blaton, M. A., and Bouchaud, C. (1995). Tumour necrosis factor-alpha induces morphological and functional alterations of intestinal HT29 cl.19A cell monolayers. *Cytokine* **7**, 441–448.

Ruckdeschel, K., Machold, J., Roggenkamp, A., *et al.* (1997). *Yersinia enterocolitica* promotes deactivation of macrophage mitogen-activated protein kinases extracellular signal-regulated kinase-1/2, p38, and c-Jun NH2-terminal kinase: correlation with its inhibitory effect on tumor necrosis factor-alpha production. *J. Biol. Chem.* **272**, 15 920–15 927.

Ruckdeschel, K., Harb, S., Roggenkamp, A., *et al.* (1998). *Yersinia enterocolitica* impairs activation of transcription factor NF-kappaB: involvement in the induction of programmed cell death and in the suppression of the macrophage tumor necrosis factor alpha production. *J. Exp. Med.* **187**, 1069–1079.

Ruiz-Palacios, G. M., Torres, J., Torres, N. I., *et al.* (1983). Cholera-like enterotoxin produced by *Campylobacter jejuni*: characterisation and clinical significance. *Lancet* **2**, 250–253.

Saitou, M., Fujimoto, K., Doi, Y., *et al.* (1998). Occludin-deficient embryonic stem cells can differentiate into polarized epithelial cells bearing tight junctions. *J. Cell. Biol.* **141**, 397–408.

Saitou, M., Furuse, M., Sasaki, H., *et al.* (2000). Complex phenotype of mice lacking occludin, a component of tight junction strands. *Mol. Biol. Cell.* **11**, 4131–4142.

Sakaguchi, T., Kohler, H., Gu, X., McCormick, B. A., and Reinecker, H. C. (2002). *Shigella flexneri* regulates tight junction-associated proteins in human intestinal epithelial cells. *Cell. Microbiol.* **4**, 367–381.

Sakakibara, A., Furuse, M., Saitou, M., Ando-Akatsuka, Y., and Tsukita, S. (1997). Possible involvement of phosphorylation of occludin in tight junction formation. *J. Cell. Biol.* **137**, 1393–1401.

Sanger, J. M., Chang, R., Ashton, F., Kaper, J. B., and Sanger, J. W. (1996). Novel form of actin-based motility transports bacteria on the surfaces of infected cells. *Cell Motil. Cytoskeleton* **34**, 279–287.

Savkovic, S. D., Koutsouris, A., and Hecht, G. (1996). Attachment of a non-invasive enteric pathogen, enteropathogenic *Escherichia coli,* to cultured human intestinal epithelial monolayers induces transmigration of neutrophils. *Infect. Immun.* **64**, 4480–4487.

Savkovic, S. D., Koutsouris, A., and Hecht, G. (1997). Activation of NF-kappaB in intestinal epithelial cells by enteropathogenic *Escherichia coli. Am. J. Physiol.* **273**, C1160–C1167.

Savkovic, S. D., Ramaswamy, A., Koutsouris, A., and Hecht, G. (2001). EPEC-activated ERK1/2 participate in inflammatory response but not tight junction barrier disruption. *Am. J. Physiol. Gastrointest. Liver Physiol.* **281**, G890–G898.

Savkovic, S. D., Villanueva, J., Turner, J. R., Matkowskyj, K. A., and Hecht, G. (2005). Mouse model of enteropathogenic *Escherichia coli* infection. *Infect. Immun.* **73**, 1161–1170.

Schesser, K., Spiik, A. K., Dukuzumuremyi, J. M., *et al.* (1998). The yopJ locus is required for *Yersinia*-mediated inhibition of NF-kappaB activation and cytokine expression: YopJ contains a eukaryotic SH2-like domain that is essential for its repressive activity. *Mol. Microbiol.* **28**, 1067–1079.

Schmitz, H., Fromm, M., Bentzel, C. J., *et al.* (1999). Tumor necrosis factor-alpha (TNFalpha) regulates the epithelial barrier in the human intestinal cell line HT-29/B6. *J. Cell. Sci.* **112 (Pt 1)**, 137–146.

Schneeberger, E. E. and Lynch, R. D. (2004). The tight junction: a multifunctional complex. *Am. J. Physiol. Cell Physiol.* **286**, C1213–C1228.

Schulte, R., Wattiau, P., Hartland, E. L., Robins-Browne, R. M., and Cornelis, G. R. (1996). Differential secretion of interleukin-8 by human epithelial cell lines upon entry of virulent or nonvirulent *Yersinia enterocolitica. Infect. Immun.* **64**, 2106–2113.

Schulzke, J. D., Gitter, A. H., Mankertz, J., *et al.* (2005). Epithelial transport and barrier function in occludin-deficient mice. *Biochim. Biophys. Acta* **1669**, 34–42.

Sharma, R., Tesfay, S., Tomson, F. L., *et al.* (2005). A type III secretion-independent, non-flagellin molecule contributes to enteropathogenic *E. coli*-mediated inflammatory response. *Gastroenterology* **128 (Supp.e 2)**, A-67.

Shaw, R. K., Daniell, S., Ebel, F., Frankel, G., and Knutton, S. (2001). EspA filament-mediated protein translocation into red blood cells. *Cell. Microbiol.* **3**, 213–222.

Shifflett, D. E., Bottone, F. G., Jr, Young, K. M., *et al.* (2004). Neutrophils augment recovery of porcine ischemia-injured ileal mucosa by an IL-1beta- and COX-2-dependent mechanism. *Am. J. Physiol. Gastrointest. Liver Physiol.* **287**, G50–G57.

Shifflett, D. E., Jones, S. L., Moeser, A. J., and Blikslager, A. T. (2004). Mitogen-activated protein kinases regulate COX-2 and mucosal recovery in ischemic-injured porcine ileum. *Am. J. Physiol. Gastrointest. Liver Physiol.* **286**, G906–G913.

Shifflett, D. E., Clayburgh, D. R., Koutsouris, A., Turner, J. R., and Hecht, G. A. (2005). Enteropathogenic *E. coli* disrupts tight junction barrier function and structure in vivo. *Lab. Invest.* **85**, 1308–1324.

Shull, G. E., Miller, M. L., and Schultheis, P. J. (2000). Lessons from genetically engineered animal models. VIII: absorption and secretion of ions in the gastrointestinal tract. *Am. J. Physiol. Gastrointest. Liver Physiol.* **278**, G185–G190.

Silva, M., Song, C., Nadeau, W. J., Matthews, J. B., and McCormick, B. A. (2004). *Salmonella typhimurium* SipA-induced neutrophil transepithelial migration: involvement of a PKC-alpha-dependent signal transduction pathway. *Am. J. Physiol. Gastrointest. Liver Physiol.* **286**, G1024–G1031.

Simon, D. B., Lu, Y., Choate, K. A., *et al.* (1999). Paracellin-1, a renal tight junction protein required for paracellular Mg2+ resorption. *Science* **285**, 103–106.

Simonovic, I., Rosenberg, J., Koutsouris, A., and Hecht, G. (2000). Enteropathogenic *Escherichia coli* dephosphorylates and dissociates occludin from intestinal epithelial tight junctions. *Cell Microbiol.* **2**, 305–315.

Sinclair, J. F. and O'Brien, A. D. (2002). Cell surface-localized nucleolin is a eukaryotic receptor for the adhesin intimin-gamma of enterohemorrhagic *Escherichia coli* O157:H7. *J. Biol. Chem.* **277**, 2876–2885.

Sonoda, N., Furuse, M., Sasaki, H., *et al.* (1999). *Clostridium perfringens* enterotoxin fragment removes specific claudins from tight junction strands: evidence for direct involvement of claudins in tight junction barrier. *J. Cell Biol.* **147**, 195–204.

Spitz, J., Yuhan, R., Koutsouris, A., *et al.* (1995). Enteropathogenic *Escherichia coli* adherence to intestinal epithelial monolayers diminishes barrier function. *Am. J. Physiol.* **268**, G374–G379.

Staddon, J. M., Herrenknecht, K., Smales, C., and Rubin, L. L. (1995). Evidence that tyrosine phosphorylation may increase tight junction permeability. *J. Cell Sci.* **108 (Pt 2)**, 609–619.

Stenson, W. F., Zhang, Z., Riehl, T., and Stanley, S. L., Jr (2001). Amebic infection in the human colon induces cyclooxygenase-2. *Infect. Immun.* **69**, 3382–3388.

Stevenson, B. R., Siliciano, J. D., Mooseker, M. S., and Goodenough, D. A. (1986). Identification of ZO-1: a high molecular weight polypeptide associated with the tight junction (zonula occludens) in a variety of epithelia. *J. Cell Biol.* **103**, 755–766.

Stevenson, B. R., Anderson, J. M., Goodenough, D. A., and Mooseker, M. S. (1988). Tight junction structure and ZO-1 content are identical in two strains of Madin-Darby canine kidney cells which differ in transepithelial resistance. *J. Cell Biol.* **107**, 2401–2408.

Stevenson, B. R., Anderson, J. M., Braun, I. D., and Mooseker, M. S. (1989). Phosphorylation of the tight-junction protein ZO-1 in two strains of Madin-Darby canine kidney cells which differ in transepithelial resistance. *Biochem. J.* **263**, 597–599.

Sugi, K., Musch, M. W., Field, M., and Chang, E. B. (2001). Inhibition of Na+,K+-ATPase by interferon gamma down-regulates intestinal epithelial transport and barrier function. *Gastroenterology* **120**, 1393–1403.

Tacket, C. O., Sztein, M. B., Losonsky, G., *et al.* (2000). Role of EspB in experimental human enteropathogenic *Escherichia coli* infection. *Infect. Immun.* **68**, 3689–3695.

Tafazoli, F., Holmstrom, A., Forsberg, A., and Magnusson, K. E. (2000). Apically exposed, tight junction-associated beta1-integrins allow binding and YopE-mediated perturbation of epithelial barriers by wild-type *Yersinia* bacteria. *Infect. Immun.* **68**, 5335–5343.

Taylor, K. A., O'Connell, C. B., Luther, P. W., and Donnenberg, M. S. (1998). The EspB protein of enteropathogenic *Escherichia coli* is targeted to the cytoplasm of infected HeLa cells. *Infect. Immun.* **66**, 5501–5507.

Theriot, J. A., Mitchison, T. J., Tilney, L. G., and Portnoy, D. A. (1992). The rate of actin-based motility of intracellular *Listeria monocytogenes* equals the rate of actin polymerization. *Nature* **357**, 257–260.

Thirumalai, K., Kim, K. S., and Zychlinsky, A. (1997). IpaB, a *Shigella flexneri* invasin, colocalizes with interleukin-1 beta-converting enzyme in the cytoplasm of macrophages. *Infect. Immun.* **65**, 787–793.

Tomson, F. L., Tesfay, S., Sharma, R., *et al.* (2005). A secreted effector of enteropathogenic *E. coli* has anti-inflammatory effects on host cells. *Gastroenterology* **128 (Suppl 2)**, A-664.

Tomson, F. L., Viswanathan, V. K., Kanack, K. J., *et al.* (2005). Enteropathogenic *Escherichia coli* EspG disrupts microtubules and in conjunction with Orf3 enhances perturbation of the tight junction barrier. *Mol. Microbiol.* **56**, 447–464.

Tran Van Nhieu, G., Caron, E., Hall, A., and Sansonetti, P. J. (1999). IpaC induces actin polymerization and filopodia formation during *Shigella* entry into epithelial cells. *EMBO J.* **18**, 3249–3262.

Tsukamoto, T. and Nigam, S. K. (1999). Role of tyrosine phosphorylation in the reassembly of occludin and other tight junction proteins. *Am. J. Physiol.* **276**, F737–F750.

Tsukita, S., Furuse, M., and Itoh, M. (2001). Multifunctional strands in tight junctions. *Nat. Rev. Mol. Cell. Biol.* **2**, 285–293.

Tu, X., Nisan, I., Yona, C., Hanski, E., and Rosenshine, I. (2003). EspH, a new cytoskeleton-modulating effector of enterohaemorrhagic and enteropathogenic *Escherichia coli. Mol. Microbiol.* **47**, 595–606.

Turner, J. R., Cohen, D. E., Mrsny, R. J., and Madara, J. L. (2000). Noninvasive in vivo analysis of human small intestinal paracellular absorption: regulation by Na+-glucose cotransport. *Dig. Dis. Sci.* **45**, 2122–2126.

Van Itallie, C. Balda, M. S., and Anderson, J. M. (1995). Epidermal growth factor induces tyrosine phosphorylation and reorganization of the tight junction protein ZO-1 in A431 cells. *J. Cell. Sci.* **108 (Pt 4)**, 1735–1742.

Van Itallie, C., Rahner, C., and Anderson, J. M. (2001). Regulated expression of claudin-4 decreases paracellular conductance through a selective decrease in sodium permeability. *J. Clin. Invest.* **107**, 1319–1327.

Viswanathan, V. K., Koutsouris, A., Lukic, S., *et al.* (2004). Comparative analysis of EspF from enteropathogenic and enterohemorrhagic *Escherichia coli* in alteration of epithelial barrier function. *Infect. Immun.* **72**, 3218–3227.

Viswanathan, V. K., Lukic, S., Koutsouris, A., *et al.* (2004). Cytokeratin 18 interacts with the enteropathogenic *Escherichia coli* secreted protein F (EspF) and is redistributed after infection. *Cell. Microbiol.* **6**, 987–997.

Wang, F., Graham, W. V., Wang, Y., *et al.* (2005). Interferon-gamma and tumor necrosis factor-alpha synergize to induce intestinal epithelial barrier dysfunction by up-regulating myosin light chain kinase expression. *Am. J. Pathol.* **166**, 409–419.

Wang, W. L., Lu, R. L., DiPierro, M., and Fasano, A. (2000). Zonula occludin toxin, a microtubule binding protein. *World J. Gastroenterol.* **6**, 330–334.

Watarai, M., Funato, S., and Sasakawa, C. (1996). Interaction of Ipa proteins of *Shigella flexneri* with alpha5beta1 integrin promotes entry of the bacteria into mammalian cells. *J. Exp. Med.* **183**, 991–999.

Wittchen, E. S., Haskins, J., and Stevenson, B. R. (1999). Protein interactions at the tight junction: actin has multiple binding partners, and ZO-1 forms independent complexes with ZO-2 and ZO-3. *J. Biol. Chem.* **274**, 35 179–35 185.

Wong, V. (1997). Phosphorylation of occludin correlates with occludin localization and function at the tight junction. *Am. J. Physiol.* **273**, C1859–C1867.

Wu, S., Lim, K. C., Huang, J., Saidi, R. F., and Sears, C. L. (1998). *Bacteroides fragilis* enterotoxin cleaves the zonula adherens protein, E-cadherin. *Proc. Natl. Acad. Sci. U. S. A.* **95**, 14 979–14 984.

Wu, Z., Milton, D., Nybom, P., Sjo, A., and Magnusson, K. E. (1996). *Vibrio cholerae* hemagglutinin/protease (HA/protease) causes morphological changes in cultured epithelial cells and perturbs their paracellular barrier function. *Microb. Pathog.* **21**, 111–123.

Wu, Z., Nybom, P., and Magnusson, K. E. (2000). Distinct effects of *Vibrio cholerae* haemagglutinin/protease on the structure and localization of the tight junction-associated proteins occludin and ZO-1. *Cell Microbiol.* **2**, 11–17.

Yamauchi, K., Rai, T., Kobayashi, K., *et al.* (2004). Disease-causing mutant WNK4 increases paracellular chloride permeability and phosphorylates claudins. *Proc. Natl. Acad. Sci. U. S. A.* **101**, 4690–4694.

Yip, C. K., Finlay, B. B., and Strynadka, N. C. (2005). Structural characterization of a type III secretion system filament protein in complex with its chaperone. *Nat. Struct. Mol. Biol.* **12**, 75–81.

Yoo, J., Nichols, A., Mammen, J., *et al.* (2003). Bryostatin-1 enhances barrier function in T84 epithelia through PKC-dependent regulation of tight junction proteins. *Am. J. Physiol. Cell Physiol.* **285**, C300–C309.

Youakim, A. and Ahdieh, M. (1999). Interferon-gamma decreases barrier function in T84 cells by reducing ZO-1 levels and disrupting apical actin. *Am. J. Physiol.* **276**, G1279–G1288.

Young, V. B., Falkow, S., and Schoolnik, G. K. (1992). The invasin protein of *Yersinia enterocolitica*: internalization of invasin-bearing bacteria by eukaryotic cells is associated with reorganization of the cytoskeleton. *J. Cell Biol.* **116**, 197–207.

Yu, J. and Kaper, J. B. (1992). Cloning and characterization of the eae gene of enterohaemorrhagic *Escherichia coli* O157:H7. *Mol. Microbiol.* **6**, 411–417.

Yuhan, R., Koutsouris, A., Savkovic, S. D., and Hecht, G. (1997). Enteropathogenic *Escherichia coli*-induced myosin light chain phosphorylation alters intestinal epithelial permeability. *Gastroenterology* **113**, 1873–1882.

Zhong, Y., Saitoh, T., Minase, T., *et al.* (1993). Monoclonal antibody 7H6 reacts with a novel tight junction-associated protein distinct from ZO-1, cingulin and ZO-2. *J. Cell Biol.* **120**, 477–483.

Zhou, D., Mooseker, M. S., and Galan, J. E. (1999). Role of the *S. typhimurium* actin-binding protein SipA in bacterial internalization. *Science* **283**, 2092–2095.

Zhou, X., Giron, J. A., Torres, A. G., *et al.* (2003). Flagellin of enteropathogenic *Escherichia coli* stimulates interleukin-8 production in T84 cells. *Infect. Immun.* **71**, 2120–2129.

Zolotarevsky, Y., Hecht, G., Koutsouris, A., *et al.* (2002). A membrane-permeant peptide that inhibits MLC kinase restores barrier function in in vitro models of intestinal disease. *Gastroenterology* **123**, 163–172.

Zumbihl, R., Aepfelbacher, M., Andor, A., *et al.* (1999). The cytotoxin YopT of *Yersinia enterocolitica* induces modification and cellular redistribution of the small GTP-binding protein RhoA. *J. Biol. Chem.* **274**, 29 289–29 293.

CHAPTER 14

Uropathogenic bacteria

Luce Landraud, René Clément, and
Patrice Boquet

INTRODUCTION

Urinary tract infections (UTI) are among the most common human bacterial infections. The prevalent pathogens are the uropathogenic *Escherichia coli* (UPEC) strains. A great deal of information concerning the genetic, virulence, and innate immune host responses against those bacteria have been obtained. Furthermore, the knowledge of uroepithelium cell biology and physiology, in particular at the level of the urinary bladder, has made considerable progress since 1995, improving our understanding of the strategies used by UPEC to colonize and invade this tissue.

UTIs account for significant morbidity and high medical costs. In 1997, the National Ambulatory Medical Care Survey and the National Hospital Medical Care Survey Report estimated that UTIs in the USA result in nearly seven million doctor visits per year, excluding visits to hospital emergency departments. The overall costs associated with UTIs have been estimated to reach upwards of two billion US dollars a year (Foxman, 2002). Moreover, nosocomial acquired urinary tract infections (NAUTIs) account for up to 40% of all hospital-acquired infections in European countries and represent the most frequent nosocomial infection (Eriksen *et al.*, 2004; Johansen, 2004; Zotti *et al.*, 2004). NAUTIs, which are frequently associated with medical procedures (the most important risk factor being an indwelling catheter), may result in significant acute morbidity, medical complications, and legal issues (Johansen, 2004). Moreover, one of the main concerns with UTIs is that they are frequently sources of recurrent infections.

Bacterial–Epithelial Cell Cross-Talk: Molecular Mechanisms in Pathogenesis, ed. Beth A. McCormick. Published by Cambridge University Press.

The severity of UTIs depends on the spread of the infection. Accordingly, two classical clinical outcomes are described:

- acute cystitis, in which the clinical symptoms (suprapubic pain, increased frequency of urination, dysuria without fever) seem to result from a limited infection of the bladder;
- acute pyelonephritis involving kidney infection together with a systemic infection (flank pain with fever).

Physicians can distinguish uncomplicated UTI from complicated UTI. UTIs can be affected by underlying host factors such as immunosuppression, diabetes, and anatomical defects of the urinary tract, which may complicate these infections. UPEC is the major causative agent of uncomplicated community-acquired UTIs and is responsible for nearly half of all nosocomial and complicated UTIs (Russo and Johnson, 2003). Other common pathogens associated with UTIs are *Staphylococcus saprophyticus*, *Klebsiella* spp., *Proteus mirabilis*, and *Enterococcus faecalis* (Ronald, 2003).

Escherichia coli strains involved in UTIs are a selected subset of the fecal flora, but UPECs differ from commensal *E. coli* by the presence of specific virulence factors, such as adhesins, toxins, and capsule- and iron-uptake systems. These specific factors are often encoded on particular genomic regions of the bacterial chromosome; such particular DNA stretches that encode multiple virulence factors have been termed pathogenicity islands (PAIs) (Dobrindt *et al.*, 2004). In this chapter, we focus on the role of adhesive factors in conjunction with the protein toxins elaborated by UPECs for signaling cross-talk with their host.

THE UROEPITHELIUM AND ITS DEFENSES AGAINST INVADING PATHOGENS

The bladder and the upper urinary tract (Figure 14.1a) are typically sterile environments. However, UPEC strains are able to penetrate into the urethra, most likely from vaginal and intestinal reservoirs. They subsequently ascend to the bladder and may affect the kidneys through migration into the ureters (Warren, 1996).

The uroepithelium (Figure 14.1b) is a specialized stratified transitional epithelium composed of three cell layers: undifferentiated basal cells, intermediate cells, and large (25–250 μm) differentiated superficial binucleated umbrella cells (Apodaca, 2004). The primary role of the uroepithelium is to prevent the entry of pathogens and selectively regulate the passage of water, ions, solutes, and macromolecules across the mucosal surface. This barrier

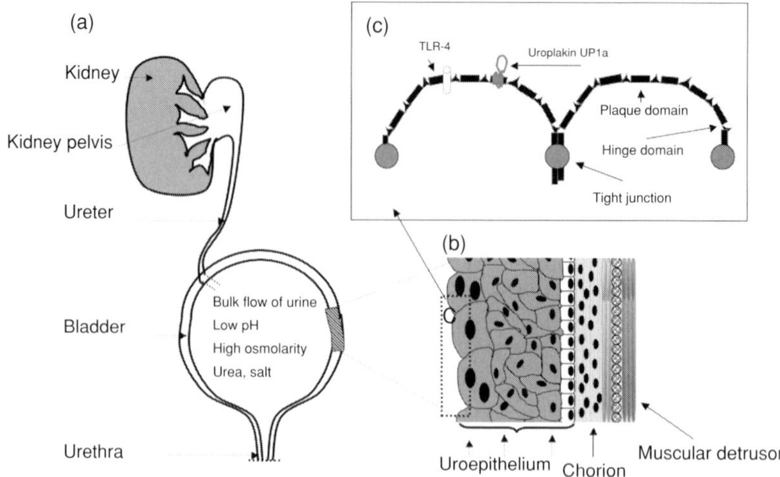

Figure 14.1 Anatomy of the urinary tract and structure of the bladder uroepithelium. (a) General anatomy of the urinary tract. (b) Schematic representation of the transitional uroepithelium. (c) Bladder umbrella cells. For details, see text.
TLR-4, Toll-like receptor 4.

role is due mainly to the presence of a structure that makes a seal between the plasma membrane of adjacent epithelial cells termed the tight junction (J. M. Anderson *et al.*, 2004). Umbrella cells in the uroepithelium display tight junctions that denote their apical and basolateral membrane domains (Figure 14.1c). Once epithelial cells have lost their tight junctions, a pathogen can easily gain access to the deeper tissues (McCormick, 2003). The apical membrane of umbrella cells exhibits extended horizontal compact structures called plaques and hinges (Figure 14.1c). Both plaques and hinges are highly detergent-insoluble lipid domains and thus assimilated to lipid rafts (Simons and Ikonen, 1997). In this respect, lipid rafts are used widely by microbial pathogens and their virulence factors to interact with host epithelial cells (Lafont *et al.*, 2004). Accordingly, it has been reported that UPEC may penetrate the bladder uroepithelium through lipid rafts (Duncan *et al.*, 2004). Plaques and hinges completely cover bladder uroepithelium and are required for its impermeability (Apodaca, 2004). Moreover, plaques and hinges contain integral membrane proteins called uroplakins (UPIa, UPIb, UPII, UPIIIa/b). Of note, UPIa is the receptor for an important adhesive appendage of UPEC, the type 1 pili (Mulvey *et al.*, 2000).

In addition to their exclusion from host tissues by the impermeability of the uroepithelium, an array of non-specific and specific defenses against pathogens are displayed by the urinary tract. The continuous flushing with urine of the lower urinary tract removes non-attached bacteria. Although umbrella cells are quite long-lived cells, they can be shed from the uroepithelium, thereby removing any attached bacteria. Several compounds act as anti-adherence factors, competitively inhibiting bacterial attachment to the bladder surface (Mulvey *et al.*, 2000). In addition, urine represents a poor medium for bacterial growth. The low pH, osmolarity, and presence of urea, salts, and organic acids in the urine can be inhibitory to bacterial growth, reducing bacterial survival in the bladder (Mulvey *et al.*, 2000).

Many epithelial cell types can sense the presence of microbial pathogens extracellularly by detecting highly conserved bacterial elements such as flagellin, lipopolysaccharide (LPS), and DNA motifs (CpG). Alternatively, inside the cytosol fragments of the invading bacterium peptidoglycan (such as muramyl dipeptides) can be detected. These recognized elements have been called pathogen-associated molecular patterns (PAMPs). Detection of PAMPs outside epithelial cells is performed by Toll-like receptors (TLRs) (Medzhitov and Janeway, 2000); in the cytosol the nucleotide oligomerization binding domain proteins (NODs) are devoted to this task (Philpott and Girardin, 2004). TLRs and NODs both activate the nuclear factor kappa B (NF-κB) transcription-dependent system for pro-inflammatory and anti-apoptotic factors synthesis (Akira and Takeda, 2004). Many observations have shown the importance of innate immunity, particularly through Toll-like receptor 4 (TLR-4)-mediated signaling, in response to the colonization of the bladder by UPEC (Anderson, Martin *et al.*, 2004). Upon UPEC infection, a rapid TLR-4-dependent signal occurs, ensuring the production and secretion of inflammatory mediators such as interleukin (IL) -8 and -6 (Godaly *et al.*, 1997; Hedges *et al.* 1992). This leads to a large influx of polymorphonuclear leukocytes (PMNs) into the lumen of the bladder, characteristic of UTIs (Svanborg *et al.*, 2001; Wullt *et al.*, 2003). In contrast to innate immunity, the role of the adaptative immunity for the defense of the urinary tract against UPEC is still largely unknown.

GENETIC SUPPORT FOR UROPATHOGENIC *ESCHERICHIA COLI* VIRULENCE

The genome sequence of the UPEC strain CFT073 has brought forth the definitive evidence that several UPEC PAIs constitute a multifactorial mechanism by which bacteria are able to colonize the unique niche of the urinary tract (Welch *et al.*, 2002). The PAIs of UPEC are among the best-understood examples of PAIs. These specific chromosomal pieces of DNA

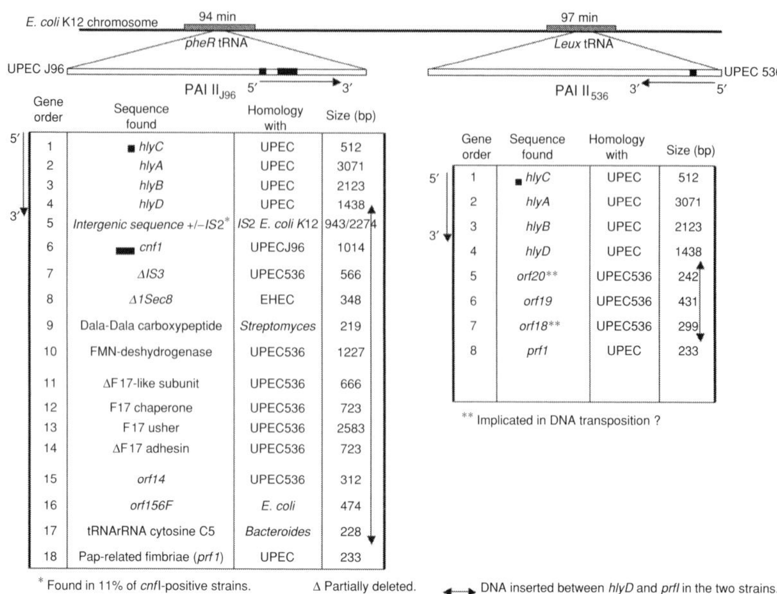

Gene order	Sequence found	Homology with	Size (bp)
1	■ *hlyC*	UPEC	512
2	*hlyA*	UPEC	3071
3	*hlyB*	UPEC	2123
4	*hlyD*	UPEC	1438
5	*Intergenic sequence +/–IS2**	*IS2 E. coli* K12	943/2274
6	▬ *cnf1*	UPECJ96	1014
7	*ΔIS3*	UPEC536	566
8	*Δ1Sec8*	EHEC	348
9	Dala-Dala carboxypeptide	*Streptomyces*	219
10	FMN-deshydrogenase	UPEC536	1227
11	ΔF17-like subunit	UPEC536	666
12	F17 chaperone	UPEC536	723
13	F17 usher	UPEC536	2583
14	ΔF17 adhesin	UPEC536	723
15	*orf14*	UPEC536	312
16	*orf156F*	*E. coli*	474
17	tRNArRNA cytosine C5	*Bacteroides*	228
18	Pap-related fimbriae (*prf1*)	UPEC	233

Gene order	Sequence found	Homology with	Size (bp)
1	■ *hlyC*	UPEC	512
2	*hlyA*	UPEC	3071
3	*hlyB*	UPEC	2123
4	*hlyD*	UPEC	1438
5	*orf20***	UPEC536	242
6	*orf19*	UPEC536	431
7	*orf18***	UPEC536	299
8	*prf1*	UPEC	233

Figure 14.2 Virulence factors contained within pathogenicity islands (PAIs) of uropathogenic *Escherichia coli* (UPEC) J96 and 536 in vicinity to the *Hly* operon. Position of the *hly*C gene in J96 UPEC was deduced from the cosmid study of Swenson *et al.* (1996) and in 536 UPEC from the work of Dobrindt *et al.* (2002). Genes found in J96 UPEC were sequenced by R. Clément, L. Landraud, and P. Boquet (unpublished data).

were first described in UPEC and diarrheagenic *E. coli* (Dobrindt *et al.*, 2004). PAIs are typically associated with tRNA genes, flanked by repeated sequences, and differ in guanine and cytosine (G/C) content and in codon usage from the core genome (Dobrindt *et al.*, 2004). All of these elements might be remnants of genes acquisitioned by horizontal transfer between different bacteria species. PAIs are present in pathogenic bacteria but absent in non-pathogenic strains, and they carry many virulence-associated genes. Encoded in these PAIs are several common genes described between different UPEC strains, such as adherence factors, toxins, and iron-uptake systems (Dobrindt *et al.*, 2004). Growing evidence suggests that virulence-associated PAI genes in UPEC strains confer not only the high capacity of the bacteria to attach to bladder epithelial cells but also their ability to invade the host cell. Invasion of host cells may indeed be a good protection of uropathogenic bacteria against the innate and acquired immune defenses and would also allow *E. coli* to persist inside host cells, thereby contributing to high recurrence rates of UTIs (Anderson, Martin *et al.*, 2004; Mulvey *et al.*, 1998). As an example, in

Figure 14.2 we have represented some genes and their organization within the PAIs of UPEC strains, J96 and 536, in the vicinity of the hemolysin (*hly*) operon.

EARLY EVENTS IN UROPATHOGENIC *ESCHERICHIA COLI* PATHOGENICITY

Adhesive appendages displayed on the surface of UPECs are pivotal for the colonization of the uroepithelium and modulate the host-cell responses towards the infecting pathogen. These adhesive appendages include the afimbrial adhesins (AfaE), and the fimbrial adhesins S-fimbriae, P-fimbriae, and type 1-pili.

AfaE adhesins are divided into two subgroups: AFA/Dr^{+} and Afa/Dr^{-}. Afa receptor AFA/Dr^{+} recognizes the Dr blood-group antigen, which is the decay-accelerating factor (DAF; also know as CD55) (Nowicki *et al.*, 1990). However, DAF is not present in the uroepithelium, therefore indicating that AFA/Dr^{+} adhesins do not play a role in UTIs. The Afa/Dr^{-} are the AfaE adhesins (termed AfaD) most frequently produced by isolates from patients with pyelonephritis and septicaemia (Garcia *et al.*, 2000). AfaD adhesins do not bind DAF but instead recognize β1-integrins (Plançon *et al.*, 2003).

S-fimbriae are associated with bladder and kidney infections. They are borne by *E. coli* strains predominately infecting newborns during delivery and causing neonatal meningitis (Bonacorsi *et al.*, 2000). The adhesin SfaS, localized at the tip of S-fimbriae, recognizes sialosyloligosaccharide residues and may aid in colonization of the upper urinary tract, particularly in UPEC devoid of P-fimbriae (see later) (Backhed *et al.*, 2002).

P-fimbriae and type 1 pili are the major UPEC-specific colonizing factors associated with UTIs (Kau *et al.*, 2005). In particular, type 1 pili are always present on UPEC. These proteinaceous appendages are encoded by a large operon of 9–12 genes, which encode three functional regions: regulation, major and minor subunits, and biogenesis and assembly via a specific secretion system (Hultgren *et al.*, 1996).

P-fimbriae and type 1 pili are fibers consisting of a rod-shaped structure of 6–7 nm in diameter. They are comprised of over 1000 structural major subunits of pilin (PapA for P-fimbriae, FimD for type 1 pili) and a tip fibrillum of 2–3 nm in diameter (composed of minor subunits) bearing the adhesin PapG (for P-fimbriae) or FimH (type 1 pili) (Kau *et al.*, 2005). The PapG adhesin of P-fimbriae confers tropism to the kidney ("P" indicating pyelonephritis). PapG binds the α-D-galactopyranosyl-(1–4)-β-D-galactopyranoside moiety present on glycolipids expressed in the kidney (Hultgren *et al.*, 1996).

The type 1 pili are the major adhesins implicated in UTIs, particularly cystitis (Hultgren *et al.*, 1985; Mobley *et al.*, 1987). The FimH adhesin recognizes mannosylated residues located at the surface of the uroplakins UPIa (Min *et al.*, 2002; Zhou *et al.*, 2001). The FimH is a bilobal molecule with a carboxy-terminal lobe involved in its incorporation into the type 1 pili and an amino-terminal adhesin lobe. The adhesin domain has an overall structure observed frequently in proteins involved in the binding of cell-surface receptors (Choudhury *et al.*, 1999). Binding of bacteria by type 1 pili prevents their detachment from the uroepithelium with the shear stress of the urine flow. The adhesin FimH is a shear-stress sensor that adapts its affinity to its host receptor by inducing a stronger binding of the bacterium when the shear-stress intensity increases (urine flushing) (Thomas *et al.*, 2002).

The different proteins required for P-fimbriae and type 1 pili building are transported and then assembled on the outer membrane. A specific secretion system called the chaperone/usher pathway is employed, in which a periplasmic chaperone binds and stabilizes pilus subunits in the periplasmic space, following membrane cytoplasmic translocation via the general secretory pathway (Hung and Hultgren, 1998). Assembly of pilin subunits is accomplished by the process of donor strand complementation (Kau *et al.*, 2005). In the periplasm, a chaperone protects the pilin subunit from prematurely coupling with other pilin molecules. The chaperone carries the pilin subunit to the large outer-membrane pore (the usher), where the pilin subunit is released by the chaperone and becomes attached to the end of the growing pilus (Choudhury *et al.*, 1999; Sauer *et al.*, 1999).

Expression of adhesin structures of UPEC are regulated at the transcriptional level. The various environmental signals encountered by the bacterium, including carbon source, temperature, osmolarity, pH, iron, aliphatic amino acids, and ion concentrations, are cues for molecular sensor and transcriptional regulators. Using an experimental mouse model of ascending UTI, it was shown that type 1 pili expression was markedly enhanced during the onset of in vivo infection and could vary during the time course of a urinary infection (Mobley, 2000; Struve and Krogfelt, 1999). More recently, the transcriptome analysis of the UPEC CFT073 strain during an experimental UTI in CBA/J mice showed that expression of type 1 pili was largely upregulated, whereas expression of other fimbrial genes such as P-fimbriae and S-fimbriae was downregulated (Snyder *et al.*, 2004). Variation of virulence factor expression, although controlled by numerous environmental signals, may allow pathogens to adapt from intestinal or vaginal niches to the urinary tract over the course of the UTI.

HOST RESPONSES TO BACTERIAL ADHERENCE

In the initial stages of the UTI, adhesive appendages contribute to the bacterial colonization of the urinary tract by interacting with their specific complementary receptors. However, this binding is followed by a host response, which can be either beneficial or detrimental to the invading pathogen.

P-fimbriated *E. coli* strains that specifically bind the abundant glycosphingolipid (GSL) receptor motif in uroepithelium and kidney tissue were isolated from symptomatic UTIs and septicemia initiated by UTI (Donnenberg and Welch, 1996). In the P-fimbria-dependent activation of the host response, a two-step model has been described. In the first step, bacterial adherence to the uroepithelium triggers cytokine production. In the second step, recruitment of PMNs and their migration to sites of infection occurs (Wullt *et al.*, 2003).

In uroepithelial cells, the specific attachment of PapG to GSLs triggers two independent signaling pathways converging on the activation of the serine/threonine (Ser/thr) family of protein kinases together with phosphatases and leading to cytokines production. In one pathway, P-fimbriae interact with GSLs and thus trigger the intracellular release of ceramides from the plasma membrane (Hedlund *et al.*, 1998). This mode of signaling mimics roles of the host immunoregulatory molecules, such as tumor necrosis factor alpha (TNF-α) and IL-1 (Schutze *et al.*, 1992). However, in contrast to these cytokines, signaling occurs independently of the activation of an endogenous sphingomyelinase activity (Hedlund *et al.*, 1998). Release of ceramides is thought to occur from P-fimbriae binding to GSL (Hedlund *et al.*, 1998), although the specific events leading to release is still unclear. In the second pathway, P-fimbriae induce the release of cytokines by the stimulation of TLR-4 without the involvement of LPS (Freudeus *et al.*, 2001), since the uroepithelium is void of CD14 (Hedlund *et al.*, 1999). Interestingly, GSL and TLR-4 were found colocalized in the cell membrane, suggesting that P-fimbriae can directly activate TLR-4 (Freudeus *et al.*, 2001). P-fimbriated strains of UPEC thus induce a strong symptomatic UTI with an overwhelming mucosal inflammatory response prevalent in patients with acute pyelonephritis, in contrast to asymptomatic bacteriuria (Connell *et al.*, 1997).

Binding of the PapG adhesin to GSLs on the uroepithelium also activates *airS* gene transcription. AirS (attachment and iron regulator sensor, also known as barA) protein is a sensor protein that belongs to a two-component regulatory system implicated in the control of bacterial iron acquisition (Pernestig *et al.*, 2000). This adhesin–receptor coupling can be a subtle strategy for the acquisition of iron by bacteria in the human host, where free iron is particularly low. With regard to this latter mechanism, P-fimbriae induce cross-talk between the bacterium and host in both directions.

The type 1 piliated *E. coli* strains have been shown to elicit the release of more cytokines than non-piliated isogenic bacteria. The adhesin FimH of type 1 pili not only mediates the attachment of bacteria to the uroepithelium but also provokes their internalization (Martinez *et al.*, 2000). Such internalization is absolutely required for the production of pro-inflammatory cytokines (Schilling *et al.*, 2001). FimH adhesin-mediated bacterial cell invasion requires localized host-actin reorganization, phosphoinositide 3 (PI3) kinase activation, and host-protein tyrosine phosphorylation, but not activation of Src-family tyrosine kinases (Martinez *et al.*, 2000). Phosphorylation of the focal adhesin kinase (FAK) at tyrosine 397 and the formation of FAK–PI3 kinase and alpha-actinin–vinculin complexes are also essential for FimH-induced bacterial internalization (Martinez *et al.*, 2000). More recently, it has been shown that FAK–PI3 kinase and alpha-actinin–vinculin complex formation requires activation of Cdc42 Rho-GTPase (Martinez and Hultgren, 2002). Interestingly, formation of alpha-actinin–vinculin complex could be induced independently from the FAK–PI3 kinase complex by activation of Rac-GTPase (Martinez and Hultgren, 2002; Martinez *et al.*, 2000). This suggests that FimH-mediated bacterial invasion is dependent on Cdc42 and PI3 kinase activation, upstream of Rac (Martinez and Hultgren, 2002). A novel macropinocytic pathway dependent on Cdc42, but not on Rac, has been shown for the internalization of glycosylphosphatidylinasitol (GPI)-anchored proteins and lipid rafts (Sabharanjak *et al.*, 2002). Taking into account the fact that Cdc42 (Martinez and Hultgren, 2002) and lipid rafts (Duncan *et al.*, 2004) seem to be important for FimH-mediated bacterial internalization, a similar mechanism of bacterial macropinocytosis other than GPI-anchored proteins (APs) might be operating by type 1 pili. Activation of Rac1 and Cdc42 by the bacteria may be responsible for the downstream activation of the NF-κB pathway and production of pro-inflammatory cytokines through a mechanism described downstream of TLR-2 signaling (Arbibe *et al.*, 2000).

Cell invasion by type 1 piliated UPEC induces the normal host-defense response of shedding of uroepithelial cells by a caspase-mediated apoptosis-like mechanism (Mulvey *et al.*, 1998). FimH-induced bladder uroepithelium exfoliation together with robust production of inflammatory mediators is not sufficient to completely clear UPEC from the urinary tract. Therefore, bacteria persist in this niche (Mulvey *et al.*, 2001). One explanation for this bacterial persistence has been attributed to a complex differentiation pathway of intracellular UPEC (Justice *et al.*, 2004). This includes the formation inside the umbrella cells of a biofilm by the invading bacteria, followed by their filamentation, which ultimately leads to the formation of so-called filamentous intracellular bacterial communities (IBCs) (Justice *et al.*, 2004). These IBCs are thought to facilitate bacterial persistence by being released in the bladder lumen (Kau *et al.*, 2005).

The afimbrial AfaD adhesin has been shown to promote UPEC internalization in cultured cells at a low level (Plançon *et al.*, 2003). Internalization of bacteria triggered by AfaD requires the presence of β1-integrins on the cells, and therefore AfaD qualifies as an invasin (Plançon *et al.*, 2003). Once in the cytoplasm, phagocytosed AfaD-UPEC stay alive for at least 72 h, suggesting that they could form bacterial reservoirs for recurrent infections (Plançon *et al.*, 2003). The β1-integrins are located on the basolateral domain of the uroepithelium cells and therefore are not accessible to AfaD-bacteria unless tight junctions are ruptured.

BACTERIAL TOXINS IN UROPATHOGENIC *ESCHERICHIA COLI* PATHOGENESIS

Major toxins

α-Hemolysin

α-Hemolysin (Hlyα) is the most common toxin encountered in extraintestinal *E. coli* infections and is associated with more than half of *E. coli* strains responsible for pyelonephritis (Donnenberg and Welch, 1996). The gene (*hlyA*) encoding Hlyα is included in an operon containing four genes (*hlyCABD*). The *hlyC* gene encodes an acyltransferase (HlyC) necessary for HlyA activation; *hlyB* and *hlyD* encode the transporter proteins HlyB and HlyD for Hlyα plasma membrane export by the type I secretion system (Felmlee *et al.*, 1985). Hlyα (molecular weight 109 kDa), released from the periplasm into the external medium via the outer-membrane protein TolC porin, belongs to the RTX (repeat in toxin) toxin family produced by several Gram-negative bacteria. The RTX toxins are characterized by a carboxy-terminal domain composed of several repeated amino-acids motifs (Lally *et al.*, 1999). The repeated domain contains a motif of nine amino acids rich in glycine and aspartic acid, which may be required for the binding of calcium and the interaction with the eukaryotic cell target membrane. Hlyα is able to form transmembrane pores of 1–2 nm in diameter in lipid bilayers in a calcium-dependent manner (Bakas *et al.*, 1998). The role of Hlyα in UTI pathogenesis was quite obscure until it was shown that Hlyα interacts with the plasma-membrane L-type calcium channels of kidney epithelial cells, inducing intracellular Ca^{2+}

oscillations and leading to the release of IL-8 and IL-6 (Söderblom *et al.*, 2003), thus contributing to host inflammation.

Cytotoxic necrotizing factor type 1

Described first as a dermonecrotic toxin upon injection into the rabbit skin (Caprioli *et al.*, 1983), cytotoxic necrotizing factor type 1 (CNF1) was next observed to be the first toxin that induces the formation of a dense F-actin cytoskeleton in cultured cells (Fiorentini *et al.*, 1988). CNF1 is found in one-third of all UPEC strains (Landraud *et al.*, 2000) and in some K1 *E. coli* strains (Bonacorsi *et al.*, 2000; Johnson *et al.*, 2001). In UPEC, the *cnf1* gene is chromosomally encoded by a single 3042-bp gene yielding a 1014-residue protein (molecular weight 108 kDa) (Falbo *et al.*, 1993). In all UPEC strains producing CNF1, *cnf1* is encoded in a PAI (PAI II) and located between the α-hemolysin operon and genes encoding Pap-related adhesin (*prs*) (Blum *et al.*, 1995) (Figure 14.2). A tight association between Hlyα-producing UPEC and the presence of CNF1 has been observed. The production of CNF1 in *E. coli* was found to be coregulated with that of proteins of the hlyα operon via the RfaH protein, a factor that allows the progression of RNA polymerase on long DNA stretches (Landraud *et al.*, 2003). This interesting observation might be a clue to the long-term enigma concerning the secretion of CNF1 by the bacterium. Indeed, to date there is no mechanism that explains how CNF1 might be secreted by *E. coli* (Boquet, 2001). The coupling between the production of proteins of the *hly* operon and that of CNF1 might be required for co-utilization of the Hlyα-secretion system for both Hlyα and CNF1. Interestingly, the C-terminus of CNF1 bears some structural homology with the C-terminal signal sequence of Hlyα required for the hemolysin to recognize its secretory apparatus (L. Landraud and P. Boquet, unpublished data).

The CNF1 toxin is structurally organized into three functional domains and belongs to the A-B (activity, binding) type of toxins (Boquet, 2001). The N-terminal third of the toxin contains the receptor-binding domain (Lemichez *et al.*, 1997), the middle domain harbors two hydrophobic helices that are involved in the CNF1 membrane insertion and translocation of acidified endosomes (Pei *et al.*, 2001), and the C-terminal of the protein contains the catalytic domain of the toxin (Lemichez *et al.*, 1997). The CNF1 cell-surface receptor has been proposed to be the laminin-receptor precursor (Chung *et al.*, 2003; Kim *et al.*, 2005). The CNF1 receptor is localized on the basolateral domain of cells (Hopkins *et al.*, 2002). Therefore, CNF1 cannot interact with an intact uroepithelium with umbrella cells due to the restriction of tight junctions. CNF1 is endocytosed by both clathrin-dependent and -independent mechanisms and is not associated with lipid rafts (L. Sisteron and P. Boquet, unpublished data). CNF1 injects its catalytic domain from late endosomes into the cytosol by a translocation mechanism similar to that described for diphtheria toxin (Contamin *et al.*, 2000). CNF1 catalyzes by deamidation the post-translational modification of the glutamine 63/61 into glutamic acid of Rho, Rac, and Cdc42 GTPases (Flatau *et al.*, 1997; Lerm *et al.*, 1999; Schmidt *et al.*, 1997), leading to the permanent activation of the GTPases.

Rho-GTPases are molecular switches for various signaling pathways and are expressed ubiquitously in eukaryotic cells (Takai *et al.*, 2001). Rho-GTPases oscillate between a GTP-bound active form targeted to specific membrane locations and a GDP-bound inactive form sequestered inside the cytosol and associated with a Rho-GTPase dissociation inhibitor (GDI) (Takai *et al.*, 2001). Rho-GTPases are activated by growth factors, where they bind their cell-surface receptors through their loading with GTP by a guanine exchange factor (GEF) and return to their non-activated form upon hydrolysis of their GTP into GDP by a GTPase-associated protein (GAP) (Takai *et al.*, 2001). Active Rho-GTPases can bind multiple effectors, playing major roles in regulation, including the organization of the actin cytoskeleton, differentiation, division, migration, and cell polarity (Moon and Zheng, 2003).

Rho-GTPases are major targets for microbial virulence factors. Thirty bacterial proteins have been documented to interfere directly or indirectly with Rho, Rac, and Cdc42 (Boquet and Lemichez, 2003). Activation of Rho-GTPases by CNF1 is followed by deactivation of these regulatory factors via their degradation by the ubiquitin-proteasome pathway (Doye *et al.*, 2002; Lerm *et al.*, 2002). Although it is known that CNF1 induces the internalization of bacteria by cells (Falzano *et al.*, 1993), the maximum bacterial phagocytosis promoted by the toxin is observed to correlate with the lowest level of intracellular activated Rho-GTPases due to their proteasomal degradation (Doye *et al.*, 2002).

Activation of Rac1 and Cdc42 can stimulate the NF-κB pathway (Arbibe *et al.*, 2000), resulting in an inflammatory and anti-apoptotic response (Akira and Takeda, 2004). Accordingly, CNF1 activates NF-κB (Boyer *et al.*, 2004), induces the formation of pro-inflammatory mediators (Falzano *et al.*, 2003; Munro *et al.*, 2004), and blocks apoptosis (Fiorentini *et al.*, 1998). Proteolysis of Rac1 and Cdc42, following their activation by CNF1, may limit the host inflammatory response together with the cell resistance to apoptotic signals, and promote bacterial phagocytosis.

Minor toxins

Toxins of the serine protease autotransporter of the Enterobacteriaceae (SPATE) subfamily have been identified in UPEC strains. Autotransporter proteins, described previously for secretion of the gonococcal immunoglobulin A1 (IgA1) proteases (Pohlner *et al.*, 1987), are an expanding group secreted via the type V secretion pathway (Henderson *et al.*, 1998) and implicated in diseases (Henderson and Nataro, 2001). Following the discovery and characterization of the autotransporter toxin Sat (secreted autotransporter toxin) in UPEC (Guyer *et al.*, 2000), a cytotoxin homologous to the Pic serine protease autotransporter of *Shigella flexneri* and enteroaggregative *E. coli*, termed PicU (Pic of UPEC), was characterized among ten autotransporter proteins encoded in the UPEC CFT073 genome (Parham *et al.*, 2004; Welch, *et al.*, 2002). Isolated in 22% of UPEC strains compared with only 12% of rectal isolates, PicU acts by degrading mucins and may confer some colonization advantage to bacteria (Parham *et al.*, 2004).

SUBVERSIONS OF HOST RESPONSES

Similar to FimH and AfaD, the toxin CNF1 may contribute to the virulence of UPEC (Rippere-Lampe *et al.*, 2001) by provoking phagocytosis of bacteria in host cells (Doye *et al.*, 2002; Falzano *et al.*, 1993). The coregulation of CNF1 and Hlyα expression suggests a synergistic activity of these toxins in the pathogenesis of UTIs (Landraud *et al.*, 2003). Hlyα might play the role of the listeriolysin O of *Listeria monocytogenes* for the rupture of the bacterial phagolysosome and the release of bacteria into the cytosol (Dussurget *et al.*, 2004), as suggested by Landraud *et al.* (2003). However, unlike *L. monocytogenes, E. coli* does not multiply in the cell cytosol (Goetz *et al.*, 2001). Therefore, it seems that there is no advantage for phagocytosed UPEC to be released in the cytosol and, thus, UPEC may reside in the vacuole. The tight coupling between Hlyα and CNF1 might be required only for the secretion of CNF1.

With regard to the activity of FimH, we can envisage a scenario in which FimH, by binding to uroplakin Ia on the apical domain of umbrella cells, first triggers the bacterial invasion of these cells (Figure 14.3). This leads to the formation of IBCs followed by the shedding of infected umbrella cells and the release of IBCs. Shedding of umbrella cells will break the permeability barrier of the bladder uroepithelium, allowing CNF1 (or AfaD-bearing UPEC) to reach their respective basolateral receptors. CNF1, by stimulating the Rho-GTPases, will promote phagocytosis of bacteria into the various uroepithelium cells but also may stimulate via NF-κB the release of inflammatory

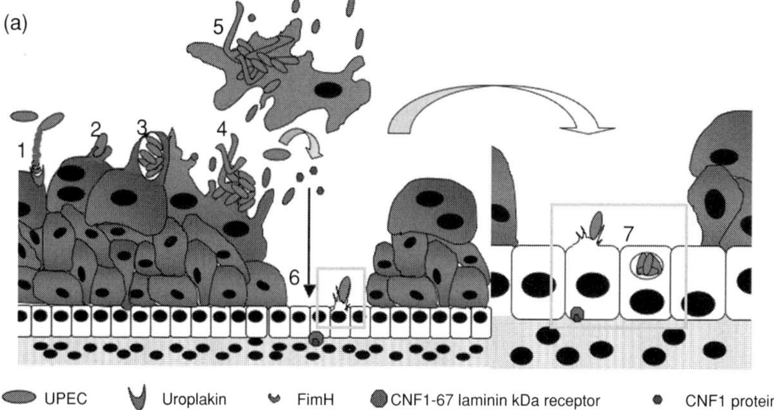

Figure 14.3 Possible cumulative roles of FimH and cytotoxic necrotizing factor type 1 (CNF 1) in uropathogenic *Escherichia coli* (UPEC) persistence. (a) (1) FimH by binding to uroplakin Ia will stimulate the phosphoinositide 3 (PI3) kinase, leading to phosphorylation of focal adhesin kinase (FAK) and activation of Rho-GTPases. (2) This will induce actin polymerization and bacterial engulfment. (3) Within the umbrella cells, internalized UPEC form biofilm-like communities and filamentation, leading to the formation of intracellular bacterial communities (IBCs) (Kau *et al.*, 2005). (4) IBCs can be released in the bladder lumen. (5) Infected umbrella cells can exfoliate and induce apoptosis. Apoptosis of umbrella cells can break open the uroepithelium, (6) leaving access to the CNF1 of the urothelium deep layers and allowing the toxin to bind its receptor. (7) UPEC penetrating in the deep layers can trigger their internalization via either FimH or CNF1 mechanisms, or both. (b) (8) Stimulation of Rho-GTPases by CNF1 (9) followed by the degradation of these regulatory molecules can induce internalization of UPEC (Doye *et al.*, 2002), but also it may limit the inflammatory process elicited by activation of the nuclear factor kappa B (NF-κB) pathway by Rac1 and Cdc42 overstimulation by CNF1 (Munro *et al.* 2004).

cytokines (Figure 14.3b). Internalized UPEC may constitute bacterial reservoirs out of reach of the host innate and acquired immune defenses and antibiotics, leading to recurrent infections. CNF1 and AfaD may be alternative virulence factors to FimH that may intensify UPEC invasion. On the other hand, during the course of a UTI, CNF1 or AfaD may play an important role in bacterial cell internalization when type 1 pili are poorly expressed.

CONCLUSIONS

Clearly, innate immunity plays a pivotal role in the development of UTIs. This host defense system presents specific characteristics in urinary uroepithelium, namely (i) an absence of CD14 protein, which could avoid

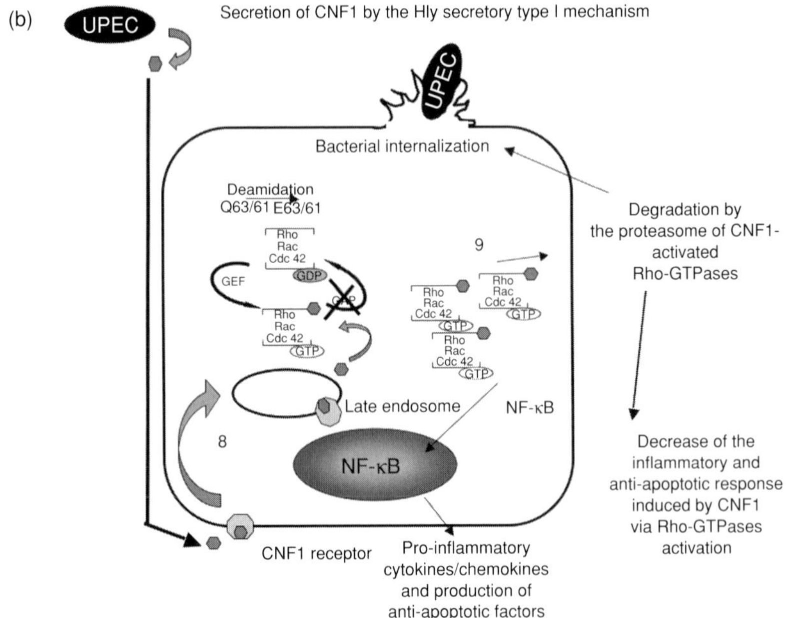

Figure 14.3 *(cont.)*

a massive response to LPS, and (ii) the presence of different TLRs, such as TLR-11 (so far found only in murine kidneys) (Zhang *et al.*, 2004), which may contribute to a specific response. However, in spite of an extremely effective innate host defense system, a significant population of UPEC does persist by bacterial invasion of the uroepithelium cells induced by FimH, CNF1, or AfaD. These bacterial reservoirs, together with the release of IBCs, might explain the high propency of UTI recurrences. Indeed, after an initial episode of UTI, one in five women will experience an early recurrence of the infection (Foxman, 2002; Hooton *et al.*, 1997; Madersbacher *et al.*, 2000). This represents a major public-health issue, leading to a high consumption of antibiotics and perhaps contributing to the occurence of bacterial resistance.

REFERENCES

Akira, S. and Takeda K. (2004). Toll-like receptor signaling. *Nat. Rev. Immunol.* **4**, 499–511.

Anderson, G. G., Martin, S. M., and Hultgren, S. J. (2004). Host subversion by formation of intracellular bacterial communities in the urinary tract. *Microbes Infect.* **6**, 1095–1101.

Anderson, J. M., Van Italie, C. M., and Fanning, A. S. (2004). Setting up a selective barrier at the apical junction complex. *Curr. Opin. Cell Biol.* **16**, 140–145.

Apodaca, G. (2004). The uroepithelium: not just a passive barrier. *Traffic* **5**, 117–128.

Arbibe, L., Mira, J. P., Teusch, N., *et al.* (2000). Toll-like receptor 2-mediated NF-kappa B activation requires a Rac-1 dependent pathway. *Nat. Immunol.* **1**, 533–540.

Backhed, F., Alsen, B., Roche, N., *et al.* (2002). Identification of target tissue glycosphingolipid receptors for uropathogenic F1C-fimbriated *Escherichia coli* and its role in mucosal inflammation. *J. Biol. Chem.* **277**, 18 198–18 205.

Bakas, L., Viega, M. P., Soloaga, H., Ostolaza, H., and. Goni, F. M. (1998). Calcium-dependent conformation of the *E. coli* alpha-hemolysin: implications for the mechanism of membrane insertion and lysis. *Biochim. Biophys. Acta.* **1368**, 225–234.

Blum, G., Falbo, V., Caprioli, A., and Hacker J. (1995). Gene clusters encoding the cytotoxic necrotizing factor type 1, Prs-fimbriae and alpha-hemolysin form the pathogenicity island II of the uropathogenic *Escherichia coli* strain J96. *FEMS Microbiol. Lett.* **126**, 189–195.

Bonacorsi, S. P., Clermont, O., Tinsley, C., *et al.* (2000). Identification of regions of the *Escherichia coli* chromosome specific for neonatal meningitis-associated strains. *Infect. Immun.* **68**, 2096–2101.

Boquet, P. (2001). The cytotoxic necrotizing factor 1 (CNF1) from *Escherichia coli*. *Toxicon* **39**, 1673–1680.

Boquet, P. and Lemichez, E. (2003). Bacterial virulence factors targeting RhoGTPases: parasitism or symbiosis. *Trends Cell Biol.* **13**, 238–246.

Boyer, L., Travaglione, S., Falzano, L., *et al.* (2004). Rac GTPase instructs nuclear factor-kB activation by conveying the SCF complex and IkBα to the ruffling membranes. *Mol. Cell Biol.* **15**, 1124–1133.

Caprioli, A., Falbo, V., Roda, L. G., Ruggeri, F. M., and Zona, C. (1983). Partial purification and characterization of an *Escherichia coli* toxic factor that induces morphological cell alterations. *Infect. Immun.* **39**, 1300–1306.

Choudhury, D., Thompson, A., Stojanoff, V., *et al.* (1999). X-ray structure of the FimC–FimH chaperone–adhesin complex from uropathogenic *Escherichia coli*. *Science* **285**, 1061–1066.

Chung, J. W., Hong, S. J., Kim, K. J., *et al.* (2003). The 37-kDa laminin receptor precursor modulates cytotoxic necrotizing factor 1-mediated RhoA activation and bacterial uptake. *J. Biol. Chem.* **278**, 16 857–16 862.

Connell, H., Hedlund, M., Agace, W., and Svanborg. C. (1997). Bacterial attachment to uro-epithelial cells: mechanisms and consequences. *Adv. Dent. Res.* **11**, 50–58.

Contamin, S., Galmiche, A., Doye, A., *et al.* (2000). The p21-Rho-activating toxin cytotoxic necrotizing factor1 is endocytosed by clathrin-independent mechanism and enters the cytosol by an acidic dependent membrane translocation step. *Mol. Biol. Cell* **11**, 1775–1787.

Dobrindt, U., Blum-Oehler, G., Nagy, G., *et al.* (2002). Genetic structure and distribution of four pathogenicity islands (PAI I(536) to PAI IV(536)) of uropathogenic *Escherichia coli* strain 536. *Infect. Immun.* **70**, 6365–6372.

Dobrindt, U., Hochhut, B., Hentschel, U., and Hacker, J. (2004). Genomic islands in pathogenic and environmental microorganisms. *Nat. Rev. Microbiol.* **2**, 414–424.

Donnenberg, M. S. and Welch, R. A. (1996). Virulence determinants in uropathogenic *E. coli*. In *Urinary Tract Infections: Molecular Pathogenesis and Clinical Management*, ed. H. T. L. Mobley and J. Warren. Washington, DC: American Society of Microbiology, pp. 135–174.

Doye, A., Mettouchi, A., Bossis, G., *et al.* (2002). CNF1 exploits the ubiquitin-proteasome machinery to restrict RhoGTPase activation for bacterial host cell invasion. *Cell* **111**, 553–564.

Duncan, M. J., Li, G., Shin, J. S., Carson, J. L., and Abdaham, S. N. (2004). Bacterial penetration of bladder epithelium through lipid rafts. *J. Biol. Chem.* **279**, 18 944–18 951.

Dussurget, O., Pizarra-Cerda, J., and Cossart, P. (2004). Molecular determinants of *Listeria monocytogenes* virulence. *Annu. Rev. Microbiol.* **58**, 587–610.

Eriksen, H. M., Iversen, B. J., and Aavitsland, P. (2004). Prevalence of nosocomial infections and use of antibiotics in long-term care facilities in Norway, 2002 and 2003. *J. Hosp. Infect.* **57**, 316–320.

Falbo, V., Pace, T., Picci, L., Pizzi, E., and Caprioli, A. (1993). Isolation and nucleotide sequence of the gene encoding cytotoxic necrotizing factor 1 of *Escherichia coli*. *Infect. Immun.* **61**, 4909–4914.

Falzano, L., Fiorentini, C., Donelli, G., *et al.* (1993). Induction of a phagocytic behaviour in human epithelial cells by *Escherichia coli* cytotoxic necrotizing factor type 1. *Mol. Microbiol.* **9**, 1247–1254.

Falzano, L., Quaranta, M. G., Travaglione, S., *et al.* (2003). Cytotoxic necrotizing factor 1 enhances reactive oxygen species-dependent transcription and secretion of proinflammatory cytokines in human uroepithelial cells. *Infect. Immun.* **71**, 4178–4181.

Felmlee, T., Pelett, S., and Welch, R. A. (1985). Nucleotide sequence of an *Escherichia coli* chromosomal hemolysin. *J. Bacteriol.* **163**, 94–105.

Fiorentini, C., Arancia, G., Caprioli, A., *et al.* (1988). Cytoskeletal changes induced in HEp-2 cells by the cytotoxic necrotizing factor 1 of *Escherichia coli*. *Toxicon* **26**, 1047–1056.

Fiorentini, C., Matarrese, P., Straface, E., *et al.* (1998). Rho-dependent cell spreading activated by *E. coli* cytotoxic necrotizing factor 1 hinders apoptosis in epithelial cells. *Cell Death Diff*. **5**, 921–929.

Flatau, G., Lemichez, E., Gauthier, M., *et al.* (1997). Toxin-induced activation of the G-protein p21 Rho by deamidation of glutamine. *Nature* **347**, 729–733.

Foxman, B. (2002). Epidemiology of urinary tract infections: incidence, morbidity and economic costs. *Am. J. Med.* **113**, 5–13.

Freudeus, B., Wachtler, C., Hedlund, M., *et al.* (2001). *Escherichia coli* P fimbriae utilize the toll-like receptor 4 pathway for cell activation. *Mol. Microbiol.* **40**, 37–51.

Garcia, M. L., Jouve, M., Nataro, J. P., Gounon, P., and Le Bouguenec, C. (2000). Characterization of the AfaD-like family of invasins encoded by pathogenic *Escherichia coli. FEBS Lett.* **479**, 111–117.

Godaly, G., Hedges, S., Proudfoot, A., *et al.* (1997). Role of epithelial interleukin-8 (Il-8) and neutrophil IL-8 receptor A in *Escherichia coli*-induced transuroepithelial neutrophil migration. *Infect. Immun.* **65**, 3451–3456.

Goetz, M., Bubert, A., Wang, G., *et al.* (2001). Microinjection and growth of bacteria in the cytosol of mammalian host cells. *Prot. Natl. Acad. Sci. U. S. A.* **98**, 2221–2226.

Guyer, D. M., Henderson, I. R., Nataro, J. P., and Mobley, H. L. (2000). Identification of sat, an autotransporter toxin produced by uropathogenic *Escherichia coli. Mol. Microbiol.* **38**, 53–66.

Hedges, S., Svensson, M., and Svanborg C. (1992). Interleukin-6 response of epithelial cell lines to bacterial stimulation in vitro. *Infect. Immun.* **60**, 1295–1301.

Hedlund, M., Duan, R. D., Nilsson, A., and Svanborg C. (1998). Sphingomyelin, glycophingolipids and ceramide signalling in cells exposed to P-Fimbriated *Escherichia coli. Mol. Microbiol.* **29**, 1297–1306.

Hedlund, M., Wachtler, C., Johansson, E., *et al.* (1999). P fimbriae-dependent, lipopolysaccharide-independent activation of epithelial cytokine responses. *Mol. Microbiol.* **33**, 693–703.

Henderson, I. R. and Nataro, J. P. (2001). Virulence factors of autotransporter proteins. *Infect. Immun.* **69**, 1231–1243.

Henderson, I. R., Navarro-Garcia, F., and Nataro, J. P. (1998). The great escape: structure and function of the autotransporter proteins. *Trends Microbiol.* **6**, 370–378.

Hooton, T. M. and Stamm, W. E. (1997). Diagnosis and treatment of uncomplicated urinary tract infection. *Infect. Dis. Clin. North Am.* **11**, 551–581.

Hopkins, A. M., Walsh, S. V., Varkade, P., Boquet, P., and Nusrat. A. (2002). Constitutive activation of Rho proteins by CNF1 influences tight junction structure and epithelial barrier function. *J. Cell Science* **116**, 725–742.

Hultgren, S. J., Porter, T. N., Schaeffer, A. J., and Duncan. J. L. (1985). Role of the type 1 pili and effects of phase variation on lower urinary tract infections produced by *Escherichia coli*. *Infect. Immun.* **50**, 370–377.

Hultgren, S. J., Jones, S. H., and Normark, S. (1996). Bacterial adhesins and their assembly. In Escherichia coli *and* Salmonella: Cellular and Molecular Biology, ed. F. C. Neidhardt. Washington. DC: American Society of Microbiology, pp. 2730–2756.

Hung, D. L. and Hultgren, S. J. (1998). Pilus biogenesis via the chaperone/user pathway: an interaction of the structure and the function. *J. Struct. Biol.* **124**, 201–220.

Johansen, T. E. B. (2004). Nosocomially acquired urinary tract infections in urology departments: why an international prevalence study is needed in urology. *Int. J. Antimicrobial Agents* **23**, 30–34.

Johnson, J. R., Delavari P., and O'Bryan, T. T. (2001). *Escherichia coli* O18:K1:H7 isolates from patients with acute cystitis and neonatal meningitis exhibit common phylogenetic origins and virulence factor profiles. *J. Infect. Dis.* **183**, 425–434.

Justice, S. S., Hung, C., Theriot, J. A., *et al.* (2004). Differentiation and developmental pathways of uropathogenic *Escherichia coli* in urinary tract pathogenesis. *Prot. Natl. Acad. Sci. U. S. A.* **101**, 1333–1338.

Kau, A. L., Hunstad, D. A., and Hultgren, S. J. (2005). Interaction of uropathogenic *Escherichia coli* with host uroepithelium. *Curr. Opin. Microbiol.* **8**, 54–59.

Kim, K. J., Chung, J. W., and Kim, K. S. (2005). The 67-kDa laminin receptor promotes internalization of cytotoxic necrotizing factor 1-expressing *Escherichia coli* K1 into human brain microvascular endothelial cells. *J. Biol. Chem.* **280**, 1360–1368.

Lafont, F., Abrami, L., and van der Goot, F. G. (2004). Bacterial subversion of lipid rafts. *Curr. Opin. Microbiol.* **7**, 4–10.

Lally, E. T., Hill, R. B., Kieba, I. R., and Korostoff, J. (1999). The interaction between RTX toxin and target cells. *Trends Microbiol.* **7**, 356–361.

Landraud, L., Gauthier, M., Fosse, T., and Boquet. P. (2000). Frequency of *Escherichia coli* strains producing the cytotoxic necrotizing factor (CNF1) in nosocomial urinary tract infection. *Lett. Applied. Microbiol.* **30**, 213–218.

Landraud, L., Gibert, M. Popoff, M. R., Boquet, P., and Gauthier, M. (2003). Expression of cnf1 by *Escherichia coli* J96 involves a large uptstream DNA region including the hlyCABD operon, and is regulated by the RfaH protein. *Mol. Microbiol.* **47**, 1653–1667.

Lemichez, E., Flatau, G., Bruzzone, M., Boquet, P., and Gauthier, M. (1997). Molecular localisation of the *Escherichia coli* cytotoxic necrotizing factor CNF1 cell-binding and catalytic domains. *Mol. Microbiol.* **24**, 1061–1070.

Lerm, M., Selzer, J., Hoffmeyer, A., *et al.* (1999). Deamidation of Cdc42 and Rac by *Escherichia coli* cytotoxic necrotizing factor 1: activation of C-Jun N-terminal kinase in Hela cells. *Infect. Immun.* **67**, 496–503.

Lerm, M., Pop, M., Fritz, G., Aktories, K., and Schmidt, G. (2002). Proteasomal degradation of cytotoxic necrotizing factor 1-activated rac. *Infect. Immun.* **70**, 4053–4058.

Madersbacher, S., Thalhammer, F., and Marberger, M. (2000). Pathogenesis and management of recurrent urinary tract infecton in women. *Curr. Opin. Urol.* **10**, 29–33.

Martinez, J. J. and Hultgren, S. J. (2002). Requirement of Rho-family GTPases in the invasion of type 1-piliated uropathogenic *Escherichia coli*. *Cell. Microbiol.* **4**, 19–28.

Martinez, J. J., Mulvey, M. A., Schilling, J. D., Pinkner, J. S., and Hultgren. S. J. (2000). Type 1 pilus-mediated bacterial invasion of bladder epithelial cells. *EMBO J.* **12**, 2803–2812.

McCormick, B. A. (2003.) The use of transepithelial models to examine host–pathogen interactions. *Curr. Opin. Microbiol.* **6**, 77–81.

Medzhitov, R. and Janeway, C. (2000). Innate Immunity. *N. Engl. J. Med.* **343**, 338–344.

Min, G., Solz, M., Zhou, G., *et al.* (2002). Localization of uroplakin 1a, the urothelium receptor for bacterial adhesin FimH, on the six inner domains of the 16 nm urothelium plaque particle. *J. Mol. Biol.* **317**, 697–706.

Mobley, H. L. (2000). Virulence of the two primary uropathogens. *ASM News* **66**, 403–410.

Mobley, H. L., Chippendale, G. R., Tenney, J. H., Hull, R. A., and Warren, J. W. (1987). Expression of the type 1 fimbriae may be required for persistence of *Escherichia coli* in the catheterized urinary tract. *J. Clin. Microbiol.* **25**, 2253–2257.

Moon, S. Y. and Zheng, Y. (2003). Rho GTPases-activiting proteins in cell regulation. *Trends Cell Biol.* **13**, 13–22.

Mulvey, M. A., Lopez-Boado, Y. S., Wilson, C. L., *et al.* (1998). Induction and evasion of host defenses by type 1-piliated uropathogenic *Escherichia coli*. *Science* **282**, 1494–1497.

Mulvey, M. A., Schilling, J. D., Martinez, J. J., and Hultgren, S. J. (2000). Bad bugs and beleaguered bladders: interplay between uropathogenic *Escherichia coli* and innate host defenses. *Proc. Natl. Acad. Sci. U. S. A.* **97**, 8829–8835.

Mulvey, M. A., Schilling, J. D., and Hultgren, S. J. (2001). Establishment of a persistent *Escherichia coli* reservoir during the acute phase of a bladder infection. *Infect. Immun.* **69**, 4572–4579.

Munro, P., Flatau, G., Doye, A., *et al.* (2004). Activation and proteasomal degradation of Rho GTPases by cytotoxic necrotizing factor-1 elicit a controlled inflammatory response. *J. Biol. Chem.* **279**, 35 849–35 857.

Nowicki, B., Labigne, A., Moseley, S., *et al.* (1990). The Dr hemagglutinin, afimbrial adhesins AFA-1 and AFA-III, and F1845 fimbriae of uropathogenic and diarrhea-associated *Escherichia coli* belong to a family of hemagglutins with Dr receptor recognition. *Infect. Immun.* **58**, 279–281.

Parham, N. J., Srinivasan, U., Desvaux, M., *et al.* (2004). PicU, a second serine protease autotransporter of uropathogenic *Escherichia coli*. *FEMS Microbiol. Lett.* **230**, 73–83.

Pei, S., Doye, A., and Boquet, P. (2001). Mutation of specific acidic residues of the CNF1 T domain into lysine alters cell membrane translocation of the toxin. *Mol. Microbiol.* **41**, 1237–1247.

Pernestig, A. K., Normark, S. J., Georgellis, D., and Melefors, O. (2000). The role of the AirS two-component system in uropathogenic *Escherichia coli*. *Adv. Exp. Med. Biol.* **485**, 137–142.

Philpott, D. J. and Girardin, S. E. (2004). The role of Toll-like receptors and Nod proteins in bacterial infection. *Mol. Immunol.* **41**, 1099–1108.

Plançon, L., du Merle, L., le Friec, S., *et al.* (2003). Recognition of the cellular β1-chain integrin by the bacterial AfaD invasin is implicated in the internalization of afa-expressing pathogenic *Escherichia coli* strains. *Cell. Microbiol.* **5**, 681–693.

Pohlner, J., Halter, R., Beyreuther, K., and Meyer, T. F. (1987). Gene structure and extracellular secretion of *Neisseria gonorrhoeae* IgA protease. *Nature* **325**, 458–462.

Rippere-Lampe, K. E., Brien, A. O., Conran, R., and Lockman, H. A. (2001). Mutation of the gene encoding cytotoxic necrotizing factor type 1 (cnf1) attenuates the virulence of uropathogenic *Escherichia coli*. *Infect. Immun.* **69**, 3954–3964.

Ronald, A. (2003). The etiology of urinary tract infections: traditional and emerging pathogens. *Dis. Mon.* **49**, 71–82.

Russo, T. A. and Johnson, J. R. (2003). Medical and economic impact of extraintestinal infectons due to *Escherichia coli*: focus on an increasing important endemic problem. *Microbes Infect.* **5**, 449–456.

Sabharanjak, S., Sharma, P., Parton, R. G., and Mayor, S. (2002). GPI-anchored proteins are delivered to recycling endosomes via a distinct Cdc42-regulated, clathrin-independent pinocytic pathway. *Dev. Cell* **2**, 411–423.

Sauer, F. G., Fütterer, K., Pinkner, J. S., *et al.* (1999). Structural basis of chaperone function and pilus biogenesis. *Science* **285**, 1058–1061.

Schilling, J. D., Mulvey, M. A., Vincent, C. D., Lorenz, R. G., and Hultgren, S. J. (2001). Bacterial invasion augments epithelial cytokine response to *Escherichia coli* through a lipopolysaccharide-dependent mechanism. *J. Immunol.* **166**, 1148–1155.

Schmidt, G., Sehr, P., Wilm, M., *et al.* (1997). Gln 63 of Rho is deamidated by *Escherichia coli* cytotoxic necrotizing factor-1. *Nature* **387**, 725–729.

Schutze, S., Potthoff, K., Machleidt, T., *et al.* (1992). TNF activates NF-kappa B by phosphatidylcholine-specific phospholipase C-induced "acidic" sphingolmyelin breakdown. *Cell* **71**, 765–776.

Simons, K. and Ikonen, E. (1997). Functional rafts in cell membranes. *Nature* **387**, 569–572.

Snyder, J. A., Haugen, B. J., Buckles, E. L., *et al.* (2004). Transcriptome of uropathogenic *Escherichia coli* during urinary tract infection. *Infect. Immun.* **72**, 6373–6381.

Söderblom, T., Oxhamre, C., Torstensson, E., and Richter-Dahlfors, A. (2003). Bacterial proteins toxins and inflammation. *Scand. J. Infect. Dis.* **35**, 628–631.

Struve, C. and Krogfelt, K. A. (1999). In vivo detection of *Escherichia coli* type 1 fimbrial expression and phase variation during experimental urinary tract infection *Microbiolgy* **145**, 2683–2690.

Svanborg, C., Bergsten, G., Fischer, H., *et al.* (2001). The "innate" host response protects and damages the infected urinary tract. *Ann. Med.* **33**, 563–570.

Swenson, D. L., Bukanov, N. O., Berg, D. E., and Welch, R. A. (1996). Two pathogenicity islands in uropathogenic *Escherichia coli* J96: cosmids cloning and sample sequencing. *Infect. Immun.* **64**, 3736–3743.

Takai, Y., Sasaki, T., and Matozaki, T. (2001). Small GTP-binding proteins. *Physiol. Rev.* **81**, 153–208.

Thomas, W. E. Trintchina, E., Forero, M., *et al.* (2002). Bacterial adhesion to target cells enhanced by shear force. *Cell* **109**, 913–923.

Warren, J. W. (1996). Clinical presentation and epidemiology of urinary tract infections. In *Urinary Tract Infections: Molecular Pathogenesis and Clinical Management*, ed. H. T. L. Mobley and J. W. Warren. Washington, DC: American Society of Microbiology, pp. 3–27.

Welch, R. A., Burland, V., Plunkett G., 3rd, *et al.* (2002). Extensive mosaic structure revealed by the genome sequence of uropathogenic *Escherichia coli*. *Proc. Natl. Acad. Sci. U. S. A.* **99**, 17 020–17 024.

Wullt, B., Bergsten, G., Fischer, H., *et al.* (2003). The host response to urinary tract infection. *Infect. Dis. Clin. North Am.* **17**, 279–301.

Zhang, D., Zhang, G., Hayden, M. S., *et al.* (2004). A toll-like receptor that prevents infection by uropathogenic bacteria. *Science* **303**, 1522–1526.

Zhou, G., Mo, W. J., Sebbel, P., *et al.* (2001). Uroplakin 1a is the urothelial receptor for uropathogenic *Escherichia coli*: evidence from in vito FimH binding. *J. Cell Sci.* **114**, 4095–4103.

Zotti, C. M., Messori, I. G., Charrier, L., *et al.* (2004). Hospital-acquired infections in Italy: a region wide prevalence study. *J. Hosp. Infect.* **56**, 142–149.

Index